Richard Halliburton and the Voyage of the *Sea Dragon*

People who have never been to sea except in a first-class cabin on a big liner can never know the full charm of the ocean, for they have seen it only under disadvantageous circumstances, where the great size of the vessel gives the life of a hotel, not the life of the sea.

—Richard Halliburton, February 27, 1922

Gerry Max

Richard Halliburton
and the Voyage of the *Sea Dragon*

The University of Tennessee Press

Knoxville

Frontispiece: The *Sea Dragon* during a trial run just off Hong Kong Harbor. (Courtesy Rhodes College's Barret Library.)

Library of Congress Cataloging-in-Publication Data

Names: Max, Gerry, 1945- author.
Title: Richard Halliburton and the voyage of the *Sea Dragon* / Gerry Max.
Description: First edition. | Knoxville : The University of Tennessee
   Press, 2020. | Includes bibliographical references and index. | Summary:
   "From Memphis, Tennessee, Richard Halliburton (1900–1939) is best known
   as a pioneer of adventure journalism. His career included numerous
   articles and books recounting his adventures, most notably his swim of
   the Panama Canal and his treks through the Mediterranean and Mexico
   retracing Ulysses's odyssey and Cortez's Spanish conquest, respectively.
   His final ploy was to sail a junk from Hong Kong to the Golden Gate
   International Exposition in San Francisco in 1939, a stunt that would
   cement his legend but also claim his life. Gerry Max's book focuses on
   Halliburton's time in Hong Kong; his painstaking efforts to build a junk
   and hire a crew during the Second Sino-Japanese War; his quarrels with
   Captain Welch; his thinly veiled homosexuality and relationship with
   Paul Mooney; and finally his death after the junk sank somewhere west of
   Midway Island"— Provided by publisher.
Identifiers: LCCN 2020026517 (print) | LCCN 2020026518 (ebook) | ISBN
   9781621905769 (hardcover) | ISBN 9781621905776 (pdf)
Subjects: LCSH: Halliburton, Richard, 1900–1939. | Travelers—United States
   —Biography. |Journalists—United States—Biography. | Voyages and travels.
Classification: LCC G226.H3 M393 2020 (print) | LCC G226.H3 (ebook) |
   DDC 910.4092/273 [B]—dc23
LC record available at https://lccn.loc.gov/2020026517
LC ebook record available at https://lccn.loc.gov/2020026518

In memory of
Kathleen Diane Williams
(1945–1953)
and our childhood encounters
with the two *Books of Marvels*

Contents

Illustrations

In the spring of 1939, just as World War II was about to begin, Richard Halliburton and a crew of fourteen attempted to sail a Chinese junk, the *Sea Dragon,* across the Pacific from Hong Kong to the Golden Gate International Exposition in San Francisco. Two thousand miles out to sea, the little ship ran into a powerful storm and perished; there were no survivors, and no wreckage was found. Besides dramatizing a confrontation with one of man's worst fears—death by drowning—the event showed that rogue adventurers, as in the classic days of ocean voyaging, could still sail off the edge of the earth and never be heard from again. News of the disappearance swiftly reached America. To Halliburton's many fans, it seemed an unfortunate yet fitting end for a soldier of fortune who had visited every corner of the globe and by his often reckless daring had cheated death so many times.

What was called the *Sea Dragon* Expedition had failed. The whole project was an artificial concoction, the caprice of one man; it didn't have to be done, nor was its conception prompted by historical necessity. Still, Halliburton never publicly doubted the merits of what would be the crowning achievement of his life. Large-scale endeavors featuring mergers fascinated him. In envisioning not many nations but one borderless *Pangaea,* the barriers separating lands and peoples seemed to him superficial. His sailing a Chinese junk across the Pacific would symbolically unite two continents and bridge what cultural and political divides existed between them.

Although he lived in the pre-jet age, when travel to distant locales was tedious, slow, and often over water, Halliburton moved faster than the wire services that reported his feats. He also lived in a pre-digital age when television was just emerging, and radio was the main medium of communication. Curious about new trends in technology, he would have thought the noiseless portable typewriter or hand-held walkie-talkie "cool," but suborbital space flights and space tourism were the stuff of dreams. His books, once long out of print, have been reissued as a generation of readers has discovered him to be not only amusing but also a worthy topic of inquiry and a pleasant escort into a bygone era. As interest in Halliburton grows, the search for materials related to his life and times has intensified. Documents buried in library archives or kept in private hands—some items repressed, others dislocated—have surfaced. Adding significantly to the record is the emergence of lost or buried letters, news clippings, newsreels, scrapbooks, sound recordings, photographs, and witness testimonials.

Several biographies of Halliburton have appeared since the publication in 1965 of Jonathan Root's *Halliburton—The Magnificent Myth*. All ably orient the reader to the life and times of their subject and vividly portray Halliburton himself. In these works, the *Sea Dragon* Expedition is offered as a last chapter in Halliburton's life, and the emergence of the craft as a personality in itself becomes a secondary player in the chapter. Little space is devoted to the Golden Gate International Exposition, even less space to the "American book" Halliburton intended to write. Sino-American relationships have served chiefly as a colorful backdrop to the *Sea Dragon* drama, while Hong Kong's role in these relationships has been seen as an incidental part of the *Halliburton* story and not the *Asian* story. For a fuller picture both stories need to be integrated with Hong Kong receiving enhanced focus. Usually noted are the tensions that arose among some crew members over the crossing's achievability in a Chinese junk—better built and equipped ships had, after all, met disaster; not noted, however, are the evening debaucheries of these same crew members, report of which Halliburton himself suppressed. The members of the crew, at any rate, seldom saw eye to eye, and brought into close proximity—and to a city about to be besieged—these strangers in a strange land were bound to clash.

Besides extended treatment of these tensions, scant coverage is generally given to the ever-widening rift between the *Sea Dragon's* captain, John Wenlock Welch, and Halliburton with Welch often appearing the brute and Halliburton the tormented angel. Theirs was an uneasy partnership. Out of boredom and by nature, Welch was designing, but not repulsive. A case could be made that he was a peeping Tom, but generally he minded his own business. A taskmaster, he was willing to be hated if harsh measures meant a common good. Halliburton was mild-mannered, poetic, childlike. A doting mother's boy, he was a devout father's son. Although he often wandered far from home, he never wandered far from his knowing their love for him, and though, in time, he learned to live without their guidance, he never overcame a need for their approval. To those who met the acclaimed travel writer, neither his manner nor his speech matched his essentially manly public image. Unknown was his homosexuality and preference for same-sex relationships. Until well-committed to the *Sea Dragon* Expedition, Welch had only a faint inkling from hearsay that this was the case. Once his suspicions were sealed, he couldn't let the matter rest and soon he concluded that Halliburton was a "fairy." If Halliburton was, as he himself claimed, a "bachelor . . . with no sons of his own," Welch saw him as a deviant who sought the company only of young men. In his correspondence, he referred to Halliburton as "the gorgeous one," "the Fairy queen," and "the sweet one," persistently dismissing him as an effeminate fop. Behind his back he called him "Hallidear" and "Halliburger" and used the pronouns "she" or "her" to further belittle his manhood. How Halliburton was able to ignore Welch's innuendos is a point of wonder. Halliburton saw at once that Welch was a conventional seaman who had been given an unconventional command where conventional nautical wisdom didn't apply. Tough-minded and paternal, he also had to be soft-hearted and have a touch of the poet. Because Halliburton is the central figure in any story by and about Richard Halliburton, it is easy to make Welch out to be the heavy—an autocratic, unbending man who was always wrong and a dream-killer who was consistently mean. Yet, to offer just one example that opposes the view, Welch's kindly reappraisal of crew member Paul Mooney, whom he called "a decent chap," proved him a man of fair play. He had the courage of his convictions, narrow though these convictions often were. To his dubious credit, Welch did not allow his prejudices to detract from his opinion that Mooney was the hardest worker of the crew or that Halliburton was at least determined in his quest. Still, while he liked Mooney somewhat, he merely tolerated Halliburton.[1]

Although challenged by vexing and often tantalizing gaps in its short history, the *Sea Dragon* Expedition is basically well documented. The "Log of the Sea Dragon," a series of dispatches, with photo insets, concerning Halliburton's and the crew's last days in China, appeared in the *San Francisco News* and other syndicated newspapers from 1938 to 1939. Nearly duplicate companion pieces from "Letters from the Sea Dragon" were sent independently to paid subscribers. Of note, the articles in the "Log" were published weeks after the events they described, while the "Letters" to subscribers reported events within days of their occurrence. Chronological vagueness characterizes the accounts, which occasionally conflate a series of events occurring over several days, weeks, or even months into a single event. Despite this, the reports are masterpieces of pictorial realism. Several of them could join the ranks of articles produced by war correspondents Stephen Crane, Richard Harding Davis, and Floyd Gibbons, pivotal figures "associated with the development of sensational news-reporting." Highly readable, these pieces were meant to feed the public's appetite for the exotic East, Halliburton's latest adventures, and, now, countries engaged in a bloody war. As records of history in the making, they are often momentous. In one instance Halliburton watched events unfold from his hotel rooftop during a nearby air raid attack, perhaps recalling in our own time CNN journalist Wolf Blitzer on the rooftop of the Al-Rashid Hotel in Baghdad during the First Gulf War. [2]

The dispatches plainly form the *public* record of Halliburton's last adventure. The three main bodies of *personal* letters—those of Halliburton, Welch, and Mooney, while they often agree, more often each contains details omitted by the others, which must lead to the conclusion that there are some details their authors may very well have overlooked. In the final analysis, one hopes that black and white, and the gray between them, comfortably co-exist. Besides write letters, Paul Mooney kept notebooks; deplorably, however, these are lost. Added to the major epistolary sources are the recollections, many long after-the-fact or incidental, of the *Sea Dragon's* crew. Aspiring writer George Barstow wrote letters home and may also have left a notebook. Survivors John Potter and Gordon Torrey left written and tape-recorded accounts of their time in Hong Kong. Various other crew members wrote letters home as well. A couple films and numerous photographs exist, and we know from letters and surviving pictures that Welch, radioman George Petrich, Mooney, and Barstow snapped pictures. While photos taken by Mooney, Chase, and Potter exist in part, those taken by Barstow, Petrich, and Welch so far remain undiscovered. Two sides to a story put together do not always equal a truthful whole, as is the case with the testimonies of crewmen Potter and Gordon. Their accounts, nonetheless, provide lively details and reflections about Halliburton's final days not found in the other major sources. Wrote Halliburton in *The Royal Road to Romance*, "There is so much material for psychologists in the association of thought and events. How much does thought influence events? How much does fear of danger encourage it, or ignorance of danger discourage?" Although he never meant to deceive, Halliburton, until his last fling as a war correspondent—an accidental one, characteristically wrote of himself in glowing terms, adding just enough winsome self-deprecation to give his narratives a measure of reality. Although he never meant to deceive, Halliburton garbed events in the glitter of romance until his last fling as an accidental war correspondent. That he wanted to write an "autobiography" was itself a confession that there was a truth to him which neither his books nor lectures conveyed. While one could wish that that autobiography existed, it is intended that the following text be in general agreement with existing sources, balance opposing views, and faithfully recreate the events and persons that form their context. [3]

While the sources for Halliburton's last days are often contradictory, partial and incomplete, those examining the adventurer image he cultivated over twenty years are of a fixed theme, telling us that he was the paragon of the well-traveled man, a romantic imbued with a sense of wonder. At the height of his fame, Halliburton in fact boasted that he would one day be remembered as the most traveled man in history. Canadian journalist Gordon Sinclair, possibly the last person to speak to him before he sailed off, called him "the most charming, the most successful, and in some ways the least understood" of the day's many travel writers. *Sea Dragon* crew member John Potter, who "watched most of the building of the junk," noted Halliburton's "humane side" as well as his "unique ability to see the commonplace as provocative and interesting." Fellow adventurer Carveth Wells also paid him high tribute: "Richard Halliburton was a most extraordinary personality," he wrote, "charming, arrogant, witty, extraordinarily impertinent, eccentric and courageous. He was full of the most original and fantastic ideas that, owing to his substantial income from books and lectures, he was able to carry out despite not infrequent official opposition." Although Halliburton earned acclaim from others, and, for the sake of livelihood, welcomed it, he was himself humble and demure, perhaps coyly so. At one of his lectures, to illustrate, the event coordinator introduced him by enumerating his many accomplishments. Trotting up to the stage, Halliburton smiled, then said, "I'm none of those things. I'm just a little boy playing Indian."[4]

Acknowledgments

To the many institutions that shared their collections, I offer my gratitude: Bancroft Library at the University of California–Berkeley for the Gerstle Mack, Paul Mooney, Noel Sullivan, and John Wenlock Welch/ Richard Albert Wetjen correspondence; the San Francisco Maritime Historical Park at Fort Mason for the Captain Yardley and other maritime documents; the Archives and Special Collections at Marist College for the Lowell Thomas Collection; Dartmouth College Library for the Potter-Torrey-Chase materials; the University of California–Los Angeles Film Archive for the Hearst *Sea Dragon* newsreel; George J. Mitchell Department of Special Collections and Archives at the Bowdoin College Library for information about crew member Benjamin Flagg; the *Memphis Commercial Appeal* for press releases; the California State Library in Sacramento and the San Francisco Public Library for the *San Francisco News*; Paul Barret Library at Rhodes College in Memphis for the Richard Halliburton Collection; Berry College in Rome, Georgia, for the Bertha Kellogg Barstow correspondence with Martha Berry; Firestone Library at Princeton University for its Richard Halliburton Collection; the Columbia River Maritime Museum in Astoria, Oregon; and the Treasure Island Museum in San Francisco.

Numerous individuals have lent assistance or discussed topics relevant to my research, and these I would like to thank: William Short, curator of the Richard Halliburton Collection at Paul Barret Library; Barbara Hunter Schultz, author of *Flying Carpets / Flying Wings*; Carolyn Treanor and Catherine Busch-Johnston of Sunflower Circle Productions; Royal Stewart, freelance Memphis historian and journalist; Erle Halliburton III, grandson of Erle I; Mike Lollar of the *Memphis*

Commercial Appeal; Edward Howell, lifelong Halliburton enthusiast; William R. Taylor, author; Charles Morris III, author; Toby Dorsey, author and attorney; Captain Wil Petrich (nephew of *Sea Dragon* crew member George Petrich), expert pilot and marine incident consultant; Patricia Suprenant, novelist, investigative researcher, and Barstow family historian; Andy Potter (grandson of crew member John Rust "Brue" Potter), who received a degree in geography from Dartmouth; Sarah Torrey (daughter of crew member Gordon Torrey); Robert Thomas Wilson, writer and sailor; Al Suildebhain, theatre technician and boating consultant; Gunnar Thompson, anthropologist and maritime historian; and Ken Kotani, Japanese World War II intelligence historian. For hosting my seminars on Halliburton, I am grateful for the opportunity given to me by Lawrence University's lifelong learning center Bjorklunden in Bailey's Harbor, Wisconsin and its director Mark Breseman to conduct Halliburton seminars. For guidance through the publishing process, I would like to thank acquisitions editor Thomas Wells of the University of Tennessee Press.

For their initial encouragement and insights, I wish to thank *San Francisco Chronicle* architecture critic Allan Temko (1924-2006) and UC–Berkeley architecture professor James Prestini (1908-1992). Several teachers from my many years in school have injected meaning into this study. One is medievalist William Chaney (1923–2013), who thought Halliburton a colorful guide into the wonders of the past and a worthwhile subject for historical inquiry. Another is classicist Herbert Howe (1912–2010), whom I thank for his account of the impact of Halliburton's books on the collegiate readers of his time. Professor Howe enlivened in brilliant fashion many aspects of the American scene of those days. His uncle Mark Antony De Wolfe Howe (1864-1960), associate editor of *Youth's Companion*, had among his many illustrious friends adventurer Richard Harding Davis (1867–1916), after whom Richard Halliburton may have been named. As a young man hiking about Greece in 1938, Professor Howe met Wesley and Nelle Nance Halliburton vacationing in Athens while their son Richard was preparing the *Sea Dragon* Expedition in far-off Hong Kong. For her encouragement, I also thank classicist Barbara Hughes Fowler (1926–2000), author of the *Hellenistic Aesthetic* who, besides recognize Halliburton's contribution to the popular travel culture of his day, applauded his "classical enthusiasms," as he himself put it, and his application of great literature to life. One may also say that the story of adventurer Halliburton's last days, besides having their elements of Greek tragedy, is somewhat a modern *Iliad* and *Odyssey* combined, about a city and a siege, a rover and a sea.

Of the people who knew Richard Halliburton, I am especially grateful to William "Bill" Alexander (1909-1997), who designed and built Halliburton's "Hangover House." Niece Elaine Hofberg (1926-2008) provided key documents and regularly offered me her recollections of her uncle and his friends. Chief among those friends was Halliburton's "secretary" Paul Mooney, whose nephews Anton Levandowsky and John Murphy Scott also shared their recollections. Retired *Washington Post* copyeditor and Princeton graduate Scott corresponded with me over many years. One of his persistent counsels was "Stick to the subject and don't chase the many rabbits that cross the path of your story." Truly, there are so many rabbits in the Halliburton story that chasing one or another of them into some dense thicket has often been irresistible.

Closer to home are other people to thank. For her patient attention to my tangents about "the *Dragon*," I thank above all my wife, confidante and best friend Carole. An avid and virtuoso reader, she has delved into works by Iris Murdoch, Joseph Conrad, G. K. Chesterton, and Jim Harrison. A student of French, she has translated Aristide Bruant's *Sur La Route*, an early paeon to the open road. Her intelligence is natural, her wisdom Delphic. A large part of our shared happiness we owe to our cats—call them psychic watchmen: Skylahr (in appearance and temper quite like *The Wizard of Oz*'s Cowardly Lion Bert Lahr), Djuna (after the author of *Nightwood*, a modernist novel whose ending perplexed Paul Mooney), and Magda (after the oldest of the Gabor sisters). I wish at last to thank my parents, Raymond and Ruby Max, who built their own home in the Midwest about the time Halliburton was building his in California. "Time is your most important commodity," both of them liked to tell me, or "You never know where you're going to end up." They introduced me at an early age to the world of books and encouraged my love for Richard Halliburton.

Kathy Williams, to whom this book is dedicated, died very young of leukemia. Her hair was blond, her cheeks dimpled and pale. She resembled a very delicate little-girl version of England's Princess Elizabeth. Miniature adults, we began kindergarten together and vowed to marry someday. We told each other stories and talked about what we would like to be "when we got big." In the evenings we might try to find the

Big Dipper and Little Dipper. During the day we might blow bubbles from a loop, mine always popping at once while hers drifted high into the air before doing so. We considered the sparkling orbs that magically formed from that loop "baby planets." Kathy had two favorite photos in *The Book of Marvels*: the nose of a dirigible touching the spire of the Empire State Building and the Great Wall of China winding toward the sky. She thought it the worst calamity to be lured into a lobster trap—I suppose she meant one of those wooden-framed ones with the entangling rope mesh around the entranceway. I have since believed that such a trap was little else than the wish for financial security, such as lured Richard Halliburton into the Orient. I now recall that Kathy kept a big seashell on a table by her bed. We would take turns putting it up against our ears. After her turn, she would smile, then hand it to me and say, "Here, it's for you," as if it were a phone call. "Put it up to your ear—they say you can hear the ocean."

Glorious Adventurer

R eads the blurb to the posthumously published *Richard Halliburton—His Story of His Life's Adventure as Told in Letters to His Mother and Father*, "The name of Richard Halliburton is synonymous with youthful adventure. During his less than two-score years he packed away more tempestuous and thrill experience than a dozen other adventurers might in twice as long. He was an originator, a trail-blazer whom countless young people have followed and are following. He epitomized their impulses and their dreams—the desire to see the world, the search for the burning moment, for the far horizon, for the unconventional life." Few would have doubted these general claims, yet over time recollection of Halliburton's name and his contributions to travel literature dimmed. Over the years his books dropped out of print one by one, with only *The Royal Road to Romance* and the *Complete Book of Marvels* surviving into the 1960s. Ironically, the large print run of his books, especially in the cheaper *Star* editions, have made his titles a mainstay in used bookstores until this day while his name more instantly invokes the oil well cementing company founded by his cousin Erle than the adventurer who, among his many other feats, scaled the Matterhorn and swam the length of the Panama Canal. Why so unkind a fate should befall a man who, with Charles Lindbergh and Amelia Earhart, once towered over an era mystifies.[1]

While Halliburton's name quickly disappeared from dictionaries of American biography, Earhart's and Lindbergh's names have been on permanent cruise control through the afterlife of fame. Radio broadcasts, newspapers, newsreels, and bookstores made Halliburton as recognizable as Bobby Jones or Babe Ruth—though it should be said that headlines chased these bigger-than-life heroes while Halliburton chased headlines. In Hollywood, great reputations could suddenly dissolve. In the showbiz world of the 1930s, too much time out of the spotlight could erase one's image from the public mind. Producers were as famous as their last successful Broadway musical, or so the saying went. A biography of the last great Antarctic explorer Hubert Wilkins, also a pilot, spy, and author, is entitled *Hubert Who*? Box office one week, an established star like Joan Crawford or Bette Davis might be box office poison a week later. A writer like F. Scott Fitzgerald might find his books best sellers one moment and out of print the next.

Accompanying forgotten people are often the events associated with them. With the surrender of Japan in 1945 and the establishment of the People's Republic of China in 1949, the war between China and Japan a decade earlier blurred. Within a few decades baby boomers could believe that China and Japan were linked by a common heritage, that China hated Americans, and, most mistakenly, that Japan, helped to its feet by America *after* World War II, had been America's ally *during* World War II. Paul Fussell's *The Great War and Modern Memory* notes how recollection of even the most catastrophic events may fade over time. Halliburton's vogue—that of the innocent traveler abroad and hawker of romantic tourism—was first broken on September 7, 1940, when the Germans first bombed London; assuredly the vogue ended on December 7, 1941, when the Japanese bombed Pearl Harbor, prompting Congress to declare war on Japan a day later.[2]

As a major figure in the development of twentieth-century travel narrative Halliburton's reputation has grown in recent decades. Opinions of him gained stature since 1940, when *Time* magazine trivialized him as "an appealing, confused individual, a US phenomenon . . . an artist and a rebel" who achieved "neither art nor rebellion," and "an innocent sort of Byron-of-his-time." That same year Pulitzer Prize–winning journalist

George Weller in a memorial feature for *Esquire* entitled "The Passing of the Last Playboy" labelled Halliburton "a good word-a-day hack, with an increasingly competent newspaper style." Thinking the "Playboy" would have agreed, he charitably added that Halliburton "considered it a great joke that anyone could write as badly as he did and still make a good living out of it." To critics who thought Halliburton a charlatan, Weller remarked, "Richard was not a phony—he had an appetite for action, and made his living having a good time." The 1940 British edition of Richard Halliburton's *Seven League Boots* notes simply that its author, "a young American, liked to live dangerously, and in the last few years before his deeply regretted death, he crowded into his life enough adventure to last many lifetimes."[3]

Cultural Ambassador

A "Voice of America," Halliburton shaped the basic way some Americans viewed their world. He had a roving reporter's eye and was respected by some newsmen as nearly one of their own. Renowned journalist H. L. Mencken saw promise in Halliburton's early efforts and said he "he would like to see" any articles about America that he wrote. As other writers of the Jazz Age and Great Depression that followed it, Halliburton had a good deal to say about the state of the world, but he thought it wise to stay mum. "Never have opinions," he advised. "A really mature person never has opinions. I'm glad I've reached that stage myself." He held points of view. As a cultural relativist, he thought the world composed of multiple perspectives, a stance that may explain his purchase of a slave child in Africa, or being led by Dyak headhunters into the jungles of Borneo. Fascinated by the "other," he was generally receptive to new ideas, at times even credulous. He knew it to be a wicked world, one of inequity and misery, but knew too of its potential to welcome moral good and self-betterment.[4]

The 1920s were famous for high-risk crossings big and small. Gertrude Ederle swam the English Channel, Charles Lindbergh flew across the Atlantic, and Amelia Earhart attempted to cross the Pacific. In fiction, Maurice Elvey's futurisitc *Transatlantic Tunnel* (1935) dramatized a technological effort to link "the English-speaking peoples."

Poet Hart Crane's epic *The Bridge* boasted of a man-made structure, the Brooklyn Bridge, linking two shores and multiple human destinies. To those who first beheld it, the 'epic' Golden Gate Bridge, completed in 1937, appeared a skyway from planet Earth to the welcoming stars. Two years later, when the Golden Gate International Exposition ("the fair") opened at Treasure Island, it welcomed sightseers to its bridges and expressed the promise of world accord.

The American Dream, as do ideas of global peace, has a history; its broad theme of freedom and equal opportunity has changed only in its mode of expression. In the 1920s that theme was rewritten. Wrote travel legend Frank Carpenter:

> In these (my) travels in Europe I shall ask you to explore a new continent. The old Europe died with the World War, and then a new Europe was born. It did not rise phoenix-like from the ashes, but is still in its swaddling clothes, sprawling about on the floor and trying to grow. Like our own dear babies at home, it shows new aspects each day and the changes are many and frequent. The social conditions are different. The new woman is rising to an equal plane with the man, and the new man thinks and acts differently from what he did in the past. The nations have new relations to us, and in spirit and fact we are fast becoming part and parcel with them, sharing in their troubles and borrowing the features in which they excel. Our social and business relations are growing closer and, with railway, steamship, and airplane joined to telegraph, cable, and radio, the world is becoming more and more one vast family where distance apart cuts no figure.

Anti-colonialist and egalitarian, the sentiments foreshadowed the "one world" concept of future republican candidate for president Wendell Willkie, a leader in the World Federalist Movement whose ideal of international cooperation would have made all roads safer for travelers like Richard Halliburton. The *Sea Dragon* Expedition inevitably bore the stamp of its times—and of its originator.[5]

Cult of Youth

Men of genius and of great promise who died young troubled Halliburton. British poet Rupert Brooke, his alter ego, died at age twenty-seven on board a ship in a foreign land. For Halliburton, this death was "one of the greatest tragedies lovers of Beauty and Intelligence and Idealism have ever suffered." Like Brooke, Halliburton's appeal was to young adults. That appeal he maintained even after he himself began to age. As late as 1938, when he was nearing forty, he was hailed "the embodiment of the spirit of youth today." His name still recalled the words that had made him famous over a decade earlier: "Youth—nothing else worth having in the world . . . and I had youth, the transitory, the fugitive, now, completely and abundantly. Yet what was I going to do with it? Certainly not squander its gold on the commonplace quest for riches and respectability, and then secretly lament the price that had to be paid for these futile ideals. . . . The romantic—that was what I wanted. I hungered for the romance of the sea, and foreign ports, and foreign smiles. I wanted to follow the prow of a ship, any ship, and sail away, perhaps to China, perhaps to Spain, perhaps to the South Sea Isles, there to do nothing all day long but lie on a surf-swept beach and fling monkeys at the coconuts (sic)." Over the course of a long career, Halliburton's message remained the same. Originally inspiring that message was not scripture but secular literature. Rupert Brooke believed "there was no place for anyone in this world over 30," and Halliburton had committed his poems to memory. Also a casualty of the Great War, H. H. Munro ("Saki") similarly believed that, "to have reached thirty is to have failed in life." Quoting Oscar Wilde's *The Picture of Dorian Gray*, Halliburton wrote, "Realize your youth while you have it—don't squander the gold of your days." *Seize the moment,* he insisted. He was the first person in modern times to admonish young people not to "throw their lives away." Don't waste your youth trying to make old age happy. Don't become the old person you will one day be telling the young man you are now how things could have been. Don't collapse into "the prosaic mold." Fear the swinging pendulum, but fear the pit even more. Shirk the "dictatorial shoulds" of the parental and public scrutiny that obstruct self-realization. Don't rush into marriage or a secure job before you explore the possibilities. Tempt fate, but hurry. "Those whom the gods love die young"—they do not *grow old.* Halliburton lamented all the young men who, martyrs to weakly felt ideals, had forfeited their lives in the Great War. He said that "life is not life if it's just routine; it is only existence and marking time till death comes to divorce us from it all." Give me "liberty or death." He chose to be a free spirit.[6]

Higher Education

Tame and even quaint by today's standards—as students now voice opinions about school curricula that are heeded and as more colleges adopt global studies programs, Halliburton's "declaration of independence," with its insistence that students abandon school to see the world, struck many educators of his day as subversive. They needn't have worried. Although he encouraged his audiences to travel "the royal road"—to board the magic carpet of the imagination and escape from the humdrum of everyday life into the realm of enchantment, he had little interest in evangelizing the heathen or spiriting innocents off to enchanted *Neverlands*. Get some education, he said, before you *ride the rails*. His royal road, after all, was not paved with cobblestones for untutored fools or rustic "boors." From a somewhat well-to-do family and himself a Princeton graduate, Halliburton traveled the high roads above the hobo campsites. "One needs the proper background for vagabondage," he wrote, "otherwise, to wander willy-nilly around the world would be a fruitless and unenjoyable task. . . . My books are nine-tenths an appreciation of the arts, not just the rambles of a tramp." Life on the road required that one be prepared emotionally and financially. Whatever its requirements, Halliburton's search "for the burning moment, for the far horizon and the unconventional life" caught on with collegians whose "impulses and dreams" he articulated, and, notably, for those among their number who could afford time-off from school life. They read his *The Royal Road to Romance* as eagerly as their counterparts a generation later read J. D. Salinger's *The Catcher in the Rye*. Good schooling prepared one for travel, but it was better if carefully monitored travel prepared one for school. Travel-abroad programs existed, but Halliburton gave them a philosophic basis. To higher education itself he offered some clarity of purpose.[7]

"Education is an admirable thing," said Oscar Wilde, "but it is well to remember from time to time that nothing that is worth knowing can be taught." In *Routes of Man*, Ted Conover speaks of college, before he learned how "to use it," as "imposed learning," and found travel, by contrast, to be "an expression of personal curiosity, of a broader education less mediated by received thought." For many of Halliburton's generation, school meant immurement, conformity, and self-abdication. Rote memorization, tedious repetition, dreary assignments, and strict obedience to authority were long-approved canons in educational policy. While approaches to learning should reflect changes in the world, often it seemed to Halliburton that they didn't attain this goal. His father, Wesley, wanted him to be a "college man," and, lured as others by the promise of material comfort and security, he too wanted to be a college man. But once secure in the role he was soon bothered by his seeing fellow classmates chained to their desks or caged up in their library carrels, buried in "a rut so deep" they could not "see over the side to the limitless horizons beyond." If they thought they were leading their lives, they were mistaken; rather, believed Halliburton, they were being led. A "disadvantage of being one of this ultra-modern generation," he wrote to his parents, is "that we are satiated before nature intended us to be." Teachers served advice, but were they wise? Had they themselves followed their own path or only one shown to them by others? Although Halliburton believed that blind devotion to academia dulled personal initiative, he proved himself a committed, conscientious student whose courses in geography, history, world literature, public speaking, French, even banking and finance, served himself throughout life.[8]

A last great exemplar of the classical imagination in action, Halliburton trusted that a sound liberal arts education combined with travel abroad produced the gentleman-scholar. He could quote whole passages from the *Iliad* and *Odyssey*. Of their many heroes, the one whom he most admired was proud Odysseus (Ulysses), cunning warrior, bold sailor, tireless wanderer. In Richard Halliburton's world, there was no afterlife, only life. He knew of the varieties of religious experience, and, though no religious creed bound him, he revered religious figures of all faiths. An advocate of the Great Man Theory, he believed extraordinary individuals through their courage, wisdom, and trust in themselves drove history. No hero himself, he emulated heroes.

Partly because the world he described has so wholly changed, Halliburton's writings today seem politically harmless. However, a deeper reading of his letters and books convinces readers otherwise. If not an outright American imperialist, he endorsed American influence abroad and its supervision over the many economically woeful members of the human community he encountered. From a country that had a dream, Halliburton came into contact with so many nations with no dream at all. A new world system had emerged from withering colonial regimes and a fast-falling British Empire, and people from the remotest regions of the earth felt the tremors of modernization. About America, he could wax patriotic, yet he knew America's roots had been sown by scoundrels and freebooters as surely as by seekers of religious freedom and political visionaries.

As a world traveler, Halliburton met many people who lived on a poverty level unimaginable to most Americans. Still, in his books native peoples in wretched straits often are presented as little more than animated features of the landscape. An actor fitted to a role, Halliburton always knew where the camera was positioned. In high profile he rubbed shoulders with the outcasts of India, shared in the miseries of the inmates of Devil's Island, and bivouacked with the hardened troopers of the *French Foreign Legion*. Although real, these demonstrations of his humanity were also temporary, and one can imagine him airlifted to safety from any imminent danger.

The World According to Richard

Halliburton's long association with the Young Men's Christian Association and the Boy Scout Movement demonstrated his allegiance to fellowship, duty, and country. The Civilian Conservation Corps, an adult version of scouting, also subscribed to those allegiances. Halliburton was not a joiner, but his ideas had some affiliation with liberal idealism. He once declared himself an enemy of fascism, jingoism, and imperialist aggression. These sentiments are to be expected from a globe-trotter, as the democratic individualism that had permitted him and other Western rovers to move easily about the world was plainly being threatened. While Halliburton's creed of personal freedom sounds naively heroic, his insistence that one could best develop one's self to the ultimate of human weal only if released from the tether of parental authority and the bonds of small-town provincialism struck a chord. As he aged, democracy had come to mean *expanded* and communism *restricted* opportunity. Thinking of group demonstrations, union leader Eugene Debs said the most heroic word is "revolution." Halliburton's was a personal revolution, but one whose unfolding would have been impossible in any but a free society. Fellow travel writer Carveth Wells noted, differently, "America is overrun with parlor pinks and before we know where we are, we may find ourselves in the thick of Communism." Lovers of a "free society" had to take note.[9]

Until the mid-1930s, Halliburton's political voice, as that of his contemporary Charles Lindbergh, if before seldom heard, now reached its fullest maturity. As social agenda in the FDR administration threatened to open the gateway to National Socialism, Halliburton blew his bugle: "Until the Statue of Liberty does a back-flip off her pedestal into New York Harbor, will Americans ever submit to despotism and such deprivation as exist under the Red Flag of Russia." As a young man he expressed a "virulent antipathy for democracy as practiced in America" and "for the laboring class" that supported people's rights. Experience only clarified Halliburton's anti-Marxist views or, rather, removed them from an American setting. When visiting Russia in 1934, he found it upsetting that the "toilers" were in control, and that displays of militarism entertained. If he considered England a model of gentility, he deemed Russia and Germany models of crudity. He appreciated what Russia had achieved in crime deterrence, prison reform, literacy growth, and women's rights, but he had to admit that "for one used to, and requiring, personal freedom to travel, to read, to enjoy the fine things that make life worth living, Russia is a hateful, tyrannical prison cell." Halliburton's "If Communism Comes To America" stirred "blistering replies from Reds and Bolsheviks in America," but a "fight about Russia" held little interest for him, and he moved on to brighter topics.[10]

Aloof and remote by nature, Halliburton realized that America was no place for an arrogant intellectual, aging maverick or enlightened egotist. Despite his appeal to the general public, he preferred specific association with wealth and privilege. He claimed to be a "liberty lover," but, often "morose" and "cynical," he admitted to having little interest "in John Jones and his local affairs." He changed, mercifully. Once contemptuous of the working classes, he saw them at last as the salt of the earth, beaten but resilient. A sense of gallantry enabled him to weather changing times with stylish grace. Until the rise of dictatorships and rogue empires of the 1930s, he generated health and happiness and functioned chiefly as an optimist in a politically muddled world. "I'm doing a magazine article entitled *The Rough Road to Reality*," he once remarked, "suggesting I've traveled this as well as Romance—and prefer Romance." To his fans and the press, he would become affectionately known as "romantic Richard." To himself he remained the modern Don Quixote with a touch of pessimism.[11]

2

Great Wall of China

Halliburton's nearly twenty years of fame coincided with pivotal events in the interwar period, including the meteoric rise of America as a world power and its increasing involvement in Far Eastern affairs. By the time Halliburton took his first stride down the *Royal Road*, the Great War had transformed America from a debtor to a creditor nation, women had gotten the right to vote, the automobile linked both ends of the continent, and electricity lit every home. Frenzied years of boom were followed by anxious years of bust, and, despite some signs of recovery from the nationwide economic crisis in the mid-1930s, "hard times" appeared here to stay—at least until the Second Great War.

Amid harsh economic realities at home, faintly seen by most Americans was the war in faraway China. Inseparably connected to that war are both Halliburton's final days and China's eventual rise as a major financial power. The dispatches Halliburton sent home helped many readers place the Far East on the map and accept it as a compelling force in their nation's economic future. By 1938, Halliburton had become more than a professional rover; he had become a foreign correspondent—with China and Japan his region of operation.

By tapping into the resources of its Asian neighbors to the west, Japan hoped to reduce and even end its reliance upon American oil, iron and other raw materials. Even while helping China to resist Japanese imperialist ambitions, the United States continued to trade with Japan. Japan even sought American backing for the industrial development of Manchuria. Although stopping shipments of armaments and munitions, Uncle Sam still sent gas, oil, iron, and other commodities Japan needed to fight its war with China. The oil embargo signed into effect by FDR in 1941 gave Halliburton and other Americans a stay of execution, but the Japanese, aiming to become oil independent, now set their sights on oil-rich Celebes, Borneo, and Java. The Philippines, manned by a limited US fighting force under General Douglas MacArthur, was all that stood in their way.[1]

At the time relations between Japan and the United States were somewhat cordial—distrustful friends but reluctant foes. While Japan's protracted war with China threatened Europe's Asian colonies and America's investments abroad, the island nation aspired to be only as a dominant regional power; it did not want to entice the US into war. Japan's attack of Pearl Harbor on December 7, 1941, was a strategic blunder undermining those aspirations. Of the circumstances that led to it, few in America had much of an idea. Writing to friend Harriette Janssen in America, Halliburton's secretary—and partner of his labors, Paul Mooney had to wonder if she knew of the calamities besetting him in the Hong Kong: "All this assumes that you know there is a war in China? It is so much farther from New York than Spain is, or atrocious Germany, that perhaps the press pays little attention. On the other hand, you may have better information than mine, on the theory that I can't see the trees. . . . My hope, after what I have seen and heard, is that Americans out here feel, after our recent senseless and short-sighted obeisance to the dictatorships. Germany's barbarism needs no further proof; and Japan equals Germany in her headlong flight back toward brutality." Halliburton and Mooney came to the Far East with the best of intentions but in the worst of times. They also traveled to places their readers back home knew little or nothing about.[2]

In plain truth, the average American had little direct knowledge of foreign lands. Halliburton was familiar with China, but most of his readers (including his parents) considered it a distant orb circling another sun, a place where children starved, gunpowder was invented, and paper money was first used. Hendrik DeLeeuw wrote in *Flower of Joy* (1939): "China is a land of paradoxes—a land where a man on greeting you shakes his own hands instead of shaking yours, where white not black is a sign of mourning, where the compass points south instead of north, where books are read backward not forward, and where women wear socks and trousers and men wear stockings and robes." The view added little to what folks back home already surmised. Comfortable in their ignorance, many Americans believed that the Chinese, with the exception of a few connivers like Lon Chaney's film character Mr. Wu or fortune-cookie philosophers like Warner Oland's Charlie Chan, were a crude bunch who subsisted solely on plants, rice, and tea. In D. W. Griffith's silent *Broken Blossoms*, Richard Barthelmess' Cheng Huan (called in the captions, reproachfully, the "Yellow Man" or, affectionately, "Chinky") was, if humane and tender-hearted, a deficient soul and social pariah.

About their Asian neighbors, Americans harbored numerous blunt assumptions. The commercially enterprising Chinese were "the Jews of the Orient." They made sacrifices to demons, worshiped dragons and fire, and consulted shamans who revered gremlins and drank goat's blood. People in China had slanted eyes, wore funny clothing, sported strange hairstyles and mustaches. They smoked opium, spoke a yo-yo language, pulled noisy, clumsy-looking pushcarts, and lived in these dwellings resembling corkscrews which they called pagodas. Major US newspapers covered the war in China, and travel magazines published extended treatments of Asian culture. Even so, most Americans thought of China as somewhere on the other side of the world; there people walked upside down, a family of five lived on food scraps which the average American left on the plate, and everyone, knees crossed, ate sitting on the floor.[3]

In popular culture Pearl Buck's *The Good Earth,* and the movie version that followed, offered a view of conditions in China that many Americans could appreciate—that of a Chinese couple struggling for survival in a rude environment. Other treatments offered clichéd pictures of Chinese manners through a generous assortment of scoundrels, beggars, smugglers, and obsequious rickshaw drivers. Frank Capra's *The Bitter Tea of General Yen* (1933) informed American audiences that a civil war raged in China. Walter Futter, who produced Halliburton's only sortie into filmmaking, *India Speaks* (1932), also produced *Hong Kong Nights* (1935). This thriller about arms smuggling featured clips of prewar Chinese civilian street life yet offered a sorry picture of

Chinese culture. Apart from such purely cinematic views, most travelers to China found the country overcrowded and busy or, less charitably, filthy and filled with slant-eyed devils. Americans were grateful a vast ocean separated them from the country.[4]

Long a force in Asian politics, the United States remained a studious if opportunistic sideline observer of the war. It did so despite its own enduring imperialist ambitions in the region and Japanese attacks on its own shipping. American interests in China visibly increased after the Boxer Rebellion in 1904 and increased even more after the death of the dowager empress in 1908. This last event led to the final collapse of the Manchu dynasty in 1917, when civil war threatened to halt further foreign investment in the country. At the time the United States was just beginning "to expand commercially into the Far Eastern markets, and several bankers and engineers arrived in Peking to discuss a railway and canal construction enterprise." The Japanese disapproved, and concerns over a torrent of foreign speculators arriving in Asia brought the two nations to the brink of war.[5]

Foreigners generally fared well. Amid revolution, war, lawlessness, and a vibrant underworld, British and American financiers successfully conducted regular business on China's shores, and American companies thrived inland. The Yangtze Rapid Steamship Company was the only American carrier to provide passenger and cargo service from Shanghai to the Tibetan hills, some two thousand miles westward. Ships bore such names as the *I Chang* and the *I Ping*. Chief administrator of this important connecting link to the interior of China was one Lansing Hoyt, who lived on Foochow Road in Shanghai. Among Hoyt's acquaintances was Walter Daub, manager of the Quaker Oats Company in China. Running a company in a hostile region was risky business, and, even with an American consulate general to oversee matters, financiers could get the jitters.[6]

In 1922, when Halliburton, in "boy-scout short pants and shirt," combed these parts—Mandalay, Bangkok, and beyond—he found "the all around-the-world tourist steamers" a commonplace sight. He saw in the jungles the "reckless, low lot, adventurers all," who "kept native women as cooks and housekeepers." He noted the American companies like Standard Oil and Shell Oil. He saw the "brilliant flowers" sprinkled about, the "cool streams" running alongside the chugging trains, and the "mosquitoes" and "lizards" in the lurch. Of political crises brewing, he took little notice, but by the 1930s, he had more than a schoolboy's passing knowledge of Anglo-Asian relations and the currents of Chinese history.[7]

Except for their eloquence, Halliburton's earliest expressed views of Asians little departed from known stereotypes. He thought that junks were unwieldy scows operated by laundrymen, and that the Japanese were just grinning idiots. Although he liked to dress in the costume associated with some cultures—he never became Far or Middle Eastern enough to think the West was strange, capitalism unfair or puritanism narrow. The Chinese, in particular, had characteristics that distinguished them from Caucasians. He lauded their advanced civilization and trading prowess, but found them dreadfully behind in material progress. Later, he saw a different China. Alongside the civilization was the squalor of everyday life. Expecting little from life, most Chinese appeared happy with what they got. Stoic by nature, being humiliated seemed of no consequence to their pride. In their willingness to please, many could be wheedling and obsequious. As trickery in a fickle world was seen as an only way to survive, many were conniving. Bound to commonplace realities, most were superstitious when it came to matters of fate and the unknown. In everyday undertakings, they could be, as Halliburton put it, "reckless." Industrious, they could also be sluggards. Indeed, Chinese work habits persuaded him that some were by nature lazy, and that, as a group, they were backward, even Neolithic. He thought Europeans remained cool under duress while the Chinese got hysterical and made funny noises. Amused by the "Chinaman," he found most of them agreeable and of good cheer. In time he learned about the place of opium in their culture and its regular use, by others, he had no qualms. Seeing the banks of the Yangtze swarming with "coolies and dogs and babies and houseboat dwellers," he concluded that no people were "more interesting" as the Chinese: "They are irrepressibly good-humored—all thorough rogues, but also good sports, so that you can't help liking them." He might just as easily have said that the Chinese 'don't think like us Westerners but expect them to be friendly.'[8]

As a herald of multicultural awareness, Halliburton receives high marks, but as a politically correct champion of social equality, his grades are lower. Strong traces of the white man's burden resonate in much of his prose. In his earliest writings, racial purity was also a concern: "I am only too fearful," the twenty-year-old Halliburton wrote to his father Wesley, "that what you say about the ultimate extinction of the white race is true. These Asiatics are slow and sluggish in progressing, but progress they must and progressing they are, and while we increase by hundreds, they increase by millions. Already their disrespect of the white, their arrogance, is terrific in the East. A vast debacle is smoldering against the Caucasian and when once these hordes begin to resist in concert, what is to stop them? One sees thousands upon thousands of half-castes in Asia. They do not tend to whiten the brown man, but to brown the white." Odd remarks, they now read as curious forerunners of a continuing debate. According to one anthropologist, "ethnic identities (rather than dissolve in the larger nation-state) have grown stronger in the modern world," and "ethnic groups have flourished and become rallying points for oppositions to governments." The fair's message, as well as Halliburton's understanding of it, was probably not so complicated or profound. Of course, one does not read Halliburton as one would read cultural anthropologists Bronislaw Malinowski or Melville Herskovitz, renowned investigators in their respective fields whose work, or the like of which, Halliburton himself read. Reading as far and wide as he had traveled, he wanted to be recognized for his substance, not his fluff, a goal he seldom achieved. Case in point: in 1935, at the height of his fame, he wrote "The Wickedest City in the World." The article explored sexual mores, marriage rituals, and burial customs, concluding that these might differ from culture to culture and perplex those who seek an ultimate truth. An in-depth look, his publishers, the Bell Newspaper Syndicate, nevertheless rejected it, insisting that he keep things light and not "spoil too many Sunday dinners." Halliburton's screenplay for the docudrama *India Speaks* gave him freer rein. Chiefly a photographic essay on how the other half lives, its cumulative impact is that hard times in America were not so desperate as hard times in the Asian subcontinent. While it is true that Halliburton's books are crammed with special-interest topics, ranging from a visit to the Inca citadel of Machu Picchu to an encounter with the head-hunting Dyaks ("Wild Men") of Borneo, these were, as he would say, "lively" pieces rather than serious ethnographic studies.[9]

Intellectually trendy, Halliburton read widely and tracked the news. He also sought the company and advice of progressive thinkers like Dr. Margaret Jessie Chung, a pioneering trans-Pacific enthusiast who advocated closer ties with China. He met media icon Pearl S. Buck and knew of Chinese foreign secretary Addie Viola Smith. More powerful and well-publicized advocates of improved Sino-American relations were *Time* magazine founder Henry Luce and his wife Clare Boothe Luce. The American public, however, was not as easily won over as such figures to America's expanding involvement with China. It was hard to overcome their lingering prejudice that Chinese Americans, called "the yellow peril," were a cheap labor source which from the 1880s had lowered American workers' wage expectations. Demands by labor parties to deport the Chinese American workforce led in 1882 to the Chinese Exclusion Act, "which (for over sixty years) banned the immigration of ethnic Chinese and denied those who were settled in America the privilege of pursuing citizenship." The law was not repealed until 1943.[10]

Halliburton knew about China's bloody past, the missionary activity there, and the seeds of Christianity long planted on its "heathen" soil. One brutal chapter from the past was the Taiping Rebellion (1850–64), which began when Christianized visionary and religious radical Hong Xiuquan rose up against the ruling Manchu dynasty supported by the British and other European powers. For nearly twenty years separatist Hong and his Taiping supporters controlled southern China and the corridors leading to Hong Kong itself. American soldier of fortune Frederick Townsend Ward headed a polyglot army of Western mercenaries and newly trained Chinese militia that won important land victories. But Britain's trained armies led by Charles "Chinese" Gordon ultimately crushed the solidly entrenched Taiping rebels. Twenty million people died during the rebellion; Hong was vanquished, but the Manchu dynasty was severely weakened. In 1894–95, during the First Sino-Japanese War, the dynasty nearly collapsed when Japanese armies invaded China. The Manchu dynasty was then saved again by European intervention only to topple two decades later. China's various regions were now separately ruled by provincial warlords who often feuded—and enslaved. The country's nearly half a billion inhabitants, according to Asian authority and cryptologist Herbert Yardley, were "enslaved to a handful of money-lenders and landowners."[11]

Travel books during this time portrayed the Chinese sympathetically. *National Geographic* and *Asia* regularly published in-depth articles on the Far East. These were starts towards cultural understanding. In *Asia at the Crossroads,* World War I correspondent Edward Alexander Powell described the complex relationship among the major Far Eastern nations, but was unable to find a link that bound them. Republic of China founder Dr. Sun Yat-sen, meanwhile, offered direction towards the goal of a unified China but died before he could complete his mission. Even before hostilities commenced between China and Japan, members of an international press corps were stationed throughout Southeast China to cover what would become the Second Sino-Japanese War. "The South China scene" was a main "focal of world attention" and the mecca of journalism's new soldiers of fortune. For many Americans, however, China was too distant and ethnically dissimilar to be important. As their apprehension of a war closer to home grew, they gave little notice to uproars in Asia.[12]

At the same time Halliburton was sending home dispatches about the war between China and Japan, another war—of which he seemed oblivious—waged between the Communists under Mao Tse-Tung and the Nationalists under General Chiang Kai-Shek. Strongest of the warlords competing for domination of China and ostensible successor of Sun Yat-sen, Chiang had the Luces' support. Chiang liked neither the Russians nor the Americans and did his best to manipulate both nations to his advantage. As he cultivated these dislikes, the Chinese rebels were in his mind posing a greater threat than were the invading Japanese. To his displeasure, the Long March of 1934-35 had relocated the opposition's Communist revolutionary base from Southeast China to Northwest China. A bitter struggle resulted between the Nationalists and Communists, leaving Chiang torn as to which enemy should receive the entirety of his energies. China's major military effort to stop a Japanese invasion of Shanghai failed, and its scorched-earth policies to check the Japanese advance cost many thousands of lives. Even so, Chiang, the titular ruler of China, would forever be criticized for the attention he placed on the Chinese Communists rather than on the Japanese.[13]

Treaties concluded in Washington in 1922 ended the Anglo-Japanese Alliance, but, to many Britons' displeasure, non-renewal was a mistake. *Oriental Affairs* editor H. G. W. Woodhead wrote the following of the naval agreements limiting warship construction by signatories Great Britain, the United States, Japan, France, and Italy: "[They] were expected to result in close cooperation between America, Britain, France and Japan in the Far East (but) they did not have that effect." Instead, the treaties "isolated Japan, and produced an undercurrent of resentment in that country."

Adding strain was the Immigration Act of 1924, dubbed the "Japanese Exclusion Act." The press gave ample coverage to "foreign affairs." In the mid-1930s newspapers covered the outbreak of war between China and Japan as fervently as they ran features about superhighways one day connecting the American coasts and robots one day assisting with household chores. Readers could travel with Ernie Pyle around America, or attend a World Fair. They could dream of new appliances for the modern home and vacations to sunny beaches even as they learned that the expanding Japanese Empire might soon jeopardize Midway and Wake Island, America's isles of defense. If fears could have been measured, the greatest one was probably a transatlantic bombing attack from Germany—not Richard Halliburton crossing the Pacific in a medieval boat.[14]

Frequent visitor to China, Wendell Willkie insisted that the Allies' more actively seek cooperation with China. He presumed that ending anarchy in China was necessary for the region's stability and for peace in the Pacific. Earlier in the decade, *Times* reporter H. C. Thomson advocated "close cooperation between the four great trading nations of the Pacific—between China, Japan, the United States and Great Britain." Militating against the ideal alliance was Japan. Halliburton deemed the country "determined" to keep China "industrially primitive in order to be conquered." He also believed Japan was keeping China "isolated from westernization," mainly by shutting out "social doctrines" arriving to it through Hong Kong.[15]

George Washington's caveat to avoid "foreign entanglements" was ignored when the United States, following its victory in the Spanish-American War, annexed the Philippines; by then the US also counted among its possessions a number of islands in the Pacific including Hawaii. In 1899, the Open Door Policy initiated free trade with China, and a waterway through Panama linking the Atlantic to the Pacific was in its early planning stages. Supporters of Theodore Roosevelt's dream to put the country at the center of the world stage could rejoice. As the tide of history predictably ebbs and flows, Halliburton himself must be seen as an envoy of later US missions in the Pacific, as well as a champion of American loyalty to its mother country. While the Englishman was not always so popular in America, Halliburton's own Anglophilia is unmistakable. His notions that a *Pax Americana* become the new *Pax Romana* and the Pacific become *mare nostrum* to Americans were equally evident.[16]

As Ernie Pyle's last days were spent on European, so Halliburton's were spent on Asian soil, a place considered by most Americans far remoter and scarier than the mother continent. If Europeans *thought* like Americans, the Chinese did not. Gold Rush storyteller Bret Harte called them "the peculiar heathen Chinee," and when the earthquake of 1906 failed to wolf down San Francisco's Chinatown, he caustically remarked that there were some things even "the earth can't swallow." In the 1930s, Charles Caldwell Dobie wrote of the Chinese in America as "a people apart," as victims of "ignorance and patronage," fear and hatred, during their early years in this country. It was known that Chinese "coolies" had helped build America's transcontinental railway and, on the other side of the world, the Great Wall. When the United States entered World War II in 1941, most Americans *still* knew little about distant China—or Japan. Harry Franck's *The Japanese Empire—A Geographical Reader* was reissued that same year to act as a primer. In newspapers feature articles about Asia soon were common.[17]

Before 1938, Halliburton acted as a sort of site visitor, arriving and quickly departing from places mostly at peace or on the brink of war. Hong Kong would become his Troy, and he its Aeneas, escaping a burning citadel soon to be captured and ransacked. Beset, and beleaguered, he built "a Chinese junk in a Chinese shipyard during a war with Japan." Seeing that America was fast becoming Britain's successor as a world power, he foresaw too that China might one day emerge as its commercial equal. According to historian Don Skemer, the *Sea Dragon* Expedition "would serve as a display of American solidarity with China against Japanese military conquest at the beginning of World War II." Still, as major headline news unfolded around it, the undertaking must be seen as a sideshow or comic relief.[18]

By 1938, Uncle Sam's involvement in the Orient had a long history. However, when Halliburton traipsed off to China, the war in the Pacific had not yet begun. A reporter of the country's stirrings, he became a late early voice in the history of modern China. These were the days before air force commander Claire Chennault and army general Joseph Stilwell both stood on the side of China against Japan. The servicemen did so at the reluctant bidding of Nationalist China and Chiang Kai-Shek, who saw another foreign presence in his country as a threat. Both Chennault and Stilwell insisted that China have an able fighting force, with Stilwell independently urging that the existing Chinese army be reformed and better trained. Stationed for a time in British Burma and India, Stilwell developed a contempt for the British high command, putting his trust in America to assist the Nationals and secure for itself a strong foothold in China. Enacted on March 11, 1941, the interventionist Lend-Lease Act, was meant to supply US allies, including Republican China, with food, gas, and equipment. Although not popular at home, it effectively ended American neutrality in the Pacific.[19]

The map of the world was rapidly changing, and its "blank spaces" were no longer so blank. By 1939, new powers were emerging, older ones declining. In March 1942, Col. James Doolittle led US aerial raids on Tokyo, and the US Navy began sinking Japanese vessels in deadly earnest. In November 1944 at the Battle of Leyte Gulf, General MacArthur destroyed the Japanese navy. In June of that same year, the Allies had landed at Normandy to continue the brutal emancipation of Europe from Nazi control. On August 6, 1945, the *Enola Gay* dropped the atomic bomb on Hiroshima. Three days later another atomic bomb was dropped on Nagasaki, and on August 15, Japan surrendered to the allies. The island nation that had inflicted so much hardship in the Far East would become America's chief ally, and, alongside defeated Germany, it would have for a time the strongest currency rating in the world. By then Halliburton was lost in the rubble.[20]

3

Why a Junk?

Before venerating junks as miracles of "beauty, grace and glamour," Halliburton confessed that, as a young man, he dismissed them as "ramshackle, unwieldly, unseaworthy scows slogging along, and manned by laundrymen." Then, during his first trip around the world—booking passage on many a shallow skiff or honking freighter, he claimed never to have seen any ship to equal the sheer beauty of a junk's "full winged sailing grace." His paeons of praise had only begun. "I spent several years in wandering by sea and land, visiting all the nice warm countries on the map. These travels at last brought me to China. And in the harbor of Foochow I found my first true love again—ships with sails. Not just one or a dozen, but scores and hundreds. The harbor was alive with sails." Here were "square miles of houseboats" and "towering Chinese junks." The sprawling, colorful community with all "its lacquered and gilded shops, ivory carving and jade shops, fans and silks" he called the "epitome of China" where "all the wildest legends and pictures and colors one has heard rumored about . . . are found."[1]

About to embark on his final journey, Halliburton spoke of sailing a ship from China to America as the basis for every travel adventure he ever undertook. "It was seeing a schooner—years ago—with its great wings spread, sailing out through the Golden Gate at San Francisco, that first made me want to go to sea. My heart went straight aboard her, and, until this day, it's never come home—for long. That beautiful, beckoning schooner disappeared down the horizon. But it gave me ideas." Over time the schooner transformed into a junk. He was just twelve when a friend gave him a "real sea-going model" of one: "With this junk my friends and I sometimes played a small-boy game," he wrote. "One side of the pond we pretended was China. A rock we called Canton; and another rock on the other side of the pond, San Francisco. The 'Chinese' merchants in Canton loaded the junk with tiny bales of leaves marked 'tea,' and with strips of red handkerchief marked 'silk.' With this cargo the junk sailed across the Pacific Ocean to San Francisco. Here the tea and silk were unloaded. Payment for the cargo—marbles—was put aboard the junk, which was towed by a string against the wind, back to China." Over the years, his recollections of Chinese merchants, marbles and tea remained vivid. The strange vessels warranted in his mind their own book of marvels. "The very thought of exploring the China Seas," he said, "and then crossing the Pacific, in such a romantic, picturesque and unconventional craft, has always excited my imagination." If Amelia Earhart had flown across the Atlantic "for the fun of it," he would cross the Pacific for the "adversity" of it. "If the junk should be small, the storms violent, and the voyage long—all the better," he said, "for if there is no hazard, no battle, where is the sport?"[2]

Years passed with hardly a murmur from Halliburton about junks, and though the idea of sailing a junk across the Pacific may have originated in childhood, what he would call "his new inspiration" actually came about, decades later, one sunny mid-afternoon in 1935. The site was Los Angeles Harbor and the place aboard a yacht owned by his financier cousin Erle Halliburton. Lovingly named the *Vida* after his wife Vida Catherine (Taber), the ship would have its own adventures at sea. Erle had a second daughter, seventeen, also named Vida, Vida Jessie. Eldest daughter Zola Catherine later remembered her romantic relative relaxed in a deck chair and his wearing loose-fitting white linens and loafers without socks.[3]

As soft waves dashed against the hull, the topic of fabulous boating excursions might have been broached while the Halliburton family lounged. The Trans-Pacific Yacht Races, begun by Clarence MacFarlane in 1906 and aimed at competition from San Pedro, California, to Diamond Head, Hawaii, might have influenced Jack London thirty years earlier. During the recent 1936 competition, the 2,200-mile run was accomplished in thirteen days, and this feat might now have inspired Halliburton to make the breadth of the Pacific his own even larger aim. There was also talk that the "Tahiti Race" in its South Pacific run would originate from San Francisco or "the Marine Exposition." New fashions in capitalist enterprise were hot topics to Erle, journeys to exotic places hot ones to Richard.[4]

With its theme of joining East and West, and promise of rekindling his career, the fair thrilled Halliburton as nothing else. And why arrive at 'the Greatest Show on Earth' in an ocean liner or steamer. How prosaic. "Ordinary sailing craft" wouldn't do; taken by "its bright-colored mat sails, its big wooden 'eyes,' (and) its fascinating history," he had to arrive in a Chinese junk. From earliest times, every expanse of water in the Far East was its domain. Junks had sailed far and wide and even gone from China to Mexico; one, the *Keyling*, reached London during Queen Victoria's reign, and about the same time a fleet of seven reached Monterey, California. Radio magnate and Odditorium creator Robert Ripley owned a junk, the *Mon Lei*, and toured the world in luxurious ease.[5]

Soon, Halliburton would have a name for his own junk. The "Sea Dragon" was a nickname given to Sir Francis Drake, the great English navigator who in 1579 landed just north of San Francisco. For several months he had sailed along the coast past what would one day become Laguna Beach and Los Angeles. So it was that both Halliburton and Drake had traveled the globe. At a Groton, Connecticut shipyard, the *USS Seadragon*, a submarine, was laid out on April 18, 1938—and launched on April 21, 1939, but Halliburton does not seem to have heard of it. Whatever the source of the name, adventurer, now scholar, Halliburton steeped himself in dragon lore. In Chinese mythology, he learned that the magical dragon— which had evolved from alligators once flourishing along China's major riverbanks, symbolized happiness, prosperity and fertility. The Chinese for their part believed themselves descended from the dragon, a wanton creature whose only demand was that it be revered. A water rather than land creature, the flames that shot forth violently from its mouth as it roamed the sea were said to be extinguished by the sun as it perpetually tried to devour them. China's greatest sibyl was the Dragon-Mother who, after her death, was carried from one shore to another by five dragons whom she had once rescued.

The only mythological creature in the Chinese zodiac, a creature that thrives on color and can readily change form, the dragon was the perfect mascot for the daring adventure Halliburton envisioned.[6]

The *Sea Dragon* Expedition, a publicity stunt plugging the Golden Gate International Expedition, was also a cultural statement announcing world unity through commerce. For Halliburton, Chinese voyages to America and European voyages to China had long been customary. He believed "Columbus' caravels had reached Asia after all, been seized by the Chinese, given a new paint job, and kept in service under the name of junks." Before Thor Heyerdahl popularized the idea, Halliburton was persuaded that the Pacific Ocean was a delivery network composed of highways, and that intercontinental commerce in ancient times was natural and commonplace. How was it done? To Westerners, these floating structures with their horseshoe-shaped sterns, odd rigging, and curving planked hulls appeared awkward, even creepy—like bats or pterodactyls on a raft. But "contrary to general belief," Halliburton told his readers, "junks are among the most seaworthy of ships." They were of many kinds: merchant junks, fishing junks, exploration junks, war junks. In addition, "novelty junks of nearly 1000 feet and metal-plated throughout called 'turtle-junks' had existed, as had ones with brothels called 'flower-ships.'"[7]

What became known as the Halliburton Trans-Pacific Chinese Junk Expedition developed gradually from its first "inspiration"—and maybe from a "small boy game" and maybe from some random talk about it aboard the *Vida*. From July 1936, Halliburton's letters home contain only offhand references to it. A junk voyage was just one adventure that another of his adventure schemes might at any time replace. Lecturer Halliburton was in New York or Washington, DC, when news of Amelia Earhart's disappearance in the Pacific was broadcast nationwide. Undisturbed, the news, illuminating Pacific adventure, piqued his increasing fascination with the Pacific and trans-Pacific transportation. At the end of August 1937, he reported that, following "several profitable conferences with Fair people," his "Chinese junk plans were getting very hot," and he had "taken practical steps forward." Distractions were many. Throughout 1937, lecture appearances, promotional events, the *Books of Marvels*, and a new house nearly smothered Halliburton's goal of a trans-Pacific voyage. He decided then to hire "a lively agent in San Francisco, named Wilfred Crowell" to take "charge of the junk plans."

A business acquaintance of Walter Gaines Swanson, Crowell was an executive at the Schwabacher-Frey Company. Halliburton's meeting with him was set for February.[8]

Meeting with Halliburton, the "Fair people"—Clyde Vandeburg and Art Linkletter, once they "agreed on the site where the junk (was) to anchor, the rent, commission, etc.," appointed him "official junk man." "*Lloyds*," Halliburton said, "will insure the investment." The maddening rush had begun, and would it culminate with him getting his testamentary affairs largely in order by the end of June. By mid-February he had met with "junk agents," capping "a furiously busy two days" including three stints on the radio and "a score of conferences." At a dinner engagement, Halliburton met the Chinese Six Companies, a group of venture capitalists whose nationalism was secondary to their profit incentive. Founded in the 1850s, the *Six*—actually consisting of over a thousand members, wielded enormous power. Charles Dobie called it a merchant league as well as "a tribunal—a government within the borders of the established government of the United States"—with a president, a reigning board, and headquarters. Another such organization was the Four Families. In descending order, the racketeering "tongs" were known to conduct shady banking operations. Representatives of the Six believed profits could be made from the junk expedition, and in short order they agreed to form Richard Halliburton Enterprises, at the time a somewhat vague business entity with little beyond word-of-mouth approvals and nothing in the way of supporting legal documentation.[9]

From the start, Wesley and Nelle Nance Halliburton thought their son's idea "hazardous and foolish." Coming to his own rescue, he told them to "think of it as wonderful sport" and "a wonderful adventure—properly safeguarded." Image mattered in a venture that would "pitch [him] out into the spotlight again and on towards another *book*." He had a penchant for positive obsession. He once confessed, "All I can think about, or write about these days, is my 'Flying Carpet.'" Now it was his "*Sea Dragon*" that would soon obsess him. Doubt, however, remained. Only after Halliburton arrived in Hong Kong did he feel the "wonderful adventure" might come to pass and thus begin writing the "Log of the Sea Dragon" articles.[10]

4

Lecture Circuit Drives Halliburton to the Sea

Simply put, job *burnout* rekindled Halliburton's interest in junks. Call it a leave of absence, or much-needed vacation, the *Sea Dragon* Expedition was the product of work-induced physical exhaustion. His switch in roles might seem, besides a desperate adjustment to boredom, a remedy more dangerous, but sailing a small wooden craft across a vast sea seemed less perilous to him than lecture-touring. Like Will Rogers and Amelia Earhart, he could talk at length about himself and get paid well for it. Train stop after train stop took its toll and soon he realized that the discomforts of irregular employment outweighed the joys. While all sorts came to see him, his audiences consisted mostly of Caucasians, Christians, young adults, collegians, middle-income people, and women from auxiliary clubs. Ardent fans—"groupies"—flocked to hear this salesman of romance. At times, the attention drove Halliburton to seclusion, as his professed need for privacy often competed with his desire for fame and fortune. Once Halliburton, writes Jonathan Root, "canceled five engagements and rested for five days but it wasn't enough." Returning to the circuit, "he gave nine-gravel-throated talks and lost his voice again in Oklahoma City where his lectures were not only mobbed, but his hotel room, too, and he could get no rest until he hired a detective to keep visitors away." The steady drama of personal appearances, book signings, and parties made him "absolutely dizzy." In one ten-day period alone, he had nearly two dozen bookings. On another occasion, his booking agent, Monica Grey, scheduled him for eight appearances in a single week. Sometimes Halliburton lectured every day, sometimes twice a day. If he wished, he could even get the lecture bureau to book him for three appearances a day.[1]

Awaiting Halliburton at any stopover were "praise and damnation, trouble and happiness." Like a victim of Ivan the Terrible, he felt alternately dipped into a cauldron of boiling, then freezing water, as a merciless schedule often brought him one day into a severely cold climate, the next day into a horribly warm one. Around admirers, he often found himself alone. "Just to mix with people for excitement," he said, "bores me painfully." His father thought him too "serious." Was he? "There's just no helping my unsociable nature," son Richard explained to him. "I've always been wearied by the company and minds and hearts of most human beings, ever since early childhood. But this in no way comes from sourness or misanthropy. I'm not in the least antagonistic—just not interested. . . . Occasionally I do find a fellow spirit, and then I'm as social as anybody." Halliburton found opportunities for intimacy and sexual dalliance in one stop or another on the tour. To "fellow spirit" Noel Sullivan he wrote, "My gland life is too scandalous to report." Fans besieged him at times, but Halliburton more often scrolled up his maps and walked off the stage to a hotel room or train alone. "Friends and acquaintances—(I have) too many," he said, adding, "All of them could disappear and I wouldn't know it."[2]

The combination of fear and excitement made Halliburton giddy. Stricken with stage fright, he might freeze momentarily before he got into the part he was to play—that of a poised showman. Dissociation from the moment helped, as did inhalation exercises. He dismissed voice projection equipment as cumbersome. In letters to his parents, he often complained of a "hoarse" or "raw" voice—by all accounts, a "pleasing tenor voice that grew in intensity, volume, and pitch as he became passionate about his presentation." In time, he canceled the demanding "two-a-days," as the first lecture might be hearty, the second one thin. Presence of mind, in any event, could be lost: "Sometimes, for no reason, I get so tense," he confessed, "my tongue goes completely out of control." On one occasion, he spoke at a metal-roofed gymnasium while it rained heavily and a freight train traveled past the building. He recalled, "Nobody could hear a word, though I yelled myself *hoarse*." Gargling salt water, or drinking lemon juice or white vinegar only helped temporarily. Lozenges and other aids in capsule form were available. For simple aches and pains, he took "aspirins." Extended rest helped, but only somewhat. For lingering ailments, he sought professional help. He stopped once at Methodist Hospital in Indianapolis for a "sinus drainage," caused by too much "roaming" in "horrible weather." At another point, a doctor examined Halliburton and told him that he had all the "symptoms of goiter." In turn, a goiter specialist in New York told him that "a marked secretion of thyroid . . . gave rise to all (his) discomfort." Halliburton characterized another affliction as the "trembles." He told the doctor, "[I suffer from] extreme nervousnous—the abnormal stimulation that kept my pulse beating and kept me so weak I could scarcely wiggle." He grew no stronger.[3]

Conclude that Halliburton, throughout his career, fought his own frailties—his a stalwart spirit joined to a fragile body. Despite his reputation for physical dare-devilry, he remained constitutionally weak over the long haul. Harsh proof of life's brevity struck him early with the death in 1917 of his only sibling, younger brother Wesley Jr. In 1915 Richard himself was diagnosed with tachycardia and confined to the Battle Creek (Michigan) Sanitarium, later nicknamed "Wellville," for four months. For Halliburton's restless nature, the sanitarium was a dull place—except for its well-articulated tips on healthful living. Owned and operated by the Seventh-Day Adventist Church, the rehabilitation facility appealed to celebrities. Once famous, Halliburton joined their ranks. His visits to Wellville and to other clinics were initially periodic, but they became more frequent. The ailments he concealed from the public during his time in China could not be concealed from his crew.[4]

Halliburton griped constantly about the frenetic pace of lecture touring, which he called "a tempestuous, nerve-whacking business." As had Mark Twain, dialect poet James Whitcomb Riley condemned lecturing as "an awful life," noting the wearisome train rides between remote bookings, and the inclement weather that might prolong those trips. Not bound to a hated nine-to-five office job, Halliburton became a "lecture-slave." The "circuit" itself he called a "grind" or "a mad gallop." Exercise, including swimming and walking, helped improve his outlook if not his general health, as did maintaining a proper diet. He relished "systematic exercise," but his "physical state" remained "an enigma" to him. Playing nine holes of golf exhausted him, and he found motoring and sailing strenuous, but he had swum the Hellespont without any difficulty. For a model of endurance and stamina on the dais, Halliburton could always recall feisty Theodore Roosevelt, who gave upward of six hundred speeches in a single campaign sweep. But the world traveler couldn't keep up. Harvard graduate Albert Shattuck, who for a time worked as Halliburton's secretary, remarked, "When he is lecturing he is in a tremendous sweat and uses up a shirt an evening."[5]

World traveler Henry Savage-Landor recalled "the horrors of the lecturing tour" and characterized lecturing itself as "tiring and monotonous, repeating the same things over and over again." Halliburton, by comparison, had a repertoire of seven stories—"old reliables." Requiring little preparation, their constant retelling became a drain. Hundreds of times he recounted in electrifying detail his Panama Canal swim, scaling of the Matterhorn, and descent into the Mayan Well of Death. Self-critical, he replayed performances in his mind. On stage he could startle and move. He also possessed a generalizing power, if only because simplification in a complex world held greater truth for him and his readers than the unconnected snarl of facts comprising reality. Information had to be pleasantly imparted as fun facts to be painlessly received. He had the common touch and was flexible. He quickly learned what approach to take with each audience. To women's groups, he had to be the boy next door; to scouting groups, 'Dick Boy Scout.' To Kiwanis Club members, he presented himself as a nice, promising young man; to young ladies' groups, a clean, neat, polite, and marriageable gentleman. With young adult audiences he might begin a lecture with words similar to those that open the first *Book of Marvels*: "As I have no children of my own, I will have to make you my children for a moment and take you to the wonders of the world." To collegians he might say, 'College only prolongs adolescence and is no substitute for the open road." Besides good cheer and gentle nudges into action, Halliburton could impart "grim, earthly, depressing things." To older audiences, people over thirty, he might say the equivalent of 'Enjoy yourself—it's later than you think.' His lectures were a sort of geographic magic show, consisting of a series of word-pictures which might feature one or another ancient wonder or modern marvel: a congested street scene in Cairo or a singular monkey peeling a banana. He was at his best when he was offhanded: 'Did you know elephants sleep while standing? . . . Ocelots like to climb in bed with you and cuddle? . . . Toucans have these huge beaks, but they're not at all self-conscious about them. . . . Here I'm swimming across Gatun Lake in Panama oblivious to the barracudas beneath me. . . . Bangkok is a fun place to be—for about two days. Did you know California was once thought to be an island? . . . Here we are at a spa on the Riviera drinking cognac from goblets used by Napoleon at Elba . . . . Here we are in Timbuctoo up to our ankles in sand—and camel excrement. Next slide, please.'[6]

Until the end of his final tour of duty as a lecturer in 1938, Halliburton seldom talked to audiences about junks or long ocean voyages By then, however, it was clear to those closest to him that the lecture circuit was 'getting to him.' Like Savage-Landor, he had aimed not so much to instruct as to fascinate, to excite, and, as travel writer Robert Byron hoped of his work, to enable his hearer "in (the) future (to) experience some quickening of historical experience." His role often made him "very blue" and threw him into fits of self-analysis. Train stop after train stop, taxi delivery after taxi delivery, bow after bow, handshake after handshake, he had to confess, "This furious motion is stimulating, but it burns up *too much*." Yet he also admitted, "I'm never so happy as when I'm busiest and have most to worry about." Still, one glimpses in his professional beginnings the pattern of existence that led him to his death. He was impatient and hated confinement.[7]

By 1935, Halliburton was spiritually and physically spent. Following his attempt to cross the Alps on an elephant—to rival Hannibal—he collapsed, spent two weeks in a Paris hotel, and overate, overslept, and, according to Root, "indulg(ed) in everything except alcohol which he still shunned." On his return to America, Halliburton's bookings and fees increased. During an appearance in Birmingham, writes Root, Halliburton suffered from "laryngitis so severe he could scarcely whisper." In just a couple years, the demands of the tour had taken over Halliburton's life just as the regular employment he had willfully averted might have done. "I've learned that I can never decide situations," he concluded, "but that situations decide me." The tour had visibly aged the Pied Piper of youthful enterprise. "Reliving and re-traveling set courses," he said, "are too much of a strain." So was speaking to hundreds of people every night and shaking their hands.[8]

Still, Halliburton could rejoice. He had successfully avoided the dead-end job, the dreaded "fixed existence, and servitude to a master. The lecture circuit had initially deluded him into thinking that he was free because he traveled from place to place and was never confined to a desk job in an office. At thirty-five, however, he hardly had the stamina or enthusiasm of the young man who had first spoken at the YMCA hundreds of lectures and small towns ago. Personal appearances, such as at the New Roxy Theatre in New York to introduce *India Speaks*, wore him out. Early in 1936, Halliburton found work on a "newspaper strip serial" that promised him "fun, and a brief respite from the grind." It also gave him an opportunity "to spare his voice." Meanwhile, there was always another lecture engagement. After a last bow, the train whistle soon blew, and off he went.[9]

Readers of Halliburton's letters cannot miss how much the touring was getting to him. "I'm so fed up with this grind, I'd chuck it for two cents," he exclaimed. Eight lectures in five days proved too much. A lecture agreement he received from Alber-Wickes—the Boston-based lecture bureau scheduling him, mightily distressed him. In a letter dated June 25, 1938, he confided to his parents, "I dread having to tell Monica I can't accept the 30 lecture dates she's booked for me." His well-being was put in serious danger by so great a commitment. "I can't sleep," he said, "without dreaming of fees and per-cents and railroad fares and circulars and contracts." Photographs show that the routine had taken a toll on his health. He nearly suffered a complete physical collapse and at least once came close to suffering a nervous breakdown. His prime of life had become but a frost of cares. The occupation that had freed him from the ordinary world of work now imprisoned him. The *Sea Dragon* Expedition would release him from the lecture circuit's demands as the *Flying Carpet* Expedition had done for him. If he was to unleash his imagination, Halliburton had to "sail away from lecture management and scheming and printers and radios and a thousand other things." His last lecture tour ended in March 1938. Three months past. Given ample time to reconsider, he had to tell his booking agent that he would be unable to accept the thirty lectures scheduled for him in the fall. By then he had far bigger fish to fry—in that other country within America: California.[10]

5

East Is East, and West Is San Francisco

In Halliburton's day, California was better known for its sunshine than its smog. Lofty mountains protected it in the east, ocean waves lashed its rocky shores in the west. Between water and rock were grape orchards and orange groves; desert terrain only added to its mystery. "Best place in the world to be is America; best place in America to be is California. Here anyone, no matter his defeats, can get a fresh start, anyone, no matter his triumphs, can soar even higher." The luckiest people lived in California. Never a dull moment. Life's awful moments of solitude were eliminated. Every wish was fulfilled, every dream realized. Inhale the cooling ocean breeze, and one's aches and pains suddenly vanished. Put all care aside. Opportunity beckoned, and, while money didn't grow on every eucalyptus tree, it grew on more trees than it did elsewhere. Contact with Hollywood people in particular bred in Halliburton a sense of his earning potential. "The money thrown around here still appalls me," Halliburton grumbled: "People of the commonest, stupidest type get $5,000, $7,000, $10,000 a week. Any actor or director or writer receiving less than $1000 feels ready to join a revolution. At the same time other people are starving by the thousands. . . . There's the worst depression in history here." If fools could make tons of money, he deduced, how might an intelligent man of vision like himself fare?[1]

By the late 1920s, Halliburton's lecture stops had shifted to the Pacific coast, and by 1935, he had assuredly *gone Hollywood*. A new culture of movie-making entrepreneurs had emerged in the mock-up studios near Griffith Park where scenes from *India Speaks* were filmed. Many filmmakers had built mansions, and Halliburton would soon build his dream house on a ridge overlooking the Pacific Ocean, forty miles south of the madding Hollywood crowd. His first home was a bungalow on the shores of Laguna Beach; atop a ridge, Hangover House, his second home, after 1937, overlooked the ocean. Laguna Beach was itself a quiet, sheltered resort community. Enjoying its breathtaking ocean views were Hollywood celebrities—some there to make movies, family tour- ists—by the car load, and cliff-side artists—many well-to-doers hailing from the east. For commutes up the coast to Carmel and San Francisco, Halliburton purchased a secondhand Dodge coupe.[2]

Through all the fuss and fury of a busy life, an adventure involv- ing California and some aspect of its history had first occurred to Halliburton in 1934. He had stood on the "then uncompleted "(Oakland) Bay Bridge" that year and wondered how to associate himself with all the publicity it would receive. The sky had held the promise of adventure a couple years earlier, but the sea now replaced it. In late February 1936, Halliburton attended the Twenty-Fourth National Sportsman Show at Grand Central Palace in New York. There, outside an authentic fishing shack in the State of Maine Exhibit, he was photographed listening to Whitney Thompson of Port Clyde, Maine rattle off yarns about the old square-riggers. Thompson, an old salt sea skipper, might have rekindled Halliburton's joy in ocean sport.[3]

Erle Halliburton had already made a name for himself in the oil and cement industry and in aeronautics. From the late 1920s, and to the end of becoming at least as financially independent as his cousin, Richard had been admitted into the circle of film and flying celebrities who congregated at the ranch of wealthy heiress and renowned aviatrix Florence "Pancho" Barnes. These who gathered at Barnes's San Marino home included noted actress (and licensed pilot) Ruth Chatterton and matinee idol Ramon "The Pagan" Novarro. Halliburton hobnobbed with members of the prestigious Bohemian Grove Club headquartered near Glen Ellen, backyard of the late Jack London. He cultivated long-standing friendships with Senator James Duval Phelan, at whose estate at Villa

Montalvo he was a frequent guest, and with Senator Phelan's nephew Noel Sullivan.[4]

Also a guest was *grande dame* of American letters Gertrude Franklin Horn Atherton, a member of California royalty who had her own special room. Not to know the now-aging state historian in residence was to proclaim one's self unknown in the higher echelons of San Francisco society. Besides a steady stream of fiction, editorials, and articles, Atherton had written books about California and San Francisco history. Of every past effort to link the fortunes of North America and Asia, she took sovereign notice. Nikolai Resanov, whose failed attempt in the early 1800s to establish a Russian Pacific Empire from Siberia to America's Northwest, in particular drew her attention.[5]

Memory of Resanov had ebbed long ago. With the 1939 opening of the Golden Gate International Exposition, a new age in California history dawned. Journalist Herb Caen was one of many who saw the 1930s as the final act of San Francisco's "golden age." As a song of the day went, "Of Frisco's Treasure Island I have dreamed." The year 1939 itself marked the start of a new century—at least the fairs, the one in New York and the other in San Francisco, would make it seem so. If the 1920s were "a precursor of modern excesses," the new decade foreshadowed doom. Halliburton's two *Books of Marvels*, published in the late 1930s, had shown that mankind could create like the gods; *Lost Treasures of Europe*, published a decade later, showed that mankind could more easily destroy like them. Against all threatening odds, the fairs heralded brighter tomorrows and championed the vague emoluments of peace.[6]

Among the many cities Halliburton had visited in his career, he liked Karachi (Pakistan), Canton (China), Srinagar (Kashmir) and Portland (Oregon) best. Yet he put San Francisco above these, saying, "It *is* the most wonderful and beautiful city in America." Another devotee of the city, Bret Harte, remarked, "East is East, and West is San Francisco." He might have added that it boasted the steepest hills and the finest natural harbor in the world. In the San Francisco of 1939, some people remembered sharing a quiet beverage down at the wharf with Jack London (1876–1916). Others even remembered mad Emperor Joshua Norton (1819–1880), who in the tattered regalia of an exiled military autocrat, roamed the city's financial district with canine companions Bummer and Lazarus and, like Jack London, marked the multitudes of extinct

Chinese junks in San Francisco Bay. Stories still circulated about the preposterously blimpish Norton and his dishing out edicts to the shopkeepers who light-heartedly honored his fake certificates of exchange. In the end, Norton did achieve a measure of civic renown—and offer further proof of his lunacy, when he decreed that a bridge be built to span San Francisco and the East Bay. In more recent times, some San Franciscans recalled "demon of the sky" Lincoln Beachey (1887-1915), the Great Waldo Pepper of his day. Dressed in suit and tie, heart-stopper Beachey performed aerial somersaults in his monoplane and said of the thirty million people who had seen his act over the years, "They paid to see me die."[7]

"City of the Future," San Francisco in 1938 no longer had the old Portsmouth waterfront look that it had in the 1850s. Gone were the ramshackle houses, dirt roads and hitching posts for horses. Fur traders had moved far north, gold miners to the east. The Asian pirates the young Jack London had bumped heads with offshore no longer seemed a threat. San Francisco's seven hills were dotted now with world-class hotels and restaurants, a streetcar network, and municipal buildings to match any in the country. Traffic hummed, and a single horn beeping here and there suggested bigger cities like Paris or New York. In the late 1930s, the city's tallest buildings were not yet so tall; the Palace and Saint Francis Hotels, risen from the ashes and debris of 1906, were among the tallest. Two luxury hotels atop Nob Hill, the Fairmont and Mark Hopkins, overlooked San Francisco and the Golden Gate, as did Coit Tower on Telegraph Hill. Some old Victorian houses seemed as tall as these hotels, and they certainly lent noble character to what could have become a shantytown. Bookshops with postcard turnstiles flourished in the busy metro district and its outskirts. Halliburton's favorite was Paul Elder's on Van Ness. Fashionable men's and ladies garment shops were located mostly in the business district, near Maiden Lane's tea shops and boutiques. Cut-rate shops stood blocks away, mostly on Market Street. With names like Geary and Montgomery, Columbus and Lombard, Kearney and Sutter, the streets and alleyways had colorful stories associated with them. Main thoroughfares were Market Street and Broadway. San Francisco still conjured up images of the pirates and black-marketers, strip joints and gambling halls of its rowdy Barbary Coast days, when the fabled likes of Lola Montez or Charles "Black Bart" Bolton still walked the streets. Notorious hangouts included, on

Montgomery, the Black Cat Café which featured offbeat vaudeville entertainment, and, on Broadway, Vanessi's Restaurant Bar and Grille, which
offered "a glimpse of old Venice." On Pacific Avenue, catering mainly to
longshoremen and writers, was former speakeasy Izzy Gomez's Café: on
one wall hung a bar-length mural with a panoramic view of sailing ships
on the Golden Gate; in the dining area fabulous steaks were served with
pasta; sometime the Portuguese Izzy Gomez himself came out to dance
a traditional gig to the accompaniment of an accordion. Hoping for a
better life, people from all parts of the world migrated here. The "Sydney
Ducks," Australian immigrants brought to America from Great Britain's
penal colonies, settled here, as did Chinese descendants of the coolies
who had survived the brutal hardships of building the transcontinental
railway. "Baghdad by the Bay" was also something of the New York of
Asia. In the same way, Asia had become something of San Francisco's
own West Coast—just as today Miami is for some the northernmost city
of South America, and for Connecticut Yankees plainly another country.
Fog regularly rolled in; a rainy season substituted for most of the rest
of the country's snowy one. Mark Twain said that the coldest summer
he ever experienced was one he spent in San Francisco; still, average
year-round temperatures were in the mid-fifties.[8]

If one were to pick the single event that determined Richard
Halliburton's fate, it was the building of the bridges. Without the bridges,
there might still have been a Fair, but without them, there likely would
not have been a *Sea Dragon* Expedition. If not an afternoon on Erle
Halliburton's yacht, credit for the idea should go to the aforementioned
Walter Gaines Swanson, public relations representative for the California Toll Bridge Authority who, as a publicity stunt, coaxed Halliburton
into walking across the unfinished bridge. Shrill winds blew that day,
Thursday, June 15, 1936, making the bridge sway like an unsteady ham-
mock slung between two trees—or like a ship pitching and yawing on
a stormy sea. Currents in the cold, turbulent water below rumbled at
4.5 to 7.5 knots.[9]

Knowing the risks but wooed by glory, Halliburton took Swanson
up on his dare. With the bridge's chief engineer Charles Hoehn leading
the way and friend Paul Mooney behind Hoehn, Halliburton inched his
way forward along the catwalks and girders, often on hands and knees.
Halliburton wore no safety harness to catch him should he fall; one misstep could have sent him barreling twenty-five stories into the tossing

waves below. Reportedly he walked only partway over the 4,200-foot span, at the time the longest in the world. On hand were photographers; so were veteran bridge workers who jeered the famed daredevil in their midst. Spry as monkeys and used to heights, a few of them taunted him with their mastery of aerial gymnastics. Lacking their agility, Halliburton looked foolish in his attempts to advance heel on toe and keep his balance, and, embarrassed, he feared he would be pitied. Sympathizing, Swanson proposed that Halliburton become the "first" man to walk across the "other" bridge as a recompense for having his athleticism so bruised. Halliburton accepted the invitation, and on Monday, June 8, Swanson and newspaper photographer Barney Peterson, both walking a step back, accompanied him. Task accomplished, Halliburton threw his arms up in triumph and panted in relief. That he was the very first person to cross the bridge is doubtful, but likely he was the first celebrity to cross the bridge. Later others would cross it on stilts, tap shoes, or roller skates—how Halliburton might have done it as a younger man.[10]

Amid great fanfare, the bridge officially opened for pedestrian and vehicle traffic on May 28, 1937. The event heralded the opening, shortly afterward, of the Golden Gate International Exposition. The Oakland Bay Bridge, another harbinger of the event, had officially opened on November 12, 1936. Across the country, on July 11, 1936, another engineering marvel, the Triborough Bridge linking the Bronx, Manhattan, and Queens, opened. It would lead visitors to the upcoming New York World's Fair, which would open on April 30, 1939. The east coast Fair's theme, "Building the World of Tomorrow," was deemed more scientific and less cultural than that of its San Francisco counterpart. The rival fair would eclipse the exposition in attendance, gate receipts, and fame, but to Swanson and Halliburton, the fair in New York was playfully dismissed as "that other fair." Besides their national and transnational appeal, both fairs were, in any case, potentially great money-makers, and Swanson conveyed as much to Halliburton. The public relations representative then brought up an adventure that he believed Halliburton would like, one that connected him to the San Francisco fair.[11]

In midsummer 1936, Halliburton's junk idea seemed fanciful. It had neither form nor substance, and the junk had as yet no name. The box office smash *San Francisco*, starring Clark Gable and Jeanette

MacDonald, released on June 26, amplified the theme of human resilience and cooperative enterprise. The Golden Gate International Exposition would be similarly successful. By February 1936, the Army Corps of Engineers had begun work on what was originally called the Yerba Buena Shoals Fill. With each step toward its realization tedious, painstaking, and dangerous, the island was literally dredged up from the bottom of San Francisco Bay. An engineering marvel in itself, some 20 million cubic yards of sand were next compacted within a 17,000-foot seawall containing 287,000 tons of quarried rock. Landscaping took months. In time hangars for the Clippers were built, trees planted, grass malls created, and waterworks put in place. At last, with San Francisco mayor Angelo Rossi in attendance, groundbreaking ceremonies commenced. Treasure Island had emerged.[12]

Already, San Francisco's citizens clamored for the city to host a successor to the California Pacific International Exposition drawing to a close in San Diego. World peace would be a theme, as would America's leading role in world economic policy. Leland Cutler, president of the fair, said that the event would depict "the art, science, industry and romance of the Seven Seas." He also noted that "it would especially portray the greatness of those nations bordering the Pacific Ocean—(besides) the Seven Seas, the Antipodes, Central and South America, the Occident and the Orient." On June 16, 1936, President Franklin Delano Roosevelt issued an invitation to the world asking all foreign countries and nations to participate in the grand undertaking. Later that year he was elected to a second term in office, and for the moment, the economic troubles of the country appeared to be lifted.[13]

Now and again Halliburton was a guest of Madame Chung, who lived atop Telegraph Hill at 1407 Montgomery Street. From his window he had a commanding view of "the two biggest bridges in the world." In a July 14, 1936, letter to his parents, he announced the "new inspiration" that vaguely owed its origin to an afternoon spent aboard the *Vida*. "As you know," he began, "there's to be a World's Fair here in 1939 to celebrate the bridges. The Fair promoters want me to hire a Chinese junk in Shanghai and sail it with Chinese crew and cargo to the fair. It would make part of a book." Known for its long rainy season, San Francisco was uncharacteristically sunny and pleasant that day, with temperatures in the sixties. Wesley and Nelle Nance Halliburton were in

Cairo, Egypt, and about to sail for Athens. Richard had himself recently visited Yosemite, marveling over the waterfalls and climbing Half Dome Mountain, whose nine thousand feet of arduous terrain had "exhausted" him. But he still thought the climb "fun." He had also produced several chapters of *Richard Halliburton's Book of Marvels*, and was preparing a series of stories for the fall lecture circuit. By that time, plans to build his house in Laguna Beach were well in place.[14]

6

Halliburtonland

If the two bridges inspired the *Sea Dragon* Expedition, the Golden Gate International Exposition crystallized it. The event was the very expression of Richard Halliburton himself. With the publication of the two *Books of Marvels*, the *Occident* in 1937 and the *Orient* in 1938, he had introduced a generation of young adults to many of the world's natural and man-made wonders. As the creations of Walt Disney would later become Disneyland, so the marvels at Treasure Island became "Halliburtonland." The exposition was "the last great event in the San Francisco Bay Area" before the United States entered World War II. Besides celebrate the Bay and Golden Gate Bridges, it paid homage to all the countries and islands touching the Pacific Ocean—a *Pax Pacifica* come true. Its causeways, resplendent courtyards, and temple-like structures, the "floating city of emerald and vermilion palaces" often called "Magic City," evoked the Aztec capital of Tenochtitlan that awed the Spanish conquistadors when they first beheld it. The fair's gardens and lagoons, princely buildings and connecting courts, colossal nude statuary and magnificent towers, evoked Babylon, sights lit and backlit by the subtly intoned colors of the newest in chromo-therapeutic technology. As Prometheus had given fire to mankind, so lighting technicians had illuminated "the diadem of the Bay." Making full use of "invisible light," the resulting color wizardry sparked everything from romance to resurrection. In scope, the cinematic epics of D. W. Griffith and Cecil B. DeMille could only suggest it.[1]

Welcoming the legions of visitors, "thirty-niners," who came to enjoy the spectacle was the colossal eighty foot tall statue of *Pacifica*, Junoesque "goddess of the Fair." "Gleaming in amber, rose, and gold," the four-hundred-foot-high Tower of the Sun loomed nearby. The plaques on each of its eight sides read in duplicate "Gentle Wind," "Cold Wind," "Trade Wind," and "Storm." The tower's summit featured a twenty-foot high golden phoenix. Each hour, forty-eight carillon bells rang from the structure's spire. Near the glittering Court of the Moon, two flanking Elephant Towers alongside "the Portals of the Pacific" opened the way to the avenues, all with colorful, alluring names and colonnaded by giant palm trees. The avenues led through terraced gardens, flowered courtyards, and Roman villa–style pools to the numerous exhibits and buildings—to Vacationland, the Court of the Seven Seas, the Temple of Music, and, an actual body of inland water, the Lake of All Nations.[2]

A reborn version—even continuation, of the Panama Pacific International Exposition of 1915, the new Exposition proclaimed, as did its predecessor, the emergence of Twentieth Century technology and a brave new world of global initiative. For the many who in 1939 still remembered the earlier Fair, the 1939 remake truly competed with it in splendor. At over $50,000,000 it certainly outdid it in cost. Enshrined were some of mankind's greatest achievements in "one of the most beautiful settings in the history of world's Fairs." While the Exposition's main themes were world harmony and cultural awareness, materialization of those themes amazed some and perplexed others. Startled by its clash of cultures, critics thought the event a heap of clotted nonsense, a hodgepodge of hokey ideas and fleeting visual thrills. Part ethnographic circus and part political statement, the garish ensemble of buildings, landscapes, and exhibits seemed a global village or sketch model of Disneyland. Gods, demigods, and heroes from some Eastern but mostly Western lands were welcomed: statues of Apollo, winged goddesses, nymphets, explorers, and Indian warriors gazed from atop buildings, graced fountains, or stood lifelike amid gardens. The Yerba Buena Club was a friendly place where visitors were given Fair information and could hear leading authorities give lectures on various topics. No logical sequencing brought visitors from one building to the next, from one exhibit or concession to the next. The Hall of Air Transportation stood beside the (new) Palace of Fine Arts. Located on the other side of the island—and adjacent to the majestic Court of the Hemispheres and Avenue of the Seven Seas, was Midget City, a "modern Lilliputia." Leading into the Court of Flowers was a Parisian-style Arch of Triumph. Visitors could grab a freshly skewered Malayan hot dog wrapped in palm leaves and cucumber slices, then go into the Hall of Miracles or watch the Transparent Man, next to whom a speaker with a pointer indicated "what happened to the human system when food was eaten, drugs and medicines were taken, or . . . sickness overtook it." The Temple of Religion and Tower of Peace stood a whisper away. A Hall of Friendship seating over two hundred persons offered a most ethereal choir of angels and mural showing the story of creation. More earthly was the Federal Plaza's "Span of Life" exhibit, an induction into the facts of life and New Deal economics.[3]

Tread forward at your own risk or reward. Near a Samoan grass-and-thatched-roof hut and a Native American Indian village was a Hall of Science where research booklets were distributed. One publication was entitled The Story of Sex Hormones. The lowest floor of the hall featured aquariums with giant squids and octopuses, while another floor devoted to prehistory contained dioramas of saber-toothed tigers and mastodons from the Pleistocene Age. Moving on to the PG&E's Electricity and Communications Building, visitors learned how far *Homo habilis* had advanced since the Stone Age. On one wall, a mural portrayed San Francisco as central to the sweep of California history from the Gold Rush to the future. Scenes included a covered wagon headed west, a sailing ship rocking with the tide, and passengers landing at a dock and some boarding a stagecoach. The Food and Cooking Pavilion introduced dream-eaters to a one-thousand-pound fruitcake. At the Court of Honor stood the Lady of the Evening holding the "evening star." The Dance of Life, a golden frieze depicting semi-clad females, could be seen between the Towers of the East. A Court of Reflections featuring two rectangular pools conveyed tranquility. One statue, The Girl and the Penguins, inspired a love for all living things, while another, Polynesian Woman, reclined in front of the Fountain of Western Water, inspired living long and well. After contemplating the statues, fair-goers could visit the

Serta-Sleeper Exhibit, lie down on a linen cloud, and dream about the latest in modern appliances. Or they could cool off at the Ice Industry Exhibit.[4]

Add it all up, the exposition was an amusing caricature of culture—not a complicated expression of cultural difference, but cultural reductionism at its most absurd. Also showbiz at its gaudiest, as some found it a jumble of leftover Hollywood sets. There was in fact a Hollywood Boulevard Pavilion with a giant motion-picture studio, but it was a small feature of an extravaganza which was really a collection of special interest topics and a potpourri of 'new worlds to comprehend.' Thrilling surprises ranged from the Lion Globe-A-Drome to the Musee Mecanique. Youngsters could enjoy the Children's Village, but the whole family could enjoy the Cavalcade of the West or the Italian marionettes at the Hall of Western States. Rides included the Elephant Train making regular runs to most of the exposition's attractions, as well as the roller coaster and two Ferris wheels in the "Joy (or Fun) Zone." Thrill seekers could also rent a Lusse Brothers Auto Scooter, board a Sightseeing Boat Ride, or hop on the Flying Scooter. For lunch or dinner, fair-goers could enter the La Plaza Area Buildings, which included the White Star Tuna Restaurant, Owl Drug Store, Dairyland Press, Continental Club, Open Air Theatre, and numerous walkways, boulevards, and rest areas. In the Foods and Beverages Palace, a wine temple celebrated California's grape and wine industry. The state's "liquid gold" flowed in steady streams of red and white at the "wine fountain." The culinary adventurists could eat at the Russian Restaurant on the Avenue of Olives, or at the Philippine Inn. Both American and Javanese cuisine was served at the Isle of Bali.[5]

Drawing the biggest crowds was "the Gayway," the fair's main avenue nicknamed the "Street of the Barkers" and "Highway of Bright Lights." An amusement park, here, "forty acres of fun" featured over six hundred concessions as well as many sideshows and music halls. Famed "fan dancer" Sally Rand's Nude Ranch brought to the zone lurid entertainment. The Folies Bergere and Ziegfeld Follies brought spectacular dance to it, and Mr. Ripley's Odditorium dazzling enigma. Swimming sensation and future film star Esther Williams made her debut in Billy Rose's star-studded water pageant the Aquacade. A less joyous experience awaited sightseers who visited the "Monster Pit" filled with unsightly ogres and rubber snakes. Elephants, lions, and tigers roamed in an "African Jungle Camp," and a South Seas exhibit displayed "a group of native villages transplanted, inhabitants and all." The "World of A Million Years Ago" introduced "moving, breathing mammals, reptiles, and men and women who inhabited the earth" in ancient times. Within the main courtyard, fair-goers could visit palaces exalting achievements in the liberal arts, agriculture, social economy, and food science and technology. At its center—all a-sparkle, was the Tower of Jewels. In the Palace of Fine and Decorative Arts, one found embroideries, handcrafted jewelry, baskets, ceramics, and the like. Portending the future of physics, the Palace of Mines, Metals and Machinery displayed an atom smasher. The Court of Palms, Court of Four Seasons, Court of Abundance, and the glass-domed Palace of Horticulture were all illusionistic. A Livestock Pavilion had grazing animals and displayed the newest in farming technology, and the Fountains of Ceres and Energy were within walking distance of the Palace of Education and Palace of Transportation. Streets throughout the grounds had names like Nanking Road and Chunking Road, Farallon Avenue, California Avenue, Avenue of Olives and the Concourse of Commonwealths.[6]

Fun as it was to get lost, it helped to have, for orientation, the Fair's *Official Guide Book*, which contained a map. Through the southern towers' lighted entranceway into the central court, enraptured sightseers found themselves between the Homes and Gardens Building and Palace of Foods and Beverages. In the 'consumption' palace was, notably, the Ice Industry Exhibit where "four (valuable) Imperial Chinchillas lived in a tightly closed air conditioned ice refrigerator." In the eight kitchens of the Heinz Exhibit luncheoneers could enjoy the various food-sampling stations. Here also was the "Pabco Linoleum Mural" depicting "The Evolution of Shelter" from cave dwelling to the typical home of 1939. At one end of the Federal Building, the "Indian court" showed Native Americans in traditional costume standing stiffly before a brilliantly embroidered tepee or dancing amid an array of handicrafts. An exhibit of the United States Secret Service of the Treasury Department focused on the history of currency and counterfeiting. Displays for the Bureau of Biological Survey and Bureau of Reclamation respectively concerned water pollution and land erosion and water shortages. The United States Weather Bureau's exhibit spotlighted the latest in meteorological forecasting. Exhibits by the army, navy, and marine corps featured history and drama in photographic displays and dioramas and proved to be huge draws. In addition, each of the nation's forty-eight states had an display at the fair. As the event's host, California provided the biggest attraction for visitors with its California Hospitality Building, (Franciscan) Mission Travels Building, Redwood Empire Building, and San Francisco Building.[7]

From these secular shrines one could be transported, suddenly, to the Philippines, Denmark, Sweden, Estonia, Hawaii, Indo-China, or Johore; all had startling pavilions. Dolls wearing the full costume of their respective countries were available for purchase. Kangaroos, wallabies, wombats, and exotic birds starred in the Australia pavilion, and panoramic maps with landscape views showed the routes leading into the country's wild interior. While the specter of war discouraged some European countries from exhibiting, England and France set up pavilions showcasing precious ceramics, tapestries, sculptures, and paintings. It was an "international event," for of the major world powers, only Germany did not participate, citing financial rather than ideological constraints. Costing over $1 million, the twelve-acre Chinese Village consisted of a walled city replete with teahouses, paddies, eating spots, and pagodas, one of which stood two hundred feet high. Inside the *Temple of Heaven*, worshipers offered prayers before images of Chinese deities and altars with burning incense. At similar cost, the Japanese Pavilion featured four thousand lanterns lighting its grounds, a fourteenth-century feudal castle, and Samurai headquarters with terraced gardens, trees, and lagoons where swan boats glided.[8]

As August came to an end, Halliburton visited the mile-square site to inspect the anchorage he had leased. At that time the fair was still a work in progress with few pavilions set up. He had of course read the promotional literature and spoken to exhibitors, but we shall never know what went through his mind, and must imagine how he took it all in. Start with "Travel." A Palace of Air Transportation used actual airplanes and airplane models to show how aviation had transformed the world. At the Bureau of Public Roads was a frieze of images called the *Highways of History* that depicted the coming of the first horses to America, the most-traveled roads, and the modern roadways. The picture gallery in Vacationland showed how modern conveyances like the China Clippers or "flying boats" had revolutionized travel, while the Chicago and North West Railway Exhibition showed how streamliners and challengers had accelerated travel. A Column of Progress reiterated the exposition's theme of industry on parade; an Arch of the Rising Sun announced the dawn of a new scientific age. Beholding its emanating colors of yellow and blue green, Halliburton (let us again suppose) strolled down the Avenue of the Court of the Seven Seas. On one side of the promenade were gardens, on the other were pylons. Sixteen in number, each pylon stood sixty-four feet high and each had on its summit the prow of a galleon representing the "Spirit of Adventure." Pink emanations of water gushed steadily from the "Fountain of Western Waters," as Halliburton (may we suppose) wended his way toward the Tower of the Sun, then drifted toward the Court of Flowers, where chubby little sea monsters spouted water into a blue pool called "the Fountain of Life." The glistening pond at the Court of Reflections recalled the Taj Mahal and its pond of roses and lilies. Halliburton could now see his anchorage. This much we know—it was tucked into the Port of the Trade Winds, just off the causeway. All manner of ships moored there, including US Navy and Coast Guard vessels.[9]

From Telegraph Hill Halliburton gazed at the land-form named after Robert Louis Stevenson's adventure tale of "buccaneers and buried gold." Would its visitors be swept away? Would what knowledge of the world they possessed now be enhanced? Ignorance of world cultures had a long tradition, and he had based his career on his perception of that ignorance. He thought little of the hoi polloi who would attend the exposition, and too much about the scissors he sharpened to cut their purse strings. He referred to fair visitors as "honky-tonk crud" more eager "to see anything that had color on it" than absorb any culture. No matter: "Once safe in the bay," he said, "I'm going to have the most exciting concession at the Fair," and the most lucrative.[10]

The Golden Gate International Exposition would open on February 18, 1939. Scheduled to run until December 2, it closed on October 29, 1939, reopened on May 25, 1940, and closed for good on September 29, 1940. Open every day from 10:00 a.m., the Gayway closed at midnight. Admission for adults was fifty cents, and for children under twelve it was twenty-five cents. Attendance figures appeared on a giant-sized National Cash Register at the top of a six-story building; the "39 Fair" averaged forty thousand visitors daily, the "40 Fair" forty-five thousand. Over three hundred thousand paid admissions were recorded in one four-day span. A parking terminal accommodated twelve thousand motor vehicles. Special rail service brought people in from nearly anywhere in California and the Western Seaboard. A train called the Exposition Flyer ran from Chicago to Oakland at great family-pack rates. The speed limit for cars was fifty miles an hour, in some places forty miles per hour. Hotels had "special" family rates. A teepee city was established in the East Bay, and large campsites were located in Marin County. To make sure everyone behaved, a "Crime Prevention" island squad watched for bunco artists, pickpockets, and other suspect characters. Banking services, a drugstore, first aid, a barbershop, and telephone and telegraph stations were also available. Tons of debris accumulated each day, and waste management teams worked throughout the night to clear it.

During the fair's 288 days, some twenty million people visited and over fifteen hundred conventions were held. In fact, it seemed like everyone then living was headed to the exposition. Attendees included ventriloquist Edgar Bergen and his dummy Charlie McCarthy, dancer/singer Betty Grable, swimmer Johnny Weissmuller, actor Henry Fonda, operatic baritone John Charles Thomas, coloratura Lily Pons, clarinetist Benny Goodman, poet Robinson Jeffers, journalist Ernie Pyle, *Music Man* composer Meredith Willson, President Franklin D. and Mrs. Eleanor Roosevelt, the Maharajah of Kapurthala, and New York Mayor Fiorello LaGuardia, who wanted to see if the upstart fair measured up to New York City's.[11]

Visitors to the fair returned home with wonderful stories to tell their neighbors. One, in particular, is preserved in journalist Joseph Henry Jackson's *A Trip to the San Francisco Exposition with Bobby and Betty*. In the fabricated story, halfway through their afternoon tour of the fair, Bobby and his sister Betty, a step ahead of their doting parents, find themselves not far from the Court of the Moon. Fountains and gardens mark their way. All of a sudden, the sight of a Chinese junk berthed at the Port of the Trade Winds takes their breath away and they gaze upon it in wide-eyed astonishment. Both move closer as their parents watch. "Celebrity" Richard Halliburton is aboard the vessel, and he gives an inquiring Bobby permission to take his picture. Bobby and Betty never doubted that the Golden Gate International Exposition was the greatest show on earth. Their parents might have supposed it a come-to-life model of Wendell Willkie's "one world."[12]

7

High Cost of Daring

Halliburton learned early in his career that traveling "the Royal Road" cost money. After experiencing a couple major investment failures—his house, at $31,000, cost him four times more to build than he expected, and the *Flying Carpet* Expedition, at $50,000, nearly bankrupted him—money, though at one time he condemned its pursuit, now figured prominently in all his undertakings.[1]

Amid other distractions, the *Sea Dragon* project with its promise of financial peace soon took precedence. Halliburton indicated to his booking agent at Alber-Wickes that he had been "killing (himself) for six months to bring (the adventure) to pass." Two weeks later he suspended his lecture engagements. His hopes for alternative employment were dashed, however, when he received word, April 12, that the "Chinese backers," fearful that "the Japs" might "prevent (the) sailing," withdrew their support. With no one showing interest, his "agent " told him that "the junk prospects look black." This was on May 24.[2]

As summer neared, the *Sea Dragon* Expedition again took "firm hold of (his) imagination." At the time, Halliburton's parents had plans of their own—a European vacation that would take them to Egypt, Greece, and Turkey. They would initially set foot at Ponta Delgada in the Azores, Richard's own "first sight of foreign land." By June 13, Richard, while his parents fussed with their schedules, reported that he was "forming a corporation and (would) sell stock to raise the necessary capital for the junk expedition." To assist him, he had gotten a business manager (film producer John Masterson), hired a publicity director (Wilfred Crowell), and retained a lawyer (J. Richard Townsend).[3]

Keeping busy, Halliburton drafted papers of incorporation, and a business plan with a statement of purpose and tabulation of the requisite expenditures. The total cost would come to $25,000. Look no further: investment capital would first have to come from the mortgage of his Laguna Beach home. Additional funds would have to come from cash reserves in one of several bank accounts and from royalties. Major sums would have to come from investors who believed in the merits of the project. On June 17, he would boast to his parents that his "junk project (had) become so organized" that he "had taken a week off and come back to Laguna." There he worked on "a fancy prospectus with maps and facts and figures" to present to potential investors. The resulting "Confidential Prospectus" included a detailed budget:

APPROXIMATE EXPENSES	
Junk	$4,000
Diesel Engine	$2,000
Alterations & Engine Installation	$2,000
SS Fare to Hong Kong	$1,000
Living Expenses Hong Kong	$1,000
Salaries for Crew	$2,000
Cargo	$1,000
Chinese Family Salary	$ 500
Provisions for Voyage	$1,500
Fuel	$1,000
Exposition Expenses ($500 Dock fee, etc.)	$1,500
Emergency Margin, general promotional expenses at Fair	$4,000
Insurance (15%)	$3,225
Total	**$24,725**

Over time the business plan was adjusted. Removed from the list, for instance, was the idea of bringing on the junk the "Chinese family" (presumably mother, father, and two children), meant to give "human interest" to the radio broadcasts he planned. Going to Sacramento, and the state Fair, seemed a good idea, but it was scratched. Expedition expenses would no doubt rise, and insurance rates would be modified. The cost to build a junk was based only upon an estimate given by the American consul. Advertising costs were not as yet tabulated.[4]

Next came profits:

APPROXIMATE RECEIPTS (FIGURES ARE ALL NET)	
Excursions and Tours	$36,000
Admissions (1/80th of 20,000,000 @ .25 cents less 15%)	$53,025
Souvenir Programs and Sales	$36,000
Sale Autographed Halliburton Books	$5,000
Resale of Junk	$4,000
Motion Picture Short	$16,000
"Covers"—(Special cacheted envelopes)	$40,000
Week in Sacramento	$5,000
Total	**$195,025**

The relationship between invested money and profit margin was eight dollars received for every dollar spent. In today's money, conservative multiples would be low at ten times and inordinately high at thirty times a given amount. A grand total in receipts reaching into the millions is realistic and, although Halliburton told his parents that he would not "daydream about profits," they often obsessed him.[5]

The Richard Halliburton Chinese Junk Expedition became Richard Halliburton Enterprises or, simply, the *Sea Dragon* Expedition. Although details regarding the corporation's operation, including its bylaws and board of directors, are obscure, individuals or groups could evidently purchase large blocks of shares. Halliburton would himself buy shares, proudly informing his parents that he owned most of them, and draw a salary of one hundred dollars a month from accruing revenues. Head-quartered at 739 Market Street, Richard Halliburton Enterprises was a small walk-in storefront office. The Schwabacher-Frey Company, which handled Halliburton's publishing needs, was located next door at 735 Market Street. Near Powell Street where the streetcars started their runs, and about five blocks from Union Square, one could see the Ferry Building at the foot of Market Street and even glimpse the Oakland-Bay Bridge. The main headquarters of the fair was located within hiking distance at 585 Bush Street.[6]

Once made known to the public, the *Sea Dragon* Expedition launched a thousand applications, as had Jack London's *Snark* Expedition of thirty years before. By comparison, polar explorer Ernest Shackleton had received over five thousand for his 1913 Antarctic Expedition of 1914. Interested parties had responded to an ad requesting heartiness of body and spirit and offering the remote promise of injury or death. Sailors, Boy Scouts, lawyers, doctors, and housewives had applied for the London adventure. While Halliburton's pool of applicants was probably similar, one must suppose that the majority came from daydreaming young men—many of them college age or older and unemployed. Young ladies, perhaps some Girl Scouts with considerable camping and home-arts experience, might have applied.[7]

The Fair Labor Standards Act of 1938 prevented minors from applying. One of the statute's provisions stipulated that children under the age of eighteen could not do dangerous work, and that children under sixteen could not work during school hours. While most applicants may have possessed presence of mind, unshakable confidence, and boundless courage, most clearly had little or no able-bodied seamanship experience. No matter; only the ability to pay tuition counted. Applicants to the *Snark* Expedition, paid no entrance fees and asked for no payment in return for their services. Although compensation was offered to them, applicants for the Richard Halliburton *Sea Dragon* Expedition had to fork out some cash. Halliburton could have gotten a few names through referrals. What interviews he conducted, mainly informal, were supposedly by telephone or through mail. Of his interviewing style, little is known. The level-headed and shrewd Shackleton, within moments of being in someone's company, could detect whether the person had true grit and a sense of romance; these qualities meant as much to him as technical know-how and mere escapist desire. That Halliburton read character as keenly as Shackleton is to be doubted. Believing that even the rawest recruits could be trained and, like himself, commit to bold adventure, he approached "lads," not veterans. Unlike Shackleton, he chose members of the *Sea Dragon*'s crew based on their financial resources and family backgrounds, again no seamanship skills required.[8]

By May 28, 1938, the day he began to conduct his interviews, Halliburton had taken up residence at the Chancellor Hotel, making it his base of operation. From the building's sixteenth floor, he could gaze toward the Far East and imagine China just over the horizon. Luckly he received walk-in applicants at the "Clipper Ship Meeting Room," a four-hundred-square-foot facility on the mezzanine above the first-floor lounge. Installed in 1937 to commemorate Pan Am clipper ship service from Alameda to the Orient, the room resembled the interior of a passenger airliner, and its panoramic aerial views of the two bridges, the ocean, and Treasure Island provided the perfect backdrop to Halliburton's spiel. On Powell and Post, the hotel was ideally located to serve all of Halliburton's business needs. Union Square was across the street, and a short walk through it brought people to Maiden Lane with its shops and diners. On Geary at Taylor stood the Bellevue Hotel, "headquarters" for those associated with the aviation industry. The advertising agency of Walter Swanson was located at 449 Powell Street, two blocks away toward Market Street. Another block over on Market was the Schwabacher-Frey Company, a West Coast leader in printing and lithography; conveniently located just a few doors down from the *San Francisco Examiner* offices, "Schwabacher" was also near Thomas Cook and Son, the travel agency which arranged Halliburton's itineraries. Outside the Chancellor's revolving front doors, people could catch a ride on a bus, cab, or one of the cable cars or streetcars that regularly ran down Powell to within a block or two of the deluxe Mark Hopkins Hotel and down to Fishermen's Wharf. From there, a ferry ride took passengers to Treasure Island in minutes. Half a mile away, at the end of Market Street and adjacent to the Embarcadero, was the Ferry Building, whose commuter services connected Halliburton to key points throughout the Bay Area.[9]

The stage now set, Wilfred Crowell urged Halliburton to win public support for the project through a series of short essays entitled "Letters from the *Sea Dragon*." Dated November 20, 1938, the first letter to subscribers was auspicious: "Let me welcome you now, right at the outset of our adventures together, into the *Sea Dragon* club," it read. "The *Sea Dragon* is the name chosen for the Chinese junk which my shipmates and I plan to buy or build on the China coast, and in which we shall attempt to sail across the Pacific to San Francisco's Treasure Island. We hope to set out the middle of January, and arrive the end of March."

His outlook brightening, Halliburton could suppose that the broadside would generate some excitement. Public interest in him had begun to wane, but he hoped that his name still had drawing power. Written in longhand, then typewritten, mimeographed on 8.5-by-14-inch sheets, and signed by Halliburton, the total of seven letters were to be sealed in envelopes with the *Sea Dragon* logo, stamped with their point of origin, and mailed to subscribers ("the Sea Dragon Club") eager for regularly updated information about the expedition. The package also included two photographs of the junk. Halliburton's hope was to sell not 8,000 but 10,000 copies at $5.00 each and pocket a third of the proceeds. China Clippers would deliver the letters to the American mainland; from there, they would go to their respective recipients. Promoted as a "rare literary treat" and "a grand Christmas gift," the letters were meant to edify young and old alike, and, as the advertising circular noted, to offer "lesson(s) in geography, modern history, literature and adventure," and by a "master artist." In addition, Halliburton's writings were to have "inestimable value to school children" and "special appeal to children in classrooms." Read the promos:

NO PARENT NOR SCHOOL CHILD WILL EVER FORGET!
RICHARD HALLIBURTON'S VOYAGE FROM CHINA TO SAN FRANCISCO
VIA HONOLULU IN A CHINESE JUNK!

The smaller print provided further information: "Early in 1939, Richard Halliburton, world-famous author and traveler, sets out upon his newest and most audacious adventure—an *Expedition* across the Pacific Ocean from China to San Francisco aboard his *Sea Dragon*—a 65-foot Chinese Junk. Leaving Hong Kong around January 1, he plans to make the voyage by way of the Pan-American Islands, used by the China Clippers, and to arrive at Treasure Island in San Francisco Bay, site of the Golden Gate International Exposition, in March. No junk has ever followed this difficult 9000-mile path across the Pacific." Halliburton said the trip would require "four months"—"one month to equip the junk and three months for the voyage." The "dramatic and exclusive story" (not ordeal) would feature "colorful and organized adventure." Promised was "a host of interest-packed events" such as Halliburton and the rest of the crew would face before their March departure.[10]

Arrival time home was as uncertain as the funding. Besides book royalties, the $6,000 Halliburton would receive from the *San Francisco News* was his only guaranteed source of income at present. "Letters, conferences over radio, engines, captain, engineer, crew, prices—keep me rushing around day & night," he wrote home on July 14, 1938—two years to the day after his idea of the expedition germinated. The "slowness of progress" irked him, but, as had happened with the *Flying Carpet* Expedition, he accepted that "delays and discouragements are part of it." The enterprise spanned over a year and a half, far longer than he had wished, and severely tested his patience. Refueling the single-engine Stearman or addressing mechanical problems when pump services, parts and able technicians were few and far between had been an issue. Still, that "scarred, weather-beaten, very worldly wise (aircraft) had returned, a veteran of many conflicts and adventures—through desert and jungle, Africa and Arabia, Himalaya and the islands of the sea." Little more than a motorized metal-bound piece of plywood in the air, the *Flying Carpet* had traveled forty thousand miles. The *Sea Dragon*, heavy lumber on the sea, had to cover only nine thousand miles.[11]

Halliburton's need for "adversity" forced him to think ahead. At the time, he was a man without a junk—having only the name for one, and a dreamer with little capital. Opportunities to go forward knocked. Brought to his attention was a "fine sea-going junk" for sale in Hong Kong, asking price, $2,500." An oil company broached the matter of his traveling throughout the West for $1,000 a week—to places like the Grand Canyon and Yosemite, and making radio broadcasts to encourage "motor travel during the Fair." Had the lucrative idea materialized, it would have delayed his departure for China by some thirteen weeks, or from September through November, which meant he wouldn't be arriving in China until December. The crew, by then "ready to sail," would have arrived there first with no one to brief them. So the motor travel idea was put aside: more to his liking was his doing radio broadcasts *from the junk* while it made its crossing.[12]

8

"The Lads"

Halliburton's many public appearances during his last months in the United States forced him to suspend his search for funding, but he soon suspended those appearances to renew the search. He had by now gotten used to the hazards it required to get people to part with their money for "romance." One potential investor was Los Angeles socialite Myrtle A. Crummer, whom he had probably met through Mr. Gay Beaman, his realtor in Laguna Beach. Formerly a court reporter and now an avid book collector, Crummer, née Kelly, had been married to famed physician, medical book writer, and fellow bibliophile LeRoy Crummer, several years her senior. In 1934, LeRoy died suddenly, leaving Myrtle with some money and the opportunity to remarry, this time to a William Ingram. While Myrtle was still married to Ingram, Halliburton offered her a stake in the project in exchange for investment capital. For weeks he gently, if persistently, telephoned and wrote to her hoping to get her into making a commitment. As well as sunbathing and swimming, he continued "plugging away on the junk without respite." The radio deal with Shell would fizzle. He stated, "The radio job boils away with no definite contract but lots of interest—from *Shell* principally." Still, he observed, "Nothing ventured, nothing gained." In addition, promised sums from unnamed sources in Santa Barbara never materialized. The two *Books of Marvels*, *The Occident*, and the recently published *The Orient*, meanwhile, were selling well—in fact, very well.[1]

While it was odd where Halliburton ultimately got money, it was also odd where he didn't get it—such as from friends Florence Barnes, Ramon Novarro, and the better-heeled members of the Hollywood community. While requesting money from his parents troubled Halliburton, it hardly troubled him to request it from his cousin Erle Halliburton Sr. or, partner in his business dealings, Erle's wife Vida. As some rivalry existed between Wesley and Erle, Richard seldom mentioned Erle in the letters he sent home. Still, on June 18, he wrote to his parents that his "conference" with Erle "turned out to be a dismal flop." Richard, meanwhile, had gotten a "high-pressure promoter who knows money people"—either John Masterson or, in an expanded role, Wilfred Crowell. Committed to the task, Halliburton drew up plans to lay before investors, but a "fierce attack of hay fever" kept him from following through on it.[2]

Vida pledged $2,500 or, turning to Erle, $5,000. Forced to withdraw from the initial *Sea Dragon* crew because of a "bad eye operation" was young Erle Halliburton, Jr.—and with him went a pledge of $7,500. Unable or unwilling to provide money herself, Zola Halliburton, a student at Wellesley, offered Richard, instead, a contact. While vacationing in Bermuda, she met twenty-four-year-old Bar Harbor, Maine resident John Rust "Brue" Potter, an amateur yachtsman, who with Dartmouth College buddies Gordon Ellicott Torrey and Robert Hill Chase and in his own 60-foot ketch piloted a voyage to the New York Sound that failed. They were lucky the coast guard rescued them. Still, what bold endeavor: Zola assumed that if the three were fool enough to sail so boldly from Bermuda to New York, they might want to sail a junk across the Pacific. Brimming with self-confidence, the tall, beach blond, self-consciously attractive Potter had a boater's full tan and gentleman's look of carefree ease. From Southeast Harbor, Maine, and a recent Dartmouth graduate, Potter had a moneyed background, and one day would marry Ann Hopkins, daughter of legendary Dartmouth president Ernest Martin Hopkins. Potter, a French major, had spent his sophomore year at the Sorbonne with plans of becoming a foreign diplomat. Importantly, Potter was offering funds for the *Sea Dragon* Expedition. Halliburton said, "[He] agreed to invest $4,000 in my project—if I'd take him along."[3]

Thrilled to make the team, Potter brought on board fellow yachtsman Gordon Ellicott Torrey, a dropout from Dartmouth, and Robert Hill

Chase, a senior from Dartmouth. The "lads," as Halliburton called them, often congregated for a quiet beverage at the "Merry-Go-Round Bar," a main venue of Boston's Copley Plaza Hotel, where they discussed the Halliburton expedition among themselves. The serious intellectual of the bunch, Torrey was in his twenties, but seemed older even though, like Chase, he was a couple years younger than Potter. Lynx-eyed and gremlin-faced below a receding hairline, Torrey seldom smiled. The least preppy of the lads and hailing from a family not as well-to-do as Potter's, he would later report that "unlike Potter, he was a paid hand and not a friend of the Halliburton family." Before Halliburton's call, Torrey had worked as a sailor aboard small craft in Maine, and as a rigger for Potter. About future plans, he hadn't a clue.[4]

Robert Hill Chase, like best friend John Potter, was drawn to elitist adventure. Having a soft air of inherited wealth, he could boast several generations of successful entrepreneurs and Dartmouth graduates. After the voyage and his graduation from college, he planned to enroll in Harvard's business school. Neatly attired in white casual wear and sporting a Harvard clip, the well-bred, blue-eyed, choir-boyish Chase exuded a cleanliness close to godliness. He was never without a friendly greeting or cheerful smile, which one could believe, remained fixed even while he slept. His parents were "distrustful of Halliburton's glib assurances about everything" and had seen to it that Bay Area attorney, and Dartmouth graduate, J. Richard Townsend would oversee the corporation. The appointment seemed perfectly okay with Halliburton. He could rest assured that with Chase and company safely committed to the expedition, his goal of $25,000 was nearly reached. He had himself contributed $5,000. The Dartmouth lads sent Halliburton a check for $14,000, or $15,000, and another $2,000 arrived from other sources. Loose ends remained. Vida had left for Europe and with her went her pledge. Insistent that her help would put him "almost over the top," Halliburton, with unknown result, cabled her while she was in Rome.[5]

With $16,000 firmly settled in his pocket and $5,000 hanging from the edge, Halliburton had still to come up with another $8,000. Piddling amounts came from here or there. "I expect Vida (Jessie) to invest $300," he said. "I have $250 from others—so I've got half the capital needed —the hardest half." For a large donation, he continued to press Myrtle Crummer. A nasty divorce from Ingram, however, had cost her, or so she said, $10,000, leaving her broke. Halliburton then turned to alternative

investors for "the missing $8,000." Some deemed the venture impracti-
cal, and others dismissed it as just plain dumb. A shrewder idea seemed
a floating offshore casino, not some wingding Chinese junk on a world
peace mission. Undismayed, Halliburton again appealed to the strapped
Mrs. Crummer, who informed him on September 4 that that she had contacted her banker in Omaha "about further investments." She evidently
offered Halliburton some investment capital—and, in the bargain, even
refused a share of the profits. It was some small triumph, and, if the day
before he had $21,000, now he had $25,000. Not done yet, Halliburton
contacted a long list of potential investors, hundreds of whom were connected to or had some financial stake in the fair: possibly, Levi-Strauss
and Company, American Express Company, Thomas Cook and Son, the
National Automobile Club, Weeks-Howe-Emerson and Peterson Clippers.
The list could go on: the Crocker First National Bank, Bank of America,
and Wells-Fargo Bank, Mission Sweater Shops, the Ediphone Company
and Caterpillar Diesel Engines Company. None responded. [6]

Halliburton's resolve did not weaken, and early in July he purchased
a one-hundred-horsepower Enterprise diesel engine. Probably last year's
model but valued at $7,000, he had gotten it for $1,000. This anchor
to Halliburton's dream was also the first crew member he hired. Because the motoring device burned fuel, it seemed reasonable that a fuel
company might put forth some investment capital. Standard Oil, which
represented Anglo-American oil trusts in China, was one such company.
Another more reliable one was Shell Oil, which had earlier provided
seed money for the *Flying Carpet* Expedition. Buick Motor Company was
approached, but company officials retreated, not wishing to be associated with a "junk" or an enterprise so far from home. Royal Typewriter
Company might have been a lead, as it sold mimeograph machines.
Another option was Enterprise, which had merged with Western Machinery Company of Los Angeles in 1924 becoming the leading maker,
repairer, and supplier of diesel engines in America. [7]

Distracting Halliburton momentarily was his parents' trip abroad.
Eager to hear about their crossing, he also wanted to know if they "enjoyed Palermo and Algiers—and how Athens looks." Once their prodigal
son, he now worried about them while they went off into the wild blue
yonder. But his own affairs, distractions from distraction, kept him
"rushing around day and night." Halliburton's main occupations included
"letters, conferences over radio, engines, captain, engineer, crew, (and)

prices." Progress was slow, and he already complained of "the delays and discouragement" that would plague him until the end. Solicitations for money continued. At first enthusiastic, fair publicity director Clyde Vandeburg, staff official Dean Jennings, and radio announcer Art Linkletter now found the project of dubious merit. Seeing Halliburton as a main headliner (as Halliburton himself had been hoping), "the Fair wanted Richard to be a part of it," comments Jonathan Root, "but not enough to justify an investment of $35,000." Years later, Linkletter said nothing about money. Mainly he recalled that the best-selling author, whom he thought a fop, was "lean, bronzed by the sun, [and] impeccably groomed and tailored," indeed, to the undiminished amusement of Linkletter, he was dressed so finely that "the stretched cuffs of his shirt protruded two inches from the sleeves" and "into one cuff (was) tucked a silk handkerchief." When Linkletter asked him if he had doubts about the mission, Halliburton resonded, "None at all."[8]

It was never too late to invest. As late as August 3, Halliburton was soliciting funds from prospective crew members. One such prospect was recent Yale graduate James Watson Webb III, a seaman, photographer, and future Hollywood film editor. Halliburton acted hastily to recruit the young prospect, whose name mutual friend and Petroleum Securities associate H. W. Dougherty had given him. Webb was furnished with ample details about the project, the route he would take across the Pacific, the time of transit, and the personnel already committed. "If the junk—for any reason at all, fails to reach Treasure Island," Halliburton said, "the investors recover every dollar from Lloyds." Just $500 more was needed to make it come to pass. Webb said no. Probably he learned what Halliburton failed to mention to prospective crew members: they were entering a war zone. As one example of the dangers that lurked: on August 24, the *Kweilin*, a Douglas DC-3 commercial airliner, carrying Chinese crew and passengers—and piloted by an American, was shot down over China by a Japanese warplane. The "Kweilin incident" might have spurred America into action against Japan—as the sinking of the *Lusitania* had spurred America's entry into World War I. Although Halliburton did not mention it, the front page news had to have drawn his attention; it may also have induced Webb to withdraw from the expedition.[9]

In the interim, Halliburton contacted Bertha Kellogg Barstow about her twenty-year-old son George Eames Barstow III and asked her to consider his participation in the group. In 1904 in a union which joined two prominent families, Bertha Kellogg, daughter of Morris Woodruff Kellogg—founder of the power-plant company bearing his name, had married George Barstow II. When her husband died of a heart attack in 1932, Bertha found herself in possession of several homes: one in Montclair, New Jersey (where George grew up), a lake house in Connecticut, a luxury hotel suite in New York's fashionable Upper West Side, and a winter residence in the Los Angeles area. "George the Third" was thus to the manor—or manors—born. Even so, Bertha found that raising a son by herself was a chore and that her son might benefit from male mentorship.[10]

When Halliburton contacted him, George was enrolled in the Night Division of Juilliard with a major in music theory. Vole-like and small in stature, he conveyed effeteness, even fragility. He did not question authority but flippantly, or impudently, defied it. George often squinted and could alternately contort his face into a snarl or expression of impish glee. He wore wire-rimmed glasses and, once in Hong Kong, where Halliburton first met him, he wore a sailor's cap. If Torrey seemed older than his twenty years, Barstow seemed younger than his. He was an accomplished pianist and accordionist who wanted to be a writer. Hardly athletic, he looked best suited for a desk job, not a gunwale post. Halliburton saw George III as a weakest link, as "super-cargo, rather a reliable funding source. But as Halliburton confided to Erle's wife Vida, "[Bertha] is allowing her boy to miss his year in college in order that he can enjoy what, in her opinion, offers a better education." Offered stock revenues for her $4,000 investment, Bertha, thinking the adventure "a great opportunity for her son," requested only that Halliburton use the money as he saw fit. The "Barstow check" now gave him $27,000.[11]

9

O, Captain, My Captain

esides funding, Halliburton needed a captain to sail his junk. He had chosen wisely when he hired Moye Stephens to pilot him around the world in the *Flying Carpet*. Cool-headed and knowing, Stephens could steer a plane and navigate as well as read maps and set courses. What would Halliburton do now? Examples from literature and life led him to consider the proper qualities for captaining a ship. As a passenger, stowaway, or deckhand on ships big and small, he had met over the years efficient but mean-spirited captains, energetic but cruel ones, bookish yet bossy ones, autocratic yet efficient ones. Most were capable.[1]

At first, Halliburton aimed high. Fifty-eight-year-old Captain George Warner Yardley, "one of the best-known Dollar Boat Captains," had been master of the SS *President Hoover* (sister ship of the SS *Coolidge*). In December 1937, his ship—after the Japanese-Chinese war forced it to alter its course into uncharted waters, ran aground just southeast of key Japanese military base Formosa (Taiwan). Passengers of the *President Hoover* were rescued from its shores and brought safely to Hong Kong, though not without bitter controversy that made the event front-page news. Out of respect for Captain Yardley, Halliburton vowed that, on his return voyage, he would stop in Formosa and view the wreck of the SS *President Hoover*.[2]

Unable to hire Captain Yardley, Halliburton tried getting one of his two sons, probably Richard Yardley, age twenty-four, who was a member of the US Marine Corps. His six years of experience at sea and rank of third mate marked him as "reliable and capable." The younger Yardley could navigate and operate a diesel engine. Halliburton contacted him straightaway, meeting him and his wife in late May to discuss the project. Richard Yardley impressed Halliburton as "very mature" and "a most likely prospect." Experienced in navigation, the young man had sailed across the Pacific "dozens of times." He had a superior knowledge of the Orient and "friends and connections in Hong Kong." In finding the right persons to man his ship and in tending to incidentals, Halliburton believed "great progress" was being made. With good luck, "the crew," he said, "would consist of Yardley, myself, (at the time) Erle, Jr. (as assistant radio operator), a Chief Radio operator, one white seaman, Chinese bo'sun, two Chinese deck hands, [and] one Chinese cook." Halliburton added, "If Erle provides the capital, we'll be away in a month." Plans to hire young Yardley fell through, however, when his father, victim of a heart ailment some believed the result of the nervous exhaustion he suffered after losing his ship, unexpectedly died. With Richard Yardley given family responsibilities that precluded other service, the $2,500 he had committed to the project was revoked.[3]

For weeks the search for a captain dragged on. At first Halliburton thought to hire a captain who was familiar with "junks and their ways" in Hong Kong, but the search returned closer to home. After the Yardley effort failed, a captain who seemed of the right mettle came somewhat out of nowhere. Forty-two-year-old John Wenlock Welch was not a "savage old man-eater," and he would not captain "scoundrels," but, like the *Octorara* captain, he was "very blustering, very profane—and very capable." Welch liked to tip a few, and got windy after a few clanks of a third or fourth schooner of ale or white mule. He spun "sea yarns" that were "endless, and endlessly exciting." The salty language of the wharf seasoned his tales: "Gord!" (God), "Dorg!" (dog), "queasy on the quay" (seasick), "glory-hole" (vagina), "flogging the dolphin" (masturbating), and "jimmylegs" (a person wobbling along due to drunkenness or nerves). "And stuff" was his high-frequency catchall for unspecified miscellanea. A colorful, opinionated character, "Almost everyone in the Colony," reported the *Hong Kong Daily News,* "has met the wild and hard (who can't be blown down) John Welch, skipper, master and the man who roars 'ready t'come 'bout."[4]

Welch had sailed on freighters associated with the United Fruit Company in Honduras, but while Erle Halliburton had mining interests in the region, no connection between the captain and financier has been found. In any event, Richard Halliburton wrote Welch a letter that has not survived. At the time, Welch, whom Halliburton understood to be a Scotchman and a sailor who could speak Chinese, was in Hong Kong with the up-in-the-air idea still of making his own junk voyage across the Pacific, aiming for Panama, then to Scotland. Welch's own letters, private communications certainly, indicate that he had given no thought to a prior voyage and didn't know much more than a few words of Chinese. Even his "junk" experience was disputable, but Halliburton evidently chose not to question it. Trying to win over prospective crew member J. Watson Webb, Halliburton told him, "[Welch has] sailed junks for years, and knows all their peculiarities." By his own admission, Welch had never sailed a Chinese junk. Although he had been to the Orient, chances are he had never been aboard a junk.[5]

A fixture along the San Francisco waterfront docks and saloons, Welch might very well have told Halliburton that he was a British navy veteran who had fought in World War I, that he was a current American citizen living in San Francisco, and that, "as a second mate aboard *United Fruit Company* 'banana boat' freighters to Latin America," he regularly sailed out from the Bay Area. He had *military* written all over him. More importantly, he seemed to know the name of every wave from Sydney to Alaska, and he "could sail a ship in his sleep." Welch probably confided little to Halliburton about his personal life. Research has found, however, that he hailed from Sydney, Australia (address 111 Peel Street, Kirribill 4 N. Sydney, NSW, Australia), and had a brother named Edward (married to Ollie) and a sister. Welch might very well have spent time in Scotland, as he mentioned the Scapa Flow, a body of water located within the Orkney Islands northeast of Scotland.[6]

To J. Watson Webb, Halliburton noted that the captain was "married." Truly Welch *had been married* from 1930 until at least 1935 to Barbara Elizabeth Bridgeford, ten years his junior, a native of Washington State, and a University of California–Los Angeles graduate. Likely Barbara Bridgeford and John Welch lived together, at one time in the Bay Area, at another time in Los Angeles. Whether she died or the two divorced remains uncertain. The couple apparently had no children. Welch's record as a sailor, rather than his demeanor as a husband, is better known. By the time he finished his teens, about 1917, he had been inducted into a life at sea. He had a number of routine shipping assignments, but his most famous one, according to the *Hong Kong Daily News*, "was a skipper of John Barrymore's yacht in the inter-club races from Santa Barbara to Honolulu." It is certain that the *Sea Dragon* command was Welch's first outright captaincy; he said as much himself.[7]

Captain Charles Jokstad of the SS *Pierce* believed Welch was once a "second officer." Gordon Torrey said, "He knew how to handle a crew and he also knew how to navigate and to rig." Welch's letters indicate that he was conversant with the tools of navigation—the compass, the chronometer, naval charts, nautical almanac, and the sextant (less certainly). Importantly, Welch knew standard maritime protocols and regulations, and he demonstrated an easy way with the language of ship design. As importantly, he knew sea lanes and the sort of shipping traffic that used them. Temperamentally, he was practical and not a dreamer. "Captain Welch isn't interested in ports-for-pleasure," wrote Halliburton, "only in getting his job done the fastest and most efficient way." What credentials the Scotsman or Australian provided him, in the end, gave way to first impressions and urgency. The low cost of securing him was also a selling point with Halliburton considering himself lucky to get so fine a sailor for only $250 salary per month, roughly $2500 to $3500 in today's money.[8]

Welch was agreeable-looking, but maybe a nudge short of handsome. Photographs of him show that he had a square jaw, big ears, a slightly cucumber-shaped nose, ample lips, a forward stance, and a solid frame that stood a bit less than six feet. His full shock of brown hair, always combed, was just beginning to recede. Appearance counted to him, and often he could be found wearing quality woolen sweaters over a white shirt and tie, pleated flannels, and boat-deck shoes or well-polished oxfords. A fitness buff, husky "Jack" swam regularly at the YMCA and did calisthenics. Now and again, he had dinner with officers from the Royal Air Force, calling them "a good bunch of eggs." Before a camera, he was mild mannered, even shy. Behind one, he could be a ham-fisted boor, venting his spleen in bursts of acid remarks and finger-wagging.

If at times a backwoods barbarian, he also had a sense of delicacy—and a way with words.[9]

He was indeed a writer, with ambitions of making a living writing for the tabloids. His "Ocean Tow," published in 1937 in the *Proceedings Magazine* of the US Naval Institute, showed he had sufficient skill. Told was the story of the *Coringa,* "a dirty tub" but "once one of the great tugs in Australian waters," and how, in 1920, through adverse weather conditions in the north Atlantic and through the Mediterranean on its way to Singapore, the tug dragged the dredger the *Mercedes* to port. Welch himself, readers are informed, stood watch on the most dismal nights and, with old Scotsman Captain Manning, co-captained the tug. Also it was said that he was the "second-mate of a full-rigged ship." Published a year earlier in the *Proceedings Magazine,* Welch's next tour de force, "Signaling and the *U.S. Merchant Marine,*" noted that "Ensign John Wenlock Welch, *U.S. Naval Reserve*" had "served in British ships during the whole period of European hostilities." The piece not only demonstrated Welch's knowledge of the International Code of Signals, and matters pertaining to safety at sea and seamanship, but also his understanding of the outreach of the British navy and its service protocols.[10]

Long-time friend popular action writer Albert Richard "Dick" Wetjen thought the captain "a topnotch seaman." He affirmed that he held "master's papers for both sailing and steamships," had served as "an ensign in the *U.S. Naval Reserve,*" and later had "experience on the Australian grain clippers that raced between Melbourne and London years ago." When barely twenty, Welch had served as the "first-mate on a transport during the World War." About Richard "Dickie" Wetjen considerably more is known than about Welch. Seafaring yarns were the specialty of the London-born writer and rover who had gotten his start in the early 1920s as an editor of *Oregon Magazine.* In 1926 he won the coveted O'Henry Prize for his short story "Command," which *Sea Stories* magazine had published. Twice shipwrecked, and for a time imprisoned by the Sultan of Zanzibar for some wrongdoing, Wetjen could match Halliburton adventure for adventure. His alter ego and most famous character was "Shark" Gotch, a seadog with "an extraordinary knowledge of men and ships" who, when "facing the gravest dangers," showed both "nerve and sang-froid." Besides having "absolutely no sense of humor," he "was utterly ruthless." While Wetjen ruled with "an iron hand," "his word, once given, could be depended on. He would cross the world for a friend and spend years chasing after an enemy. He could be neither cajoled nor forced from his purpose. . . . It was wise when dealing with "the "Shark," to be "utterly on the square." Seaport bookstores everywhere carried the magazines that told of the devil-eyed rover whose "name (was) feared and respected throughout the whole of the South Seas."[11]

Although superficially resembling "the Shark," more is known of the person John Wenlock Welch himself. For openers, Wetjen said that Welch was "a technical director for sea films in Hollywood for five years." No hint of such work, however, is expressed in Welch's own extant letters. The *Hong Kong Daily News* noted that he had served as the consultant on a couple Hollywood nautical adventure films—in number "two," according to Jonathan Root. In any case, fellow "mateys" and the "sailing fraternity in the United States," found the captain and his tales of the sea amusing. Richard Halliburton respected Welch. When, just days before the SS *Coolidge* departed from San Francisco, Halliburton's mother Nelle Nance met Welch, she found him charming and well spoken. The ship's chief officer, Dale Collins, called Welch "a personal friend" and would better get to know him during the "trials and tribulations" of building the *Sea Dragon*. "Old man" Karl A. Ahlin, the ship's captain, also considered Welch a friend.[12]

Halliburton called Captain Welch "Jack," and this "Jack" may have reminded him of Captain George Anderson of the *Octorara*, whom he called "an old seadog but a gentleman." Halliburton also believed that he and the captain would "get along in great shape." Welch, however, called the characterization of him as an "*old* seaman" in Halliburton's first newspaper article rudely far-fetched. The captain smoked and from time to time enjoyed a fine cigar. He drank, often to excess, and he was rumored to be a womanizer. He had a booming voice and talked a tough line. He could throw tantrums and did not easily welcome compromise. Capable of reversing first impressions about people, he got to appreciate the Chinese carpenters whose work habits he originally deplored. He also came to think differently of Paul Mooney, whom he originally dismissed as a slacker. "During the time it took to construct the *Sea Dragon*," writes James Cortese, "Welch proved a headache for Richard, [Welch] flying into uncontrollable rages and bullying the young collegian sailors." Eventual crew member John Potter thought the captain a "martinet" and "not a very good sailor." Jonathan Root affirmed that Halliburton would soon find Welch "totally incompetent," but, in the beginning, he must have thought him competent enough. In hiring Welch, Halliburton had, in any case, made a decision and had to be positive about it. The captain's character would unfold variously, but for now, he was simply a "veteran sea-dog" and "master of all manner of sailing ships." Contract details seemed okay with Welch. Said pay now set at $250 a month, should he lead excursion cruises once the *Sea Dragon* arrived at the exposition, his salary would increase to $300. Also there was the possibility of a commission from motion picture companies wanting to use the junk in films and for shorts on China.[13]

10

The Black Magic of Machinery and the Wizardry of Radio Communication

A captain in place, now all Halliburton needed was an engineer. To
fill that spot, he chose quickly and well. "The Engineer is the best
possible—a German who has had six years with Diesel engines in motor
sail boats," said Halliburton. He also remarked that Henry Von Fehren
had "complete control of the black magic of machinery." Importantly,
the engineer had "crossed the Pacific a dozen times in charge of a Diesel
motor." Crew member John Potter would recall Von Fehren as "a true
professional." According to the *Hong Kong Daily News*, Von Fehren knew
expertly the workings of the one-hundred-horsepower Enterprise diesel
engine that would be fitted on the junk. Besides operate the engine,
Von Fehren, by profession an electrician, would be "in charge of radio
and lights and things mechanical." Halliburton also said of him, "[He]
has been aboard an ocean-going yacht for several years, and is thor-
oughly sea-wise." The "ocean-going yacht" was the famed *Zaca*, at the
time owned by financier, sportsman, and explorer Charles Templeton
Crocker, himself a member of the prestigious Trans-Pacific Yacht Club.[1]

Born in Germany but now a naturalized American citizen, Von Fehren
had neither flaws of character nor ability of which Halliburton was
aware. The *San Francisco Telephone Directory* for May 1939 listed his
address as 1746 Bryant Street, the light industrial part of San Francisco,

and his telephone number, ominously, as HE-mlock 6169. Six feet tall, he was muscularly lean and lanky. His dexterous hands were notably large. His eyes were gray as iron. A captain's cap covered his blond hair. A pipe often hung idly from his mouth. Retiring and quietly diligent, he was also softhearted and personable. If at times gruff, he more often cracked a well-wishing smile. Thirty-four, Von Fehren was married, to Hannah, 33, an office clerk also from Germany; the two may have had children. He was of course fluent in German, but the little English he knew was heavily accented. When vexed, he murmured a string of German expletives that somehow became him. His relationship with Captain Welch rapidly soured. Their disagreements, mainly over technical matters related to ship operation, often flared into sudden, near-violent quarrels. When this happened, the engineer invariably walked off in a huff, and, still grumbling, went about his tinkering. By the end of February 1939, Halliburton noted that Von Fehren had proved himself "a faithful friend, capable, serene, dependable."[2]

Evidently Halliburton never visited San Francisco's Mackay Radio and Telegraph Company in person, though, to establish a main lifeline, it would have been wise for him or representative Wilfred Crowell to have done so. Now that he had engaged a captain and an engineer, Halliburton's next order of business was to engage a radio operator. Because his English was poor, Von Fehren chose to forego the position. Radio communication is a key safety issue in any perilous travel adventure. Halliburton had to have had some inkling of Amelia Earhart's well-publicized indifference to specialized radio equipment such as the Bendix direction finder, and about her subsequent failure to establish contact with ground crew during her last flight. Indeed, the best radio communication was of paramount importance to him.[3]

Halliburton was wise to seek sound counseling here as radio communication perplexed him as much as it had Amelia Earhart. Remarking, on July 14, that "the details & complications are infinite," he was well-advised to have a professional do the job. At the time he supposed that he would be leaving for Hong Kong by September 1. The busy days passed quickly. By August 1, however, he had hired a well-qualified radio operator in one George I. Petrich—described as "head of the radio department in the coast guard boat (the *Itasca*) that stood by, in mid-Pacific, to help Amelia Earhart." Halliburton saw the dark-haired, short, taciturn Petrich as "a veteran" and a man he wanted on his team. Originally from

Tacoma, Washington, the thirty-five-year-old Petrich was a skilled sailor and photographer. While it mattered little to Halliburton, he could also hold his own in a fight, having once been a Golden Gloves champ in the bantamweight division. His main job now was to maintain shortwave radio contact (600 meters) with all merchant ships traversing the same waterways as the *Sea Dragon*. "Our radio is going to be wonderful," Halliburton wrote to his parents. "We can radio you messages every day or so—& you can wireless me—almost (as) good as (a) telephone." Success of the communication system of course remained to be seen. Like Von Fehren and Welch, Petrich was married, or had been married, no fewer than five times. He would join the team in Hawaii, but would not arrive in Hong Kong until sailing time.[4]

There was still unfinished business. At first, Halliburton figured he would need seven people to sail the junk, then figured a couple more might help. J. Watson Webb had declined to be the eighth member of the crew, so why not add Paul Mooney in his stead? Mooney had proven writing and photography skills, and he had once applied for a seaman's certificate. As Halliburton's secretary, he would assist the him with the "Log of the *Sea Dragon*" and the "Letters from the *Sea Dragon*," but would ostensibly serve as the expedition's official mimeograph operator. "He's in a rut in Laguna—drinking and smoking too much," Halliburton explained: "This violent change will stir him up and give him a new interest in life and work." In July, Mooney had had no idea he would be joining the crew. In August, when he received the invitation, he had his bags packed and was headed for Taxco, Mexico, a haven for expatriate artists, and a place where he had friends. Halfway out the door, he stepped back into the house the moment Halliburton called his name. It was an awkward change of heart. Before committing, he tried talking his friend out of the venture as he tried talking him out of his last one, riding an elephant over the Alps, an ordeal begun, but not finished.[5]

Occupants for Hangover House, friends Mary Lou Davis, her daughter Dorothy and son Tommy, had already agreed to rent the house. While Halliburton said that Mooney was "overjoyed" about joining the crew, Mooney had misgivings: "Only God really knows why we're doing it," he said. "Four of the crew paid $17,000 for the privilege of coming along," he indifferently elaborated, "(of) suffering seasickness and working on deck at all hours: so there must be something in the idea. I naturally am getting paid for it, which is some excuse anyway." Tired of job uncertainty, Mooney could breathe a sigh of relief as he now joined the ranks of the gainfully employed. Times for him had been tough. In a letter to a friend, noted Cezanne biographer Gerstle Mack, Mooney confessed that "sometimes the strain of not finding a job to stave off chronic undernourishment" had depressed him. Following a referral from realtor Gay Beaman, he had nearly gotten a job writing a book about "flying," but the vague undertaking had fallen through for reasons unknown. Mooney never lived in one place for long, occupying a tenement here, other people's rooms there. In the 1920s and 1930s he spent his time mainly in New York and Los Angeles, and to a lesser extent in France and Germany. These years had been largely lost ones for the freelance journalist. "No one," he declared, "has suffered more than I (have) from the Wall Street Crash, which gave everyone the ready reply, 'Well, you can always find government relief work.'" Mooney referred to himself as "steadfastly a ragged individualist, marvelously lazy but knowing the laziness." Little changed for him between September 12, 1934, when he had written this description, and September 12, 1938. The resort community of Laguna Beach, within view of Catalina Island on clear days, had become his new home, but he seemed at times as purposeless there as Napoleon on Elba. On September 9, 1937, Mooney summarized his recent days to Mack: "The only place I ever travel to is San Francisco, surely my favorite city; it gives a breath of life after the moribund days in Laguna. (Even the nights are moribund, for I am growing old)." Mooney was then thirty-three. [6]

These were, at any rate, the concerns of the last person in America to join the *Sea Dragon* crew. "My crew is all fixed," Halliburton beamed. "(Paul) will leave on the '*Coolidge*' with me—we are *all* going Tourist class—$200." Besides Mooney and Halliburton, "all" included Captain Welch and Henry Von Fehren. They would leave San Francisco on Friday, September 23, stop in Honolulu on September 29, and arrive at Yokohama, Japan, on October 7. On October 18 the group would be in Hong Kong. The *Coolidge* had stowage for no fewer than two hundred passenger cars; among them the ponderous diesel engine would ride. [7]

But, except for the dim promise of some money and vainglory, why go at all?

11

The Royal Road to Romance in America

oet Walt Whitman proclaimed that a passage to India would be more than a passage to India. Similarly, Richard Halliburton hoped that the *Sea Dragon* Expedition would be more than a crossing of the Pacific. Now that funding and a captain and crew were in place, a higher meaning needed to be attached to the enterprise, one that gibed with the bigger aims of the Golden Gate International Exposition. To this end, Halliburton courted advice from friend the noted physician Dr. Margaret "Mom" Chung, a main voice in championing greater cooperation between America and China. Like Commodore Perry, whose 1854 expedition to Japan opened that country to the West, Chung intended "to develop a community of aims between two self-respecting nations."[1]

A trailblazer in trans-Pacific aviation, Dr. Chung had organized what became known as "Mom Chung's Fair-Haired Bastards," a contingent of Pan Am Clipper pilots who regularly flew from San Francisco to Hong Kong. Halliburton may have been introduced to Dr. Chung as early as 1936; by May 1938, he evidently knew her well. While informing Halliburton of her views on establishing stronger ties between the US and China, Dr. Chung may also have instructed him on the basics of proper etiquette and communication with Chinese businessmen, whom he knew to be shrewd. "There is no disaster so great," he would write, "as to extinguish the Chinese trading instinct." Little did anyone realize then, as did Dr. Chung with her crystal ball, that China would one day become a global economic superpower.[2]

For a fuller mission statement, Halliburton turned to "Dame" Gertrude Atherton. During his last days in San Francisco, he met with Atherton regularly at the Blue Lagoon Restaurant on Maiden Lane, across the street from Union Square and near his hotel, the Chancellor. Chatty and given to dropping names, the eighty-year-old Atherton might have spoken to Halliburton of her run-in with Oscar Wilde—he died at age forty-six in 1900, the year Halliburton was born. To Halliburton, Wilde represented intellectual emancipation, but to Atherton the playwright personified "the decadence, the loss of virility that must follow over-civilization." Her esteem for Russia's early empire builder, the said Nikolai Resanov, however, was immense. In her opinion, he was "by far the finest specimen of a man the Californians had ever beheld. . . . With an air of highest breeding and repose, he looked both a man of the great world and an intolerant leader of men." Besides women's and author's rights, both pet subjects of Atherton's, the Resanov experiment could have been a topic of conversation.[3]

Within these supposed contexts, Halliburton likely mentioned his next book. Smitten with "the epic impulse," he yearned to do something grand and meaningful, to write a national anthem or an odyssey for Americans who seemed to him without a common bond. As had Whitman in *Democratic Vistas*, Halliburton set adoration of America's picturesque landscapes, magnificent achievements, and indomitable spirit as his mark. "I never want to be without a book-in-the-making the rest of my life," he wrote home. "The next (book) must definitely be on America," he remarked, "(and) the junk trip will be a grand spring-board. All my creative thoughts for the next two years will be toward the America book." Just as his swimming of the Hellespont metaphorically bridged Europe and Asia, Halliburton's crossing of the Pacific would symbolically link Asia and America. Like the letters and notes that formed the basis for Mark Twain's *Innocents Abroad* and Halliburton's own books, so too would multiple sources comprise this new *Royal Road*. The book was to be an epic of America, its narrative ranging from the remotest geological evidence accounting for the country's wondrous landscapes to the routes its earliest explorers traveled and thence into modern times. The *Sea Dragon* Expedition, providing a pre-Columbian chapter of America's "discovery," would serve as the work's prow.[4]

Halliburton's writing was taking different turns. He could no longer think about freedom without also thinking about the responsibility it imposed. He now thought in bold strokes; as the Augustan poet Virgil had celebrated imperial Rome, so would aspiring poet Halliburton celebrate America. Just as control of the Mediterranean Basin inspired Republican Rome to build its empire, so control of the Pacific would inspire America to build its empire. Another chapter in the great China trade was opening. That *Manifest Destiny* should transcend the Continental United States and extend over every island between California and China seemed a measure of human progress.[5]

Entitled *The Royal Road to Romance in America*, the book would crown him as the cultural ambassador of America, a nation whose motto, echoing that of the Founding Fathers, read, "Alliance with all nations, compliance with none." A travel version of the "Great American Novel," exploring what Thomas Wolfe called "the all-engulfing wilderness of America," and Halliburton called "the American book," would "capture completely this immense and new society of ours." The work would cover engineering marvels like Boulder Dam and Grand Coulee Dam and natural wonders like Niagara Falls and the Grand Canyon. Besides numerous magazines and newspapers, Halliburton scanned numerous books, one on Mount McKinley, another on Yosemite Park. Other sources focused on the lives of famous Americans: George Washington, Abraham Lincoln, Nathan Bedford Forrest, George Armstrong Custer. Lesser-known figures such as "Death Valley" Scotty and John Hunt Morgan also merited attention, as did advocate of improved workplace safety and childhood labor laws Illinois Senator John Peter Altgeld. Vicksburg, and those Civil War soldiers who fought there, would also be a subject. Contemporary adventurers, such as Seymour Gates Pond and Admiral Richard Byrd, were omitted from this gallery of high achievers, as were illustrious women and minority figures.[6]

As any flirtation with radical politics by a man of his vaunted ideals could draw censure, Halliburton honey-coated his views of America. Congressional committees had periodically convened to investigate supposed subversive (Communist and, later, Nazi) activities within the United States. Halliburton had recently traveled to Russia. Once safely home, he condemned the country's Communist government as an affront to democracy, and he continued to put as much distance between himself and the principles of the Soviet regime as possible. In all events, Halliburton's new *Royal Road* was to be a flag-waving *Yea*-sayer—Irving

Berlin's "God Bless America" sung not by Kate Smith but by Richard Halliburton: Hail America, right or wrong. While today Halliburton's tone sounds insipidly patriotic, overblown, even cornball, in 1939 it was *au courant*. His deepest fear, recalled Wesley Halliburton years later, was "that the U.S.A. was too close to home to be romantic." Curiously, Richard also feared that "he was too old to write any more adventurous, romantic travel books." Another concern should have been the spate of books currently in print on the same subject. The Chinese connection would, in any event, be the main element which set Halliburton's book apart. The American flag would wave from the mast of a Chinese junk and herald US economic and political ascendancy in the years to come, or so Halliburton supposed.[7]

Responding to a "demand for clean, robust entertainment," exposition pitchman Art Linkletter produced *The Cavalacade of the Golden West*, billed as "romance, drama, [and] action." Later expanded into *America! Cavalcade of a Nation*, it featured colorful tableaux of "Washington's Inaugural," "Lincoln's Gettysburg Address," Napoleon seated in his bathtub signing off the Louisiana Territory, the joining of the transcontinental railroad ("Meeting of the Rails"), and a "Gay Nineties" tableau. Other scenes were of Christopher Columbus landing in the New World, the Puritans praying (to note religious tolerance), General Washington wintering at Valley Forge, and Patrick Henry delivering his "Give Me Liberty or Give Me Death" speech. A key stop was the famous Arrow Rock Tavern (a replica), where Kit Carson and other western-bound pioneers once gathered. Attractions emphasized "romance, historical adventure, sturdy humor, patriotism, respect of God, humanness and wholesome living." Last but far from least were the American murals of Thomas Hart Benton and the Pacific Rim panoramas of Miguel Covarrubias. These were matched in contemporary music by Earl Robinson's patriotic cantata *Ballad for Americans* and Ferde Grofe's orchestral suites, like the one celebrating the Grand Canyon, inspired by American scenic wonders.[8]

The Royal Road to Romance in America never reached completed form. Discarded sections or episodes from the two *Books of Marvels* might have served partly as text. As noted, the book would begin with the *Sea Dragon* Expedition, preliminary studies for which were the "Log of the *Sea Dragon*" and the "Letters from the *Sea Dragon*." Chief Officer of the SS *Coolidge* Dale Collins believed that, had he lived, Halliburton would have written "another book . . . as interesting as Richard Henry Dana's *Two Years Before the Mast*."[9]

12

You Never Die in Your Dreams

His plan to sail across the Pacific nearing realization, Halliburton told his parents that he was "very happy over (his) project—and over the calm way (they were) accepting it." It was his roundabout way of making the other side of the world seem next-door and the voyage itself an afternoon walk in Overton Park. There might be some showers and a wish for warmer clothes, but he'd be home by dark.[1]

Of course, traveling from Memphis to Charleston by car or bus was one thing, sailing from Hong Kong to San Francisco on a tiny wooden boat was another. Images of Amelia Earhart and her plane lost in the Pacific or of the *Hindenburg* exploding in New Jersey showed people that fame and fortitude are weak enemies of cruel fate. Besides, a war raged in China, Japanese gunboats aimed deadly missiles at any suspicious-looking ship passing by, and cutthroat pirates roamed the shores. The region was in utter turmoil. Why should anyone want to put himself in harm's way? To shift their attention, Richard told his parents about successful earlier junk voyages. They knew, however, that his travels in the Far East had only exposed him to "ships with sails." He was knowledgeable about trans-Pacific nautical history, but he had next to no knowledge of practical seamanship.

The crossing "will be tough, I know," Halliburton told reporters, "but plenty of fun." He advised those who wanted an easy crossing of the Pacific to book passage on the SS *Coolidge,* the biggest ocean liner of its day, and enjoy such amenities as air conditioning, telephones, a swimming pool, a gymnasium, a dance floor, and an Otis elevator. Halliburton had a schoolboy's acquaintance with ships, and through reading he had gained a docent's knowledge of Oriental maritime craft. He *knew* about scows, bateaux, dhows, dories, feluccas, shallops, sloops, catboats, skiffs, yawl-boats, and junks; he had sailed in some and could ask intelligent questions about others. Still, as late as November 21, 1938, when the *Sea Dragon* was about to be built, he had to admit, "Boats are a new world for me, so I've much to learn, and am sometimes troubled when I have construction decisions to make."[2]

Told by cousin Richard that he planned to cross the Pacific in a Chinese junk, a startled Juliet Halliburton brought up the dangers of so fanciful an undertaking. Brushing care aside, he assured her that the expedition would be less perilous than his earlier flight over the Sahara in a single-engine open-cockpit airplane. In any case, his "new inspiration" seemed less risky than the 1896 crossing of the Atlantic by two Norwegians in a rowboat, or the 1915 crossing of the Pacific by a junk named the *Ning-Po*. What Richard failed to tell Juliet was that it had taken him a year to recover from the two-year flight around the world. That was seven years ago. Years before, aboard a fast-moving train, he enthused, "Whewww!—This furious motion is stimulating, but it burns up too much (energy); however, I'm never so happy as when I'm busiest and have most to worry about." He had been younger then, and the train rested firmly on metal tracks going in one direction, not turbulent waves whose direction might suddenly change. It was some concession to age, however, that he should tell Juliet, as Amelia Earhart had told the press prior to her around-the-world flight, that this would be his "last trip." Wesley Halliburton, despite his son's assurance that "never was any expedition so carefully worked out for safety measures," wondered about the project's merits. Evasive, Richard responded that he had "a wonderful captain and engine and engineer"—what could go wrong? He also mentioned the huge profits from "common stock revenues" that he expected. Truth be told, he took a chance when he decided to forego the certain income from his scheduled lecturing for the uncertain profit from his proposed adventuring. Yet he believed that income from the adventure would exceed what income he lost as a lecturer. In this best-laid plan for financial success, common sense was the first fatality. Halliburton held fast to his plan, however, and, to help get it underway emotionally, he arranged for his mother, Nelle Nance, to come out to California. A most thoughtful son, he did what he could to make her trip fast and easy. Preoccupied with business matters at home, Wesley was unable to join her.[3]

A month before, in early August, Nelle Nance and Wesley had returned from Europe. The threat of a war throughout Europe had nearly preempted their trip. "You're safely home—and this is to greet you with wide open arms," Richard wrote them. "We've been so lost from each other for so long! I never knew where you were, and you didn't know whether I was in America or China. Well, you're in New York, safe and sound—and I am still in San Francisco just where you left me." Booked trips to Istanbul, Athens, Budapest, Prague, Vienna, and then Berlin hardly bothered Richard as much as their intention to book a trip to Leningrad, which he thought a troubled place. With Hitler's armies invading one country after another on apparent impulse, he was as much worried about his parents' safety and well-being as he knew they were about his. "I groaned over your passport difficulties in the Balkans," he wrote them on August 31. "I had the same experience in going from Vienna to Istanbul—and I know what you had to face." Mrs. Halliburton was coughing, and Richard hoped that the ailment would "be dissipated by the sea voyage home." In a summer "full of struggle," he was preoccupied with his parents and his continuing quest for "25,000, a figure he seemed stuck upon. According to Root, Halliburton pretended the $21,000 he raised was "ample," so "to reduce the chances of financial disaster, (he) mortgaged his house (in Laguna Beach) for $4,000." On September 10, he told his parents, triumphantly, "I have (with the Barstow check and mortgage) more than enough money—$27,000."[4]

These were vintage moments, of joy alternating with thrilling expectation. Through it all, Halliburton had managed to stay in shape—exercising and keeping to the dietary regimen ministered to him years before at the "wellness" institute in Battle Creek. Still, the intimations of mortality stirred up by recollections of childhood illness vexed him as grippingly as fears that he might rot in a detention camp as a prisoner of war, or be struck by a stray bullet or mortar shell while traipsing about some killing field. He had entered middle age and it frightened him. No longer that little boy playing Indian who had wowed audiences with his tales of far-off places, he was beginning to show the stretch marks of a life spent mostly on the road, in rumbling train cars and low-budget hotels. His celebrity was fading, as weathered and discolored as an old circus tent pressed into service again and again.[5]

Tediously slow to unfold, the *Sea Dragon* Expedition got momentum with financial backing, a fixed crew, and permission from both the navy and Pan Am to use its bases across the Pacific as stopovers. Even as he rejoiced that things were "moving rapidly and splendidly forward," Halliburton complained of "delays." His boyish enthusiasm usually triumphed. Writing home from the Chancellor Hotel, he called his voyage "the talk of the town" "and the envy of everybody I see." On Sunday, September 18, joined perhaps by friends Gertrude Atherton, Margaret Chung, and Noel Sullivan, he held a farewell party with Nelle Nance as guest of honor at the Blue Lagoon Restaurant. Author Charles Caldwell Dobie was invited at the last minute, but he might not have come. That last week Richard seemed full of pep and cheer. Valedictions poured in from all over the country; full-page press releases offered him a spirited send-off. Meanwhile, radio deals, promotions, investors, legal matters, bills, a new house, and correspondence "swamped" him.[6]

Through it all the *Sea Dragon* Expedition emerged alive and well. Halliburton had even sent his parents a picture he had drawn of the kind of junk he planned to buy; he called the ship's fiery designs and curving lines "lovely." He also enclosed a photo of Welch, Von Fehren, and "General Manager Dick" leaning against the rail of the SS *Coolidge*. As a gesture of faith in her son's latest adventure, Nelle Nance asked that he bring home gifts: "Yes, Mother dear," Richard assured her, "I will bring you a *Mah Jongg* set from China!—the nicest one I can find. I plan to load the junk down with Chinese things." Nelle Nance would later request a "red Chinese jacket" which son Richard also made a point of getting for her.[7]

13

Columbia, the Gem of the Ocean

Those final few days in Laguna Beach, then San Francisco, had left Halliburton "bewildered and dizzy." Yet he said those same days could not have been "any busier or happier." Nelle Nance hid her worry. She had accumulated many farewells from her son, beginning with that tearful one twenty years before when he boarded the *Octorara* bound for Europe. Resigned then, she was resigned now. She had met Captain Welch and Henry Von Fehren, both of whose apparent strength of character assured her that her son was in capable hands. "Seeing what a mature, experienced captain and engineer I have, she can feel at ease," Richard concluded. "With this engineer, & (add) radioman, & captain I'm safe as can be." He foresaw discomfort but no peril, and "plan(ned) to be back in San Francisco on schedule." His mother's visit he called "a great success," one of "fleeting moments" that were "very sweet and very happy." Although her son's last-minute preparations had kept them apart, Nelle Nance "was patient not to get too restless during the best part of the day when she was alone, just waiting and waiting for (her son) to 'come home.'" She had survived the heavy press coverage and believed her son had selected a "mature, experienced" captain and engineer. The connection for her return trip home—a confessed "long, lonesome ride," prevented her from waving her son off.[1]

On Friday, September 23, 1938, Halliburton and company boarded
the SS *President Coolidge*. A week later, British Prime Minister
Neville Chamberlain would give his "peace in our time" speech. In Europe, Hitler's army had just marched into Czechoslovakia; soon after,
Czechoslovakia reluctantly ceded the Sudetenland to Nazi Germany.
Six months earlier, Austria had been annexed to Germany. Watching
events unfold, far-sighted observers felt confident that Hitler intended
to annex all of Europe to the Third Reich. In Asia, the isles were restless
and a war raged. Many believed Japan had designs on China similar to
those of Germany in Europe. As cameramen stood by, fans crowded the
dock to cheer Halliburton and the *Coolidge*. Alongside Halliburton—
smiling and waving back from the ship's rail, were Paul Mooney, Henry
Von Fehren, and John Welch.[2]

Captained by Karl A. Ahlin, who had made the crossing innumerable times, the SS *Coolidge* was a marvel of shipbuilding engineering.
Associated Press journalist Violet Sweet Haven called the vessel "an
island city" and "a village on the high seas." Getting lost aboard the
ship, the largest all-electric passenger liner and merchant ship in the
world, was easy. Over 600 feet in length, with an 81-foot beam and
52-foot depth, the steam-propelled liner could reach speeds of over
20 knots. Each trans-Pacific crossing usually ran from San Francisco
via Hawaii to Japan, Hong Kong, and the Philippines, then back again.
Luxury staterooms carried all the modern amenities, including private
baths in each. As other ships of the line, the SS *Coolidge* was noted for
its "smoothness," an attractive feature to passengers prone to seasickness. A floating city, the ship's 'downtown' comprised "Peacock Alley,"
literally a shopping mall with boutiques and dining concessions. For
investors, there was a brokerage counter. There was also a dance floor
and swimming pool. Besides halls for lectures and musical events, there
was at least one massage parlor. Comfortable enough to ramble about
in any outfit at hand, Halliburton might have hankered, now and then,
for a much-needed rubdown. Over 350 experienced seamen and 600
outbound passengers, from first-class to steerage, were aboard any one
of its regular trans-Pacific crossings. Besides fish and poultry, 45,000
pounds of prime meat, tons of other comestibles, and 2,300 tons of fresh
water were loaded aboard by crane, for the "total of a hundred and forty
thousand meals which (went) to make up a round-trip to the Orient."
Thousands of cartons of cigarettes and hundreds of cases of wine were

also stowed aboard the liner, which 42,000 barrels of fuel were required to power. When Haven conducted her own tour of the ship, she stopped to observe the chief engineer "tending the boilers, warming turbines, watching temperatures and the pressures of the various pumps."[3]

Evening spread out across the sky. An hour before the ship left, a gong sounded. The railed gangplank was raised, the engines of the giant liner hummed and droned, and whistles blared. The ship backed off into the bay. "Once the *Coolidge* got underway," wrote Haven of her later shipboard experience, "the engines had to produce and consume more electricity than is used by the city of Honolulu." As it would for Haven, a bell now rang for Halliburton. Moments later it was "full steam ahead." As the ship roared further from the dock, Halliburton gazed for the last time at Treasure Island. A neon torch, *Exposition Tower* sent off beacons of light. The *Court of the Hemispheres*, with its giant world globes, gave a Godspeed. Under now starry skies, the *Coolidge* now headed out of the bay and beneath the Golden Gate Bridge into open waters.[4]

Halliburton retired to his quarters: "It's late—It's late—I'm tired," he told his parents. "You know I love you all—and embrace you, 5000 miles away. Good night—Richard." He soon plopped into bed and slept till noon the next day. Instead of sheep, he counted dollars, endless numbers of them marching into his bank account one by one. "I don't let myself daydream about the profits I'm to make out of the venture," he remarked, "but it may be considerable." He added, "[I'm] dreaming of the story I'll do about my *Sea Dragon*." He then noted Paul Mooney's increasing mastery of the mimeograph.[5]

From his office in San Francisco, Wilfred Crowell printed the circulars. Ten thousand had gone out on October 1, and, if profits mounted, another ten thousand would be issued. "With letters, envelopes, *Bell Syndicate* stories, my *Sea Dragon* and its crew of ten," said Halliburton, "I'll have my hands full—which I'll like." Halliburton was in the midst of life and the sirens sang. Actually, the *Coolidge* did have an orchestra which, day and night, played soothing melodies. On her later trip, Haven recalled them do "Wishing," one line of which was, "If you wish long enough, wish strong enough, you will come to know . . . W-I-S-H-I-N-G will make it so."[6]

Once out to sea, and fully rested, Halliburton almost seemed merry. Just beyond the next wave was a pleasant layover in Honolulu. Already he was thinking about the return voyage home and the most practical

means to accomplish it. The ship's second-in-command, Chief Officer
Dale Collins, said that Halliburton and Welch "spent many hours on the
bridge studying . . . charts and discussing the best probable course they
would take and the weather they might expect." Perhaps with Collins's
help, they also observed the relationship of the currents to the stars
and marked land-points. Collins knew that Welch had spent time on
sailing ships and been an officer in the United Fruit (shipping) Line.
"He had a very likable, dynamic personality," Collins said, "and was
always bubbling over with vitality and fun." During the voyage, Welch
told many "frank and amusing anecdotes" which "kept everyone entertained. (His) lusty tales were always well-seasoned with a seaman's
phraseology." The captain might eye the ladies rakishly behind their
backs, but to their faces, he was "as courtly and gallant as any knight
in King Arthur's Court." Collins added, "Although he was extremely
popular with both sexes, he was a man's man imbued with the love of
life and the sea."[7]

If trans-Pacific voyaging was routine for the SS *Coolidge*, it would not
be so routine for the *Sea Dragon*. Not yet sure what route to take back
home, he predicted that the crossing, with stopovers, would take ninety,
not thirty, days. "We must leave Hong Kong—go south of Formosa—then
north with the Japanese current," Halliburton told his parents, "but
(we will) not go as far north as Japan—turn left at the 30th Degree
latitude (crossed out, 'longitude') and follow this to Midway. This is
4000 miles between ports—but it's quick—with following winds, and
mild season—The 30 degrees is about the latitude of Los Angeles and
Charleston, South Carolina—so we won't suffer from cold." He failed
to mention the currents, namely the rotating currents known as *ocean
gyres* above whose southern arc, "the great circle route," the junk was to
follow. Liners regularly traveling that route could track their progress.
No landfalls apparently stood between the main stopovers of Midway
Island and Honolulu, both of which seemed immense in the context of
smaller landfalls. But how immense? From the air, a pilot, as his plane
prepares for landing, has difficulty at first seeing even the Hawaiian
Islands. About 1,150 miles west-northwest of Hawaii, what might a
pilot see of Midway Island from the air? Magnified in size by its proximity to much smaller land sites, it was in fact an atoll measuring only
two square miles. A US possession since 1867 and a stopover for *Pan
Am* air-traffic beginning in 1935, the tiny island—chiefly a sanctuary

for birds and a dumping ground for ocean refuse, was considered the United States' first line of resistance against Asian aggression. Few people lived there, but most who did were ship service technicians or postal officials. The San Francisco–Manila mail route included regular stops there. Besides being a key station for *the Commercial Pacific Cable Company*, the appropriately named Midway was a crucial refueling stop for navy ships.[8]

From the moment he laid eyes on Halliburton, Welch with sidelong glances and prolonged stares sized him up. On trial for him was Halliburton's image as a playboy, the heartthrob of women young and old, "the *enfant terrible* of travel," and "good-looking Gulliver" who had "met fascinating women" but whose "beau ideal (had) eluded him." Welch's suspicions grew that Halliburton was a powder puff—not one cut out for the grim realities of seamanship. Had Halliburton not characterized the voyage as "a picnic" and categorized "hurricanes" as calamities of little concern? Likely Halliburton talked to Welch about the "perfectly gorgeous color scheme" he wanted on the junk. It was maybe odd to Welch that the world traveler had never been to Australia, equally odd that the 'most eligible bachelor' should have a male secretary whom—what a laugh, Halliburton called an "experienced seaman." And who were these collegians? Sure, they had sailing interests, but had they any *proven* sailing skills? Handle a rowboat in a pond? Maybe. But could they handle a ship on the open seas? Doubtful. Welch soon got the unsettling feeling that he was to be a drill instructor, whose task it was to mold a middling bunch of raw recruits into an effective combat unit.[9]

Halliburton, Mooney, and Von Fehren, meanwhile, sized up their fearless captain. Von Fehren, sensing Welch's disdain for his klutzy mannerisms and Germanness, instantly disliked him. Mooney found Welch repugnant. Even though Welch and Von Fehren feuded—mainly about the diesel engine and its need aboard a small sailing ship which had ample rigging, Halliburton told his parents that the four of them were getting along "beautifully." By attacking Von Fehren, Welch indirectly attacked Halliburton who brought the engine and hired its operator. Of course the engine's unstable temper and odious "groaning and coughing in the hold" was music to few ears. For even small craft on a long voyage, Jack London thought a diesel engine indispensable. Besides helping a ship to motor in and out of port and move forward regardless of prevailing wind, the motor provided the electrical power required to ensure a well-working radio communication system. On what he thought would be "almost entirely a sailing trip," one with "little to do but work and read," Mooney saw the obstreperous contraption as a necessary nuisance, but one that would hasten their crossing if resorted to sparingly. Seeing the diesel engine as a hard-driving member of the crew, Halliburton, in disputes over its necessity, tended to side with its operator, "real gentleman" Von Fehren. He at first said little about Welch. Unable to overlook it, he hoped only that the growing animosity between him and Von Fehren would pass. But even after attempts at mediation, Halliburton told his parents that the men "still hate[d]" each other.[10]

Several days passed; weeks, to Halliburton. "The voyage has been perfect—hot sun and calm seas—and it will continue on to Yokohama," he told his parents. "I'm very happy over my project—and over the calm way you are accepting it." To Monica, at Alber-Wickes, he reported that the "harassments" had so far outweighed the joys: "I'm too bedeviled with engines and permits and salaries and radios and contracts and crew and a hundred other worries to know whether I'm really happy over or not over my success." As a diversion, he read Gertrude's Atherton's "entertaining" (and bulky) autobiography. He also wrote perhaps a thousand words each day. Collins described Halliburton as "quiet, unassuming, and a trifle shy." Other crew also assessed the celebrity: "Our attention," said Collins, "was irresistibly captured by his slow, soft-spoken tones, his flawless manipulation of the "King's English, and his inexhaustible vocabulary." He related then a curious event involving the author. It so happened that one evening a ship's watchman found him wandering about the passenger quarters in "slacks, open shirt, and an old jacket" and not in the customary dress-wear stiff-collared shirt and tie. Straightaway, the "efficient and rather breathless" watchman reported him to the officer on deck as a possible stowaway (a 'stowaway' who was maybe stargazing). Taken to his cabin, Halliburton identified himself and that seemed that, but, "for the remainder of the voyage," he "received considerable good-natured kidding about (the episode)."[11]

If not on the deck, proper attire was an unwritten requirement in the dining hall, whose menus offered the finest cuisine Halliburton and company would see. Appetizers customarily included a seafood cocktail with lime twist, stuffed eggs *Strasbourgoise*, yacht-club sardines, and canapé Monte Carlo. Start out with a bowl of *tomato monegasque*. Entrees customarily featured saddle of Belgian hare *sauté financiere*, *tete de veau en tortue*, braised tenderloin of beef with natural gravy, poached medallion of chicken with remoulade sauce, halibut *flamande*, Petaluma capon with celery dressing and cranberry sauce, and prime ribs of Kansas. Available were many sides. Potatoes, for one, came in a choice of snowflake, château, or baked Yakima in the jacket. Desserts included Yorkshire pudding, buttercream cakes, and different flavors of ice cream or Jell-o. An assortment of cold cuts served with potato salads—"*avec la cuisine sont plusieurs vins classiques de France et Amerique*"—was always present for the asking, as was around-the-clock room service.[12]

Later Halliburton described the voyage to Hawaii as "perfect," with "hot sun and calm seas." One first saw Molokai, then Oahu, and then surfing mecca Waikiki Beach and Diamond Head. Putting aside Atherton's autobiography, he was again reading Somerset Maugham's "autobiography" *The Summing Up*, with its many strange admissions, ranging from its world-renowned author's self-identification as the best of the second-raters to his wish to be born with a better brain. A confessed skeptic, Maugham spoke at last about the importance of religion in one's life. He had turned sixty; once one reached that age, he said, *death is no longer remote*. As the ship neared Hawaii, a brilliant sun burned the glistening sands. Halliburton stepped closer to the rail to inhale the balmy air, as the roaring surf beat on the shore. He said little about the stopover or the welcome he received, but customarily, wrote Violet Haven, when ships arrived, "nut-brown Kanaka boys shouted for coins and dived deeply into the harbor as dimes and quarters came spinning from the decks above," shouting. "Friends (of those de-boarding) stood on the pier with fragrant leis of ginger, gardenia, carnation, *maile* and *pikaki* over their arms, waiting for the ship to dock." Halliburton was a celebrity arrival, and a hive of photographers gathered around him the moment he stepped off the walkway. Joining the team in a photo shoot was radioman George Petrich who, as noted, would remain in Hawaii until the junk was ready to sail from Hong Kong. Reporters from the *Honolulu Star Bulletin* next drew Halliburton aside for a brief interview. While the *Coolidge* was docked, he relaxed. Photos show him sunning on Waikiki Beach, where, as poet Rupert Brooke (there in

October 1913) had written, "new stars burn into the ancient skies, Over the murmurous soft Hawaiian sea." In no hurry to leave, Halliburton, after ten lovely hours in Paradise, re-boarded his ship. Compared to the stretch from San Francisco to Honolulu, he found that from Honolulu to Yokohama "pleasantly dull."[13]

14

"Japanese, If You Please!"

On October 7, the SS *Coolidge* arrived in Yokohama, with Halliburton complaining that fifteen days aboard the ocean liner was "overlong." By comparison, he had to wonder how ninety days imprisoned in a rickety old junk—truly a less hospitable ship and one that didn't serve chicken *haliburt flamande*—would affect his spirits. The distance to be covered also troubled him, as did the privations to be endured. "Lord knows," he wrote to his agent Monica at Alber-Wickes, "I'll need all the prayers and blessings I can get. I'm coming across in mid-winter—9000 miles in a 65-foot (length by water-line) boat and I don't know any more about a boat than I do about heaven. However I've got a smart professional captain who does know and a good engineer and a good engine. I'm taking a big gamble financially, plus a little item, (another gamble) known as my life." Standing alongside Captain Welch, he continued to observe the currents and weather patterns: "I have noted that the wind has been strong against us this trip—Hope it remains so on our return. Jack (Captain Welch) says that if we have the same wind in December we'll be across in *thirty* days." Wishful thinking was never more wishful.[1]

Yokohama was a bustling port. Here in 1890—a date midway between Hong Kong's founding and Halliburton's final sailing from the port city, stunt journalist Nellie Bly, on her newspaper-sponsored effort to beat Phileas Fogg's around-the-world record of eighty days, stopped for some 120 hours. Bly called Japan "the Land of the Mikado." Finding Japanese culture in many ways superior to her own, she reported with some joy that the Japanese used what they adopted from the West to better advantage. "If I loved and married," she said, "I would say to my mate . . . (I would) desert the land of my birth for Japan, the land of love—beauty—poetry—cleanliness." About the Japanese, she remarked, "[The men have] good-natured faces, at least the *jin ricksha* drivers," and the women are all "charmingly sweet." Of Japanese geishas she commented, "[They] are the only women I ever saw who could rouge and powder and not be repulsive, but the more charming because of it." The globe-girdling reporter had come to the Far East in 1890 and had found much of it colorful and quaint.[2]

Arrived then the twentieth century and the last days of Meijan social reform and modernization. "Time alone must have made some changes," said journalist Violet Sweet Haven in 1939, adding that the war had also brought change. After an absence of only eight years, she had returned to Japan. Customs officials poked around her person and looked through her purse, but she seemed not to care. For her, being in Japan "was fun." The quaintness of everything was to Haven most arresting— "the kimonos and the colorful obis, (and) the Japanese women wearing wooden shoes and wafting down the streets under paper umbrellas." in so regulated a society, she looked novel, but was generally left alone and unobserved. "The hundreds of Japanese on the railway platform looked at me only casually," she said. "The Kamakura express whizzed to a stop, not a moment early, not a moment late, as usual. Just as punctual as the Kona rains." The manners of the people, most relaxed and levelheaded, little bothered Halliburton, but the "officialdom and red-tape" by bureaucrats who were high-handed and vain boiled his blood. He had first visited Japan in 1923, and spent several months there. "Every moment of my stay (had) increased my delight in being there," he wrote. "The people were so extravagantly courteous and quaint; their bustling paper cities so bizarre; even the natural surroundings had been shaped in the Creator's most whimsical mood." Once, and in winter, he had scaled Mount Fujiyama, an active volcano revered by many as the very embodiment of nature, and by the Buddhists as an entryway into another world. Nearing Yokohama, murky weather blocked his view of the legendary peak. "Revolutionary" changes on the island nation blurred other things as well.[3]

Friday the seventh in Yokohama was a fine October day, cool, crisp and clear, with temperatures in the mid-fifties. Besides seven hundred bags of mail, the *Coolidge* dropped off a number of prominent passengers. These included assistant commercial attaché to the American embassy Paul Steintorf (with his wife) and US Marine Corps colonel J. C. Fegan, both little noticed. Cameras flashed, however, the moment Halliburton and company stepped ashore. Smiling and dressed casually, the sportsmen might have been mistaken for insurance agency executives or a barbershop quartet had agent (and publicist) Wilfred Crowell not warned the press of their arrival. The caption beneath their photograph in the *Japan Times* identified them as just four of a projected crew of ten who would accompany "author, lecturer, and adventurer" Richard Halliburton on his latest mission.[4]

Asked how the idea of a trans-Pacific voyage had come about, and how he would accomplish it, Halliburton responded, "(I have) no special purpose in making the trip other than it (would) provide me and my companions with adventure and sport and incidentally material for the next book which I shall write in the near future." He added that "the *Bell Features Syndicate* and several American papers" had "commissioned him to keep a record of the trip." Halliburton said he initially intended to take the route of the Clipper ships. After consulting with "certain persons . . . well acquainted with the Pacific," he decided that it was better to "sail from Hong Kong to Taiwan or to southern Japan and then head for the Midway Islands and then across the Pacific." He said the trip would take "ninety days," an estimate now etched in his brain. That tally included fifteen days for stopovers, originally at Manila, Guam, Midway and Honolulu. The changed route, however, might have them spending more time at sea. He also mentioned that he intended to go to the American embassy to obtain official permission from Japanese officials "to sail to Japan." Captain Welch was also interviewed: "There is not the slightest doubt in my mind," he said, "that a junk cannot make the voyage we have in mind. The junk has been used by the Chinese for a thousand years or so and it has stood up under all kinds of weather. The only difficulty we might encounter is that none of us have had any experience handling a junk and we have been told that they cannot be run like ordinary sailing vessels such as are known in the west." Soon after meeting with the press, the group divided. "Jack Welch," reported Halliburton, went "overland to Kobe," where he would make "purchases of ship stores and go on to Hong Kong direct." Welch might have looked over a junk while in Kobe, one based on a tip given to him by Dick Wetjen. The eight-year-old vessel cost $8,000, and would have made a perfect 'Sea Dragon.' A deal was never struck. Henry Von Fehren and the engine would stay aboard the SS *Coolidge* and reach Hong Kong on October 18, Halliburton also expected to be in Hong Kong on that day.[5]

September 10, Halliburton wrote to his parents, "I'll be in Japan five days, seeing the Jap Navy." Taking the British navy as its model of efficiency, Japan's navy had launched torpedo attacks on vessels in every sea separating the Japanese islands from mainland China, especially to the north. Sealing a better fate for himself, Halliburton sought assurances from the Japanese government that he and his crew would receive safe passage through Japanese-occupied waters. To this end, his first task was to contact US ambassador to Japan Joseph Clark Grew, who was currently stationed in Tokyo. If a generation older than Halliburton, Grew was very much like him, at least superficially. The ambassador was an Ivy Leaguer, an avid outdoorsman, an admirer of Japanese culture, and a figure as well known in Japanese society as Halliburton was in American society. Of special note, Grew's wife was the great-grandniece of the man who opened Japan to the West, Commodore Matthew Perry. Their daughter, twenty-three-year old Anita Grew, outdoing Halliburton, had swum "the entire length of the Bosphorus in five hours." Halliburton said that he telephoned the ambassador's office; if he reached Grew, the Ambassador might have told him of the dangers that lay ahead and the shaky relations between America and Japan.[6]

Attentive to world news, Wesley and Nelle Nance Halliburton worried more about their son falling into the hands of the Japanese military than about him meeting expenses for a trans-Pacific journey in a Chinese junk. Richard dealt constantly with the public's conviction that he would never return to America alive. Precedents of survival under circumstances as dire were available if they chose to explore them. A famous one centered around war correspondent Jack London, who in 1904 was hired by the Hearst paper the *Examiner* to cover the Russo-Japanese War. While his fellow reporters, under the spell of Japanese hospitality, languished in Tokyo, London accompanied the Japanese Imperial Army through parts of Korea. Denied the use of his camera, he wrote the best front-line correspondence of the war. Three times he was arrested, twice for spying and once for striking a Japanese official. Only the timely intervention of fellow war correspondent Richard Harding Davis and President Theodore Roosevelt secured his return to America. London's time abroad, captured in the best-selling *Sailor on Horseback* (1937), included a sailing adventure. For over six days through freezing rain and lashing headwinds and with three Koreans, none of whom spoke a word of English, he had sailed a junk across the Yellow Sea, an inlet of the East China Sea; during the ordeal, the mast collapsed and the rudder broke loose. London was himself a "physical wreck," said one witness, and looked deathly ill. Yet he persisted, he survived. Halliburton could take heart.[7]

The Chinese, meanwhile, feared the Japanese, whose island nation lay less than three hundred miles from their doorstep. By 1938, Japanese influence extended throughout the western Pacific. The nation's armies had successfully invaded China in the mid-1930s and occupied many of its inland and shoreline communities. Shanghai had fallen to the Japanese in 1937; the end of the year witnessed the infamous "Rape of Nanking," which saw the ruthless slaughter of over three hundred thousand Chinese civilians. Meeting little resistance, the Japanese next captured Canton, southern China's political capital, the oldest of China's mainland cities, and the Yangtze River's outlet to the sea. Japan's forces earned a reputation as butchers by committing atrocities that shocked the generation that had experienced the war to end all wars. Reports circulated of soldiers raping mothers and daughters, and mutilating and decapitating public offenders. It was also said that fathers were ordered to shoot their own sons, then ordered to shoot themselves. Civilians were shot or bayoneted, their bodies then booted into burial pits they had dug themselves.[8]

The Japanese government exercised complete control over its citizenry. Told how to think and how to spend money, they were also told to cheer loudly "for Japan's alliance with Italy and Hitler." Halliburton believed that in America "the Japan-Germany-Italy alliance" was not taken seriously. What Americans failed to grasp was obvious to him, however: "When Italy and Germany go to war with France and England, Japan will seize Hong Kong, Singapore, and Australia, and take the Philippines and French Indo-China in stride." Because France and England would "be fighting for their lives in faraway Europe—and because Russia can't fight, and America won't," Japan could accomplish this easily. Sent from Shanghai on November 20, the letter provided a short history of Pacific crossings by Chinese junks and by so doing invested the *Sea Dragon* Expedition with another higher calling—assist a weakened China and check the advance of an evil empire, Japan.[9]

Since Halliburton's second and last visit to the island nation in 1932, a short one to Kobe, Yokohama, and to Tokyo, changes had taken place. He wrote a series of impressions of these changes with sad, often narrowed eyes. Muddy roads cutting past ramshackle huts and villagers in rags led to modern Western-style metropolises like Osaka and Tokyo. Here buses, automobiles, taxis, and electric streetcars moved noisily down paved thoroughfares, and factory workers mindful of the clock shuffled to their daily routines. Unlike industrializing Japan, agrarian China had not Westernized, and while a city like Shanghai had modern elements, the nation as a whole had paid a steady price for its backward stance. The Japanese once revered the Chinese, but, after defeating them handily in the First Sino-Japanese War, they viewed them with contempt. Few people—unless they were "officials equipped with the official phrases"—had much of an idea of the current war in China. Some even believed that Japan was trying to "save China." Having no idea "of the real nature of the actions by which Japan condemns herself in foreign eyes," said Halliburton, the average Japanese had even less idea of the "mass murder of China's civil population."[10]

Diplomatic ties with Japan and the United States were strained but not yet severed; the surprise bombing attack on Pearl Harbor was still three years away. For Halliburton, Japan stood beneath the bridge of Asia as a troll awaiting the billy-goats from the west. He would have liked to skirt Japan all together, but the "junk-sailing expedition" required that they travel through its sphere of authority. Many of Japan's officials were hard-core militants—anti-American, anti-intellectual, anti-literary. One in particular epitomized these qualities. Asking Halliburton to show his passport, "the little man with the sword and the rubber stamp (an immigration officer)" saw that his profession was "writer." "For what newspaper do you write, please?" he inquired. A "deadly polite" period of questioning then followed. Halliburton cringed. "Had the passport described me as a professional thief or an indigent leper," he said, "my welcome would have been warmer. Writers are suspect in Japan these days. Only grudgingly did the (fellow) admit that he couldn't actually prove my criminal intent, and—still bristling with suspicion—allow me to land." Once ranked above generals, writers, called "scribblers," now ranked below drummer boys. A book was considered "as dangerous as a bomb," and, though Halliburton carried with him "nothing more subversive than Webster's *Dictionary* and (his) latest *Book of Marvels for Children*," the inspector gave them a "careful" going-over.[11]

Breezing through "one of the most densely populated areas in the world" and past "a continuous stretch of tall (munition-factory) chimneys breathing smoke," an electric train took Halliburton and Mooney to Tokyo. Once a quaint hallmark of Asian sweetness and light, and since arisen from a devastating earthquake and rebuilt, Tokyo now wore an iron mask of modernistic conformity. Halliburton barely recognized it. What he remembered as "God's little creation," a place where everyone behaved happily was now a paradise lost. The failed coup known as the February 26 (1936) Incident, precipitated by disaffected elements within the military, had had violent repercussions and further militarized the civilian government. The new order "might prove strange indeed," Halliburton feared, and so might the people it ruled.[12]

High-profile visitors such as Halliburton were condemned as spies before they were cleared as tourists. Once when he was a naïve kid and photographed the gun placements at Gibraltar, he was summarily arrested and thrown in jail. The US Consulate, the assistance of a friend, who paid his fine, and an understanding judge got him off a charge of espionage. The memory of being thus detained lingered and Halliburton knew that he had best exercise the gravest vigilance in the new Japan. He needed to remember that his main purpose was to contact the Japanese Foreign Office and explain to them that "several weeks hence" he would be sailing away from China in a Chinese junk manned by ten Americans. He knew that the Japanese navy had blockaded all the Chinese port cities and used torpedo boats to sink thousands of junks along the coast. If waved before their eyes, the safe-conduct papers would have had no meaning to Japanese torpedo launchers, as the *Sea Dragon*'s "shining new sails," said Halliburton, were "a target too alluring for the Japanese navy to resist." The three over-sized American flags on his junk would make little difference, In fact, it probably would have been better to hoist a German flag, as osteopath Dr. E. Allen Petersen would do in his effort to cross the Pacific by junk.[13]

In Tokyo, Paul Mooney evidently spoke with 'Air and Navy'

attachés at the American embassy, and, in turn, with several secretaries of the Japanese Departments of War, Navy, and Communications. At the Japanese foreign office, he met several times with "the Spokesman," a Mr. Tsuchiya, "one of their best." Mooney described Mr. Tsuchiya as a "hydra-headed creature" and "most politely appalling." Tsuchiya might have been the very immigration officer who, reproving Halliburton for his professional status, issued him the safe-conduct papers. Believing the stamped documents were sound, Halliburton clung optimistically to the belief that he would receive safe passage; moreover, the Japanese Foreign Office in Tokyo told him that he and his party had nothing to fear. Assurances of the kind meant nothing, as failures of communication and accidents happened. Halliburton was armed only with paper and assurances. No fool, he knew that the torpedo boats had an awful habit of sinking junks that entered the "war zone" and apologizing for it later. With the papers tucked in his pocket, he could say, "(I) will carry them with me—even to the bottom of the China Sea."[14]

15

Pictures of a Floating World

M ost Japanese, in Halliburton's view, "seemed busy and prosperous and unconcerned with war." The few who spoke of the conflict actually believed that Japan was "trying to save" its Asian neighbor. Except for Germany and Italy, the world at large was disgusted by Japanese brutality, but the average Japanese citizen, left in the dark, appeared unshaken. Japanese in the hinterlands seemed to care little about the war in China. Citizens in the capital of Tokyo seemed to care even less about their country's involvement "in a great war for the subjugation and annexation of the most populous country on the globe." Two things led journalist Haven to believe Japan was at war: "the scarcity of young men on the city streets" and "the absence of taxicabs cruising in the streets." Gasoline rationing had begun, and buses in Tokyo were reconditioned to burn charcoal instead of gas. Soldiers were "little seen and less talked about."

Incredulous, Halliburton had to rub his eyes. "Despite rumors of impending financial collapse," he said, "wartime Tokyo has never seemed so prosperous: hotels are over-flowing, theatres packed every performance, and so great is the department-store business that the stores stay open until eleven o'clock each night." "Thousands of neon signs, which lined the railroad, rivaling Broadway itself," signaled a thriving commerce. The "dazzling display of color and fantasy" advertised soap, beer, and shoes. "Tokyo after nightfall had the hectic color of fever, but surely not war fever: nothing could have appeared gayer or brighter." A fascinated Halliburton paused to note "the happy, orderly throngs—in the theaters, along the Ginza, in holiday mood at the national parks of Nikko and Nara". Said he, "It was hard for me to believe that people who are so well behaved at home can be such savages in China."[1]

Kept informed about events in Europe, Halliburton inferred that Japanese and German aggressions were concerted actions. "The September crisis in Europe," he wrote, had given Japan "her chance," and, with war looming, the Tokyo government "rushed 300,000 soldiers, supported by a good part of her navy, to Formosa—and kept from there, ready to pounce upon Hong Kong the moment German bombers set out for London." Hostile Japanese military forces were at Hong Kong's front door, and searchlights mounted on the Peak illuminated their battleships and raised guns at night. Halliburton would soon see for himself "the terror of light" and "the gloom of the grave," but had he settled in Japan to write of these things, he would have been arrested. If pens were to the Japanese immigration authorities poisoned darts, cameras were unpinned grenades. "Anti-camera ordinances" were "insanely funny" to him. Photographing "any building above the first floor" was strictly forbidden, and permission from the foreign office was required even to "take a train to the national shrines." Protocols were to be followed, rules obeyed. Official business with the Japanese authorities often stretched from a one- or two-day affair to several days, but tarrying gave Halliburton the time he needed to observe casually and form impressions from "everyday personal contacts." He noted that foreigners discussing "the stock exchanges of the world" were particularly distressed by "Japan's dwindling gold reserves." Asked the perplexed, 'Where did all the gold go?' Halliburton had a theory: "The instant the first Japanese grinned at me I knew the eternal Japanese smile was just one continuous gold tooth."[2]

Prohibitions were many, the majority of which struck Halliburton as absurd. Opening and closing hours for retailers and wholesalers were regulated, as were the times when citizens could attend to their devotions (called "'spiritual thought and patriotic activities'") and go to work. Colorfully costumed geishas had performed traditional dances for visiting sailors in the olden days, but no longer. Tokyo's pace was hectic, even chaotic, and its subways and skyscrapers gave it the look and feel of a "modern American city." Formerly but a few city dwellers had dressed in the current European style; now most did. For Halliburton, this practice, not so regrettable for men, was a "disaster" for women. They had previously worn "graceful, flowered kimonos," and with "their gentle, intensely feminine manner," they were "the most appealing gals in the Orient." Now they wore "middy blouses and tailored suits." In addition, many had "a most unfortunate fondness for unnecessary horn-rimmed spectacles, worn mostly by students and business men, and unnecessary gold teeth!" On occasion "an old-fashioned Japanese girl" strolled by, still clinging "to the color and grace once cultivated before Japan set out to conquer the world!"[3]

Seasoned traveler Halliburton had grown used to having military authorities take him aside for questioning and borne enough "spy-phobia" for one lifetime. Then came "National Spy Week," topper to it all: "Posters and editorials announced it," he wrote. "Lectures with movies were held in public auditoriums; school children wrote essays on how to detect and capture foreign agents. 'One day on my way to school, I saw an old man dressed as a fish-monger, peeping in the window of an airplane factory . . . ' begins one of the published prize winners. There is for me something terribly depressing in the thought of thousands of well-drilled school children, all heroically counter-spying at once." Almost as a conditioned reflex, the Japanese "smiled incessantly," leading Halliburton to believe that they had forgotten how to break out laughing, especially at the absurdity of their new pretensions. Noting that Germans and Italians were immune from Japan's "intense suspicion," he soon formed an opinion as to why: "Bound by 'cultural' alliances, the three nations emulate the famous monkeys that are carved over the doorway of a shrine at Kikko: they see, hear and speak no evil—of each other. Such an exchange of compliments, such a waving of each other's flags. The Italians rate pretty high, but the Germans rate so *very* high that the Japanese have managed to ignore everything which Hitler, in his *Mein Kampf*, says about the yellow races. (It is nice, of course, to know that Japan can accept an apology as easily as she gives one). For the time being, these three Fascist powers are treaty friends." About this time Violet Haven was gathering notes for her book *Gentlemen of Japan—A Study of Rapist Diplomacy*. The Americans, she said, had been "duped" into trusting the Japanese, whose warlords made a mint on the opium trade in China. Japan to her only *seemed* a peace-loving nation. Once war broke out between Japan and the United States, the bumptious reporter accused the Japanese of long-running "nation-wide perfidy": household servants were coached in espionage; villagers living on apparently "innocent outposts" of the nation's dominions were trained in shortwave radio communication.[4]

Introducing one of its many totalitarian controls, the Japanese government decreed that citizens must greet the country's alliance with Germany and Italy with roaring applause. Showing no opposition, theaters showed German films and feature revues "full of gondolas and *O Sole Mio.*" Remarked Halliburton: "The largest theater in Tokyo—where the entire cast, including the 'male' players, are always very pretty girls—was having great success with one of these curious cultural efforts." Of course the show was a propaganda vehicle: "On the stage appeared tableau after tableau of girls in Nazi uniforms, more girls in Italian uniform, still more girls in Japanese uniform. During a "finale " that was "a blaze of mingled Italian, German and Japanese flags,"they marched and counter-marched to German music; they sang the Italian national hymn." Halliburton watched the hours-long drama in puzzled amazement.[5]

Halliburton had ample opportunity to roam about Tokyo, and the Japanese government granted his requested letters of transit, as well as permits to enter certain Japanese-occupied territories in China. A few hours or days later, he would request and receive safe conduct through the Formosa Islands. The stopover troubled Welch, who at this early date (October 29) thought it would delay passage home. Although Halliburton poked fun at Japanese officialdom—whose representatives he thought preposterous caricatures of military pomposity, he was wise to keep it to himself as he sought their cooperation. If not in word or deed, he sinned in thought. Upon departing from office, George Washington, admonishing his successors, thought it America's "true policy to steer clear of permanent alliances, with any portion of the foreign world." As a private citizen from a free country, Halliburton believed it his right to act in the reverse—to intervene and lightly meddle with the "foreign world." Accordingly, he met a number of heads of state during his travels: the president of Andorra in Spain, Peruvian dictator Augusto Leguia, Ibn Saud of Arabia (the "Mohammedan Pope"), Greek revolutionary Eleftherios Venizelos, Ethiopia's Emperor Haile Selassie,

Mrs. Vladimir Lenin ("First lady of the Land of Russia"), Henry Pu Yi (last emperor of China), King Feisal al Husain of Iraq and his son the Crown Prince. On this stop, however, he did not visit heads of state—just boorish underlings.[6]

Paul Mooney had little good to say about Japan, writing to friend Alice Padgett that it "was a god-awful place for an American. If you aren't German or Italian," he continued, "the Japs make themselves a nuisance—opening your luggage secretly, opening your mail, tailing you on the street and being generally obnoxious. White visitors are so rare in Japan nowadays that those who do come get more than their share of official attention. You don't begin to know how offensive such attention can be either, until you've had a flock of the grinning little monkeys asking—very courteously—every possible private question." Roving reporter Mooney was happy to *escape* Japan—even if the escape was to shell-shocked Shanghai. Although he told his parents that he "loved his few days in Japan," Halliburton also had to admit that whatever bonds of affection had once linked him to that nation of silk dresses and welcoming faces were now severed. With Mooney alongside him, he boarded a German ship bound for the China coast. "The last thing I saw," he said, "was an enormously fat Japanese gentleman, who bulged over the railing of the pier, waving a Rising Sun and a Swastika, and mingling *Banzais* with *Heil Hitlers*." As Halliburton sailed away from Kobe, no Madame Butterfly sang melodic arias; no painful remorse weakened a saddened heart.[7]

16

Shanghai Pen Pal

S afe-conduct papers in hand, Halliburton, leaving Japan with "Captain Welch, engineer Von Fehren, and seaman Paul Mooney," headed for mainland China. Although he told subscribers that the four stayed together, elsewhere he indicated that the team had *divided*. He and Mooney lingered in Japan. Von Fehren, and *the diesel engine*, continued onward to Hong Kong aboard the SS *Coolidge* to Hong Kong. Captain Welch went overland to Kobe to get "supplies," then on his own reconnaissance boarded the *Suiang* to Shanghai, where the search for "the *Sea Dragon*" began.[1]

Nowhere did Halliburton indicate a plan to rendezvous with Welch on the Chinese mainland. Changes of plan often occurred with the *Sea Dragon* Expedition, so, after he met up with Halliburton and Mooney at Kobe, Welch may have gone with them to Shanghai. By himself, he may have proceeded southward on the *Suiang* and stayed briefly in Swatou, a prosperous port city with a British and an American consulate. The city was also a large shipbuilding center that produced its own style of junk. Welch hated the place: "The Japs were over (there) but they didn't drop any bombs," he wrote. "Just wanted to see what was going on and how many they could make drop dead from fright. My rickshaw boy was foaming at the mouth when I made him take me back to the ship while the sirens were blowing. I didn't fancy having my guts spread out on the dirty streets of Swatou and nobody knowing anything about it for a year later. Not that it would matter a great deal, you understand, because, so what, but it's a nasty mess and I have some pride in where my guts is to lie." His guts intact, Welch yet suffered a "bout of dysentery" that stayed with him even after he arrived—by October 21, in Hong Kong.[2]

Just months before, Dr. E. Allen Petersen and his Japanese American wife, Tani Yoshihara, had sailed off from Shanghai in a junk named the *Hummel Hummel.* They exited a theater of war; Halliburton and company entered one. Halliburton considered purchasing a junk in Shanghai and, like the Petersens, sail down the coast to Hong Kong, in Hong Kong have the junk further outfitted, then cross to America. Shanghai, for now, was quiet as a graveyard. The Japanese, said Halliburton, wanted "to seize the Settlements and the vast foreign investments concentrated there," but so long as "the USS *Augusta* stood guard in the harbor, along with other nations' warships, they held back."[3]

Shanghai proper had undergone change. A few years earlier, newspapers, book publishers, and schools flourished, and so had print culture in general. As in any major American city, billboards along the waterfront advertised medicines, tobacco products, and household goods. Ordinary citizens could travel to China for adventure and even conduct business. In 1936, for instance, accompanied by a convoy of assistants (some armed), American socialite Ruth Harkness, against all advice, had gone deep into China's Szechuan Province toward mountainous Tibet. Going by steamer and on foot, she was determined to find a giant panda and bring it back alive. Had she waited another year, Harkness on her trek into China would doubtless have found herself in greater danger. The Battle of Shanghai in 1937 had left the city in a shambles, and the subsequent uneasy truce between the two warring parties stipulated that Japanese troop barracks be placed throughout the urban ruins. If Japan had been to Halliburton politically stifling and oppressive, at least it had law and order. Shanghai had neither law nor order; it boded only death and destruction. Despite its transformation from golden pavilions to department stories, Nellie Bly's fabled "Land of the Mikado" was civilized compared to Shanghai which Halliburton called "a wasteland."[4]

It is at this point, when reading the "Log of the *Sea Dragon,*" that Halliburton seems to have forgotten the junk search and made reporting the war in China his chief reason for coming to the Far East. As a mirror of that war, Shanghai took center stage, with the bravery of the resisting Chinese extolled and the brutality of the invading Japanese condemned. For Halliburton the ruination of the city's university and its library, its temple of learning and literature, symbolized a defeat for civilization. No longer a travel adventurer, Halliburton was now a war correspondent composing dispatches about a devastated China to send home.[5]

Ideally located for international commerce, Shanghai was a main turnstile into China's heartland. Halliburton quickly learned that Shanghai was also easy prey to military opportunists and another 'wickedest city' in the world. In her book *Shanghai—The Rise and Fall of a Decadent City*, Stella Dong tells the story of a city that had evolved from a "wilderness of swamps" to Asia's "Sin City," a "modern-day Babylon redolent with the sickly sweet smell of opium, teeming with illicit sex, crime, and poverty, rife with corruption and glamorous wealth." Over 750 miles due north of Hong Kong, roughly the distance from Seattle to San Francisco, the city was close to the Russian port city of Vladivostok, a spy mecca and vacation retreat Halliburton had visited in 1922. Port Arthur, on the Laodong Peninsula, was also close by. In the earlier Russo-Japanese War, the deep-water harbor had been a bone of contention between Russia and Japan. "Shanghai was never beautiful," wrote Halliburton. "As commercial as Chicago, and showing some signs of rapid growth, it spreads over a plain as the Illinois prairie. Not a single hill gives it distinction. Scarcely a tree ever graced its teeming streets. In the commercial center, tall buildings house the trade of every nation on earth and most of these buildings are ugly. In the small shopping centers and the residential sections, the buildings are less tall, but no less ugly." In contrast, Dr. Petersen called the city "the Paris of the Orient." To Halliburton, who had been there numerous times, Paris rather meant "grace and loveliness," qualities Shanghai clearly lacked.[6]

Attempts to beautify the city included a "fine new road" winding drearily some four miles from the business section to a civic center. Nearby, "superb modern buildings rose, styled after China's historic architecture" which "symbolized the national order and patriotism to which China aspired." Regardless, the Japanese heavy artillery started pummeling the structures until they fell. Although "pounded unmercifully," the civic center yet stood, becoming a stalwart symbol of Chinese resilience. The Broadway Mansions, a hotel built in 1934 to announce Shanghai's entry into the world of twentieth-century high-rises, was proud no more. It had been converted into a Japanese military head-quarters with a Japanese flag flying from its lofty summit. Among the building's many bureaus was one on the fifth floor devoted to opium regulation. The French Concession extended along the waterfront; the well-established British companies of Jardine, Matheson and Company and Butterfield and Swire operated here. Behind the warehouses, docks, and sampans were modern skyscrapers and boulevards—all run down, squalid, and bleak. If Shanghai was "never beautiful," as Halliburton had said, still it bustled with a sort of muted commerce and, even now, trailed only New York, London, Liverpool, Rotterdam, and Hamburg as a major world port.[7]

Well-to-do mercantile families, by their grand public works and importation of foreign culture, had long ago laid the foundations for Shanghai's Westernization. Halliburton remembered when fashionable English ladies walked about with parasols or were driven in carriages along tree-lined thoroughfares and past Saint George's Hotel, with its popular garden theater. Once a boardinghouse, the Astor House Hotel had become deluxe; an orchestra played in the courtyard, and guests were required to wear formal attire for dinner. Legitimate businessmen and known crooks wandered through its lobby and attended its parties. Furthermore, saloon cultures mingled; a French club was located near an American club. The city was home to the prestigious Shanghai Race Club, as well as numerous other clubs devoted to every sport associated with the West, from golf to rugby, tennis to cricket, football to field hockey. A short drive away were the Shanghai Museum of Natural History and the Cathedral School for Girls, which Royal Ballet star Dame Margot Fonteyn had attended. As did other headliners from the West, Fonteyn now performed in Shanghai's high-end overture centers. Open-air theaters showed newsreels and feature films; the foreign-language ones were dubbed or subtitled. Almost from the time Hollywood took its lead as the film capital of the world, Chinese residents in Shanghai had "discovered American cinema with a vengeance." A film industry emerged, and full-length films were shown at such gala indoor theaters as the Majestic. Shanghai even had its own leading film star in Ruan Ling-Yu (*The Goddess of Shanghai*), a champion of women's rights whose tragic suicide in 1935 at age twenty-five occasioned the grandest funeral procession in the city's history.[8]

Solemn rituals of course could not conceal Shanghai's civil warfare. Halliburton cast distressed eyes on the humankind that poured into the fallen city, and the "sheer anarchy (that) had been loosed upon the world." The city's mixed ethnic population included many western Europeans, some of them adventuresome Englishmen, some of them descendants of the freebooters who had come to China during the Opium Wars of the mid-19th century. Since the formation of the Soviet Union in 1922, White Russians and Russian Jews had fled to the city by the thousands, adding to the high number of Russians already settled there. Besides Eastern Orthodox churches and those of other Christian denominations, Shanghai was home to Moslem temples, Buddhist shrines, and Jewish synagogues. These were the worshiping places of many eastern Europeans, most of them traders, who had settled here, and whose earthly remains would one day be deposited in an exclusive Christian or Jewish cemetery.[9]

Soon after their ship docked at this isle of gloom and enchantment, Halliburton, with Mooney, was joined by a friend, now a resident of the International Settlement. Said to be a young American and unmarried, he might very well have been *Chicago Times* Far East correspondent William Montgomery McGovern whom Halliburton had met years earlier in Tehran. With little ado, the unnamed friend volunteered to show the two around. Undaunted, they drove in his car down Edward VII Avenue "through barb-wired barriers into the Japanese-controlled areas." At every point, the three had to show permits to move forward. The Japanese invaders only controlled the city "up to the midpoint" of one of its main bridges; they did not as yet control the International Settlement. However, they controlled all traffic going to and fro. Skirting impassable roads, Halliburton, Mooney, and their guide continued their drive into "the badlands," an infernal place infested with Chinese cutthroats and bandits. Land mines were hidden on the streets or under bridges, and lawns "pitted with bomb craters" surrounded the toppled buildings. Halliburton remarked that the "earth was sour with the dead." Scuffles might break out. But for now, gunfire could no longer be heard riddling the fleeing Chinese troops and civilians, many of whose bodies lay decaying in Soochow Creek, the main shipping route into China's interior. Next, they drove into the downtown district known as Chapei. Now "a wilderness, overgrown with weeds," Chapei's landscape had featured "mulberry trees and rice paddies, lotus ponds and orchards" only a short time before. What remained was "nothing but thistles and weeds, with a few mounds of bricks to show where houses had stood," including that of Halliburton's friend and guide. In a state of picturesque semi-dilapidation, the district was eerily still. Halliburton found "not a human in sight" and "no danger in any form." He remarked, "More than 17 months had elapsed since the Japanese 'saved' Chapei from its Chinese defenders, and all was peaceful, as the dead are peaceful."[10]

An "ugly" city that once thrived, Shanghai now had "bulwarks and sandbags" lining its drab, muddied streets. The American, British, and French cruisers that roamed the harbor challenged Japanese incursions along the coast. US Marines had set up barricades to stop the Japanese from advancing farther into the city. The indigenous population fought fierce battles among itself just to determine who would control a cross-walk. "A few shop signs still dangled from shattered doorways," but "whole buildings had exploded." Japanese soldiers stood about. Here and there a Chinese woman or child wandered through the rubble. Apart from these stark images, "there was no other life." Quitting the Japanese-occupied section of Shanghai called Hongkew, Halliburton came to the famous Astor House Hotel and to "Garden Bridge." The latter, connecting Shanghai's east and west bank, led to the city's International Settlement. "Crossing this bridge," wrote Halliburton, "was like going from a ghost city into the living world. In the conquered part were deserted streets and shuttered shops. Beyond the bridge were crowds of people, and free-moving traffic. On the Japanese side is the dummy 'New China' on the other side, all that is left of real China in Shanghai, huddling for protection in the settlements of the powers which are more foreign than Japan, but apparently preferable." The Japanese now kept to their side of the bridge but policed its traffic. Motorcars, buses, and pushcarts were not allowed to cross, and neither were streetcars and rickshaws. People in Shanghai customarily took a rickshaw to the bridgehead, left it there, and proceeded on foot to the other side of the bridge, where they might hire another rickshaw. Beyond that point "only the middle of the roadway" was open—the "sidewalks (were) a tangle of barbed wire." Japanese sentries were stationed along the road, and those whom the guards addressed were to follow the proper etiquette.

"Handbills were distributed" with imperious directives to Shanghai residents: "Remove your hat, hold it in your hand, smile cheerfully, bow, say good morning pleasantly, and present your pass. When it has been accepted, thank the sentry, and cross the bridge at a seemly gait." The sentries drew a distinction between the ill-mannered Americans and English who ignored these directives and the hated Chinese who, at the slightest hint of insubordination, were tied to a lamppost, slapped, and prodded with a bayonet.[11]

The heavily populated International Community was both a refuge for the misbegotten and hotbed of calamity. "[Here began] the Shanghai of trade and intrigue," wrote Halliburton, "(the Shanghai) of shipping and sin, of East meeting West at the lowest common denominator, which is the joy of adventure writers and the despair of missionaries." Bullets whizzing by their ears made some people feel intensely alive, and for certain battlefield sightseers bombs striking a golf course, resort, or racetrack were pyrotechnic amusement. Rather than manage crisis, many in this "paradise of adventurers" found that crisis "added to the garish pleasures of the city" and brought "the gaiety of the Settlements to fever pitch." If only the dead knew Brooklyn, only the dispossessed knew Shanghai. "Because it lacked the protection of passport barriers," the lusterless port had long become the "haven of a good part of the world's riffraff." Fugitives from justice given immunity by Shanghai's laws fled here. White Russians and Jews from Europe immigrated here in droves. Halliburton wondered how those in this "metropolis of the lost" survived, and concluded that they did so only "mysteriously and deviously."[12]

During the influx and exodus of many of the city's occupants, nightlife continued to flourish in a *danse macabre*. The neon lights of main thoroughfare Nanking Road revealed the sly looks of foreign adventurers, gangsters, artisans, missionaries, and merchants. Love affairs amid the hideous flare of excitement began and ended in sleazy cabarets and piano bars. From Asia, Europe, and every corridor of the earth, soldiers of every flag and of fortune, sailors, refugees, and riffraff packed into a city that managed its catastrophes in a manner that, looking back, frightens, befuddles, and invites eerie wonder. Chinese terrorists and bandits prowled as spies lurked. Army checkpoints separated one quadrant of the city from another. Squadrons of police whose source of authority was unclear maintained some semblance of order.[13]

Drawn into Shanghai's red-light district, Halliburton and Mooney
sauntered down a dingy block-long street called Blood Alley. This
transplanted version of Hogarth's "Gin Alley" in London was "so very
notorious" as to inspire "shame-faced pride" in the city's residents.
The street's chilling rhapsody of laughter and tears made itself known
"on gala nights," when one could "hear Blood Alley two blocks away."
Amusements in Shanghai were offered "on every social level from top
to bottom," remarked Halliburton, "broadening considerably toward the
bottom." At one drinking establishment he saw how, through guile and
lubricity, dance-hall girls from Siberia and parts unknown separated
customers from their money—just for "simply sitting down, sipping a
drink and listening." Some told tales of hardship and woe, living "from
night to night by spinning fancy yarns." Serving Halliburton at one
nightclub was a "lissome young thing of 17," who, presumably knowing
some English, told him "the whole tragic story of her flight from the
Bolshevik revolution." Halliburton classified the many "Russian host-
esses" as "Scheradzades" who "lived from night to night by spinning
fancy yarns." Although skilled hustlers, they "had nothing on Shanghai's
slim Sonias," Chinese girls of whom Halliburton said, "[They] reigned at
most of the dance halls and must share honors—or something—with the
Russians." Halliburton was amused. He found all the women "charming
and gay, and sometimes extraordinarily pretty," and, echoing his earliest
theories of miscegenation, concluded, "When Chinese blood and foreign
blood are mixed, especially if the foreign blood is Russian, Portuguese
or French, the devastating result is something to write home about."[14]

Patronizing the bars at night were rowdy soldiers from America,
France, Great Britain—and "whatever other nations happen[ed] to have
Army or Navy forces in town." All of the servicemen flaunted "tomorrow we die" attitudes. As an unwritten policy, payment for drinks in
these dives was deferred, ultimately assumed by whomever. Groups of
French soldiers and occasionally of American marines gathered at the
New Ritz, Blood Alley's most raucous bar. Italian soldiers, whom no one
at the time liked except the Japanese, might join the crowd, and tempers
quickly flared when they did. Halliburton described the cause of a riot
that broke out one night, recalling, "One of Mussolini's men insulted
a French soldier, who, since Blood Alley is in the French Concession,
properly felt he was being attacked on home grounds." The combatants
threw punches, flung bottles, and swung heavy brass buckles on the

end of military belts. Controlling the scene required "all the police in Shanghai." Said Halliburton, "The war was on in earnest—with America in for the duration!" Initially dismissed as "only a tempest in a teapot—or rather, in a gin bottle," the story was cabled to news stations in Paris, London, Berlin, New York, Washington, "and especially . . . Rome." The riot became an international incident, "for sharply and unmistakably, this spontaneous combustion in Blood Alley showed whose soldiers were ready, of their own accord, to fight for—and against—whom."[15]

Soon after, Halliburton bade "good-bye" and "Arrivederci" to Shanghai.

17

Toward the South China Sea

~~~~~~

The search for "the elusive junk" renewed.

The passenger ship, probably British, with Halliburton and Mooney aboard bravely chugged south down the China coast through one of Asia's many "torpedo alleys." Japanese gunboats were a constant menace—as were pirates hiding in every cove. Overpopulation concealed the pirates, extreme poverty added to their number, and inland coves where fishing boats abounded offered them safe retreat. Lines of jurisdiction were unclear as pirate bands often formed cartels in league with law enforcement agencies. With some apprehension then, ships proceeded steadily and watchfully.[1]

Halliburton's ship moved southward, muscling its way through the shoreline slime and pungent stench of dead fish. Temperatures warmed while sanitary conditions worsened. Treacherous shoals fronted the rocky coasts and sandy beaches. Under the rising and falling hilltops, a few boating villages emerged ghostlike through the fog. Halliburton's weary eyes roamed the often blockaded coastal communities for the ideal junk. He later said he had searched extensively for a junk "big enough and strong enough" to sail from China to San Francisco. Before the war, thousands of the ships had been huddled in floating villages off the mainland; the fittest ones now seemed on the verge of extinction.

Scouting about for two whole days, Halliburton learned only that the best junks had been sunk by Japanese torpedoes or pilfered by Japanese troopers or pirates. Many owners had fled south with their junks to safer coves. The vessels left behind were ramshackle fixer-uppers and outright wrecks, both of which commanded high prices. To his queries, perhaps done through an interpreter, owners underplayed their bad points and overplayed their good ones. Halliburton so far had surveyed junks of every breed and caliber, but his "Sea Dragon" was not among them. As a joke on himself, he purchased five miniature junk replicas to send home to his family.[2]

Halliburton's ship continued southward. When the ship docked (if it docked, as he did not record stopovers), he likely didn't climb aboard what junks were alongside it for close inspection. Rather, from afar, he relied on first impressions or the report of others. These sources of information he then crafted into cogent essays for his subscribers. At the once-bustling port of Wenchow, his ship almost certainly stopped, but (I suspect) Halliburton did little more than scan the waterfront from the deck or, stepping ashore, talk to harbor officials. "Wenchow junks with their white and scarlet hulls," he said, having only read about them, "are famous both for beauty and for speed." Yet he saw or heard about none that fitted the high estimation. Most were over ten years old, some over a hundred; nearly all of them were irreparably disabled. He looked about, again and again shaking his head at the sight of so many broken-down junks. He was told, meanwhile, of an American captain named Nicholas who had purchased a junk (maybe a Wenchow-style one), and headed for America with his wife and a crew of seven. Three days out, they ran into a typhoon (in Asia, the terms *typhoon* and *hurricane* are interchangeable) off the coast of Formosa. Captain Nicholas and his party were rescued, but the junk was totaled. Later, Nicholas wrote Halliburton and told him that had he "possessed an engine, however small, he could have escaped the rocks." Later he would hear about another junk heading from Shanghai for America that would run into a "gale," have its mainmast "snap," and be pushed backward. The ship had an engine installed in port, then restarted its voyage. Halliburton of course had an engine; he just needed a ship. His next stop was Ningpo, where he saw not the "finest junks in China," about which the "dead port" once boasted, but rather so-so junks "quietly resting in their moorings."[3]

At Swatou, Welch reported that he had inspected some locally built junks marked for sale. Although 'sticker' prices varied, they could be steep. Welch drew the limit at $9,000 for an eight-year-old junk he thought was "a good one." That asking price would have sounded high to the budget-minded Halliburton. By comparison, Dr. Petersen had paid $250 for the *Hummel Hummel*, a thirty-six-foot junk that *did* cross the Pacific despite some overhaul needs. In his "Confidential Prospectus" he estimated a $4,000 price tag for a junk; he later upped that figure to $5,000. The Swatou-style junk eliminated from his consideration, other junk inspections followed. Most junk owners, by now alerted to the cash-and-carry American in their midst, hiked up their asking prices accordingly. The junks they promoted were neither big enough nor strong enough to hold heavy cargo and cross the ocean. "Few fishing junks (were) sturdy enough," remarked one observer, "to stand the weight and vibration of a heavy auxiliary motor unless very extensive alterations were made." Such timely and expensive alterations were for the moment of zero interest to Halliburton.[4]

Halliburton might have gotten off ship in Amoy (Xiamen), the British-treaty port where he had originally planned to purchase a junk. Here, a cartwright volunteered to build him a junk. Halliburton thought the three-month construction period too long; it meant that the vessel would be ready by the end of January 1939. Hindsight argues that he should have taken the deal, as Amoy junks were reputed to be of the highest quality. These robust ships were built to endure crushing waves and persevere in the worst weather conditions. He knew of one named the *Amoy,* a tough little junk which, following "the great circle route," had in 1922 carried "a Dutch captain and his Chinese wife, with three Chinese seamen" across the Pacific. "For stores and repairs they put in at Dutch Harbor in the Aleutian Islands," wrote Halliburton, "Later, having visited Vancouver and San Francisco, this brave little ship sailed into the Atlantic via the Canal, and is now tied up for display near New York City." Rejoicing at their success, Halliburton left Amoy without regret—and without a junk. The next port was Foochow and it had junks—notably carved and painted ones—and these were better suited for his purpose. Once a smuggler and a ship that had borne the scars of Asian wars, the *Ning-Po* was built in Foochow during the early 1750s. Presently it was on view in San Diego Harbor, and again on view on Santa Catalina Island where Halliburton could have seen it.

But if a Foochow junk seemed the right type of ship for his expedition, Halliburton knew that Foochow itself hardly seemed the right place: "There was such a lack (again due to the war) of tools, timbers, gear, canvas, paint, provisions—all vitally necessary for a long sailing trip— that I dared not embark from there." War and upheaval continued to form the backdrop of Halliburton's search. In "normal times," he could easily have found a suitable junk, but "having had a taste of the war raging over all China," he sensed the odds against doing so. By his own count, he had inspected some fifty junks. None passed routine inspection, and none seemed within his price range.[5]

Halliburton's ship now entered the South China Sea, one of the most heavily trafficked shipping channels on earth. The foreground of villages and paddy fields along the mainland coast looked distinctly Chinese, though, for visitors like British naturalist Graham Heywood, the area's background "mountains and valleys" invoked "the (Scottish) Hebrides." Many lake districts greeted ambitious trekkers who traipsed off into the New Territories, or beyond the range of hills stretching from Tai Po to Mirs Bay in the east. As dawn quietly settled into day, Mirs Bay looked "its loveliest, with its nacreous blue waters and its queer-shaped islands and with banks of white mist creeping along the northern shore." Villages came into view with fishing boats in their harbors. Waterways were crammed with sampans. One heard voices murmuring, oars striking and splashing the water, fishing poles lightly tapping the water, and fishermen, seated in their scows, shaking their nets empty of a recent catch. Hillsides and vast wooded areas were filled with colorful flowers and exotic birds. In some locales, villagers hunted deer and wild boars as well as clawing monkeys and poisonous snakes. As they had to skirt foraging Japanese soldiers, people explored these outlying areas at their peril. Junks were "anchored two by two" closer to Hong Kong, near Hang Hou at Junk Bay. As it was not so crowded now, bathers gathered at the beach at Hang Hauy just beyond Razor Hill and a short ferry ride from Hong Kong. Clearwater Bay, the most beautiful outdoor recreational spot in Hong Kong, was located nearby.[6]

Sailing through a confusing maze of islands that also defended the port, Halliburton's ship neared Hong Kong, whose name literally meant "fragrant harbor" or "harbor of fragrant streams." With Paul Mooney alongside him, Halliburton might have observed how one breeze or cloud cover, one curling wave or lofty ridge reminded him of his two previous trips to Hong Kong—and of Laguna Beach. He might have noted how small the world had become since his first trip to Asia in 1922 and his return trip in 1932. The heart of a chain of seaports "still free from Japan's strangling blockade," Hong Kong was Halliburton's stated "last hope" for finding his junk. Known as "the Colony," the city was the flagship of Great Britain's Far Eastern trading posts. Along with Singapore, it was also the brightest gem of the British imperial crown. When Halliburton was a boy reading about the port in picture books, it seemed somewhere on the dark side of the moon. He later wrote, "[Getting to Hong Kong] has been the goal and many a time I looked at the map of the world and wondered was it physically possible to reach this tremendously faraway port." Declaring it "one of the most interesting cities in the world," he arrived there "at one of the most critical moments in its history." This, "the back door to home," would be Halliburton's last earthly stop on the royal road.[7]

18

See Hong Kong, Riviera
of the Orient, and Die

First-time American visitors to Hong Kong—"the Colony," landmark hub of international commerce, knew at once they were in a foreign country. Indeed, some wondered if they had landed on another planet. Set at the foot of a steep hill nearly two thousand feet high and separated from the mainland by scores of tiny islands, the city appeared sheltered from invasion. There were few tall buildings; the natural peaks surrounding its harbor were its skyline. Rows of granite-faced buildings, some turreted, some with bell-shaped parapets, many with railed balconies and Venetian windows resembling Canaletto's rococo paintings of the Grand Canal, stretched along the port's bustling waterfront and wharves. So Hong Kong looked just before World War II. The Japanese occupation, slum clearance, redevelopment projects, and the skyscraper boom all transformed the open port city. When Halliburton arrived there, a person who stood atop Victoria Peak, some 1800 feet high, could look across the harbor and not have tall glass buildings obstruct his view or see expressways circling the harbor.[1]

In the early 1890s, just before the outbreak of the First Sino-Japanese War, reporter Elizabeth Bisland wrote enchantedly of Hong Kong: "The verdure is magnificent; the town is submerged in it, and flowers are everywhere. . . . The climate she likened to that "of Eden," where the "airs of Paradise wave through the splendid tropical foliage. . . . The sun is pleasantly hot at midday, and the morning and evenings are dewily cool." Bisland's contemporary Nellie Bly saw this but also saw "wharfs crowded with dirty boats manned by still dirtier people, and its streets packed with a filthy crowd." A generation later, Bly's views still applied. East and West cordially met. Called "Little China," the Colony still embodied all that was distinctly Chinese. Some alteration was inevitable. News now traveled faster; radio-wave messaging proved quicker than dot-and-dash transmitting. British interests in the region had yielded to American ones as foreign entanglement appeared inevitable for the rising commercial superpower. By the 1930s, the old still blended with the new. Colonnaded mansions, royal residences, diplomatic embassies, and elegant hotels in the grand Victorian style still stood as a reminder of the Colony's British heritage amid the many ramshackle yet quaint buildings lining its winding streets and shadowy corridors. The "smells of opium," which Bisland had reported, still filled the air, along with the scent "of the dried ducks and fish hanging exposed for sale in the sun, of frying pork and sausages, and of the many strange repulsive-looking meals being cooked on hissing braziers in the streets and in doorways." Welcome to the Colony.[2]

Like Shanghai, Hong Kong had become a repository of the world's derelicts, adventurers, and political dissidents, of all that reeked of white mischief and Asian mayhem. Call it the 'Casablanca of the Far East.' Wealthy British traders and diplomats employed servants to attend to their every need and lived in hillside villas with terraced gardens and porticoes. These homes were far from the riffraff, but not so far from its smells—generally disseminated odors of soy, sorghum, gasoline, and fumes from factory furnaces and kilns. Spiceries, barbershops, bars, sex-toy shops, coffin factories, pawnshops, and repair shops filled the winding streets as rickshaws, palanquins, and two-wheeled trucks wove through the throngs of pedestrians. Shoddily attired "human burden-bearers" and "beggars of dreams" trundled alongside gentlemen in white three-piece suits and ladies carrying parasols and wearing bright dresses. Cobblers, fullers, rice and baked-goods salesmen, fishmongers, kerosene vendors, knife grinders, cloth vendors, and tobacconists hawked their wares on every other street and corridor. Shopping districts were many. Queen's Road offered the fullest selection of foreign and domestic goods, including porcelains, sandalwood wares, shark-fin products, tiger claws, and elephant-tusk ivory. Wholesalers were every-where, and adventurous shoppers might run into anything. Pharmacies selling Bayer Aspirin and five-and-dimes offering toy figurines stood a couple doors down from the money changers and jewelry shops. Here, as on the equally busy Wyndham Street, English-language product signs hung beside Chinese-language ones.[3]

At the Jade Market shoppers bargained for precious and semiprecious gems—buyer beware. On Bird Street, one might buy a macaw, mynah, or Daurian jackdaw. Kiosks flourished: some sold snake meat and snake oil, others sold fruits, vegetables, and sacks of rice, still others sold skinned rabbits, and wild boars. Shoppers could find "fast-food" stands and newspaper kiosks around the corner from a bookstore or tobacco shop. From one street vendor or another, people bought beaded bags, exotic perfumes, opium pinches, cigarette-rolling papers, matchbooks, rice wine, flower bouquets, ceramic Buddhas, and teas—Won Hop Gon Jin was especially recommended for its internal cleansing properties. In gambling parlors, patrons played traditional games of fan-tan and mah-jongg alongside such modern games of chance as roulette and twenty-one. Opium dens were everywhere. The Dragon Boat Race and Happy Valley Horse Races were suspended but would reopen, sometime. While Hong Kong with little stretch could be called beautiful, quaint, pestilent, or squalid, the Japanese believed the thriving city only en-capsulated Western arrogance. Halliburton observed that Japan coveted "the great port" as much as it hated it, seeing Hong Kong as a key source of military support to mainland China that had to be destroyed. They already had Shanghai "at their mercy," and now Hong Kong was to be brought to its knees.[4]

The bubbles of the scum had long ago risen to the surface. Legit-imate and illegitimate commerce thrived, but neither could obscure the Colony's most flourishing underground enterprise—spying for a price. By the late 1930s, Hong Kong had become a leading mecca for information merchants. Every other person, regardless of appearance, class, or occupation, seemed engaged in espionage, black-marketeering or smuggling. Having a disguise worked as effectively as having no disguise at all. Shop-keepers, jobbers, rickshaw drivers, sampan oars-men, the plodder on the street, all kept an open ear; all, eager to sell what information they gathered to any interested party, liked the quick buck. Into the mix of intrigue arrived the tourists, businessmen, foreign dignitaries, and entertainers, as well as "trainee diplomats, journalists, medical missionaries and teams of 'economic investigators.'"[5]

About the physical geography of Hong Kong, Halliburton spoke in general terms. When he mentioned Kowloon (Jiulong), he meant a separate northerly location on the mainland. When he referred to Hong Kong, he meant both Kowloon, a peninsula, and Hong Kong, an island. For many travel writers, "Hong Kong," measuring some four hundred square miles, was a catchall for the over two hundred islands that comprised it. Those place names were also a catchall for Hong Kong's three major parts—Hong Kong Island, Kowloon, and the New Territories. Halliburton and other writers often designated Victoria Harbor, which separates Kowloon from Hong Kong Island, as the "harbor," the "main harbor," or "Hong Kong Harbor." From terminals with towers holding Big Ben clocks, a ferry, usually filled with tourists and commuters, ran regularly from Kowloon, near Kowloon Point, to Hong Kong Island and the city of Victoria (after Queen Victoria). A tram or funicular ran to Victoria's steep summit.[6]

The peninsula's many mountains, hills, and peaks, many in the New Territories bordering Kowloon, all had distinct, often colorful names. The terminus of the trans-Siberian rail link and portal to Hong Kong, Kowloon roughly means "Nine Dragons," nine the sum of eight mountains and a Chinese emperor. Rows of wharves, one after the other, lined the shores of both Hong Kong Island and Kowloon, which was connected to Hong Kong by rail or ferry. A complicated tangle of waterways, channels, and inlets extended from its harbor west to the South China Sea. A series of winding roads led from Kowloon and the New Territories to the mainland and to Kwangtung Province, whose hills form the backdrop of the island. Located offshore from Kowloon Peninsula was Stonecutter's Island. Macao, a gambling mecca that Halliburton called "the Eastern Monte Carlo," was forty miles to the west just past the Pearl River Delta. In 1922 Halliburton's once-pirated and "now famous" steamer *Sui An* docked there in Macao. In 1938, British artillery protected the "mountain-bound harbor," thought by Halliburton "the most spectacular on earth" apart from that of Rio de Janeiro. Amid the chaos surrounding it, the strategic military and commercial outpost of Hong Kong continued to harbor ships from throughout the world. Steamships, luxury liners, tour boats, merchant vessels, and yachts cruised into port unharmed.

Along its shores fishermen "merrily" went about their business, often "thumbing their noses at the Japanese."[7]

Following his lack-luster trip down the coast, Halliburton found Hong Kong and its many wondrous junks a "glorious sight." He sparkled with enthusiasm, knowing his search had come to an end. Hundreds of junks with raised sails and sturdy hulls met his dazed eyes, and he noted "their foremasts leaning rakishly forward, their mainsails a vast brown web of matting formed like the wing of a bat, their perky little fan-shaped mizzen-sails stuck high on the up-swelling poop." In this hatchery of junks, the ships were born and bred, then multiplied and spread. From a mile or so out to sea, one could literally walk from boat to boat to the shore without getting wet. Halliburton found himself in a boat-buyer's heaven. "Perhaps it was the morning light on the waves and on the gold-blown sails," he enthused, "perhaps it was the gorgeous tropical scenery in the background, perhaps it was the cheery way the fishermen shouted at us as we wound our way through the fishing fleet, that made the Chinese junks seem strikingly beautiful and appealing, all over again." Suddenly erased were his recent setbacks in procuring a seaworthy vessel: "I rejoiced, anew, that I had resolved to possess myself of one, and live aboard it, and sail it out across the biggest ocean in the world." Besides countless junks, a myriad of sampans and barges littered the harbor. As they had for centuries, tens of thousands of Chinese spent their entire lives in these floating prisons of reality. "(The) harbor is filled with ships," Captain Welch said: "(There are) about five cruisers and seven destroyers, ten subs, and a plane carrier. A couple of American ships and a Frog. Oh yes, and one small Portuguese gunboat that looks like a garbage wagon to the fleet." Clearly, Welch did not share the glorious adventurer's sense of romance.[8]

The moment they stepped off the boat, Halliburton and Mooney were joined by Captain Welch and engineer Von Fehren. Eagerly reporters and camera-men followed as the four made their way to Gin Drinker's Bay. There, they saw "a magnificent junk" with a ninety-foot hull that was "built like a battleship." The vessel's "deck was made of polished teak; the mat sails, dyed scarlet, (were) heavy enough to withstand almost any wind; the cabins, occupied by thirty people, were lacquered red." Halliburton characterized "the carved joss shrine, filled with gilded gods" simply "as a work of art," with its "wooden 'eyes,'" at a yard across, just the perfect size. "With its graceful crescent hull and its scalloped sails" the "admiration of sailors in any country," the craft "could easily have crossed the Pacific in fine style." This might have been the very junk Dick Wetjen had told Captain Welch to check out. Welch was mum about its cost—triple what Halliburton wanted to pay. Hindsight again argues that Halliburton should have gotten the ship, but he didn't. He had "better luck" in Joss House Bay, where he found a junk nearly as beautiful as the one at Gin Drinker's Bay. It also had a "castle poop (that) swept thirty feet above the water" and, along its deck, touted "five beautiful iron (probably lantanka swivel) cannons"—ample protection from pirates. "The price was right," but before the deal could be sealed, it was discovered that the "hull planking" was only an inch thick. In the quest for his version of the Golden Fleece, he often ran into trouble.[9]

After a quick once-over of other junks, Halliburton registered at the Kowloon Hotel on 503 Canton Road, first checked into by Captain Welch, who arrived in the Colony on October 14 or 15. Its location was nearly ideal: only a few blocks away, a tall, spired tower with a conspicuous clock-face served as a compass needle directing people to the wharves and the terminal for rail service to and from Canton. Commuters to the Island parked their vehicles in an adjacent lot, then boarded the Star ferry that took them to their jobs. Canadian journalist Gordon Sinclair described the Kowloon as "a cheap little back-street hotel," adding that "the place was respectable and clean enough but the rates (by comparison) at the leading hotel, the (classy) Peninsula, were far from high." Halliburton informed his parents that he and Mooney each had a separate room; in fact, both shared room number four. From his window he could view, pedestrian traffic, honking foreign cars, and double-decker buses pass. The hotel, he said, provided "good food" and "clean water." By "good," did he mean tasty, nutritious, germ-free? By "clean," did he mean potable? China had no housing codes or inspections. The Kowloon might have offered laundry services, but it likely did not delouse its beds. Plumbing was crude, privacy limited. If not in each room, down the hall there might have been a bath, sink and flush-toilet. The thin walls were dirty, and the floors creaked. If one did not see vermin, perhaps one heard them. Halliburton thought some things best. "Only the war refugees are sickening," he told his parents. Mooney added that these "sickening" individuals were actually "wealthy Chinese refugees" and "a handful of impoverished German Jews who came lately." The hotel had also become "a notorious center for German agents doing Tokyo's dirty work in Hong Kong." Unmindful, Welch said that the place was "mostly a Chinese hotel," one "no different from any other save that the house rules are funny." Read one rule: "This is a respectable hotel and prostitutes are not allowed. If any gentleman guest is found to be *companioned* by such, a charge of ten dollars will be made for the extra guest."[10]

While Welch had no pressing need to delve into Asian history and culture, Halliburton did. Tucked in his mind were countless images of China that he presumed readers of his articles didn't have at the ready and should now receive. He believed that most Americans, though reasonably well-informed about *recent* events in Europe, knew next to nothing about China's history and culture. He had arrived in Hong Kong within a year of the anniversary of its founding. As some claim that Rome had been founded by shepherds and whores, so had Hong Kong, a sanctuary city, been founded in 1840 by refugees—"a group of Britishers from Canton"—and pirates. "In a short time, the refugees had pirated the island from the native pirates," wrote Halliburton. "Then, in quick order, forts, offices, banks, houses of God, and jails began to rise as fast as Scots, Irish and Englishmen arrived to use them. Hong Kong, from the start, gave promise of becoming a colonial imitation of a London suburb." By 1900, the little outpost created from the First Opium War of 1839–1842 had become one of the largest shipping bases in the world. "(The many Asian and European) commercial transportation companies," writes junk expert Hans K. Van Tilburg, "required not only large steamships, but the associated dry docks and repair facilities, engineering shops, cargo boats, coaling stations, railway heads, fresh water systems, depots, wharfs, breakwaters, boiler shops, deepened channels, and other infrastructure." A mix of Asians and Europeans, meanwhile, had poured into the city. In just a few years, Hong Kong's population, ordinarily some 900,000 souls, had swelled to over 1.5 million, making it "probably the second largest (city) in the British Empire."[11]

In 1938, Asians outnumbered "whites a hundred to one." The majority spoke Chinese and many spoke English, German, or French. What with the influx of refugees, most of whom slept on the street or in alleyways in an already crowded community, cramped living quarters were to be expected. It was also to be expected that these conditions created a breeding ground for infectious diseases, a factor which might have lent, as will be seen, some urgency to the *Sea Dragon's* departure. Ignoring disease, Asian residents, long conditioned to expect little personal space, were less troubled by the overcrowding than were their European or American neighbors, who, accustomed to greater operating room in their daily rounds, found street life oppressive and perilous. Even as Britain's power in Hong Kong appeared on the wane, the British moved proudly through the crowd. "The refugees come here for safety," wrote Paul Mooney, "and think they'll have plenty of time to starve on the sidewalks (where many of them slept on mats). But they haven't talked with members of the American Club, or with the British residents—all of whom are certain the Japs can take the Colony whenever they wish. Officially almost smug, privately the city is waiting for the showdown. British prestige is nearly at zero." Ironically, people in the Colony seemed invigorated by the very desperation Mooney found alarming. Despondency energized them. If congested in the 1930s, Hong Kong also pulsated with life. As the flames of war surrounded it, the city seemed protected by these very flames, and even spurred to prosperity.[12]

For generations China had a legislature, one based ostensibly on the British parliamentary model. "All property-owning citizens, regardless of sex, color, or race, could vote," wrote John Powell, who often had to inform the US Congress of America's present and future role in China. The country's array of laws proved confusing, and a main issue was whether offenders should be tried under Chinese or British law. A police department comprised of Europeans, Chinese, and Indians enforced the laws or kept the peace despite their laxity. On paper, at least, visitors to China in 1935 could feel safe and well protected. The acting commissioner of police had "12 super-intendants, one police cadet, 270 European, 843 Cantonese, and 296 Weihai constables, numerous Indian constables and a special force (including some Russians) of anti-piracy guards" under his authority. A "District Watch Force" assigned to predominantly Chinese neighborhoods worked "in some degree of cooperation with the police." Criminal activity thrived. Pirates frequently attacked junks, creating an "uncomfortable amount of gunfire, robbery and death." Journalist Violet Sweet Haven noted that while all this activity was occurring, "Chinese river junks plied back and forth laden with rice, bamboo poles, spices, tea, and mysterious bales of other cargo." Customs officials kept watch over Hong Kong's docks as they did in other ports, regularly confiscating contraband and surveilling illegal activities.

Often crammed, courts held trials and heard litigants. "In Hong Kong's courts," wrote Halliburton, "piracy trials are as frequent as cases of opium-smuggling, child-slave holding, and overtime parking."[13]

Illiteracy, notoriously high throughout China, was less so in Hong Kong. As in any European and American city, however, many types of schools existed throughout the Colony. Teacher-training programs existed in both Chinese and English. A Queen's College and a King's College mainly served privileged Chinese men. The University of Hong Kong had a College of Medicine and a Technical Institute and drew students from all over Europe and Asia. An arts program was later added to the university. The Young Men's and Young Women's Christian Associations offered room and board and tutoring in reading and writing to many who needed it. American children, often the sons and daughters of Protestant missionaries, grew up in China and attended its parochial schools. The country's hospitals were as numerous as its schools, and they included government and military hospitals, religious and civilian hospitals, and some teaching hospitals. After 1928, a director of medical services saw to these facilities' quality, as they often lacked the necessary services, drugs, rooms, and beds to be rated efficient. It was best to go to Queen Mary's Hospital, also called the "British Military Hospital" or "Bowen Road Hospital." Located at Pok Fu Lam, the facility overlooked the main harbor.[14]

Halliburton spent some of his vacant time at the Hong Kong Hotel, a convenient trysting spot situated at the heart of the financial district and near the major shopping and theater centers. The hotel's gift emporium carried magazines and newspapers from America. Other options for his leisure were the exclusive Hong Kong Club and Victorian Recreation Club. For travel advice, he might visit Thomas Cook and Son, a firm in the same district and just a block away, going down Connaught Road to Queen's Wharf. From there he might saunter across the street to the Colony's main post office. As banking services and English-operated banks were plentiful, Halliburton could conduct financial business at the Chase National Bank next-door or the First National City Bank of New York on the same block. The financial district had a Wall Street look about it with the Hong Kong Bank and Shanghai Bank operating from windowed monoliths that were especially imposing. Hong Kong was also home to the Overseas Chinese Bank, the Bank of East Asia, and the Farmers Bank of China. Some Chinese banks modernized, installing air conditioning and replacing pen-and-ink ledgering and the abacus with current bookkeeping equipment. The Hong Kong Bank and Shanghai Bank occupied the most imposing structures on the island. Policies for improving services were constant concerns. In 1938, the Hong Kong dollar bearing George VI's picture became China's chief monetary unit. Other currencies, notably the US dollar, were also in circulation. As the number of urban proletariats and foreign nationals grew, so did the number of foreign investment groups. Duties on imports and exports added money to foreign coffers; many of the officials at the Chinese Maritime Customs office were foreign nationals. The Chartered Bank of India and the Bank of Taiwan were located along the Bund. The latter institution was under Japanese supervision.[15]

Every form of conveyance was available in Hong Kong, ranging from horse-drawn carriages, rickshaws, and pushcarts to autos, scooters, and ercolinos—bikes had long been in use. Travel outside the Colony in times of relative peace posed few threats. Travel by sea was swifter than travel by land, and, if one could afford it, travel by air was better than both options. The Hong Kong Flying Club was established in 1929. By 1935 Kai Tak airport flew regular commercial flights, and by 1937 two flights a week, averaging ten days with stops, connected Hong Kong with London. Making the same trip on the Trans-Siberian Railway through Russia took eighteen days. Making the journey by steamship took about five days. In 1937, writes Nigel Cameron, "Pan American linked Hong Kong to Manila, and it was then possible to fly from the colony to San Francisco in six-and-a-half days."[16]

Since Halliburton's last visit in 1932, Hong Kong due to war and natural disaster had changed. On September 2, 1937, a typhoon had devastated the island and its vicinity, largely wrecking the sights a visitor would have seen in 1935 and presenting a new landscape to a visitor in 1938. The splintered remains of houses and boats floated in the grimy, oil-filled waters along some of the Colony's shores. Looters made their way into the dilapidated structures before salvage operations could begin. Hundreds of vessels, both large and small, were ripped apart or lost. Scores of buildings collapsed; over eleven thousand people lost their lives. Halliburton held faith that Hong Kong would emerge from the catastrophe as a modern metropolis, and not regress into a collection of shoddy hovels and huts. Unfortunately, it would not happen soon enough.[17]

19

Rats, Lice, Morphine, and Misery

Chapter Eight of Charles Darwin's *Origin of Species* begins, "I will now describe the struggle for existence in a little more detail." For Halliburton, that struggle had just begun.

Writing home, he said he was in "a beautiful and interesting town." Paul Mooney agreed that Hong Kong was "beautiful, but thought it looked best from Victoria Peak, not close-up. He said nothing of the awning of fear that hung over the city. Captain Welch, however, saw the fabled port as a festering wound on the butt of humanity, a ladder in a stable, each rung caked with dung. He called Hong Kong a "goddamn place" and a "lime-hole," meaning too many British limeys were around to suit him. Only the "movies" distracted him. Townsfolk dressed up to see films, and the crowding into the theater was maddening. Adding his own complaints, Welch remarked that the port was "a lousy place to be stranded in spite of all they say about China." "Shanghai was dirty," he observed, explaining then how Hong Kong compared to it: "[It] was cleaner, but so goddamned slow that it hurts. After a movie the whole street is deserted save for the police patrol." At least Hong Kong was safer than Shanghai, but that made no difference. Whether on the street or alone in his room, Welch hated all things Asian. "I have never liked the Orient," he said, "and like it less now. These people nauseate me to no end. You can't look upon people as human beings who let their own live as they live in some of these streets. If anybody ever tells me that the Orient is a great place to have a good time in, I would like to tell them what I think of the whole goddam place. Nothing here at all. The whole place closes tight at midnight sharp, and it takes an act of Congress to be able to talk to anybody without a proper introduction." Welch could, of course, bring a bottle of booze to his quarters and, through the smoke of a Corona, write letters home. During the day he could go to the golf course or race track, or maybe sit in a teahouse, but such amusements held no interest for him. He preferred clean ports with strong European traditions, places like Antwerp in Belgium. A city of "refugees" to Halliburton, Hong Kong was to Welch an open cesspool whose odor rose and spread throughout all of Asia. Exotic-sounding diseases abounded. As for the women, they looked sickly and flat-chested, smelled "ricey," and carried germs. "If these women are good lays," he said, "then I'm willing to let some other lad use up the (hospital) beds on the hill. I still need my legs to walk about with, and will not limp." Reading through his letters, one quickly suspects that the discriminating captain spent his entire time in Hong Kong without a woman at his side. Complaint was his current companion, and a constant one: "I wish to hell I had never started on this goddam trip" became an old refrain.[1]

Romantic in the abstract, the 'Royal Road' often proved degrading in reality. Years of travel had diminished Halliburton's expectations and, grown accustomed to a wide range of sanitary conditions, many foul, he no longer recoiled. He had walked down streets roiling with filth and debris, had waded into streams turbid with sediment and grime. In ports throughout the Mediterranean, flies grazed conspicuously on the backs of stray dogs and nestled comfortably in their wastes, then settled atop the food laid out in outdoor eating places. Animals were slaughtered in alleyways, their innards routinely tossed into the gutter and soon claimed by some starving addict or drunk. The struggle for existence excluded any kindness for animals. Westerners saw them as pets to be cuddled, but in Hong Kong animals were more often seen as food. Rats as big as small dogs prowled through alleyways and were often caught, cooked, and eaten with a bowl of rice and weeds. As for etiquette, grunting, picking one's nose or teeth, and flicking gnats off one's food were dinner-table okays. To be sure, cleanliness standards in Hong Kong differed from those in mainland China where refuse littered the embankments of waterways already filled with sludge. Dead fish, cats, and dogs lay in the canals. In many boarding houses and hotels, drinking water tainted with oil, urine, lead, and zinc salts might only look clear in the glass. Rusted pipes carried water to the tap and were often clogged. Finding it salty and repellent, Halliburton called it "brackish." The Colony, with no wells, depended for its drinking water upon rain that accumulated in its four reservoirs during the short rainy season. As a result, demand exceeded supply. While a septic tank leaking into lake water was a catastrophe in an American city, it was a minor issue in many an Asian city. The poor and downtrodden, ignorant of its perils, drank waste-water. If over time they developed good immune systems, many still died young. For travelers from abroad visiting one or another Asian port, and just off the ship, it was best to quench immediate thirst with a pint of Jack Daniels or Sneaky Pete, or go to one of the better hotels where some guarantee of clean water was given. Other problems concerned the routine washing of kitchen pans and chamber pots in the nearest river. In many households, the stench from butchered animals conflicted with the blunt taste of earthy food. Sampans dropped garbage overboard, and fisherman squatted on the shore to relieve themselves. Birds and land animals frolicked and scavenged while bathers waded in the shoals and local people in shabby clothes bent down to capture water in their hands and lift it to their mouths. The Chinese let little go to waste, cultivating and eating what grew both above and below the ground. Hong Kong, if itself congested and "filthy," was far less so than many an inlet or bay further north or west. For the better-off there were fine restaurants—ones often connected with exclusive hotels like the Peninsula, Gloucester or Hong Kong, and these served excellent European and American cuisine. Certain higher-brow Chinese restaurants also were known to cook food agreeable to the American or European taste.[2]

The air one breathed in many parts of Hong Kong had to have been putrid. Surfaces had to have contained contaminants both seen and invisible. Inadequate sanitation and pervasive human refuse created repulsive odors. One could purchase civet and castor to go with snuff and other inhalants. Visitors couldn't help but observe the animal waste littering every street and alley and the slaughterhouse by-products filling every creek and waterway, but what of the invisible, noiseless pestilences all about them? For professional rovers like Halliburton becoming ill was an occupational hazard. To the Colony, his current stopover, refugees brought beriberi, cholera, malaria, and even leprosy. Ships from the world's ports also brought illnesses, so it was best to live out of reach on a hilltop. If the California sun infused well-being, the Hong Kong sun incurred sunstroke and fever. Urinary tract infections, food poisoning, and constipation competed with the indifferent sun, but dysentery shined brightest. Even after the slum clearance projects undertaken in the 1920s that reduced the frequency of its outbreak, the threat of plague remained.[3]

Fortunate to have a small appetite, Halliburton thus had a greater aptitude for abstinence than would his crew. Over the years, he had learned that people amid material deprivation and hopelessness can weather the most repellent living conditions if given time—and so could he. "Insouciance," comments Jonathan Root, was one Halliburton's "trademarks." A seasoned traveler, he accepted the occupational hazards of his chosen profession. Just as someone who cannot handle routine exposure to human waste, the shattered limb and blood should not become a nurse, someone who cannot handle routine exposure to disorientation, discomfort, and delay should not become a traveler. Halliburton spent eight days (with rests aboard a small rescue boat) swimming across the slimy waters of the Panama Canal. In swimming across the Hellespont, he suffered excruciating sunburn that was of little dire consequence. Emulating Robinson Crusoe, he lived for nearly a month in a stockade while a menagerie of largely domesticated animals ate and defecated around him. He slept with an ocelot named Tommy and chummed with a monkey named Nino. He rode a grunting elephant over the Alps, shimmied down the musty Well of Death, inhaled the unhealthy breezes of vermin-packed slums, and supped with the hard cases of the French Foreign Legion. If Halliburton found his soul purified by the trip to Devil's Island and a multitude of caravansaries in the Middle East, along with swims in the Sea of Galilee and the Nile, he also found his resolve toughened. Now in Hong Kong, the better hotels might guarantee safe food and water, but were they really safe?[4]

Despite his pampered needs, life in all its rawness appealed to Halliburton. His precise allergies were not published and his aversions, including hand-shaking, were mostly concealed. He might very well have felt most alive when faced with what most repulsed him. He observed the insulted and the injured, even breaking bread with them. Stories emerged of missionaries bringing their belongings in coffins when sent to the malaria-infested regions of Africa. With a life expectancy of two weeks, the missionaries believed the coffins would be purposed to bring their bodies home. Knowledge of such things titillated Halliburton, who once remarked that he wanted to experience and to feel everything. Although Halliburton believed that romance, like civilization, was ultimately the distance one put between one's self and one's dung, his tolerance for the despicable was high, for the displeasing even higher.[5]

Whether Halliburton pondered the emoluments of Asian medicine as had osteopath E. Allen Petersen is uncertain. Open-minded, Halliburton accepted alternate lifestyles. Tried-and-true *tribal* aids to healthful living, he also supposed, might be better than new enforced *synthetic* ones. In *Seven League Boots*, he remarked that a Soviet initiative to impose modern therapeutics like vaccinations, toothbrushes, and pink soap on its centenarian population in Georgia could prove ruinous, observing, "These rough and hardy mountaineers [have] grown quite content in the company of body-lice and fleas." Was it so in China? Available were numerous time-honored healing herbs to address every ailment from the common cold and constipation to chronic back pain and indigestion. In 1938, as now, numerous diseases existed in Asia and elsewhere that were unidentified or, identified, were untreatable. Some were toilet related; many of unknown aetiology were idiopathic. If they existed, hygienic laws were impossible to enforce, especially in war-time. Epidemiology in an acutely scientific sense was itself little regarded by Halliburton, but of chronic diseases spreading throughout a community he was keenly aware. Partner to disease-carrying bacteria were other human-generated diseases. In major commercial hubs, like Hong Kong, inoculations for cholera, typhoid and smallpox were available, and willingly Halliburton submitted to these. Years of global wandering had surely acquainted him with folk remedies—and with the knowledge that symptom pools peculiar to each culture might require different cures and that certain diseases might resist recognized diagnostic (Western medical) labels.[6]

Captain Welch knew that all men are mortal, especially "Captain Jack Welch." From experience, not from books, he knew what illnesses he could catch. As a precaution, he had been vaccinated for the "military gout." Although Halliburton and Mooney in their correspondence said nothing about it, Welch, less governed by matters of taste, solemnly admitted to Dick Wetjen, "We have all been sick as hell with dysentery." The inflammatory disease often had Welch running to the toilet ("the throne"). During its cruel but brief reign, of "four days," he lost ten pounds, and soiled garments might have compelled him to ask his friend Wetjen to send him some new clothes. A new food regimen followed, and the captain swore off Chinese food *for good*.[7]

In port but a week, Welch by his own admission had spoken only to the rickshaw boys who took him to and from the harbor or a bar. The captain watched the traffic from his hotel window; a sign on one streetcar read, "In case of this tram getting caught in a typhoon all windows must be lowered and not to be opened until the typhoon has passed." On his jaunts he observed that the "great universal gesture" was "the open palm," which he explained as " a form of salute, like the Hitler gesture so to speak." Welch talked to squadron officers, likely about sea and air operations with some hometown sentiments thrown in. The officers might have grumbled about how poorly the dated fleet of airplanes their crews commanded would fare against their Japanese opponents in an all-out war. Welch needed to be kept busy. When not busy, he succumbed to bouts of griping, mostly about squalid Hong Kong. The public services and conveyances frustrated or confused him. The huddled masses nauseated rather than thrilled him. "Outside of all this," he grimaced, "I am in the pink." Even the most innocent discharge of civic emotion riled Welch. A dance hall (possibly the Queen's Dance Hall) just across from the Kowloon Hotel occasionally drew a crowd. Once, a military band came there to play old tunes "Dardanella" and "Sympathy." He wanted to scream." To cope, the captain enacted a set routine. Mornings now saw him trundling off to the "broker's office," maybe to see "old Swede" Captain Anderson, then to the American club "for a drink and a yarn with some of the chaps there." In the interval between one or another social diversion, he went to the YMCA, perhaps finishing off a regular swim with some calisthenics. He also wrote letters.[8]

Paul Mooney's routine varied. "There is just one way of passing the time during a total blackout," he said, "but here in China that (one way) isn't scandalous and it costs just thirty cents." Scandalous? In China's major ports, gambling and opium dens were everywhere. In Shanghai the Special Service Section of the Japanese army created the "Shanghai Supervised Amusement Department" (run by Chinese and Japanese underworld figures) to regulate the narcotics trade. Heroin production in Japanese factories in Tientsin (Tianjan) and Dairen (Dalian) grew, and, as the need for the substance increased, factories sprouted throughout Asia to serve it. In America, concern over narcotic and alcohol use had risen toward the end of the nineteenth century. Under the administration of the US Treasury Department, the Harrison Narcotics Act of 1914, whose chief proponent was prohibitionist secretary of state William Jennings Bryan, limited the supply of opiates coming into the United States. As the supply dwindled, the number of addicts declined. Noted Indian ethnographer James Mooney, Paul's father, was a voice in the Peyote Controversy of 1916. To gain an intimacy with his subject, he regularly participated in religious rituals that required the use of peyote. Along with many Chinese physicians, Paul, like his father, saw no crime in exploring the therapeutic and psycho-active benefits of pharmaceutical substances thought taboo. Paul, unlike his father, was not on an anthropological field mission; he simply liked 'to get high.' In Hong Kong the manner of opium consumption varied. Called *chandu* or *maddak*, a bite-sized chip of opium resin or hashish was easily obtainable and affordable.[9]

To Mooney's delight and joy, Hong Kong attached no stigma to the inhaling, ingesting, and absorbing of narcotic substances. Of course, in the Hollywood community and among his and Halliburton's wider circle of acquaintances, smoking marijuana would have fallen beneath the level of venial sin. Halliburton himself claimed to smoke tobacco only on occasion—maybe for a magazine ad in which he noted, as did famed athletes, cigarettes' salutary effects. Films show the travel writer smoking cigarettes as though he did so customarily. He knew "the pungent odor of opium" when it filled the air and did not recoil from it. Day laborers in the various "third world countries" he visited used stimulants to relieve pain and distract them from repetitive job tasks. Although known among foreign visitors as an alcohol community, Hong Kong was better known among its longtime residents as a narcotics one. Halliburton *at first* took small notice. Besides some over-the counter aspirins and a pack of unfiltered cigarettes, the only drugs he seems to have taken were the few remedial ones licensed medical practitioners prescribed to him. Drugs branded illicit little appealed to him.[10]

Without elaborating, Halliburton reported to his parents that the food was "good" and the water was "clean"—at least in the hotel where he stayed. The store of food supplies, however, dwindled, as such provisions could only be brought into the Colony by ship. Also, there was so little water that the pipes were cut off for six hours every day, leaving the four reservoirs "surrounded by glaring white stone dams, (that made) perfect targets for bombers" as the only resort. Yet Halliburton maintained that the Colony was "normal." Mooney expanded on "normal," noting that "big guns and planes and squadrons of French, American and British ships (loaded) the harbor." He also noted the wide variety of people filling the streets and pubs: "Americans, Irish, French, Scots in kilts—soldiers as well as sailors, Moslems with turbans, Sikhs, and whatever else England can dig up in her Empire." Japanese patrol ships, meanwhile, lurked "just over the horizon." Ports were seized, and Japanese sentries blocked entry to any region fewer than fifty miles north of Kowloon. Unless aboard a warship, an American or Englishman could not reach Canton. Boats only carried Orientals, making some exceptions for foreign journalists. Big ocean-liners and the Clipper planes remained the only safe links with the outside world. The report of bombing raids only forty miles away sent the city of Hong Kong into a near panic. Evening blackouts hid the city from bombers but didn't stop them. Halliburton told subscribers that Hong Kong was "defenseless" against the Japanese despite its "apparent strength as a fortress." What with the population swelling "to the bursting point by the tremendous influx of destitute refugees," fear spread that the city, unable to support these individuals, would soon be starved into submission. The British had antiaircraft guns mounted on every peak and a hundred airplanes on the ready. But the troops took no offensive action against the Japanese and their stranglehold on trade. Esteem for the British soon plummeted. Observers agreed that the Japanese could capture the Colony "whenever they wish[ed]."[11]

Captain Welch, for his part, was resigned to the city's fate: "The story here is that the Japs will take Hongkong in less than six months. If you ask one of these Empire builders, he looks at you as though you wanted to sleep with his aunt, but that's the story and I believe they can take it if they want to." Halliburton did his best to see a bright side, but, even so, soon realized that the *Sea Dragon* Expedition was fast becoming the *Sea Dragon* Evacuation. He predicted that he would be home by March 12, but as things now looked, he would be lucky if he left port by then. The dramatist in him saw Hong Kong as a city besieged, as Homer's Troy,

as Gordon's Khartoum, with mayhem and murder, rape and pillage in
the offing. Halliburton saw himself as a tragic hero in a comic role.[12]

Not too long before Halliburton had high hopes for a speedy entry
and exit from the scene. Within a week of arriving in Hong Kong, his
intention of sailing a junk across the Pacific hit most front pages with
bravado: "This morning we were blasted out of our beds by the morning papers," he wrote, "which gave a highly glowing account of the
famous author, Richard Halliburton who is here buying a junk to sail on
a dangerous and hazardous voyage across the Pacific Ocean." Captain
Welch "blew up" over the news. Were potential sellers to get wind of
Halliburton's business, their asking prices for junks might soar. For
days he sulked, perhaps not wanting others to steal from the same
pot he stole from. Needless to say, it would have been hard to keep
Halliburton's plans "dark." Welch, nevertheless, saw reporters as gnaw-
ing pests. One, "a callow, pimply-faced limey," showed up after breakfast
to query him. Welch divulged only that he and Halliburton had abandoned the idea of getting a junk in Hong Kong and had set their sights
on "one already ordered" in Foochow.[13]

Aiming to settle Welch down and to gather additional needed supplies, Halliburton took him to Portuguese Macao, an island measuring
about ten square miles and located fewer than forty miles due west of
Hong Kong. The chief corridor to the Pearl River Delta, major commercial artery into mainland China, its famous port, once the busiest in the
region, was now war-ravaged and overrun by Japanese troops. Welch
and Halliburton were almost certainly accompanied by Paul Mooney,
and maybe by Henry Von Fehren. If Captain Welch loathed Hong Kong,
he detested Macao, repository of knaves and gambling mecca of the
Orient. "Of all the lousy, stinking holes that place takes the bun," he
said. The food made him wince, and while he pitied the Chinese who
consumed it daily, the smell they exuded repelled him. To share his own
joy in the culture, Halliburton dragged Welch down a winding trail to
some old ruins, almost certainly those of the Church of Saint Paul the
Apostle adjoining a fortress. Built in the 1620s by Japanese Catholics in
the service of Italian Jesuit missionary Carlo Spinola, the church was
largely destroyed in 1835 by a fire, aftermath to a typhoon. The original stone masonry was still visibly crumbling, and the church's spire
trembled as a frightened ghost. Welch loathed the whole excursion.[14]

Once back in the Colony, Welch thought he would journey to

Foochow, perhaps to get such "supplies" as were unavailable elsewhere. Halliburton was to accompany him in a renewed effort to purchase a credibly sea-worthy junk. In the meantime, both Halliburton and Welch met for dinner with "lovely fellow" Mr. Scott, a ship broker. Scott applauded Halliburton's mission, and Halliburton briefly—and secretly, considered firing Welch and putting Scott in charge of the operation, but when it later came out, Welch waved his contract in front of him so the matter was dropped. What mattered now was getting a good seafaring junk and at a good price. Discussed with Scott were the merits and availability of a Foochow junk. Nothing came of it. The Japanese currently occupied Foochow, so why bother to go there? Scratch Foochow. Try Canton, ninety miles to the north. For Halliburton, Canton represented the apex of Chinese civilization, and an apt starting point for his cross-cultural mission. But with British arms at rest, the Japanese presence there would have made working conditions frightfully difficult. Japanese fighter planes had recently pummeled the ancient city. Furthermore, Welch heard nasty things concerning the Japanese soldiers: "Talked to a chap at the American Club, and he said they were cocky as hell: slap the whites across the face as a friendly gesture—Nice little bastards aren't they?" Chinese and Allied resistance to Japanese incursions had tentatively reopened entry to Canton. On Sunday, October 30, Captain Welch drove northward to Canton with Halliburton. Just miles from their goal, they for some reason turned back. Welch attempted another trip there, this time on a riverboat from the Canton River Company skippered by "a reckless sort of bastard—hard as nails" who was married to a Chinese woman. Nothing came of either effort, and for now Canton was off limits.[15]

20

The Battle of the Books

On October 24, soon after he arrived in Hong Kong, Welch wrote, "Halliburger and I get along alright." His neutral regard for his employer soon withered. His slurs to Halliburton's character became sharper and more frequent. Welch was ashamed of his association with the adventurer and embarrassed to be seen alongside him. Being in the Orient rattled Welch's nerves; being in Halliburton's company made his condition worse. Given an opportunity to work on a ship of the Canton River Company, Welch did consider it. While he parleyed with the ship's captain, Halliburton, out of earshot, stood nearby. "Take him too?" Welch asked his prospective boss. Running his eyes up and down "Halli," the "hard as nails skipper" said, "No, just you come along." Welch declined the offer. The money probably wasn't good enough, and the position was far less exalted than his current one. Apart from a wide divergence in fame, what separated "Jack" from "Halli" was that "Jack" had served in World War I while "Halli" had not, that Jack had at least co-managed a ship while Halli had only sailed on one.[1]

Halliburton seemed unaware of Welch's disapproval of him. In public he never stood apart from him but showed great pride in stepping forward to introduce him to reporters. Such courtesy made Welch dislike Halliburton even more: "I can't chase about all day with the gorgeous one, listen to her talk business, or rather ask everybody else what they think of our business, eat lousy food three times a day, drink tepid beer, then be expected to look with lustful eyes on Chinese women as flat as boards and who don't like the white anyway." At bottom, what troubled Welch was how others might perceive his relationship with "the gorgeous one." "She always introduces me as 'my captain,'" he said, "and after a while the chaps (probably at the American Club) look at me as though they are wondering a bit." Halliburton proved hard to shake off. "I stay to myself as much as possible," Welch said, "but I have to spend most of the day with her, but at night, after dinner I get the hell out of her way."[2]

Welch had to bow to Halliburton; he needed the job and the measure of fame it promised. For weeks the matter of Halliburton's hiring "lovely fellow" Mr. Scott in his place had him brooding. He knew of at least one instance in which employers had broken a contract with their skipper and faced lawsuits as a result. Both parties stewed. Then and there, Halliburton might have replaced Welch with someone less contentious—perhaps the said Mr. Scott. But time ruled, and getting a capable helmsman at this late date could be difficult. "The China Coast is no longer the home of the renegade and heavy drinker," said Welch. "The jobs are too scarce and there are plenty of ships laid up." He felt he had Halliburton cornered, but the contract protecting him also bound him.[3]

As early as November, Welch wished with every sinew binding body to soul that he hadn't come to China. Now more than ever he wanted to retire from life at sea and become a writer like his friend Dick Wetjen. Welch had already produced a few short stories, including "Ocean Tow," published in the US Naval Institute's *Proceedings Magazine* in 1937. *The World's News*, headquartered in Sydney, Australia, later published his story "Paradise Regained." His "Farewell to Sail" appeared in Sydney's *Man Magazine*. Richness of specification and colorful incident distinguish these stories which also show a gift for racy narrative and lively dialogue. Like Wetjen, he had the lecherous eyes of the pulp writer and, peculiar to himself, a gift for bored observation. That Halliburton and his "secretary" occupied the same stateroom on the ship to China got him to wondering. When Halliburton and Mooney again occupied the same quarters at the Kowloon Hotel, Welch was confounded. Details of the incident are lacking, but the captain peeked through an open door into Halliburton's and Mooney's room and beheld what to him explained everything. "I saw the secretary (?) *wife* running about in his room the other day naked," he wrote, "and I think I have the answer to the whole thing. All same horsey [*sic*]." If before Welch only suspected that Halliburton and Mooney were thick, now he knew.[4]

Halliburton now became a sort of hobby to him. As Welch told the story, shortly after arriving in the Colony, he bumped into Halliburton at a bookstore (possibly Swindon Books near the ferry) that carried titles in English and various foreign languages. Halliburton had come to see which of his own books, if any, were available in the distant port. "He (Halliburton) found one of his (own) books in the store," Welch told Wetjen, "and didn't like it too much when I hauled yours out to show him saying that I knew you well." Halliburton, not knowing quite how to respond, wondered only why the captain would raise the issue. Welch continued to bait Halliburton and might have told him that Dick Wetjen's wife, Edith, was a fan of his books. The comment implied that women, not men, read him.[5]

As a writer for the pulps, Albert Richard Wetjen courted a readership different from Halliburton's; in renown he was hardly a competitor. The 1926 winner of the coveted O'Henry Prize for short fiction, Dick Wetjen probably introduced Welch to the profitable tabloid marketplace and gave him tips on how to produce salable copy. The cigar-chomping, back-slapping Wetjen, a pudgy little firecracker, seldom wrote a story he couldn't sell. Adult males, with a craving for hard-core adventure featuring hairy-chested protagonists, tough talk, lusty jungle maidens, and a scuzzy story-line, were his reader base. Hunting, fishing, safari, and shipboard romance were his provinces of intrigue. His colorfully ribald books were found in book-stands throughout the English-speaking world. Monthlies as diverse as *Adventure, Argosy, Action Stories, Blue Book, Colliers, Cosmopolitan, Grit, Saturday Evening Post, Everybody's Magazine, Storyteller*, and *Wide World Adventure* often bought his stories sight unseen. His prose was fast-paced, his titles, "The Lion Goddess" or "Wild Girl Taming," "Six-Feet and Horse-Faced," and "One-Man Road," lurid.[6]

While a case for *Treasure Island* author Robert Louis Stevenson could be made, Wetjen's true mentor was "sailor on horseback" Jack London, a writer devoted to a creed that exalted manly virtue, action and daring. Like London, Wetjen wrote exciting adventure tales set in the South

Seas and believed that biology as much as psychology accounted for a character's actions. He recommended that stories be rich in specification and waste no opportunity for conflict. As did London, Wetjen drank heavily. Fellow pulp writer Ed Price described him as "a querulous drunk, and at times a nuisance" who "when sober, was most likeable." Both London and Wetjen were generous and kindhearted. In addition, London's life paralleled Halliburton's in some ways. London also built a magnificent home, Wolf House, and a remarkable ship, the *Snark*; the former burned down, and the latter made it to the Orient and back. London died at age forty in 1916. Halliburton had to dissociate himself from rumors that he resembled London, yet resemble him he did. Both men sought adventure, wrote about it, and believed success in some sport the highest "personal achievement."[7]

By October 24, the date of his first surviving letter to Wetjen, Welch had developed a negative fix on Halliburton, whose success he considered unmerited. Iago-like, he would ridicule him subtly and weave his petty plots. "Halli told me tonight that she has put me in her first newspaper article," he told Wetjen. "I am an old seaman it seems and the others are just playing at it." Ranting on, he told his friend, "You lug, me makes the money and you could too." "He [Halliburton] told me that he has a contract with the Bell people to do the trip." Halliburton had no idea of Welch's wish to pursue a literary career, but Welch, by hinting that he intended to write a "yarn" about the "trip" from a mariner's point of view, baited him. *San Francisco Examiner* columnist Herb Caen, just starting his career, might have prodded Welch into sending him information so he could write a story about the expedition—and Welch thought to do one himself. "Just wrote to Herb Caen," Welch said, "and asked him to go easy on the story that we are all here for fun and adventure." Once at sea, the captain would keep "two logs." Of any independent writing apart from these logs, he told Wetjen only that he would do "(his) best to get a good article or story together for you when I get back; I'll work on it all the way across." Welch might have told Halliburton that he was "keeping notes" during the trip and, as he told Wetjen, that he would keep a "very accurate log," maybe *two*: "One for the ship in which there will be the bare essentials and the other in which I will do all my sights on one page and all the in s(?) that might be helpful on the other. I'll keep three track charts, one for you, one for (writer) Jacland Marmur (he asked for one) and one for myself. This trip will fill in my other chart very nicely too. Getting a blueback for the fancy work." "Fancy work?" Whatever the story's theme, Halliburton was sufficiently upset that Welch was writing one at all. Unpleasantly surprised—at first unsure how to pursue the matter, he asked the captain to tell him more. Why shouldn't he? Naturally, Halliburton worried about his reputation; Welch could dispute any of the adventurer's claims about what happened to them in Hong Kong. He had also contracted with the *San Francisco News* for exclusive rights to the story. Halliburton needn't have worried about Welch stealing his audience. In creative outlook, the two men were seas apart: Halliburton edited the mundane to fit ideals while Welch edited the mundane to fit reality. No imposter, Welch was as accomplished a wordsmith as Halliburton, but devoid even of a modicum of Halliburton's fame. Anxious for additional information and perhaps concerned that the captain's work would trump his own, Halliburton learned only that Welch would tell his story "purely [from] the seaman angle." Except for a letter or two to Wetjen, Welch had written nothing so far about the expedition.[8]

Halliburton did not address all his captain's grievances, but, sticking to business, raised the possibility of his having for his own play-time the *Sea Dragon* "during the [Treasure Island] Fair (1939) at a straight salary of three hundred a month, and a commission when the motion picture companies [took] her over for use in pictures and shorts on China, etc." Another idea involved his taking the junk to the Sacramento Fair by sailing up the San Joaquin River and mooring on the Sacramento riverfront. The viewer market was solid. Sacramento was home to McClellan Air Force Base, a supposed link to the outside world. Also, the city's West End had gambling parlors. Welch understood that the most money was to be made in Sacramento, but he hardly cared: "My interest ceases after I am anchored in the Marina or tied up alongside of the Fair Dock." Indeed, Welch's writing career beckoned. He would soon bid a "fond farewell to the bloody sea and all its works"—and test *new waters*.[9]

Welch of course had schemes of his own tucked inside his shirt. Like Chaucer's Shipman, he fully knew the "craft" of navigation and could "reckon well the tides," but "of nice conscience he took no keep." Mission statements meant nothing to him, and he dismissed Halliburton's first "Log of the *Sea Dragon*" article as "a vain search for beauty in China." As Welch was no stranger to the criminal culture that operated on the docks, he now contemplated profitable ways to enliven the voyage.

Besides the British, he knew the shipping industry and its underworld. He cut deals while carrying out his assigned duty of gathering supplies for the junk trip. The morally creative Welch defined "cumshaw," or easy money from kickbacks, as "all good clean money," and stated that "the British authorities, now colleagues in his secret commerce, "have been very decent to me all through this because it has been out that I am from Australia." A common usage for Welch, "cumshaw" was among a constellation of terms associated with waterfront knavery and a criminal lifestyle at once exciting and perilous. Cunningly Welch knew that a basic salary of $250 a month was only a foot in the door to unlimited largesse. Why not add to it "for the fun of it?" Profits by speculators from kickbacks were so well known that local gazettes didn't even think them news. "The engineer (Von Fehren) told me today," he informed Wetjen, "that he *heard* that somebody was getting fifteen hundred dollars *cumshaw* on the building of the junk alone. I acted very surprised, but I wondered how the hell he heard that or if he was taking a shot in the dark. Funny thing that a man can't make a little change without everybody getting mad." "Funny thing," the captain had a "store list" of his own, one comprised of "nice things" and had made "a very handy deal with the *compradore* of one of the bigger stores." Welch worried what "those bastards of Chinese" acting as middlemen might impound, but he was sure that he would be "well paid for . . . (his) foolishness in coming all this way to sail a junk when (he) might well have stayed home and had . . . a nice time over the holiday."[10]

The "nice things" on Welch's list were likely tangible items—lacquerware, porcelain, jade, amber, maybe ivory. Other items such as perfumes, rugs, and pigskin suitcases were a steal in what many visitors saw as a shopper's paradise. In a ship the size of the *Sea Dragon,* such "nice things" could be easily concealed to elude customs inspections and duties—and Halliburton's not-always-watchful eyes. Although the cumshaw would arrive late, Welch behaved as though its arrival was imminent. When Anderson (the ship surveyor) gave him a gift pass to the "swanky" American Club, the captain lamented the high price of the establishment's drinks but hoped to be *sitting pretty* at the bar there soon.[11]

Time was passing too slowly . . . and yet too quickly.

21

The Master Shipbuilder and the Shipyard by the Peachy Garage

Halliburton and company had been in Hong Kong less than two weeks. They had gotten on one another's nerves, but had not gotten a junk. "I had hoped to shop around in a sampan," Halliburton wrote, "select some junk which was already in service, bargain for it, perhaps change a bulkhead or two, add a coat of paint, and set sail." By now he had "looked at 50 junks" and "hundreds of worm-eaten sea-goers." While some were "superb examples of Chinese ship-building, most were either "too big, too expensive, (or) too battered." The remaining vessels were "too old and too ugly." Japanese blockades had limited the search for a suitable 'Sea Dragon,' as had time constraints. By November 1, "when it became obvious that no ready-made junk suitable for a Pacific voyage could be had at a reasonable price," Halliburton decided to build one from "keel up." He wanted it "designed, constructed and decorated in strict accordance with native custom," and he further insisted that "the work must be done by natives." In step with the idea, Welch hoped operations would soon get underway. "We'll have a junk all right," he said—this was on October 29, a few days after he arrived in Hong Kong and a few days before building of the Sea Dragon actually commenced.[1]

Who would build the junk was the question of the moment. Likely Captain Anderson from the American Bureau of Shipping or shipbuilder

Mr. Scott offered Halliburton a small list of reputable names and from that list emerged a standout with the "singular name" of Fat Kau. "Hong Kong's best junk builder," he was "recommended above all others." The "worthy Mr. Fat Kau," a shrewd operator who had amassed a small fortune leasing and building ships, exercised some clout in the community. Employing a contingent of workers, he owned a shipyard headquartered on the outskirts of Hong Kong, to the north and beyond the metropolitan Wan Chai district. Once a meeting with Fat Kau was arranged, Halliburton and Welch—possibly Von Fehren and Mooney as well, led by a male interpreter, walked to the Kowloon dock from where the harbor ferry brought them to the Island. There they boarded a tram that "carried (them) along the mountain-framed shore, away from downtown Hong Kong, past the British barracks, farther and farther into the squalid Chinese slums." The tram halted to the clank of a gong in a less forbidding region off the harbor where "thousands of families lived on boats."[2]

Once off the tram, the company paused to get their bearings, proceeded uncertainly toward an apparent wrecking yard "opposite a high-smelling soy-sauce factory and between the *Peachy Garage* and a *Gentleman's Parlor for Beauty*." They soon reached their destination, the place where, as Halliburton recalled, "junks are born." But rather than elated, he was speechless. Unlike work sites he had seen in America and Europe, this one was ghastly and primeval. Working conditions at once struck him as "wholly, hopelessly Chinese," muddled, "cramped," and unordered, an affront to logic and the clarity of the Aryan mind. "It sprawled along the waterfront," he expanded, "in a wild confusion of timbers, bamboo poles, babies, old cannons, cooking fires, carpenters, and wives. Hens roosted on the big two-man saws. Tailless cats prowled, slant-eyed, among the logs of teak and camphor wood. Cached in every cranny were the workmen's bedrolls—for all the workmen sleep at the job at night. 'Job' and 'home' are synonymous in the Chinese laborer's lexicon." The shipyard could be easily mistaken for a landfill operation. "Every hour or so," wrote Halliburton, "another smelly fishing boat is hauled ashore on the ways by Fat Kau's ancient engine, for besides building new junks, Mr. Kau services old ones." Lighting the yard, joss sticks, set on each of the slipways leading to the water, burned day and night.[3]

Brought into Fat Kau's living quarters, Halliburton confronted "domestic bliss." In attendance were, along with other extended family members, Fat Kau's four wives—"each with a baby to care for." Actually the shipbuilder "had five wives," Halliburton learned, "but of these the youngest, aged 16, the day after the wedding had run away with one of the shipwrights." Fat Kau was left ashamed and humiliated, and his other wives called the "errant bride" "a minx and a hussy." The past left behind, Fat Kau "graciously" welcomed his American visitors; he had been expecting them. One pictures the junk-builder as a Mandarin priest seated on a raised pillow and wearing a pillbox hat, slippers, and a richly embroidered silk tunic. Halliburton instead saw the man as "a middle-aged, good-natured Buddha (who) was very fat, very bald, and very rich." Captain Welch otherwise described Fat Kau as an "old chink" who was "leather-faced" and had cheeks that narrowed into deep furrows the moment he "squinted."[4]

Congenial, the shipwright gave Halliburton and his captain a tour of the shipyard. Halliburton thought the place would be "overflowing with work," but Fat Kau was experiencing "hard times" and had received little business. Half of the junks that used to come to his yard for repairs had been "sunk or burned by the Japanese" and "over 10,000 of the fisher folk (a great part children and women) killed or drowned." As Halliburton (through his interpreter) spoke of his needs, Fat Kau eagerly listened. "We were clearly a group of foreign idiots," he later summarized the event, "wanting to build a junk and name it the Sea Dragon and sail it to San Francisco, 9000 miles away; but we were cash customers, so the generous man agreed finally to accept only twice what we offered. When we departed we had a contract, beautifully sealed in red with Kau's private chop, and the Sea Dragon—we imagined—was as good as built." Halliburton had expected to pay $2,500 to have the junk built, and Fat Kau, writes Jonathan Root, "quoted him a price of $5000." What with the need for scarce commodities in wartime, both parties could expect the cost to climb.[5]

Halliburton knew how to greet and bid adieu in Chinese, but most of the time it was 'point and grunt' for him. Likely Fat Kau spoke both the Cantonese and Mandarin dialects, and likely he knew a few words of English, or "Hong Kong English" as well. For lengthier discourse, the shipbuilder and his client relied on an interpreter—perhaps the one Halliburton brought, whom Welch eventually took under his wing. Born in Burma of a Chinese father and Burmese mother, the translator was, in the captain's generous opinion, "smart as hell." Immediately after he appointed the fellow a shill in his little con games, he asked that he coax "the old chink" into giving him ten percent clear of any deal struck for material; hard-bargainer Fat Kau countered with an offer of five percent. Was the translator double-dealing? "Lived in Liverpool" was all Welch said about him. "I pay him well too. He is going to be messboy on the junk." While Fat Kau and the interpreter discussed money in Chinese, Halliburton might have seen only smiles and nods and understood nothing. He said simply, "We made a deal." A ceiling cost for labor and materials amounting to perhaps $13,000 was presumably agreed upon and evidently put down in writing. A forty-five-day completion time was also stipulated or about half the time given to him by the Amoy carpenter.[6]

Fat Kau had won the contract. Although the reason for its sealing cannot be rightly determined, Welch received a kickback of $1,500 for his efforts. Side deals (cumshaw) were second nature to him. To cut corners, he likely acquired low-cost timber for the hull and deck from decommissioned junks as well as acquired other basic materials, including clothing, food allotments, and "extras," from competing wholesalers whom Halliburton himself never even met. While the "extras" perplex, the source of the side money is clear—5 percent markups as payoffs to Welch, a sum of $400 for $8,000 of "stores." The messboy, and interpreter, procured for Welch some "bawdy prints," perhaps of *shunga* erotica. These Welch intended to give to Dick Wetjen, whom he presumed had, as he, "a depraved mind," and, as he, needed visual stimulation as an aid to "hoist the mainsail."[7]

Halliburton watched the calendar—the hours and days trickled by; so did dollars and cents. Oddly, he didn't quibble with Fat Kau about the doubled cost of his services. He admitted ignorance about junks but plainly told the builder to make "a *Ningpo* model, 75 feet long and 20 feet wide." He specifically requested "a beautiful junk," one that was "big, but not too big; colorful, but not garish." Captain Welch requested for himself "two small houses on the poop"—one for use as the chart room, another to serve as his living space. At the time, he described the poop, sight as yet unseen, as "a measly twenty five feet up and the bow a modest nine." When Halliburton asked Welch where he was to have his own quarters, the captain snapped, "Down below with the others." Halliburton then suggested that the lads bunk on the upper deck near the forecastle, and that "a small space (be) set aside for himself and his secretary." Welch looked quizzical and uncertain. "In the same room?" he asked pertly. "Yes," said Halliburton, "with berths and washbasin together." He wouldn't let himself be cowed by Welch. Rarely, however, did he stand up to Welch, though once he asked his captain "not to swear too much." As early as November 2, he privately branded him "dictatorial and evil-tempered," but added, "he's a good sailor, so I don't care." In coming days, Halliburton would offer the rankled captain a number of interior design ideas. One one occasion, the SS *Coolidge's* Captain Ahlin, dropping by, took Welch's part, and solemnly rebuked Halliburton for "interfering with the master."[8]

Assuming it his prerogative, client Halliburton continued to interfere with the project. During the effort, Welch did not upset him as much as the impassive Fat Kau did. It was hardly impertinent, for instance, that he should offer "to supply [the shipbuilder] with all the necessary blueprints and drawings," ones Welch had himself drawn up. When the interpreter told Fat Kau what Halliburton wanted, the shipwright was invariably mum. Besides a rough sketch, Halliburton might have shown Fat Kau one of the small junks he had purchased in Shanghai to serve as a model for his *Sea Dragon*. He might have pointed to a junk under construction—possibly the *Pang Jin*, which would sail to the New York World's Fair, or a junk near completion. He was soon led to wonder, however, if any of it mattered. Finished products meant nothing to the shipwright. Blueprints certainly left "no impression" on him. All he wanted to know was the desired length and width of the junk. Like other shipwrights whose skills spanned generations of trial-and-error experiment, Fat Kau saw the junk inherently as a phenomenon of nature that just needed hewing and shaping. Sharks and dolphins inspired the slender design that accounted for their speed, bats the sails that steered them, ducks the grace that glided them, and creatures of myth like the dragon the furious look that made them bold. So it was that master shipbuilder Fat Kau created from life a lifelike form.[9]

Fat Kau's pedigree as a professional went back many centuries—to the Song dynasty and the polymath Shen Kuo, who wrote about ships. The *Sea Dragon's* shipwright almost certainly belonged to the "Guild of the Sails," whose members subscribed to the highest standards of the shipbuilder's art. For these craftsmen, as for the Italian violin makers of Cremona, written blueprints served little purpose alongside those already stamped indelibly on the tablets of an efficient memory. Despite his own deep respect for Chinese shipbuilding tradition, Halliburton portrayed Fat Kau as a sort of Asian Luddite, a man suspicious of novelty, of innovation, and, importantly, of Western technology. He remained respectful, however, of the shipbuilder's position. "For 4,000 years China has been building these strange craft," he wrote, "and after 40 centuries of trial and error has learned to build them so that they handle remarkably well in all kinds of weather." The shipbuilding techniques of these masters were not committed to writing. Rather, they were transmitted orally from generation to generation. In defiance of these traditions, Fat Kau would at times bend to his client's whims.[10]

Amused by his client's *progressive* ideas, Fat Kau thought the drawings and blueprints Halliburton laid out as useless as dinner-table doilies. He needed no pictorial assistance. His was an ancient art practiced for millennia, kept in the head and not put on paper. He knew instinctively that a keel, small on junks, serves as the ship's backbone or spine and assists the rudder, a partly submerged blade that, with a twelve to fourteen foot shaft, is notably large on junks. Functioning as the ship's stabilizer, it controls balance and direction, movement and coordination. It gives lift to a vessel and directs its forward motion. A ship without a keel can founder or list, just as a fish without gills or weak in the fin will turn to one side and die. Fat Kau saw in his mind's eye pictures of a keel and a rudder, not the intangible words *keel* and *rudder*. And *helm*? After hearing the basic English terms that pertained to his craft, Fat Kau perhaps attempted to repeat them. He knew that *hull* was his client's word for the frame or shell of the ship, *beam* denoted its width, *fore to aft* its length, and *gunwale to keelson* its depth. While a shipbuilder from the West might explain that sails provide the ship its propulsive force and, combined with the stays and mainstays, constitute the ship's rigging, Fat Kau, not given to abstract theorizing, rather saw them as the ship's wings. Halliburton appreciated the junk as a wondrous freak of nature: "Their three bat-wing sails, made of shining yellow mats, drove them over the water with the ease of a flying seagull." In the end, Fat Kau and Halliburton must have seen "eye to eye." "Hesitantly, a little fearful of such casualness," said Halliburton, "we signed the contract."[11]

22

Mr. Halliburton Builds His Dream Ship

Work on what would become the *Sea Dragon* began at once—perhaps as early as November 2. The carpenters wore sandals or went barefoot. They clothed their smallish bodies in loose-fitting work trousers and buttoned shirts. Most wore drab wide-brim fedoras, bucket hats, or army caps. Some crouched down before their tasks, while others stooped or stood. After work, they stayed aboard the ship or nearby. Materials not on site had to be ordered. These included woods such as camphor, cedar, and sandal, all of which, said Paul Mooney, exuded a "heavenly" smell. Teak, a tropical hardwood resistant to water damage and rot, was the principal wood used. The laying of the keel, itself "a magnificent log of *jacal-wood*, 60 feet long," represented the first step in the junk's construction.[1]

Public interest in the goings-on in the shipyard were immediate. "Within days after the keel for Halliburton's junk had been laid in Fat Kau's shipyard," writes Jonathan Root, in a simile worthy of Homer, "the pirate tipsters came to gaze upon it much as a panther would regard a suckling pig in a flimsy corral." Guards, unwatched themselves, were posted. The work progressed without incident. All would be accomplished peacefully and in an orderly fashion, or so Halliburton wrongly assumed. "Once the keel was laid," he reasoned, "one might suppose the ribs would follow, and then they would begin to make the hull. In any sensible shipyard this would be the sequence. But not in Fat Kau's!" The planks, suspended from bamboo ropes and propped up with bamboo poles, were set in place before the ribbing. When asked about it, Fat Kau explained that the carpenters couldn't tell where to position the ribs unless the planking was there to guide them. Thinking the reasoning "demented," Halliburton groaned: were they commissioned to build a house, these same carpenters, given Fat Kau's absurd directive, would probably have started with the roofing and worked down to the foundation. Amused, Fat Kau told his client that construction had been done in this manner for four thousand years, long before the Great Wall was even conceived. The reliance of the carpenters upon hammers, chisels, and saws, which were of the Stone Age variety at best, also appalled Halliburton. Although the keel had been laid in "24 hours," the piece had first to be "hand-cut," shaped and smoothed. While a sawmill might have "squared it off in 20 minutes," the Chinese carpenters spent a week working with their "little axes" and saws. The measuring devices Halliburton expected the craftsmen to use—had they relied on blueprint specifications—would have included tapelines, plumb bobs, and possibly rulers, but these were never in evidence. And why cut wood with an ax or bore holes with a manual drill when an electric saw and drill were near at hand? One could only wonder. Captain Welch found the methods "primitive" but conceded, "You take it as it comes and say nothing."[2]

Still, what next? For Halliburton a momentary dream could suddenly turn into a dark nightmare. One day, on his way to the shipyard, that nightmare became reality as he saw, from afar, wisps of smoke curl up from one of the work areas. Fearful that the three-inch teak planks he had purchased at great cost were in flames—or, worse still, that some crazy arsonist had torched them, he rushed to the scene. There he found stretched between two sawhorses those very planks with smoke rising from them. He nearly fainted at the sight. 'What on earth!?' The unexcitable Fat Kau explained through an interpreter that the planks had to be heated so they could be bent into the desired crescent shape. "How else could it be done?" Fat Kau asked Halliburton. "Why bother to complain?" To end what was becoming an old refrain, 'It had been done this way for thousands of years, so why change now?'[3]

A dozen carpenters started operations. That number grew as the carpenters' wives began to lend their assistance. The noise from the pounding and chattering was often unbearable to those seated in the front row. Harmonic music to Fat Kau's ears, Halliburton had to wonder if he had hired a "lunatic" to fulfill his present dream. Providing constant drama were Fat Kau's four wives. Belligerent and scolding, "the eldest wife," observed Halliburton, "tyrannized over the other three—and over Fat Kau too." He saw that "each wife had a special job: one cooked the rice, one sold the chips, one collected the money, and one (the youngest and prettiest) just sat around, languidly berating the workmen." The ludicrous distribution of job tasks teased a smile from Halliburton. "All four looked alike and dressed alike,' he said, ever more amused, "and as each carried a baby on her back, we had a hard time trying to tell which was which." Halliburton also observed that each woman saw herself as of a higher military rank than the others: "We were forever mistaking the general or the colonel for the lieutenant, and as the wives were intensely jealous of their rank, outraged feelings were the result."[4]

What Halliburton called "complications" started multiplying. As the workers appeared to be making progress of some kind, he at first said nothing. Each morning he put on overalls and brought a hammer with him to the shipyard, where he tried to be of help. He spent the afternoons going over "construction details"—that is, handing out unsolicited advice. At cross-purposes with his client, Fat Kau told him outright that he—that is, Halliburton himself, "did everything backwards," and "asked (him) to go out and play till his own carpenters had the job done." Captain Welch, finding Halliburton meddlesome, partly agreed.[5]

Despite hammer, chisel and saw, by November 16, much of the junk had taken form. The keel and the keelson had in fact been laid down and buckled, and the frames set up. Planks for the hull and interior had been cut and curved and were ready to piece together. Wooden ladders were set against the hull, and poles or planks acted as buttresses to cradle the ship. So far so good. Wisdom, however, dictated that iron bolts, not the wooden pegs the Chinese customarily used, were needed to keep the ship together during a nine-thousand-mile voyage across a "raging winter sea." The hand tool carpenters used to bore holes in the junk's many planks seemed to Halliburton a sort of repurposed weapon of war. He described the implement as "a gadget like an Indian bow, with the string wrapped a couple of times around the shaft of a sharp-tipped arrow which, when rotated, digs a hole." Halliburton groaned.[6]

A method deemed speedier troubled the workers. To their way of thinking, completing a one-week task in a single day would cost them six days' wages. The labor theory of value expressed here became only too clear to Halliburton—the craftsmen feared starvation if he rushed things. He had a ready grasp of the hourly wage, incentive pay, stipends, barter, and the slave wage. But he also knew that saving time was saving money. With reporters present, Halliburton donned a suit and tie and demonstrated the use of speed tools for the carpenters, applying an electric drill to an eight-by-eight timber. Later, Henry Von Fehren went to the shipyard and, astonishing the Chinese carpenters with the miracle of tool technology, drilled hole after hole into a row of timber. When Halliburton gently suggested they use the drill, the shipbuilders balked and again insisted that the tool would "ruin their livelihood even quicker than the sawmill." Halliburton gasped: if he subscribed to these crude economics, he would be in San Francisco in ten months rather than in ten weeks. His deadlines meant nothing to the workers or to Fat Kau, their boss, whom he called the "Buddha."[7]

Fed up, Halliburton stormed over to the "Buddha's" office "to complain, to threaten, to curse—as well as one can curse through an interpreter." His wives in attendance, Fat Kau, remaining "perfectly calm," invited Halliburton into his quarters, where he hoped he would be "more reasonable." Indignant, Halliburton told the shipwright he wanted a *finished* ship—and *soon*. As the two conversed, Fat Kau's eldest wife, "who bossed the other three," served an exotic dish of rice, shrimp, and duck. Once his guest was "besodden and subdued with food," Fat Kau proposed having a mill-sawn keel, but insisted that Halliburton "pay the bill, and pay wages to all his carpenters who would be deprived of their labor." Halliburton, "in desperate haste to get on with the job," acceded. By then fifty men were working on the ship, a labor force that one would think could accomplish a good deal and fast. Halliburton didn't quibble. He said he would trade seven days' pay for one day's work or, as he phrased it, "pay wages to all his carpenters who would be deprived of their labor." Although agreeable to Fat Kau, this solution confused the work crew. One crisis averted, another quickly arose. Compensation issues appeared resolved, as wages had been paid. But client Halliburton had neglected to throw a party, thus exhibiting "the worst possible manners." In retaliation, the workers thought "a sit-down strike would teach [Halliburton and his crew] the customs of the country." Why so? "Every self-respecting workman in China," explained

Halliburton, "expects the employer to give two big parties per job—one at the beginning and one when the work is finished."[8]

And what might it cost to learn these customs? Halliburton's interpreter figured that thirty Hong Kong dollars, or about nine American dollars, would cover "a whopping big blowout." He was told not to give the money to "Wife Number One," who would throw a twenty-dollar party and pocket the rest, but to give it to the "less grasping 'Wife Number Four.'" "Less grasping" meant that this wife would keep only five dollars for herself. Smart in "cajoling, intimidating and generally wearing down the (local) shopkeepers," Fat Kau's fourth wife was worth her cumshaw. She supplied the shipyard party with "barrels and baskets of Chinese food" at a good price and enough "rice wine" to "to float [the] *Sea Dragon.*" Also on hand were "girls and music." Soon everyone "was tipsy." They achieved an even higher elevation when opium was introduced. Said an astounded Halliburton, "Without opium, so I was told, no party was worth being invited to." Grinning uneasily, he watched from the sidelines as each of his "50 guests" smoked his full quota. The party lasted all night. As dawn approached, the exhausted workers with their spouses "climbed up onto the unfinished ribs and scaffolding of the *Sea Dragon* and went to sleep." Next day they "returned to the work with redoubled energy, and the pounding, sawing and nailing could be heard all over the neighborhood." Captain Welch thought the workers industrious. "They look slow as hell when they are working," he said, "but when the day is done and they all squat down to their rice and fish you can see that plenty has been done for that day." In a happy ending to the episode, the carpenters started to like the electric drill which at first had frightened them, and even competed to see who would use it next. Progress was made. Captain Welch could rejoice that they were "building a junk after all," and even Halliburton noticed some slight progress.[9]

For a time harmony reigned. It festered in Welch's mind, however, that the "construction details" Halliburton had discussed with Fat Kau— consisting mainly of talk about "the gorgeous color . . . on the stern," were pure nonsense. He did concede that Halliburton "sign(ed) the bills for the material," but otherwise paid him little attention. "All in all he leaves me alone," who busied himself mainly with the "lanyard rigging and iron rigging." Conferring first with Halliburton—probably about cost, he ordered canvas sails, presumably from a naval-supply merchandiser. Either he or Halliburton then ordered "Chinese (or rattan) sails." Resembling giant bat wings, these could be scrolled up and down *like venetian blinds*. Heavy and cumbersome, however, it was intended that these sails be substituted during the actual crossing for lighter canvas ones, and then be brought over in a liner to the Farallon Islands near San Francisco. Here they would meet up with the liner and replace the canvas sails and, given the authenticity of true junk sails, the *Sea Dragon* would proceed triumphantly into San Francisco Bay—and the fair. Next, the sixty-foot mainmast and the foremast, rising fifty-two feet, would be "smoothed off" and soon put in place. The main boom was forty feet long with a hoist of fifty feet. A poop would rise nearly as high from the deck. Next, the "outside planking" of the hull would be joined to the ribbing posts, and, by November 30, the bulkheads and deck beams be put in place. A credible vessel was beginning to emerge—a junk with a horseshoe-shaped stern and curving planks, the whole of it about seventy-five feet long and twenty feet wide. Installing the engine came next.[10]

Given steady work at last, Captain Welch seemed pleased. He "got along well with all the (work) crew" and even wanted "to take some pictures of them" before he sailed off. Of the routine he wrote, "Leave here (Kowloon Hotel) at eight in the morning, get to the yard at nine, leave there about six and get home in time for dinner." Out of bed early and knocking off late kept him jovial. Further lifting his spirits was the materializing hope of a fast exit from beleaguered China, "a lousy place" where he felt "stranded." Steady work eased the captain's troubled mind. He admitted he was "feeling pretty good these days" and that "the dead feeling (had) worn (away)." He had never worked on a junk, which to him did not seem to be a ship at all, much in the same way an ostrich might not seem to be a bird. Intercultural communication was not his strong suit; despite the presence of an interpreter, often his directives were not understood. Left to their own insane devices, the carpenters did okay. As far as he was concerned, everyone else in Hong Kong was supremely "nuts," in particular the harbor's boat dwellers, whose celebratory rituals were to him eerie and unearthly. "About every hour or so there is a blast of firecrackers from one of the junks," he said of their behavior. "Somebody died, a baby is born, a god to be propitiated for something. When they are hauled on the ways for cleaning, there are firecrackers for burning the weed off, more for launching. Joss sticks burn night and day at every slipway in the yard so that nothing will happen. The Chinese laugh, but they still do it." Welch found these "humorous moments." He liked having full command of operations in the shipyard, and he even found good things to say about the carpenters whom he found "smart as hell" and capable of teaching him "what to do with a piece of lumber." Yet he held doubts about the mission at hand, wanting to embrace it one moment and scuttle it the next. "I wish Halliburton would slip once," he said of his commitment. "I'd be home by plane!" Evidently only Halliburton stood between him and freedom. "I hate this place with all my heart," Welch ground his teeth, "and the thought of living here for another month or six weeks drives me nuts. They can have the Orient." Again, he wished that he had never left San Francisco.[11]

November 2, just when construction of the junk began, Halliburton predicted that he and his crew would "set out about January 1." In the appointed forty-five days, on December 16, the ship would be finished. Halliburton and the crew would be on their way, traveling via the east coast of Formosa to Midway, then to Honolulu, and then, "about March 10," arriving in San Francisco." Cynics said a crossing by junk couldn't be done; they thought it a "tough trip." Welch disagreed: "These things can sail like a bitch." Having test-ridden three such vessels his first week in port, he concluded, "They are swell" and "[They can] come about on a dime and cut the water like a yacht." He vowed to "push this cow [the *Sea Dragon*]" to its and his own full endurance. "With any decent break in wind at all," he added, "we should be in Midway by the end of January; after that I shall push the guts out of her because she is new, as heavily built as ships twice her size and will stand it." The thrill of added money, or what Welch called his "pigeon" business, also counted. "That is all that matters to me." *That* and all the intrigue and excitement of making shady deals: "I make all the contracts," he said, "get them signed and arrange for the payments to be made—after a little private talk with *the other side of the fence*." Evidently, Welch's contract with Halliburton did not forbid him from conducting other business. For investment capital he might have counted on the Halliburton Chinese Junk Corporation, or such petty cash as Halliburton himself carried. Welch had bilked the adventurer in the past—why stop? Customarily Halliburton carried large sums of cash, including a $1,000 bill on this trip. Welch wanted that bill, which amounted to four times his own monthly salary, but he had competition. "I never saw so many wolves in all my life on Hallidear's trail as soon as she showed [that bill] in the offices uptown. Now all is well. I outsmarted them all. I have her for mine." How the "outsmarting" was accomplished is not known, but purchasing needed supplies cheaply and selling them dearly is a reasonable guess.[12]

23

Flight of the Bumblebee

Once he saw that side money could be made from the *Sea Dragon* enterprise, Captain Welch strengthened his commitment to it. "I know that every moment I put into it brings me nearer to sailing day," he said. "I have never been so interested in a thing in my life as I am in this junk." Wearying of the daily ferry ride to the shipyard, he and the others quit the Kowloon Hotel on November 5. and moved to "Hong Kong proper." Where the others dropped anchor is uncertain, but Welch dropped his at the shipyard. The change of location, however, brought him no closer to home. "Drop Me a line like a good guy," he asked his friend Dick Wetjen. "I need a letter so badly. The mails are so few and far between that it hurts to wait for a letter. I'll keep you posted on events here. Some of them are very amusing but those that keep me here are almost tragic to me." While Halliburton's faith in Captain Welch tarnished, Captain Welch's faith in Halliburton reached new lows. Both in his jealousy and contempt for him, Welch often bristled. Halliburton knew at least vaguely of Welch's shady dealings. Yet Welch, knowing that Halliburton was falsely proud, hardly cared. When opposed, Halliburton liked to remind his tormenter, "I *am* Richard Halliburton," uttering the name like a king legislating from his throne. The design features of the junk lay at the heart of Welch and Halliburton's brewing cold war. Originally, Halliburton was to room on the starboard side with Welch and George Petrich, while Paul Mooney joined "the common herd in the main cabin," but insisted on having what Captain Welch called a "special cabin for herself and her secretary," one with the sybaritic luxury of "a French bed." Given for his own use two small cabins on the poop, Welch still snarled: "I am doing it for her" because "she needs company at night."[1]

Although the Chinese did "everything backwards and everything slow," Halliburton conceded that "boats (were) a new world" for him and that he had "much to learn." He welcomed good news, agreeing with "local people" who called his and the carpenters' "progress astonishing." By November 24 the "entire skeleton of the junk" was nearly finished. "Unless we have some major set-back," he said, "I still plan to leave Hong Kong January 1." He wanted to leave sooner. The next day Halliburton learned that the *Pang Jin*, a rival junk, had begun its voyage westward to the New York World's Fair. As work on the *Sea Dragon* continued, he learned that the *Hummel Hummel* (roughly, *Bumblebee*), captained by one E. Allen Petersen, had completed its trans-Pacific voyage. "The very day my shipmates and I sailed from San Francisco," Halliburton said, "a 35-foot junk named the *Hummel Hummel* sailed into San Pedro. With an American captain, his Japanese (American) wife (Tani), and two young Russian sailors, it had crossed the Pacific from Shanghai." It stood to reason that, "If they could cross the ocean in a 35-foot boat, surely we would be able to cross in one twice as big." While Paul Mooney thought the competing voyage "stole our thunder," Halliburton openly "rejoiced in [the rival ship's] success," hailing the crossing as "the most incredible of all." Why feel upstaged? After all, the *Hummel Hummel*'s aim was *any* landing in the Western Hemisphere, not the Golden Gate International Exposition. Moreover, the *Hummel Hummel* had crossed the Pacific in midsummer, not in midwinter as the *Sea Dragon* would.[2]

Welch couldn't understand the "fuss" over "the *Hummel* thing." Junks can cross a large expanse of water, so why be surprised if one does? The voyage nevertheless illuminated any attempted crossing of the Pacific by junk. Of course, Dr. Petersen was not sailing to the fair in San Francisco, but, as others, was escaping a battle field, as Japanese invaders had forced much of the foreign population of Shanghai to flee by ship, often to Hong Kong. Unable to board a commercial liner (for unstated reasons) and believing that Hong Kong had become too over-crowded and expensive, Petersen decided to take his chances with the sea. He purchased a three-year-old thirty-six-foot river junk from a Mr. Emmerman for $250. Subsequently, he filled the ship's narrow hull with plenty of food, a "new primus stove," a portable pressurized kerosene burner, a pistol, a rifle, and a sextant he barely knew how to use. All looked well. At the moment "unbeknown to Richard," the *Hummel Hummel*, Petersen's "dream ship," began its "great adventure" from Shanghai on April 15, 1938. After a harrowing trip down the murky, swirling currents of the Yangtze River, Petersen and crew sailed into a violent storm During the ordeal, he had an additional run-in with a cutthroat band of junkmen "with shaved heads and gold rings in their ear lobes," and narrowly escaped bombardment by Japanese warships. Steered by the inexperienced Dr. Petersen, the badly leaking junk (all junks are prone to leaks) drifted aimlessly for miles. Led to port by some good maritime Samaritans who appeared so cutthroat that Petersen at first thought they were bandits, the *Hummel Hummel* eventually hobbled into Shanghai.[3]

Recaulked and refitted, the *Hummel Hummel* started off for Yokohama. On its perilous way, the ship labored and at times bobbed about "like a cork." Thirty-three days out, in late May, Petersen reached Japan, where he docked at a remote lighthouse port near Yokohama, his ship needing vital repairs. Arrested as spies by Japanese customs officials, the sailors were interrogated to exhaustion. The *Hummel Hummel*, meanwhile, was ransacked from "bow to stern." Passports, books, and letters were confiscated and brought to an inspector who spoke English ably enough to inquire about what unfamiliar expressions the documents contained. Next, the Japanese secret police again detained the sailors, this time for five weeks. Tani, who was Japanese and could converse with her interrogators, endured the "longest" period of questioning. The Russian crew members, twenty-five-year-old Nick Perminoff and twenty-one-year-old Victor Ermoloff, endured the shortest one. They were released without further ado. Funds from Los Angeles arrived, meanwhile, to help prepare the *Hummel Hummel* for its long voyage. The junk, returned to them, was again recaulked and given upgrades. Its bottom seams were stripped with zinc and its exterior painted black. A ton of scrap iron and half a ton of rocks were dumped into the hull for ballast. Besides five hundred gallons of fresh water, "three cases of eggs," sacks of onions, potatoes, and garlic, and numerous canned goods were loaded on board.[4]

On July 12, 1938, while Halliburton was in San Francisco and complaining about the red tape delaying his own trip, the *Hummel Hummel* sailed off on a voyage that became famous. The ship was soon greeted by a gray dawn breaking over a sullen sea. Ahead lay five thousand miles of ocean. Throughout July, Petersen and the crew advanced nearly sixty miles a day, and after forty-five days at sea, they were halfway home at the end of August. Compass or no compass, they pushed ever eastward along "the great circle sailing route." As it had on its way to Yokohama, the junk labored. Storms and thrilling rides over high-cresting waves came and went and became routine, but occasional sightings of a sea creature—a whale, a school of dolphins, or a seal—offered delightful distraction. Porpoises were most often sighted; Petersen dubbed these "happiest" of "all earth's creatures" the "kings of the ocean." When land was finally sighted, captain and crew could wearily rejoice. A bearded Captain Petersen told the press corps covering the golden moment, "One thought kept uppermost in my mind: we had crossed the North Pacific in a thirty-six-foot junk. They called it suicide in Shanghai and Yokohama but we had proved that man can rise above the threat of death: that defeat can be turned into victory, if the will is there to go ahead." *Life* magazine's account of the adventure provided additional suspense: "Bobb(ing) into San Pedro, California harbor 85 days out of Yokohama, Japan . . . the crew had a nightmare voyage. Drawing only 2 ½ feet," (the vessel) "bounced like a cork over the waves, fog lasted for a whole month and once a party of whales swam along with the boat, almost knocking it over." Crew member Nick Perminoff would later say, "Sailing on this boat is like walking on the water, you are so close to it." For Captain Petersen, the experience was not always so implicitly religious, but rather a torturous deterrent to a sound sleep. Sleep-deprived and undernourished, the ship's passengers were also "groggy, wet, [and] miserable," just as they had been before the voyage started.[5]

Like Dr. Petersen, Halliburton supposed that hands-on experience would hone his limited navigational skills. During his voyage Petersen remarked, "My navigation was improving." In time, he would call himself an "experienced junk sailor." That much Halliburton understood about his 'rival.' He seemed little interested, however, in the design features and seamanship that enabled the *Hummel Hummel* to accomplish its mission. Besides some superficial similarities between Halliburton and Petersen, closer similarities existed between their respective vessels. To its advantage, the *Hummel Hummel* was small and streamlined, and it lacked a high poop. All these qualities guaranteed it better ocean-going success than Halliburton's own gaudy monstrosity. While the *Sea Dragon* was made of several wood varieties, the *Hummel Hummel* was "built entirely of camphor wood" and "heavily constructed." At the time of its purchase, Dr. Petersen had not inspected the ship's bottom, but when a low tide revealed the badly bruised planking, he quickly saw to its reparation. It is not noted whether wooden pegs, iron bolts, or cording bound the ship's planks. By comparison, the *Sea Dragon* was *ultimately* held together by iron bolts. While the *Sea Dragon* curved toward its keel or spine, the *Hummel Hummel* "was practically flat-bottomed with but a six-inch keel." The measured thickness of the *Sea Dragon's* exterior planking is unknown, but known is that the "smooth bottom planking (of the *Hummel Hummel*) was an inch and a half thick." Also, "starting from just below the water-line, hand-hewn-logs seven inches thick made up the sides." "Serving as horizontal ribs, these (same) timbers gave the junk tremendous strength." At last, five three-inch-thick wooden bulkheads divided the interior"; these "proved to be an important safety factor on more than one occasion." Halliburton put safety last when he dealt with his own ship's bulkheads.[6]

The *Hummel Hummel*'s Russian crew members, Nick Perminoff and Victor Ermoloff, underplayed the horrors of the voyage—what Halliburton saw as "reeling decks, cold beans for supper, and sea-drenched turns at the tiller." The two men were "White Russians," down-and-out sailors who had fled from injustice in their native country. They asked for no pay other than sanctuary in America. Down on their luck again, the two sent word to Halliburton that they had enjoyed the crossing and would like to do it again. "You are a fool not to climb Fujiyama once in your lifetime," Halliburton replied, quoting a Japanese proverb, "but you are a worse fool to climb it twice!" Yet he would have second thoughts. "The time may come when I shall regret having refused the offer of these veteran junkmen," he said, "for we are not altogether an expert crew." Soon after, Nick and Vic disappear from the pages of maritime history.[7]

Following word of the *Hummel-Hummel's* success, Halliburton next received word that another junk was "sailing from Wenchow up the coast." He also heard a rumor that the ship's captain had made "some deal with the fair in case [the *Sea Dragon* arrived] too late to participate." The news may have forced Halliburton to act in haste and bypass Honolulu to make up for "lost time." Although he did not mention to Captain Welch a race or prize given to the winner, he did tell him that, should the *Sea Dragon* make an appearance at both the Golden Gate International Exposition and Sacramento State Fair, financial opportunity awaited.[8]

Other potential competition? Even before he left San Francisco on September 28, Halliburton heard that "another big junk (was) leaving China October 15 for San Francisco." As he had the diesel engine, the news of the ship's two-month "head-start" did not disquiet him; so high was his initial optimism, *on September 28*. The *Tai Ping*, another junk making the trans-Pacific voyage, *did* have a diesel engine. Skippered by merchant seaman and retired Yangtse River pilot John Anderson, the forty-one-foot three-masted ship was manned by crew including Anderson's wife, Nellie, and four Europeans (a German, a Dane, and two Norwegians). The junk would sail off from Shanghai on April 7, which was later than planned. Both the *Tai Ping* and the *Sea Dragon* were bound for Treasure Island. Only a couple hundred miles into open waters, a worsening storm forced the *Tai Ping*'s retreat to Yokohama, where it was overhauled. Recovered from its wounds, it embarked from Kobe on April 22. Anderson later claimed he had spotted the *Sea Dragon* enter that same storm, telling Commander G. C. Jones of *Coast Geodetic*, "[We trailed Halliburton] until we were separated by a typhoon." But if Anderson commenced his first trans-Pacific attempt on April 7, a couple weeks *after* the *Sea Dragon's* last radio message, it must be concluded that he was mistaken.[9]

Earlier, still another junk, the *Adventure,* had departed from Shanghai on November 12 headed for San Francisco. But a violent storm in the Formosa (Taiwan) Strait near the Pescadores or Penghu Islands just west of Formosa crippled the ship. "The other junk is a total wreck," Welch said. "The gang went back to Shanghai with everything lost. Tough luck." Remarked Halliburton: "[The captain's] unfortunate misadventure may cost him a month, and give me ample time (with my engine) to get there first." Halliburton felt relieved: "For him to have arrived just before me," he said, "would have been disastrous." Welch made no mention of a "race," of losing the concession or of receiving a cash reward, but Halliburton's lack of remorse for fellow sailors nearly lost at sea bitterly angered him. "Halliburton cheered," Welch said. "The bastard. I mentioned that it might happen to us also." The captain of the *Sea Dragon* supposed that the rival junk, whether it had a diesel engine or not, had met hardship because of its unprepared crew—would-be sailors like the ones he himself captained. "I met one of (the crew members) in Shanghai when I was there," Welch remarked, "and he was a fine specimen to take a trip like that." The captain facetiously wondered "why . . . the adventurers all look alike"—meaning preppy and pampered. "I can pick them out a mile away," he added: "There is a grand point of psychology for you if you look." In the end, race or no race, the Industrial Revolution had introduced modern technology and the aforementioned deadline—a work-incentive concept that even Fat Kau was learning. Ahead lay "the fury of the open sea," as Dr. Petersen recalled, and "wind, the rain, (and) the white-crested waves."[10]

24

Pandas and Other Distractions

The success of the *Hummel Hummel* only fueled Halliburton's passion to complete his mission. November: Hong Kong, "a city on the edge of the tropics," was generally bright, and November was "still a hot and sunny month." The tramway or funicular carrying passengers from the harbor communities to Victoria Peak—home to the governor's residence, a cathedral, and a women's and children's hospital, seemed built for reflection. Walkways curled about into gardens and near palatial homes, making the war seem remote to visitors. Located half way up the peak on 41A Conduit Road was the Foreign Correspondents Club. Here Halliburton and Mooney could relax on the portico overlooking Victoria Harbor, or hobnob with members of the press corps reporting the war in China. Alternative topics of conversation were Halliburton and his junk. Shielded by a newspaper or his back turned, Halliburton overheard more people talk about his chances for failure than for success. Needing better-informed odds, Halliburton, or so it seems, met at Kai Tak airport with pilots from the Clipper ships and captains at the docks. Often accompanied by Mooney, he also visited an orphanage in Kowloon run by a woman from New York. Writing also occupied him: "In nearby waters," he summarized recent days, "pirates are a very real and constant menace," and wintry storms "may play us cruel tricks."

Halliburton further stated, "[All is] anti-climactic compared with the state of perpetual crisis which we have endured while building our Chinese junk with Chinese labor." He courted adversity but wed damnation. For anyone wishing to be "driven rapidly and absolutely mad," he recommended "building a Chinese junk in a Chinese shipyard during a war with Japan."[1]

When not mimeographing or writing, Mooney visited the open-air curio shops, hiked off into the outlying regions with his camera, or rode the funicular to the top of Victoria Peak. Like Halliburton, he enjoyed heights and grand views. Although building the *Sea Dragon* was Captain Welch's chief occupation, that task seemed of only incidental interest to Mooney. Welch knew that the Japanese were sinking ships off port, raping women on the mainland, and dropping bombs everywhere, but he paid less attention to threats outside the shipyard than did Mooney. Nearly every day the crumple of artillery could be heard, and nearly every evening rockets illuminated the night sky. Although they landed a distant forty miles away, the bombs seemed right outside the gate. A week earlier, Japanese aircraft had flown near Hong Kong's border only eighteen miles away. Kodak in hand, Mooney hastened to the scene and saw "refugee peasants and merchants by the thousands" scurrying to the safety of British-held territory while the well-armed Japanese troops chased the pitifully armed Chinese troops into the hills." Most of these Chinese forces, dropping their weapons, would soon be "interned on a ship in Hong Kong harbor." Japanese troops were nearing the Colony, their occasional bombs now landing in British territory. They had even captured a British military post. Mooney tried to get a better look, but "barriers of barbed wire" blocked his passage. Still, he reached the British half of a place called Sha Tau Kok, "right up to the border barricade." From there he could see that the "Chinese half of the city was boarded and abandoned." He returned to Kowloon that night.[2]

Mooney's next act of mischief was a trip to Canton. He wanted a story, but, a lapsed Catholic, he may have been prompted by religious impulse. Famed sixteenth-century missionary Saint Francis Xavier set out for Canton after being expelled from Japan and establishing an underground Christian community. He died suddenly, however, before reaching the Chinese mainland. Saint Francis Xavier's letters about China are said to be the best account of the country since Marco Polo's, written two and a half centuries before. Whether Mooney read the missionary's account remains unknown. Reaching Canton was, in all events, dangerous, and the Japanese presence severely restricted access into the city. While thousands of refugees were fleeing south into British territory, "in the opposite direction, truckloads of English soldiers pushed their way through the mass of children and carts and bullocks being pulled along by the fleeing farmers." The British detained and disarmed battalions of Chinese soldiers who wanted to get back into the fight. "For three weeks I've been trying to get just seventy miles up the river to Canton," Mooney wrote, "and there is still no hope of getting there." He noted that "a few war vessels (had) run the blockade," but the spectacle was "not for the benefit of tourists"—or reporters. Also, the train that started off near Mooney's hotel traveled only thirty miles before a dynamited bridge blocked its progress. With Japanese pressures mounting, Hong Kong would soon be put "in a hopelessly difficult position."[3]

Hong Kong now conducted its first practice air raid drill. In pitch darkness, sirens sounded and planes blinked their lights to signal dropped bombs. Citizens lit candles and hid behind drawn blinds. During the two hours of blackout, Mooney downed a few drinks at a corner pub. When the lights returned and rickshaws "creaked through the streets again," he started reminiscing, mostly about his residence at 14 Gay Street in New York's Greenwich Village and the friends both he and Harriette Janssen had known. Even as he spoke, the Japanese stranglehold on the Colony tightened. Blackouts and drills soon became commonplace. As Francis Scott Key had watched the "rockets' red glare" over Fort McHenry, Halliburton, with Mooney alongside him, went to the roof of his hotel in Kowloon to watch "the blaze of lights" looming over the harbor. He described "the scenic effect of these blackouts" as "weirdly beautiful" or even apocalyptic, with "utter darkness descending." The Golden Gate International Exposition could hardly compete with such pyrotechnics. Halliburton (or Mooney writing for him) elaborated on the blackout experience: "While a million and a half people creep lightless through the streets, sit in unlighted houses, feel their way in boats lightly across the harbor, a score of British army planes rise from the airport and begin to roar overhead in military formation. Instantly, from the heights which hedge Hong Kong on every side, great batteries of powerful searchlights go into action, lighting the clouds with their blazing rays. They find the 'attacking' planes, and hang on as the planes dive and dodge to escape their implacable beams. The anti-aircraft guns bark in realistic fashion. One by one the planes signal, by dropping flares, that they have been 'brought down.'" As the Second Coming was said to be preceded by a day of darkness, Mooney could rejoice: "So Hong Kong is saved—and a million lights, in one blinding flash, come on."[4]

Driven to distraction, Halliburton could be found most often at the Hong Kong Hotel. Guests arrived there from all over the world, and it was possible to run into anyone. Thus it might have only been a coincidence that spy coordinator and fellow Bobbs-Merrill author Herbert Yardley came to the hotel in November 1938. By his own account, Yardley sat in the establishment with his interpreter (one Ling Fan), "drank Scotch," and "watched the beautiful and well-dressed Chinese women in their long, colored silk gowns, some slit above the knee." Outside the hotel, as the sedan chairs scurried past with their fares, he saw "an English officer and a blonde in a low-cut evening gown" step from a taxi. Like Yardley, the brooding Halliburton liked to 'people-watch.' Seldom was he alone. According to one report, he sat hour after hour in the hotel lounge "in the company of an eighteen year-old boy, a tall and gangly youngster with sleek black hair, an olive complexion and exceptionally slender, tapered hands." While they sipped grenadine and soda, Halliburton seemed "preoccupied and bored," the boy indifferent. At first Halliburton thought "the dark-skinned Latin" might be named "Miguel or Manuel." The adventurer's understanding was that the "boy," whose name was Patrick Kelly, "had been born in Canton of an American father and a Portuguese mother." Kelly owned an American passport but had never seen America. After "warning him of the terrors ahead," Halliburton invited him aboard the *Sea Dragon*. As Welch had a Burmese messboy, why shouldn't he have a Portuguese messboy?[5]

Others, besides Kelly, were offered a chance to see America. Among them were some furry and cuddly candidates. As Paul Mooney told the story, a friend of his and Richard's, said only to be a West Point graduate and an adjuvant with Chiang Kai-Shek's army, had brought a couple baby pandas to the Colony from the bamboo forests of Northwest China, near Tibet. Recounted Mooney, "They travel with three nurses, and native boys, who sleep in their cages and comb their long fluffy hair all day long and feed them condensed milk in punctured cans. I spent a whole day playing with them while they were in Hong Kong waiting shipment to England. One is a baby; another is pregnant, so there is no way really to say how many pandas I did play with." Mooney soon became their proud papa: "They are sleepy things," he said, "but (are) full of fun when you get them roused—say, by bribing them to keep their eyes open: sugar-cane and Eagle Brand are the best bribes. When they become annoyed with each other, they bark fiercely and show their teeth. They lift their paws to strike: then the paws halt, waver, droop again. The fight is over—too much trouble! Big and strong enough to do damage, with claws like needles, they only sometimes forget to be gentle; then, when you yell at them roughly, they sit back and look sorrowful. The first time someone asks you to a panda-party—*go*." Within a confined army barracks, a corps of local militia guarded the pandas. At an appointed hour each day, the militiamen led the bears to a play area where Halliburton and Mooney socialized with them. Once he had lavished his affections on one "amiable giant-panda," Halliburton passed him (or her) over to Mooney. The other cubs had to wait their turn.[6]

While Halliburton's frolics with animals were well-known from his books, Mooney's were not. Yet his love of animals had a long history. When a boy growing up in Washington, DC, he liked to bring home stray animals he had found—once a baby alligator, often a cat or turtle. He could hardly contain his enthusiasm for the pandas, and snapped multiple photographs of them at play. Overjoyed being with "the only pandas in captivity except for one in America," and little minding their tendency to salivate and drool. he gently cradled, stroked and hugged them. As late as the 1930s many believed pandas were purely mythological creatures in the phylum Chimera or Hydra. They were products of a fertile imagination and not of this verdant Earth. While the pandas should not have been taken so drastically from their element, public curiosity about them had to be fed. The panda mania of the late 1920s was as great as the King Tut mania of a few years earlier, and, before 1930, only an estimated four "white" people were known to have seen a giant panda.[7]

Given "a chance to purchase" his own cub and make it "a prospective mascot for the *Sea Dragon*," Halliburton weighed seriously the pros and cons of actually bringing one or two back to America. A pair of pandas might make the perfect gift to President Roosevelt to symbolize his watch over China. Exotic animals had served as gifts to potentates since the days of Solomon and Sheba. In 1927, the mayor of Johannesburg,

South Africa, had presented two lion cubs to President Calvin Coolidge. Although he took an instant fancy to them, Mr. Coolidge concluded that the Washington Zoo would be a better home for them than the White House. Unlike lions and giraffes, pandas languished in captivity. In the wilds, while a giant panda was rarely seen, hunters had acquired dead specimens—that is, before Manhattan dressmaker and socialite Ruth Harkness came along and brought back two very live ones.[8]

In her 1938 book *The Lady and the Panda*, thirty-seven-year-old Harkness provided harrowing detail about assuming the life's work of her late husband, noted explorer Quentin Young. Harkness at last became a mother to an apparently abandoned bear found nestled in the crotch of a tree. The panda was so delicately small that it could be cupped in one's hands. Brought to America in 1937 from China's mountainous Szechuan Province, the panda was heralded as the greatest discovery since Stanley found Livingston. Harkness faced many dangers in her rescue mission. A strange woman in a strange land, she was surrounded by clashing armies and roaming bandits. Determined, even possessed, she and her guides ventured far into the jungle, past many random deadly booby traps and the grim, tortured faces of war casualties. If lucky in her first mission to rescue the panda Su Lin, she was a little less lucky in her second mission to rescue Mei Mei. The horrors of war made her ill. During one leg of her journey back to safety, the sight of so many mutilated and rotting bodies drove Harkness into an alley where she vomited. She had glimpsed into the abyss but escaped before it looked back at her.[9]

Calling their plight "pitiful," Mooney continued to photograph the refugees pouring into and out of Hong Kong. He snapped photographs of the *Sea Dragon* on the same roll of film, thinking of the ship's impending launch and his own flight from the city. He sent everyone the best Christmas wishes. Letters provided Mooney "a great deal of pleasure," but he feared he would not receive another one until he and the crew reached Honolulu "around the beginning of March." To quell his internal struggles, he had begun reading poet William Blake again, including his *Prophetic Books* and *Songs of Innocence*, particularly the poem about "the little boy lost in the lonely fen, led by the wandering light." The words made him wonder why he had come on this improbable journey. "Halliburton, and eight other Americans and I," he concluded, "are all about to die for the sake of another *Royal Road*." Death by Japanese gunfire or torpedoes was one concern; a more remote one was death by shipwreck, a fate that had befallen his namesake Saint Paul. Theatrics and Hamlet-like soliloquies often colored Mooney's discourse, notable especially in the summation of his life in a letter to Harriette Janssen: "I was born in November, 1904," he began, then continued his story in this manner: "Not until 1916 was I expelled from school. In 1918 the world war ended and my voice changed. Scarcely six years passed before I found myself hutched in Greenwich Village. Very few stores would give me credit. In 1928, in France, I learned that the French merchant has much the same mentality; it was there I took seriously to drink. In 1930, I died in Hollywood. By 1937, I had managed to erect a magnificent mausoleum at Laguna Beach. But due to terrific upkeep on the Tomb, it proved necessary for me to Walk, toward the fall of 1938. So here I am in Hong Kong. Still haunting the world, and materializing every so often to you. Ectoplasmic egotist, I!" As far as is known, Janssen didn't write him back.[10]

Mooney's expected departure time conformed to that of Captain Welch, who told Wetjen on November 25, "We'll be clear of here between Xmas and New Year," sailing "south of Formosa and across to Midway keeping north of thirty." A course was set: "Should make the run to Midway in fifty days, if all goes well. It is forty five hundred miles. I anticipate plenty of gales from here to Midway and from Honolulu to San Francisco it will be toughest. I remember that last winter too well to have any illusions." Welch kept his hopes high. On December 8, the *Coolidge's* Dale Collins, seeing the junk, supposed it three-quarters completed; when he returned from Manila a week later, he supposed it fully completed. An over-anxious Captain Welch hadn't set a "sailing date," but on November 30 vowed that he would "be out of here by the 24th of December if all goes well." Despite their having a monetary value of over $20,000, the pandas, however, would stay. Prone to seasickness, they could die on so long and bouncy a trip. As for Su Lin and Mei Mei, they succeeded in making the crossing to America on a big ship, but both would die in captivity. Adventure in her soul, Harkness herself would have another harrowing adventure, this time into the Pangoan jungle of Peru.[11]

25

Big Men on Campus Join the Fraternity

From late November into early December, the weather in Hong Kong remained sunny and warm. Under clear skies, the *Sea Dragon* had taken form slowly, almost imperceptibly. Despite setbacks—dismissed basically as 'normal abnormalities' by Halliburton, November proved a productive month. The estimated arrival time in San Francisco was still set for mid-March, and that was good. A letter from Halliburton's publicist Wilfred Crowell informed him that the "lads"—collegians John Potter, Gordon Torrey, Robert Chase, and George Barstow—were on their way and had arrived in San Francisco in a "blaze of publicity." In his letter, Crowell enclosed a clipping from the October 20 edition of the *San Francisco News*. The news item noted that the lads were about to go off to the Orient to become members of Richard Halliburton's *Sea Dragon* crew. The accompanying photograph showed them clean-cut and in suits while they sipped drinks at the top-floor lounge of the deluxe Mark Hopkins Hotel, with its startling views of Treasure Island and the many ships bound for the Orient. All the young men appeared game for adventure. Crowell assessed the lads as "fine big boys, very good looking and quite strong." He evidently took them on a tour of Treasure Island, and possibly to the China Clipper Cocktail Lounge and Café near the *Sea Dragon* berth, where they were showered with further media attention. Besides high-end accommodations, the lads were given a three-day travel package valued at $250 which included transportation to and from the fair and admission to its grounds. According to Crowell, they "had a great time in San Francisco" and presumably at the fair.[1]

When interviewed, the foursome seemed chipper and gung-ho. All admitted that the work would be physically demanding and that three months of confinement on a tiny ship with a dozen people would be "hard on the nerves." But the young men believed these "hardships" would give them "stamina." Sight unseen, Captain Welch called the lads "enthusiastic hot shots." Yet they saw themselves as "adventurers" wrought from the same action-hero mold as glorious adventurer Richard Halliburton. Sport and the sheer joy of being associated with a celebrity travel writer brought them west. John Potter spoke for all of them when he told reporters, who thought the foursome "blasé," that he wanted to "purge (himself) of wanderlust." He had read one or more of Halliburton's books, or knew of his reputation as an adventurer. Like the other lads, he had a high pedigree; indeed, Potter, Chase and Barstow in particular were progeny of the old guard of New England's financial elite. All the collegians were "able-bodied seamen" in theory. Still, they were but "the greenest of green crews" paying a hefty tuition to earn their seaman's stripes.[2]

Potter would serve as the junk's "first mate," which meant he would superintend operations above and below deck, answer to the captain, and take command in his absence. Torrey, at first the ship's cook, would instead be appointed "second mate." Shipping and international petroleum consulting were in his future, but at this junction in his life he had only a dim idea as to a career calling. Chase would serve as a general purpose deckhand and lookout. All the lads would serve as gaffers to untie tangles in the rigging and act as grips to pull ropes. As for Juilliard School of Music student George Barstow III, whom Crowell singled out as "the youngest and nicest" of the four, his part had to be written; for the time being it was probably enough that he boost crew morale with his accordion. In all, the lads were a composite of the Halliburton persona, weakly prepared for the worst, amply prepared for the best, alternately naive and brave.[3]

After reading the news feature Crowell had had sent Halliburton, Welch, shaking his head and turning livid, growled at the very sight of these white males born to privilege seated smugly at the Hopkins;

to him, they appeared to be vacationers, not toilers of the sea. Softies who never did a hard day's work in their pampered lives, they would require fresh linens daily and likely expect room service as well. In truth, the lads hadn't the foggiest notion how long it would be before they again slept on a soft bed, how long before they feasted on sirloin and ice cream, saw dry land, bathed in hot water, or conversed with a lovely lady. Welch had hoped that he would not have to train anyone, but instead would command a crew that could hit the ground running—crack troops, not raw recruits. Instead, he had to get a bunch of wet-behind-the-ears dandies in shipshape in a hurry. The *Coolidge's* Dale Collins noted that Welch was the only man aboard the *Sea Dragon* who had "previous practical experience in sailing ships." Collins added, "The handling of a large three-masted junk in heavy seas should certainly require a thorough knowledge of seamanship from all hands as well as the master." He acknowledged that the lads were "a fine group of young men," but added, "[The] fact still remains that it certainly takes more than 6 days, or 6 weeks, to train landsmen into competent deep-water sailormen." School-ships existed that did train and graduate able-bodied seamen. Associated with the YMCA, the Seaman's Institute in Hong Kong itself apparently offered practical advice on ship-handling. In his *The Making of a Sailor*, mariner Alan Villiers offered novitiates how-to suggestions accompanied by photographs of "cadets converted into crew and land-lubbers into worthy exponents of the seaman's art." Guides like Richard Henry Dana's *The Seaman's Friend* were also available. Although published in 1863, Dana's work offered still-relevant instruction on setting sails and tying knots. Besides define the various roles and duties of an assembled crew, it provided a glossary of nautical terms and a section on merchant-service customs, usages, and laws, which subsequent editions expanded. So what if there were teaching tools around? Welch didn't want *his* ship to be a "school-ship." Nor did he want to be an instructor. Even so, he would "keep (the) crew going all day at different jobs such as mooring lines, gun frames . . . varnishing the new masts and other things to be ready." Welch would show the lads how to work the rigging and handle the steering, but it remained to be seen how well they learned all these skills in so short a time.[4]

When Halliburton showed Welch the letter he had received from Crowell, the captain was instantly livid. While the lads were "all fine boys, very good looking and quite strong," Barstow, however, would need some "mothering," recommended Halliburton. Welch, shrugging angrily, insisted that the lads would get no special treatment from him. Halliburton, however, would not let the matter rest. Behind closed doors, the captain raged. "I understand I am *to mother* (the Juilliard lad) for a few months until he learns the ropes," he told Wetjen. "Who is to take the wheel while he is being mothered? I know goddamn well I am not." Welch himself had not been a bleating babe gently introduced to a life at sea: "I got plenty (of mothering) when I was 18—(that) I can tell you!" Later Halliburton told Welch that *all* the lads, not just Barstow, needed "mothering"—instating the dictum that 'boys went out, men came home.' Shocked and disgusted, Welch, thinking their being over twenty made them adults and beyond a mother's care, threw his arms in the air. There were additional concerns. "All very big chaps," they looked able and hearty, but would they come to work or to play? Would they consider the *Sea Dragon* a floating frat house or a hard-knocks school of life? Halliburton called the collegians "green hands who are amiably planning to mutiny if the work bores them." On November 15, two weeks before he even met them, Welch vowed, "I'll mow the bastards down if they get gay with me"—that is, if they got chummy with him or misbehaved. "I'll heave the log book at them as soon as (we) get to sea," he grit his teeth. "I made up the plans for the junk a week ago," he said then. Indeed, what plans—and how did they relate to the lads?[5]

November 1 or 2, when Halliburton learned "the four boys were enroute," the construction of the *Sea Dragon* had barely gotten underway. He spent the following days nervously hoping Potter, Torrey, Chase, and Barstow would linger in San Francisco, then Hawaii, and finally Japan before coming to Hong Kong. He wanted to show off the ship in all its glory. If they arrived "too soon," Halliburton thought, "they would be another responsibility." At the time, his intention was to "rig and train [the] crew" over a two-week period starting December 15, or the day the ship was completed. He would then depart on January 1. Although they tarried, it appeared the lads would arrive just after November 21— almost *too early*. A passenger freighter, the *Matsonia,* took the team to Honolulu. Color films taken by Robert Chase or John Potter show the two vacationing at Hawaii, where they spent nearly three weeks. They told reporters in San Francisco that their next stop would be Kobe, Japan; from there, they would board another freighter to Hong Kong, estimated arrival time December 1. Their plans must have changed. Welch assumed that they would all arrive together and by Clipper, not by plane. Torrey, a guest of Ambassador Grew, lingered in Tokyo. He also mapped in his mind the waterways pertinent to the *Sea Dragon's* likely path out to sea. Years later, he wrote that he had come from Shanghai to Hong Kong in "a passenger liner," observing firsthand the "kind of seas" one might encounter "in the Formosa Straits." Halliburton noted that Torrey came down from Japan. He might have spent some time in Manila, perhaps to job hunt. Then, boarding the *Philippine Clipper* on Friday, November 25, he arrived in Hong Kong the next morning. Chase and Potter, whose films captured Hong Kong as their plane descended, arrived together.[6]

Earliest arrival George Barstow had come to Hong Kong from Yokohama or Kobe aboard the giant *Empress of Japan*. Captain Welch noted the day, adding, "The others come in a few days by the next clipper. They stayed in Manila to *look see*." Welch also noted that one of the lads, identified only as a "seaman" but probably George Barstow, had shown up in Hong Kong as early as November 20, then sped off to Singapore to "look it over." The young man said he would be gone three weeks, but Captain Welch hoped his absence was permanent. Barstow wore a sailor's cap plopped somewhat pell-mell over his thin, elfin head and a gooney grin on his pimply face. Pressed and new, his midshipman pantaloons looked borrowed from a Hollywood costume studio. Captain Welch folded his arms and winced. Was this crinkly-billed shorebird even remotely seaworthy? "Barstow is a mess," Welch wrote. "The first thing he ask(s) for was how many bottles of whiskey he could take." Next, he asked, "How about a couple of women?" If Welch thought Hong Kong a stink hole, soda fountain Romeo George thought it an aspiring rake's paradise. The ungovernable, impish "George III" loathed taking orders. He vowed to sue a crew trainer, one of Welch's prospective hires, who threatened to get tough with him. When pressed, Barstow might have routinely resorted to the corporate-style threat. His training was in music, not navigation. According to Halliburton, the lad brought along an accordion because his chosen instrument, the piano, was too big to bring. On the ship, Barstow could only be given an undemanding role.[7]

Captain Welch, though troubled by Barstow, took an instant liking to the Dartmouth lads—namely, Potter and Chase—who arrived on December 4. "They are good able-looking lads," he wrote, "and it was a pleasure to see them." In Welch's opinion, Torrey was the most manly of the new crew members and by far the best sailor among them. He also seemed more regular and less entitled than the others. So much for first impressions. Within a few weeks of contact with the collegians, Torrey alone escaped Welch's maledictions. All the lads except Torrey had what the captain called "lisping voices that shriek with impotency when the going gets tough." Their seemingly precious manner of speech probably had more to do with a prep school upbringing than with an over-secretion of the feminine hormone.[8]

As December approached, Captain Welch's spirits rose. An eager crew had arrived, the junk would soon be launched, and all the hours and effort put into the enterprise seemed worth it. Work commenced early and continued until ten each night. Even with the end in sight, Welch found it all "a bit hard on the nerves." Some observers thought him "nuts," but he said, "With all the caulking done, the sheath oiled today, the masts lying alongside all ready with the rigging spliced, I don't think I am." By December 5, when all the lads had safely arrived, the *Sea Dragon* had evolved from a singular junk design based on a traditional model into a perplexing marine oddity. Still under construction, the ship lacked its masts, paint job, diesel engine, water tanks, and other appurtenances. In its present state, it looked like a smaller version of Noah's ark, or like the titular character in *The Little Engine That Could*—something not up to par with its purpose. Meeting Halliburton for the first time, the lads were stunned by how delicate and finespun he looked. Although his commitment to the enterprise seemed at once clear, his surviving on a gallon of water a day and tasteless food for three months seemed to them a joke. Paul Mooney was cheerful and had a mordant sense of humor. Welch, through a half- smiling face, was stern and appraising. Henry Von Fehren was a nice fellow who kept to himself. [9]

At this early date, the lads gave little thought to the *Sea Dragon's* chances at sea; they had come to work the ship, not build it, to enjoy port life, not be led at once into the drudgery of regular employment. Once aboard the *Sea Dragon*, reported the *Hong Kong Daily News*, the four would doff their "dinner clothes" and don "tennis shoes, sweat shirts and sail trow." Misapprehensions were sown long before they they could be reaped. Conceding that the voyage would demand enormous energy, Gordon Torrey said that it would also be "fun." Years later, he would say that the junk was an accident in the making. John Potter diplomatically remarked that Halliburton had a knack for making the "unusual" become "exciting and romantic."[10]

26

The Dark Side of Laughter

Welch didn't put the lads to work immediately. Little did it matter as they were more interested in exploring the uncanny wonders of area living than in adapting to the hammering sounds of a shipyard. Before put to work, they roamed fancy-free through the crowded streets. Campus-garbed in white leisure suits and foulard ties, "Brue" Potter and "Bob" Chase gaily snapped pictures of the sights. Seeking camaraderie, George Barstow tagged along while wintry-eyed Gordon Torrey operated, it seems, as a lone wolf. They sought lodging, possibly, at the Gloucester Hotel which catered to well-heeled Americans. And, as it reminded them of the Merry-Go-Round in Boston's Copley Plaza Hotel, they patronized its bar. But the scene there was dull, so they took to slumming in the night clubs and gaming parlors of the Wanchai entertainment district just blocks from Victoria harbor. Flashing money and winning smiles, the lotharios were soon known fixtures in the Colony.[1]

The lads' formal training commenced on December 12. This was just days before the junk reached finished form. At times Welch seemed to like the scoutmaster role, and, as best he could, kept the lads busy, getting them accustomed to the rigging and managing the rope lines and pulleys. Difficult to handle efficiently was the eighteen-foot tiller whose proper use was crucial to the *Sea Dragon*'s operation. "Junks do not have steering wheels," explained Halliburton, "but rather clumsy tillers manipulated by block and tackle." For its mastery, strength and dexterity were required. "By the time they have wrestled with that baby for two months," a grinning Welch remarked, "they'll be able to wrestle an elephant." On reflection, Captain Welch might teach the lads a few practical things, but about abstinence he could only advise. He told them that "they couldn't wash every day," and let them know that "the water supply is going to be a gallon a day and one gallon per man for the cook." In response one of the lads asked outright if Halliburton would be pulling his weight aboard the ship. He would "take his trick along with the rest," Welch replied. When told this, Halliburton (said Welch) "grew very quiet."[2]

December was a month of great expectation, and Hong Kong was a "lively" place to be if one could say good-bye to it. Welch was sad, meanwhile, that he hadn't received word from home or from family and friends living in Australia. The few letters Halliburton received from his parents heartened him. To Welch, Christmas simply meant the end of another calendar year. To Halliburton—wherever in the world he might be, Christmas was special. This year, to his parents, he sent "a "beautiful Mah Jongg set in a lacquer box with tray and cigarette box and racks to match." Three packages would arrive in Memphis on December 10—eight blue and eight white china horses, small herds of carved elephants and carved horses, a pair of mat-wrapped black lacquer tables (to use for cocktail parties), several ebony heads with stands (for the house in Laguna), and another set of blue horses for family friend Mary Hutchison ("to present to the VIII grade as a Xmas present"). Five junk models were his to dispense with as he saw fit. Economically self-conscious as ever, he converted all the presents' total cost in yen or pounds to thirty American dollars. Paul Mooney conducted his own shopping spree. At the Jade Market he purchased figurines of a fisherman, a philosopher, and a dragon. Along with other items, he sent these to his mother in Washington, DC in a teak chest with words from Confucius's *Analects* carved in front: "When young, beware of fighting; when strong, beware of sex; when old, beware of possessions."[3]

By the end of the first week in December, work to get the junk into finished form reached an accelerated pitch. More workers were brought aboard; the number of caulkers increased from three to ten, and the number of joiners from four to ten. While thirty workers had begun building the *Sea Dragon*, seventy-five were now engaged. To Captain

Welch, the extra manpower meant a faster completion time. "I will be out of here by Xmas day," he said. By his own count, he consulted with the shipwright hundreds of times each day, often grudgingly communicating to him Halliburton's design wishes. To these Fat Kau would only squint "from (that) face like leather" and shake his head as if to say, 'What utter nonsense.' Welch had to laugh: "I fear we are to arrive with a pansy for an ensign," he told Wetjen. "We are almost finished as the pictures show, but this damned fool almost drives me nuts with the continuous changes and the everlasting plaint about carvings and gorgeous colors."[4]

"Ensign" Halliburton's throes of creative endeavor had just begun. He "still thinks it's going to be a picnic," said Welch who was not amused. Nearing a breaking point, the captain suggested that Halliburton paint butterflies on the hull so captain and crew could be flown home. Thankfully, Halliburton didn't contract local artists to paint the masts. He liked it that they were orange, but, if butterflies were to grace the hull, he wanted the masts "shaped like a butterfly's wings." Recoiling, Welch reported that Halliburton also wanted to cart along "a lot of gorgeous sea birds, booby birds." No sooner did Halliburton give up one outlandish idea, he subscribed to a next. There was no end to it—Halliburton creating and Welch carping. "Now she wants seven cannons sticking over the side," griped Welch, "all the same size, painted black with red trimmings!" If he put in the cannons, where would he put the ship's stores? So much aggravation—Welch started to wonder, now more deeply, whether the world he lived in was crazier or less crazy than the one other people lived in: "The mad ship-building program goes along on its merry, screwy way," he remarked. "I have a bet with myself that I'll go nuts and go to the house on the hills for screwballs just before it's finished." Welch was referring to the nearby asylum at the end of Centre Street. "I am not sure whether I am still sane or not," he expanded, "but I feel goddamn funny up top." Halliburton was sympathetic. "The cramped little shipyard," he said, "has been a sort of private madhouse for John Welch, our captain; for Henry Von Fehren, our engineer, and for me."[5]

The exterior to his pagan temple completed, Halliburton, whom Welch now called "the gorgeous one," turned his attention to the interior. The poop deck was to have two cabins, as earlier mentioned; the captain's cabin would serve as his sleeper and chart room, and

Halliburton's would serve as his daytime quarters. However, cabins had "the wheel and sails . . . close by" to facilitate "constant command." The four lads would "occupy a four-bunk cabin below and forward—the two Chinamen (cook and boy) in the bow." Four additional bunks, originally meant for Von Fehren, Petrich, Mooney and another recruit, were In "the (dining) saloon—16 X 20 feet." The Sistine Chapel of the ship would be Halliburton's own quarters, which required some redoing of the existing bulkheads and planking. Welch fumed; "the marvel of China and the *Hope of the Fair*" was becoming *Halliburton's Folly*. He recalled Halliburton's earlier request for "a special cabin all private like for her and her secretary." Two bunks were to be installed and a desk put alongside "in case they really did want to work sometimes." Once he overheard Halliburton and Mooney quarreling "about writing articles or something." The next thing he knew a (pale green) curtain was put up to separate "Halli-dear" from "the common herd." Halliburton's personal quarters were in fact "painted black, with a ceiling of bright red." "Enthroned in a carved and gilded shrine" was god of sailors and fishermen Tai Toa Fat. "Chinese artists," meanwhile, had gone "to work on the side walls," and "woodcarvers . . . on the (remaining) bulkhead." Halliburton's "decorative schemes," like his moods, were seldom fixed. "Halliburger vacillates like a debutante trying to decide which dress she shall wear for the coming out party," said Welch. "One day she's all enthused, the next she has a brand new set of ideas." When Halliburton asked that the entrance to the cabin be made "lovely," Welch finally "blew up." The adventurer reportedly told him that fair-goers, whom he called "honky-tonk crud," would "pay $2.50 to see anything that had color on it." For sheer Orientalism, it must have rivaled Dowager Empress Cixi's throneroom. Of it, little is known. A photo shows near one corner of the decorated cabin a Taoist, or Buddhist, priest (alongside the scroll painting of an Oriental lady on her veranda) blessing the shrine of *Tai Toa Fat*. That is the main, perhaps only, visual record. A friend to the crew, twenty-eight year old journalist and magazine distributor left the only known written description of the cabin as a whole: "The living quarters," he said, "were like entering a Chinese temple as the hull was one vast space like the inside of a cathedral, and no thought was given to her for cargo space, which would have given her stability. . . . Bunks were arranged along the side of this vast cabin. Each bunk was decorated according to the rank and station of the occupant. . . . At the end of that huge cabin was a miniature Chinese altar, exact in every detail to the famous temple in Peking. The other end of the hull was built like the ante-room to that noted temple, and would be the entrance for visitors when the ship was tied up at the San Francisco exposition. If you stepped out of this false entrance you would have ended up in a bit of the old briney." Captain Welch called the ship a "hooker," meretricious rather than austere, decadent rather than romantic. Of all things to push it over the edge of aesthetic joy, the *Sea Dragon* featured a fiery dragon painted on the main wall—and at a cost of $500.[6]

Of course Captain Welch was a pilot, Halliburton a showman. Besides fulfilling his promise to fans and fair officials that the junk be beautiful, Halliburton wanted some home comforts for himself and the Dartmouth lads. This meant putting a "bunk in the galley" so "one of the lads" could have "coffee at night." It followed that a cook would have to be on hand at all hours to serve crew members. Irate, Welch insisted that the lads must "make (their) own coffee in the night watches." He further stated that "neither the cook nor the messboy were going to be personal valets for him or anybody else." Halliburton also wanted special treatment given to "messboy" Patrick Kelly. The sight of the two together made Welch cringe; Kelly strutted about the shipyard like an exotic shorebird while Halliburton, whom Welch believed "in love with him," cooed at his side. Fearing he would soon be surrounded by "pansies," Welch told his friend Dick Wetjen, "[I am] a little worried in case all the lads make passes and nobody will be able to handle the tiller."[7]

By December 21, close to the estimated time of departure, Captain Welch, instead of rejoicing, slumped into the doldrums. Now more than ever temporary relief from his worst fears depended on any word he received from family and friends. He often ran to the ships that had just arrived from America or Australia to see if mail had come for him, but he had not gotten "one goddamn letter" for weeks, and it made him "mad as hell." Welch explained to his friend Dick Wetjen, from whom he expected some word, "This letter is not intended to be a sailor's lament, but only to give you an idea to what depths I have sunk." Letter writing for the captain had become therapeutic. He would start a letter, walk away from it, then return to it. Sometimes his thoughts were garbled; at other times he slobbered—perhaps drinking was to blame. At times his letters show him to be full of bluster. In an important issue, did he really tell Halliburton off, or only fantasize about doing so?[8]

As Welch's regard for Halliburton continued to plummet, his affection for the junk grew. He said of this type of ship, "Sweet lines that slide through the water like a yacht, they sail without wind or at least move as though they were all equipped with motors." Of some note, Welch used common naval designations for a ship—*she, bitch, cow, hooker*. As did Halliburton, he also called the junk a *debutante*, a name he also attached to Halliburton. He never called the *Sea Dragon* by its *birth name*. Towards the innocent child, however, he acted like a doting father assisting in its birth, a first child no less whom he wanted to have a decent chance of surviving beyond the womb. Junks had inherent defects—leaking and leaning to starboard. "Still," said Welch, "they are marvelously able craft, husky and strong, so I shall certainly drive the jesus out of it [the *Sea Dragon*] when I get it outside of Formosa." Elsewhere he said, "These things are the best sea boats I have ever seen." Amid all the strife, a singularly odd thing had happened—Welch at once fell in love with the *Sea Dragon* and appreciated his role in its creation. At the late date of December 6, it also dawned on him that he was the ship's captain. Oddly, that the junk was 'good-looking' and not just 'fair' drew his fancy: "I have a ship," he said, "and she is a sight to see." Now and again his affection for 'her' was effusive: "I am happy to know that I built the bastard and will sail her if I first don't go nuts. . . . She is in the hands of the little angel who guides the destinies of fools, parasites, drunk and trusting sailors." Welch wished he were with Wetjen at Izzy Gomez's Café and could relate his enthusiasm in person: "I feel proud of the bitch (ship) as she sits on the blocks for I know that every bolt and every plant is what I put into her," he said: "The masts are really something to see and will try and get you a snap of her when the mizzen is up and rigged." Welch made no attempt to conceal his emotions from friend Dick Wetjen. He could laugh, and be filled with childlike wonder. "Funny, but, until this last week, I haven't given much thought of what it will be like to have the command for the first time really," Welch said. "I have taken over when the old man got sick in sail and thought nothing of it, but now, with the time drawing close and the ship getting to look like a ship, I am giving it thought and feel very happy about it all." The "old man" might have been Welch's father or a revered captain from his past; Welch wasn't that good with names.[9]

Halliburton said that the junk had evolved by "spurts and whims." In truth, Halliburton edited the ship as he edited his books, adding, removing, and rearranging until he obtained a result true to himself and accessible to his readers. He acted like the producer of an over-budgeted film who, fearful that it would fail at the box office, kept futzing with the script, at one time wishing to fire the director (his captain) and replace him with someone easier to control, and at another time wishing to dismiss amateurs and replace them with professionals who knew their lines. "I've still another unused bunk," Halliburton said, "and we'll add a movie-man in Honolulu." Before Welch could yell, "Stop," Halliburton had already mandated, "Cut and process." "The junk's peculiarities are *hers*," claimed Welch. "*She* demanded the high poop." Welch let contention over the matter drop, but later regretted doing so.[10]

Much has been made of this "added stern" or high poop, which, although it provided the perfect surface for decorative motifs, added eight feet to the height of the poop and seemed a leaning tower that might at any moment topple over. Read the survey report: "The owner (had) raised the height of the poop house 8 feet which makes the vessel difficult to handle in heavy weather, masts being too high, not stepped right and sails are too large for dimensions of the junk." As the stern had to have "gaudy dragon pictures," reasoned Halliburton, so the deck had to have an "up-soaring castle"—or *poop*. Purists might contend that the high poop violated the principle that junks be sleek and low. Even though the junk Halliburton inspected at Gin Drinker's Bay had a poop as high, it was more of a houseboat than an ocean-going ship. Required of every vessel whose owner sought insurance coverage was a survey certificate. The one issued for the *Sea Dragon* recommended that the ship's high poop be removed. Alternatively, he could have built a cabin on the deck, beam to beam and toward the stern, an architectural extravagance he considered. When, for the sake of more leisure space, he removed one or more of the bulkheads, he committed yet another transgression against traditional Chinese design. The bulkheads, built with crossbeams or ribbing, were safety partitions: should flooding occur, they kept the junk afloat or forestalled its sinking; also they better secured the masts. "From the Chinese perspective," comments Van Tilburg, "both the *Sea Dragon's* symbolic and physical integrity had been breached by incorporating an engine and propeller shaft, the inclusion of which necessitated the removal of the watertight bulkheads and a redesign of the hull and stern area." What modifications were made to the ship's interior also had the potential to upset its center of gravity.

Said Paul Mooney, "There isn't any sort or shape of ship more easily managed, and more unsinkable, than a Chinese junk." Unsinkable the junk might have been, but only *as long as it remained a junk.*[11]

Welch did not like Halliburton's "construction decisions" any more than he liked his aesthetic ones. Whatever Halliburton conveyed to the Chinese carpenters, he attempted to reverse, screaming into their uncomprehending ears in ever louder Aussie-English. Embittered, the effort also frustrated him: "I never took an hour off for two months from the building yard," he said. "Late at night, all day, shouting and yelling at these awful bastards, driving them, breaking everything but my promises to them to get them cumshaw when it was over, then when it was finished I had to stand and see this Fairy Queen spoil the ship's lines with her screwy ideas. Now what was a decent looking junk looks like a Spanish galleon" or "one of Columbus' caravels gone Chinese." To make space for that engine and the tanks that fed it, the junk's interior was "torn up" a drastic *fourth* time "where the main cabin had been." In the rush to complete the ship, a few jobs were hurried with workmanship suffering as a result. Strict adherence to safety regulations also slackened. When the survey man (either from the Crown Colony, or from Lloyds Insurance Company) inspected the junk, he found a loose plank leading to a gap in the hold. Rather than further reinforce it, Welch secured the plank with an extra bolt. If true that misfortune aids invention, here invention aided misfortune.[12]

Halliburton claimed that the building of the *Sea Dragon* had transformed him from a stoic entrepreneur into a "mental wreck." For the delays in the *Sea Dragon*'s construction and for the ship's defects, he specifically blamed "the superlative perversity of Chinese carpenters." Captain Welch, for his part, blamed the junk's problems on the superlative perversity of Halliburton. John Potter chose to blame Fat Kau, who had sole responsibility for the ship's construction and repair. Halliburton agreed. "Our carefully drawn paper plans made not the slightest impression on Mr. Fat Kau," he said, "so we abandoned them and added or subtracted to our heart's content—as far, that is, as we could without altering the ship's authentic junk lines. Cabins, storerooms and lockers were torn out and rearranged three times. The rest of the ship went through much the same evolution." As the *Sea Dragon* grew—was reconfigured and remodified, it became, says Van Tilburg, "not a combination of Chinese coastal styles but a mix of eastern and western technologies." Indeed, the vessel was an "untested hybrid design" resembling a Spanish galleon, British frigate, or Dutch trader—a 'freak ship.' Historic connections linking East and West fascinated Halliburton; he imagined that Columbus had reached Asia, and that junks with high poops and with bowl-shaped hulls were actually repurposed caravels. Western influences aside, Halliburton wanted the ship to have a distinctively Asian look. He preferred its underbelly to have the curving shape of the speedy Malayan bedars, its bow to have the spoon shape of the spiffy Soochow river junks, and its overall appearance to resemble a sleek Macao-based lorcha. Elements of the heavy-duty Samoan and Tongan sailing crafts—"the Dreadnoughts of the Pacific"—were also incorporated into the *Sea Dragon*'s design. A taxonomic conundrum, a Frankenstein's monster or bride, the ship ultimately defied classification. The Crown Colony's "Survey Report," prelude to one issued by Lloyds, listed the *Sea Dragon* as a junk of the Wenchow type. This classification was based on the vessel's deck plan and sail placements. With a copy of it before him, Dale Collins iterated the report: "The *Sea Dragon*," he wrote, "was an auxiliary junk of the Wenchow type with a length of 75 feet, a beam of 20 feet, and a molded depth of 9 feet. She was well constructed, having a keel of yacal [*sic*] 10' X 10' and a stem of teak 8' x 8,' frames were of hardwood 4' X 5' spaced 14' apart in the engine-room and 18' apart forward and aft. Three watertight bulkheads were provided, one forward 4 feet from stem, one 2 feet forward of engine-room, and one in way of stern-post." While these specs are reliable, the ship's heightened stern could just as easily have classified it as a ship of the Foochow pole type. A hodgepodge, certainly, of traditional Chinese fishing and seagoing vessels, it most resembled the Ningpo-style junk (minus the high poop). Described as "duck-shaped rather than fish-shaped," sleek rather than cumbersome, Ningpos were "able to plunge into the trough and to rise again on an almost vertical wave with far less danger of being submerged than any Western craft double its size." Halliburton told his subscribers that the *Sea Dragon* was indeed a Ningpo-style junk.[13]

If a camel appeared a horse designed by a committee, the *Sea Dragon* appeared a floating circus wagon designed by a lunatic. The comic, garish mix of styles and technologies weighed fifty tons when complete. And blueprints or no blueprints, it still resembled a junk, though some squinting of the eyes was necessary. On the bright side, the ship had been completed in a record time of forty-five days; construction started

November 1 and ended December 15. With the *Sea Dragon* finished, a stunned Halliburton had to wonder how it had been achieved. "Now that the construction is almost over," he reflected, "we realize that it was needless for us to worry about Mr. Fat Kau's disdain of blueprints. The final job is all we could have asked for. We have as sturdy and as beautiful a junk as was ever seen in Hong Kong." The time had come. "The *Sea Dragon* is now ready to sail," Halliburton rejoiced, "and I have no doubt either that its construction violated every accepted rule of shipbuilding known to man."[14]

Eager himself to 'get the show on the road,' Captain Welch wasn't so sure the junk was "ready." "I know this one is going to be a cow in a running sea," he said, "and I know too well that there are going to be plenty of running seas waiting for us outside of Formosa. I have sixty fathoms of four and a half inch manila, already spliced on to a cross sixteen by eleven feet with a cover of two cloths of zero hemp canvass and a wire jackstay. I have two thousand gallons of fuel oil in the tanks that I will pump over the side." Welch had at least one contingency plan: "If it (the tiller) gets funny," he said, "I'll lash the tiller and send the boys by-by: I have no intention of trying to run before a sea that will tip me over if I broach." By December 12, 1938, when the first of the "Log of the *Sea Dragon*" articles appeared in syndication—with dated news—Halliburton believed that he and the crew might leave any day. "The masts go up tomorrow, the engine and water tanks by the end of the week," he wrote his parents. "A lifeboat *with sails* is provided. It's a great thrill watching the ship grow in beauty. Everybody is working hard, and eager to set sail." By the time they received his letter, Halliburton assured them that he would be ready to sail. "Happy New Year," he signed off. "I'd hoped to be gone by today—but the usual delays will keep us here till about January 10."[15]

27

Preparing for the Real World

Rain had not fallen for weeks. Autumn had been sunny, breezy and comfortably warm, with little hint of the overcast skies and cooler days to come. Harbor temperatures now reached the sixties. December 16, as the sun glowed lustrously overhead, the *Sea Dragon* dove smoothly into the murky harbor waters. Halliburton and the lads were having tea, probably with shipwright Fat Kau, and watched from a window or loft. A *Hearst* cameraman and one of the lads each captured the action on film. To their silent b/w and color frames Halliburton's independent narrative could serve as a soundtrack. Obedient to the centuries-old custom that the eyes be opened and Tai Toa Fat paid homage, a dutiful Halliburton had, "on the day of the launching," recruited from a temple near Fat Kau's shipyard a priest (met before) to christen the vessel. "He appeared dressed in white and yellow robes and wearing a curious square hat," marveled Halliburton. "In one hand he carried a tom-tom. Walking about the unfinished deck, he rattled the tom-tom loudly and fixed paper prayers to the mast. Then he bathed the eyes in Chinese *samchu* (rice wine), strung up a string of firecrackers three feet long, and instructed us to 'let her go.'" Next one of Fat Kau's wives smashed a bottle of water or wine against the bow. "With the great eyes dripping wine, and firecrackers popping," the junk "slid into the bay." With a gentle splash, its spoon-shaped nose dipped into the water. The craft swayed for a moment, then settled peacefully. To ward off evil spirits, "the finest, jolliest, fattest statue of Tai Toa Fat that could be found in Hong Kong."[1]

After the crew stepped on board and took to their stations, Fat Kau, his several wives, and his retainers climbed aboard. With reefed sails and driven by engine power alone, the *Sea Dragon* left Hong Kong's "mountain-bound harbor" minutes later. As the ship neared open waters, the engine was stopped and the "brilliant orange foresail raised." Then "five men at the capstan" hoisted "the huge white mainsail." Moments later, "the little scarlet mizzen was stretched aloft." All boded well: "The rainbow canvas caught the wind," Halliburton wrote. "The *Sea Dragon*, shining with its fresh red and white paint, responded, and—slanting in the breeze, tilting over the waves, with her dragon flags flying against the blue sky—was proving herself at last. The wind was brisk that day, as we had hoped it would be. The waves, compared to our little ship, were mountainous. The debutante *Sea Dragon* gaily climbed up and down the water, leaning before the wind until her rails were almost buried." The *Sea Dragon,* set free from its moorings, looked at home *in the calm wild.*[2]

Putting "her through her paces violently," Captain Welch subjected the ship to "every possible strain," and soon learned that "unlike automobiles, which drive off as much alike as peas from the assembly line, every ship has a temperament of its own." At first the ship "responded sensitively to the rudder," but "the rigging creaked" and " the timbers groaned." Although "she sailed well," the *Sea Dragon* moved "slower than hoped." Seen as a liability, the high poop had the ship "heeling in a puff of wind that wouldn't blow [a] hat off" and "running up in the wind." Luckily it did not dip or dive: "The deck wasn't even wetted with spray, so well (did) she (ride) the swells." Even so, the *Sea Dragon* "leaked" "like a bloody basket," and "so heavily built" that it sat too far below the waterline. "The deck slanted and heaved," reported Halliburton. "Very soon our guests, the Chinese, began to regret their excursion. Building a junk was grief enough, said Mr. Fat Kau afterwards. But sailing in one . . . never again!" The ship rumbled upward and downward, riding high with the waves one moment and slamming down hard on its belly the next. Many got seasick. "(Fat Kau's) third wife succumbed soonest," recalled Halliburton. "Wrapping her head in her shawl, she moaned and groaned. Then the painters and the carpenters and the Chinese pilot's son followed suit. This youngster, a child of twelve, baffled us all by being actively seasick without once varying his cheerful grin. Our cabin boy went below to die in his bunk, after thirty minutes on the bounding main. Paul Mooney, at the tiller, looked grimly resolved not to give in to a mere ocean." He admitted to being "the world's worst sailor" and said he "lay prone on the deck, praying for a wreck, a pirate attack, or for Doomsday—for anything, in fact, that would end this oscillation." Chase had "some experience in small boats," but "piano student" Barstow and "journalist" Mooney, the three of us, Halliburton said, "scarcely know port from starboard" and it showed. Barstow found the situation comical. "During this crescendo of distress," said Halliburton, the Juilliard lad "went around with his movie camera, taking movies of the rest of us as we hung over the rail . . . We could gladly have killed him, only we didn't have the strength." Tormented by seasickness, many were hunched over and barfing over the junk's side. Weakened in body, they were also dulled in spirit." Happily, Halliburton found his crew resilient: "Our greenest hands managed to hold (the ship) to her course." He held fast to his belief that Captain Welch "could sail a ship in his sleep" and called Potter and Torrey "sound" sailors.[3]

A good report, instilling crew harmony, some of it was even true. "Incidentally, we've had angry battles and clashes of temperament among ourselves," he wrote to his subscribers. "Each of us has vowed, at least once, to leave this blankety-blank *expedition* and take the next boat home." In a New Year's Day letter to his parents, Halliburton remarked, "The crew is all at each other's throats—of course—but it was never otherwise on such an *expedition*." Mooney in his assessment was less charitable:

Each member of the crew has some element of his nature that's hard to bear: the engineer is stubborn—the captain is a tyrant—the radioman uses dreadful English—But nothing fatal. The company isn't entirely compatible. [We have] a blustering extrovert for a captain; Dick [Halliburton] for a boss; myself, temperamental as hell; the engineer, a very proud and stupid German; one boy a Connecticut Yankee of the worst [sort] (probably Chase); another longing for his habitual background of zebra stripes at the El Morocco (Potter); still another fresh (how!) from Dartmouth and full of anti-Roosevelt jokes. The radioman (Petrich) is tattooed from head to foot, the cameraman (Barstow) is odious in every way. The two Chinese waiters will take the real dirty work; for the rest of us, work will be merely hard, endless, and at first unfamiliar. But the enforced companionship, in crowded quarters, with a collection of people in whom I haven't the faintest interest, physically or mentally, is going to be a real trial. Not one of us knows anything about sailing.

By now the bitter reproach felt by Welch and Von Fehren for each other was old news. Almost certainly the crew member characterized as "full of anti-Roosevelt jokes" was Gordon Torrey, ranting non-stop about every aspect of American foreign policy from protectionism and free trade to colonialism and the new imperialism. Later employed by an American company as a petroleum consultant in Asia, Torrey had already learned that "the U.S.A. is the most hated nation on earth." "The Press," he expanded, "fills (people at home) full of hot air about the U.S.A.'s help and sympathy for the Allies and China, the British, French, Dutch, Belgians, Norwegians, Czechs and Chinese think we're windbags and the cause of all this trouble due to former diplomatic meddling and Roosevelt's big mouth. (Truth be told) the Chinese hate the hell out of us for talking a hell of a lot about helping them while we slyly feed Japan all she wants to keep her out of the Dutch East Indies and the Philippines."[4]

Halliburton entertained like views but seasoned them to his readers' taste. While the others were not so politically antagonistic, they feuded nevertheless. First mate John "Brue" Potter thought the crew a bunch of "good guys" who "didn't exactly epitomize able seamen." With all the "quarrels," Halliburton observed rivalries he feared would stay. Rivalries emerged that stayed. His own personal preference for Potter and Chase over Torrey and Barstow clearly showed; "The two boys are jewels," he said. To the point, Wesley Halliburton cautioned his son "not to play favorites," and Richard assured him that he would "keep detached from everybody in order to act as mediator." As a kid, John Potter had read *The Royal Road to Romance,* and, in 1988, claimed to no disappointment upon meeting its author. He thought Halliburton "an indomitable man" who never got "angry" and a man whom he "never saw . . . or heard complain." Detecting a streak of Scotch puritanism in the adventurer, he thought it to his credit that he never resorted to foul language. Once, though, "when he dropped the coffee pot over the first day out," he muttered a few swear words.[5]

Halliburton put great faith in his crew. "Once at sea," he said, "we'll

PREPARING FOR THE REAL WORLD

all be too busy with ship and sails for the cultivation of animosities."
He considered himself a "better than average mediator" and believed
he could quell any snit, feud or riot. What mediation skills he possessed
were, of course, never questioned—maybe because, as soft impeach-
ments, they were too ineffective to be noticed. Optimistic, he hoped
that the example of his "own even temper" would "be a big help," and
that commitment to a shared goal would bring "harmony" and sow camaraderie. Many mariners, however, would dispute that living in close
quarters over time breeds harmony, but rather produces the worst from
even the best of friends. Aboard a tiny ship at sea, crew members become subject to what explorer Thor Heyerdahl called "expedition fever."
Heyerdahl, who was all too familiar with the ailment, defined it as "a
psychological condition which makes even the most peaceful person
irritable, angry, furious, absolutely desperate . . . until he sees only his
companions' faults." Mooney, for his part, wished for more distance
from his fellow crew members. "A hundred times a week I determine to
return by the first comfortable ship," he told Laguna Beach friend Lee
Hutchings, "instead of spending three months in the human churn we're
building here." Besides "a human churn, one painted up like Jezebel,"
the *Sea Dragon* was for him a "crazy cockle-shell."[6]

When critics told him of the many perils he faced in this "cockle-
shell," Halliburton's desire for accomplishment only grew stronger. He
tried to remain lighthearted toward crew members, but he succeeded
only in appearing moody. John Potter said that Halliburton "hardly ever
smiled and almost never laughed." Journalist Gordon Sinclair, said that
Halliburton had "been accused of being anti-social and high-hat because
he shunned conversation with the average person whom he met." An interviewer for the *Hong Kong Daily News* thought otherwise, saying that
Halliburton possessed "a keen and amusing sense of humor" and communicated "terrific curiosity" as well as a sense of "fun." Gordon Torrey
disputed the notion that Halliburton had "a keen and amusing sense
of humor." He in fact found Halliburton preposterously self-centered
and dangerously delusional. Like Potter, he had read one or another of
Halliburton's books. Like many who attended Halliburton's lectures,
he found that the man did not measure up to the image he projected
of himself in his books. He said he had expected to meet a "fun-loving,
humorous" fellow. Over time, however, he found "the real article" rather
"driven, un-funny, [and] unreasonably stubborn." Torrey didn't think

Halliburton tortured like the deranged Captain Ahab or tyrannical like the bookish Wolf Larsen. His assessment of him was oblique; to have Halliburton "as the center of our universe in a very alien environment," he said, "was one of the biggest disillusionments of my life."[7]

Writing home, Halliburton said, "[The] clashes and quarrels will pass." But disputes concerning the merits of the *Sea Dragon* Expedition itself were not likely to pass. There emerged those one or two who opposed the voyage or whose confidence in it wavered, and those who fully supported it. "They are trying to ditch the lad Torrey," said Welch, "who is by far the most manly of them all." Not one to mince words, Torrey believed that the *Sea Dragon* "could have been heavy-lifted on the deck of a San Francisco bound freighter and carried across the Pacific and still have been a tremendous success at the *World's Fair*." The trial runs sealed the bitterness that developed between Torrey and the others. Seeing how the junk had struggled in the harbor and how ineptly the crew worked as a team, Torrey concluded that the whole expedition was an invitation to death. Nothing remained but for him to defect from the enterprise, and soon he "tried to talk Brue (Potter) and Bob Chase out of going." These mutinous actions then "led to un-pleasantries and (he) became dissociated from the others." Even so, Welch still liked Torrey and thought him the only competent sailor among the bunch.[8]

Reporters and members of the sailing fraternity generally liked Welch. They at least found him amusing. Halliburton liked the captain or at least thought it impractical to dislike him. For his part, Welch detested Halliburton simply for being Richard Halliburton, a publicity seeker. The Dartmouth lads were privileged, spoiled thrill seekers unacquainted with tough work. As for Mooney, Welch was beginning to see him as a "decent chap" who was no stranger to hard work and who pulled his small weight. Respectful of Captain Welch and Henry Von Fehren, John Potter was convinced that they were the only crew members "who had ever been aboard a sailing craft." Von Fehren was "a true professional," by trade "an engineer," but he was no sailor. Believing Welch to have been a "freighter captain," Potter called him "an irascible guy" who "seemed to carry around a chronic grudge against the world" and "could possibly have been the original Captain Queeg." Endorsing the opinion, the captain, he noted, was not at all chummy: "He always ate alone at a single table while the rest of us were at one large table." Although emotional exposure to relative strangers made Welch uneasy, his distancing had a motive studiously kept from the crew. In a letter to Wetjen, he communicated that he deliberately dined alone, thinking it good policy to keep his crew at a distance, an instructional technique apparently lost on Potter. Welch also chose to live aboard the junk while the crew for a time stayed at the hotel. Curiously, he seldom, if ever, called anyone by their proper name; crew members were *work units* activated by nearby or remote command. The captain was not mean-spirited, and his willful isolation did not spring from an evil heart. A passage from another letter he wrote to Dick Wetjen suggested the gist of his management philosophy: "The fact that it is not going to be a picnic after all is slowly seeping into the minds of the autocrats and it is a painful process," he said. "I like them personally but I am afraid that I am going to have to give them the works before we are many days at sea. . . . It is a slow business but very sure and certain. I try to be tolerant but my life has not made for tolerance in any form. I have learned the bitter truth about people in a bitter school and I am afraid it sticks too close to easily fade." A bold teacher, Welch was willing to be hated in order to instruct.[9]

Years later, in 1987, Gordon Torrey recalled that "none of the *final* crew had practical off-shore sailing experience and only Welch and two others (possibly recent recruits Ralph Granrud and Ben Flagg) had seamanship backgrounds." From the moment he met Halliburton, Torrey's small faith in him deteriorated. "During the short, initial sea trial a few miles outside of Hong Kong," his faith in the *Sea Dragon* dwindled equally rapidly. Torrey recalled his reaction to seeing the craft struggle atop the settled waves of the harbor: "It became apparent . . . how, in the kind of seas I had seen in the passenger liner, solid water could come right over that low spoon bow, wipe out the sky-light trunk and fill the lower deck section with enough water to enable that great heavy diesel engine to take her to the bottom." Welch adjudged Torrey as "by far the most manly of them all." Even after Torrey had defected, he saw him as "a good gutsy sort of egg." Only Torrey had sailed through the perilous Formosa Straits, and only he could visualize the misfortunes so untested a vessel as the *Sea Dragon* could encounter.[10]

Torrey said nothing, Welch little, about diet and nutrition, food safety and consumption. Late, on December 16, James N. Sligh, part Chinese and originally from Los Angeles, was hired as ship's cook. Welch called him a "white cook"; Halliburton called him "a fine cook." Scuzzy, sullen-faced, Sligh's sordid history began with the abandonment of his wife and two kids for no higher purpose than a life at sea. He had apprenticed as a cook and, after years in the galley, could boast of preparing meals "on the ships of a dozen nations." As a reporter for the *Hong Kong Daily News* said, "Ships and their people are second nature to [Sligh]." Since proper food and water management meant a fit, hearty crew, good cooks made for good voyages. The *Sea Dragon* would hardly be a floating Palace of Food and Beverages, but a greasy spoon on water that featured food 'unfit for human consumption but good enough for sailors.' A hardworking crew's nutritional needs, in any case, had to be addressed practically; for strength, and to prevent vitamin deficiency, it was important to keep plenty of beef and citrus on hand and limit sugar and salt intake. Note that the human body burns about 2,600 calories a day on average, and that one ounce of jerky or bully beef produces just 45 calories while an 8-ounce ribeye steak produces 660 and a 10-ounce perch 330 calories. That said, a good ship's cook, like Sligh, parcels out and properly distributes food and water so that supplies did not run out. To assist Sligh, Halliburton hired a "cook's boy" from China when John Rowland declined the job.[11]

Before reporters and tourists Halliburton tried to look cheerful, but the stress showed. The *Coolidge*'s Dale Collins saw him as "pale and haggard." On December 29, a couple days before he hoped to sail, missionaries Dr. and Mrs. J. M. Lapp photographed Halliburton in front of the *Sea Dragon*. The resulting image shows him wearing a beige overcoat, a white shirt and tie, a suit jacket, and sharp-creased dress pants. He is shying away from the camera, and his appearance suggests the elderly man he would soon become rather than the young man he had been. Halliburton had met the doctor while traveling in India during the adventures that fill *The Royal Road to Romance*. Lapp, "a veritable white maharajah," had worked with his wife to establish a mission in the province of Dhamtari. Twenty years had passed since his initial meeting with Halliburton. No longer was Halliburton the 'live-young-forever' rover he remembered, but a toppled figure, weary of both himself and the world. On January 9, Halliburton would turn, as most books show, thirty-nine or, as he himself believed, forty.[12]

28

The Whole Town Is Talking

January would prove the bitterest and longest month for Halliburton. He remained in port, the crew continued to feud, and the *Sea Dragon* had flunked its trial runs. His increasing number of critics had him wondering about the merits of the adventure. Still, he kept a brave face and, ever friendly with passersby, had "lost none of (his) enthusiasm and confidence." The defects of the junk corrected, the "last vestige of uneasiness" he felt about its "seaworthiness" faded. It now ran "perfectly." Moored "alongside the main dock," it needed just a few "final touches."[1]

The final delay gave Welch and Halliburton time to discuss, besides navigational details, a best route across the Pacific. Welch preferred a northern, Halliburton a southern route—"via Manila, Guam, to Midway, and Honolulu, then directly to San Francisco." Manila seemed a reasonable first stopover. An American military base was stationed there, and Halliburton had landed the *Flying Carpet* on one of its airstrips in 1932. Just before Christmas he told Captain Welch that he wanted to make a week-long trip to Manila; whether he did so is uncertain. From Manila, the junk could link up with Japan's north-flowing Kuroshio Current, and track what cartographers called "Urdaneta's route," an arcing, bow-like course that touched Midway Island and led to the California coast. Longer than the northern route, this one was best taken in summer.

Now that it was winter, Halliburton supposed that seasonal winds coming from the east would only blow his ship backward, so he dismissed the southern route as "impossible" and, believing "the typhoon season [would] be over," opted for the northern route. Keelung (Chilung), on the northern tip of Formosa and some seven hundred miles from Hong Kong, would be the first stop on the *Sea Dragon*'s journey. Enroute would be "a rather rough sea, but from there on, the prevailing winds are more favorable than if we had kept to our original intention and gone by way of Manila." For the moment, the weather concerned Halliburton less than the pirates—and the Japanese gunboats which for over a year had been "sinking ships at the rate of one a day." Against the pirates he of course had the four shotguns and three rusty cannons; against the Japanese he had the letters of transit and "three easily visible American flags." What with "a splendid radio" and "a fine crew—and a big medicine case—and lots of food and 2000 gallons of oil & 2000 of water," why should he worry?[2]

Provisions loaded on board, rumor spread throughout the Colony that the junk was set. Those among the crew who were seasick were now homesick. On New Year's Eve, Paul Mooney wrote to his friend Alice Padgett, "Why did I ever leave Laguna???" Still, he trusted that the *Sea Dragon*, leaving in the next two weeks, would soon bring him there. He told Padgett the ship was headed to "Keelung in Formosa, Midway Island, Honolulu—and then home." Amid the epic grandeur of the *Sea Dragon* Expedition, pressing business related to water-assessment notices on his properties in Laguna Beach drew his attention. Offhand, he hoped things there were "going on well enough." Still, with not a single drop of rain falling in Hong Kong, he missed what rain he imagined falling in Laguna. Mooney also mentioned Southern California's notorious earthquakes, noting Hangover House's vulnerability to such rumbles. Little considering the *Sea Dragon*'s vulnerability to storms at sea, he rather had his mind set on eating chilis at El Padre. Enroute to that goal his mailing address would be *"Care of American Express, Honolulu—Hold For Halliburton Junk Expedition."*[3]

While Mooney's letter simply *announced* the upcoming adventure, Halliburton's January 18 letter to the *Sea Dragon* Club subscribers *celebrated* it: "Our junk is soon to sail!" he exulted. "It is lying at anchor this moment in Hong Kong harbor, shining with red and white paint, its dragon flags flying, its yellow sails furled but ready, its big 'eyes' staring toward the sea. We hope to leave within the week—as soon as we can clear the decks of the riggers and carpenters and painters and fitters and photographers, who, during these last frantic days, have turned the junk into a floating crazy-house." Only ten days separated this letter from his next one to subscribers. "Dear friend of the *Sea Dragon*," Halliburton began. "A few days ago I wrote you that the *Sea Dragon* was about to sail. Well, it is still just about to sail. We've fixed a new date for our departure, and we hope, this time, really to get away. Our first 'sailing' date came and went a week ago, while the *Sea Dragon* stayed tied up at a wharf, sulkily enduring the last-minute ministrations of carpenters and caulkers." There was also rough water reported outside the harbor to discourage immediate sailing.[4]

Clearly, better judgment was a frequent casualty of the *Sea Dragon* Expedition. At these crucial moments, and in an action that endangered his very mission in China, Halliburton decided to take a trip. "While the rest of the crew were at work under the Captain's orders," he announced, "I took advantage of a very special opportunity, and ran *off* to Canton." He did so for no other reason than "to get (his) souvenir envelopes stamped." In October, bitter conflict in the region had prevented him from reaching the beleaguered city. "Japanese airplanes had methodically bombed Canton," leveling "railroad stations and freight yards" where munitions were believed stored. Canton fell, and over ten thousand civilians were slaughtered. Chinese armies retreated as columns of Japanese soldiers advanced. The British protested, but their threats of retaliation rang hollow. In their own sense of retaliation, however, many Chinese had adopted a scorched-earth policy and set fire to their "homes and shops." Soon the "fires swelled together into one vast, terrifying conflagration, roaring across the city." Looters by the thousands roamed the streets while local police and troops took no action to stop them. All that was left of Canton was "a ghost town" occupied by "the halt and the blind, cowering in the ruins." A mass exodus ensued. "Stoically, with scarcely a backward glance," wrote Halliburton, "a million citizens evacuated their doomed town, and disappeared into the country-side. They opened the jails, released the lunatics, emptied the hospitals. Flowing out the main avenue, leading west, for three days and three nights, there was a solid river of humanity." Paul Mooney made the effort, but only Orientals could travel up the Canton River to the broken city, and, though some Caucasians could venture forth on

English and American warships, trains and commercial ships were not allowed to pass. Japanese troops ransacked the land where the fleeing Cantonese had settled. Japanese gunboats, meanwhile, roved the channels, often attacking and maiming American and British cargo ships. Pirates infested every coast, and brigands roamed the countryside.[5]

From the relative safety of Hong Kong, Halliburton noted the depredation. Since he first tried to reach Canton in October, hostilities in and around the brutalized city had relaxed. The Japanese attempted to establish a sort of *era of good feeling* whereby at least some foreigners were conditionally welcomed. Occupation brigades cleared the streets of fallen debris, repaired the bridges leading into the city, and opened food kitchens to beckon home those who had fled. "Your city is good as new," the Japanese soldiers in command announced. "Come back to your happy homes." When it appeared that most of the evacuees would not return, the Japanese fumed. "Tokyo had counted on a rich income from taxes and trade in this former center of commerce," wrote Halliburton. "So far the only reward for the huge expenditure of money and material has been ashes and hatred."[6]

Halliburton had come to believe that "no disaster (was) so great as to extinguish the Chinese trading instinct" or wanton thievery. Indeed, "a few Chinese, more mercenary than patriotic, slunk back into Canton and opened a market where the looters and the Japanese (could) buy and sell the property stripped from unguarded homes." Any day or night, "one could buy Leica cameras, seized in a camera shop, for a dollar. The traders had no idea what they were. New electric refrigerators went for three dollars; grand pianos, sitting out in the rain, for their value as firewood." The Japanese held out hope that those Cantonese less driven by gain would return to their homes. In time, thousands did return, but many fled elsewhere, often to Hong Kong. In going to Canton, Halliburton might have had aims besides procuring stamped envelopes. Although legitimate and black-market businesses still operated, procuring needed supplies was not his purpose. That he was scouting out the viability of US offshore business investment in the region remains only a possibility. As his birthday was January 9, the trip might also have been a gift to himself. Or (romantic that he was) it was his way of honoring the patron saint of travelers, Ming poet and philosopher Xu Xiake (1586–1641). Apart from his stay in Hong Kong, Canton was Halliburton's last fully realized trip to a major foreign city.[7]

After some wire-pulling, Halliburton got himself aboard the *Mind-anao*, an American "river boat" bound for Canton. Making the 110-mile ride with him were about twenty "mostly medical missionaries return-ing to their wrecked schools and hospitals" and a small corps of inter-national press correspondents sent to cover the war. Journalist Gordon Sinclair, who had insulted Halliburton in print years before, sat next to him and was now congenial. Halliburton was as congenial, but focused mainly on the traffic beyond: "During the 10-hour ride up the river," he reported, "we passed fully fifty Japanese ships—cruisers, destroyers, patrol boats, and freighters carrying supplies." Expect brigands and freebooters to be a threat as well. As the *Mindanao* approached what had been Canton, Halliburton saw beauty even in its ruins. Bounded in the north by mountains, which included several inactive volcanoes, and fronted by numerous small islands, channels, narrow inlets, and estuaries, Canton had been well protected from invasion until modern times. Pictures of how the city once looked flooded Halliburton's mind. "I was in Canton back in 1922," he wrote. "Then, the city's riverfront swarmed with sampans. Canton had the biggest floating population of any city in the world. But now, as we approached, I soon saw that the sampan dwellers had followed their dry-land brothers. There are per-haps a hundred boats left, where once they numbered twenty thousand." Astonished, soon Halliburton would be appalled.[8]

Mid-afternoon the *Mindanao* docked. Once off the ship, the author of *The Glorious Adventure* and *New Worlds to Conquer* "crossed to Can-ton proper, on the mainland." As the Far Eastern cradle of civilization, Canton had still represented for the sentimentalizing Halliburton the old China. Now, however, it was a garrison with Japanese armed guards standing side by side and Japanese troops conducting exercises in one open space or another. Halliburton's thoughts of the "wasteland" that had been Shanghai returned. "At every step, Japanese pillboxes and sentries confronted (me)," he said. "A special pass" permitted him to move on. Once "proud of its wide avenues and of the dense care-free throngs that filled them," war-devastated Canton was now a "derelict, tragic city!" Its quaint buildings and shops were now "charred and blackened ruins." Scorched by fire, the trees lining the city's streets were "lifeless and black." There was "scarcely a single hint of life"; the city's million souls had simply "gone with the wind." Once "one of the busiest cities in the world," Canton was now a junkyard of "homeless dogs and blind beggars." What Gordon Sinclair called the "queen city of South China" was now the "city of death." The high humidity made the stench from "rotting bodies in the canals and lagoons" pungent. Careful where he stepped, Halliburton slowly walked past bicycle and automobile wrecks, scattered bricks, fallen buildings, chards of bric-a-brac, animal carcasses, and the bones of dead children shrouded in the tattered remains of shabby clothes. It was so eerily quiet that he heard only his own footsteps. "There's something frightening about this stillness," he would write, "this ghostly, deathlike stillness. I've had a little of the same feeling on the streets of New York, at dawn of a winter morning, when, on looking down a side street, I've seen no living thing. But in New York, life was only sleeping. In Canton it's just not there." The images of desolation and loneliness never left him.[9]

If such hardship could befall Canton, what lay in store for Hong Kong? From the beginning of the China war to the fall of Canton, Chinese armies were supplied chiefly through Hong Kong with help from the British. But traffic to the mainland appeared halted. Demolished bridges and railway trestles only encouraged the resilient Chinese armies to rebuild. In Halliburton's opinion, Imperial Japan felt a mixture of "covetousness" and "hatred" toward Hong Kong. Despite the British presence and outcries from Britain's Western allies, Japan remained grimly determined to conquer the Colony. Since the fall of Canton, "trade (had fallen off) sixty-eight per cent" and promised to further decline to an 80 percent decrease. Prices rose on most goods, including shipbuilding materials, oil, and food.[10]

Halliburton did not say how long he was in Canton—maybe a day. Amid so much that had appalled him, the *Sea Dragon* returned, as an analgesic, to his thoughts. Before their trip to Canton, Gordon Sinclair reported that Halliburton "was sipping innumerable cans of tomato juice and taking a few vitamin pills for his nerves." During the trip, one "interlaced with troubles, he [Halliburton] talked far into the night about his beautiful bachelor home, his plan to write a great travel book about the United States, and his belief that the junk *expedition* must and would succeed." To and from Canton, he would tell Sinclair how proud he was of his ship, describing the *Sea Dragon*'s beauty rather than enumerating its shortcomings. In his fifth "Log of the *Sea Dragon*" article, which appeared in the *San Francisco News* on January 23, 1939, he told subscribers that the *Sea Dragon* would be ready to sail by the time he returned from Canton to Hong Kong. On January 23 he also wrote to his parents, whom he had not contacted for three weeks. "Storms will probably be unpleasant," he said, "but will only delay our arrival." The end of April was now set as his arrival time at the fair. He assured his parents that he was not "in trouble," he was "just delayed." Every hardship he endured was "worth it" for the "perfectly gorgeous ship" that resulted. Word from halfway around the world was notoriously slow; when he wrote those words, Nelle Nance actually believed her son was already at sea.[11]

"Romantic Richard" Halliburton as he appeared on the back cover of the first edition of *The Royal Road to Romance* (author's collection).

Halliburton about the time he swam across the Panama Canal in 1928 (courtesy Rhodes College's Barret Library).

"Hollywood" Halliburton in Los Angeles in 1930 with actress and comedian Dorothy Lee (1911–1999) while preparing his trip around the world in a single-engine Stearman called *The Flying Carpet* (author's collection).

Halliburton arrives at Los Angeles Airport in 1933 to promote his film *India Speaks* and is welcomed by fans Frances Drake and Eve Ellman (International News / author's collection).

Halliburton plugged a number of commercial products on radio and in magazines and newspapers. This is among the most visually witty of his advertisements (author's collection).

Flyer introducing Halliburton as speaker for the Alber-Wickes agency and noting sponsorship by the Ypsilanti, Michigan, chamber of commerce (author's collection).

Halliburton shows Explorer Scouts Bill Hart (left) and Jack Green places of interest on the globe, ca. 1937 (author's collection).

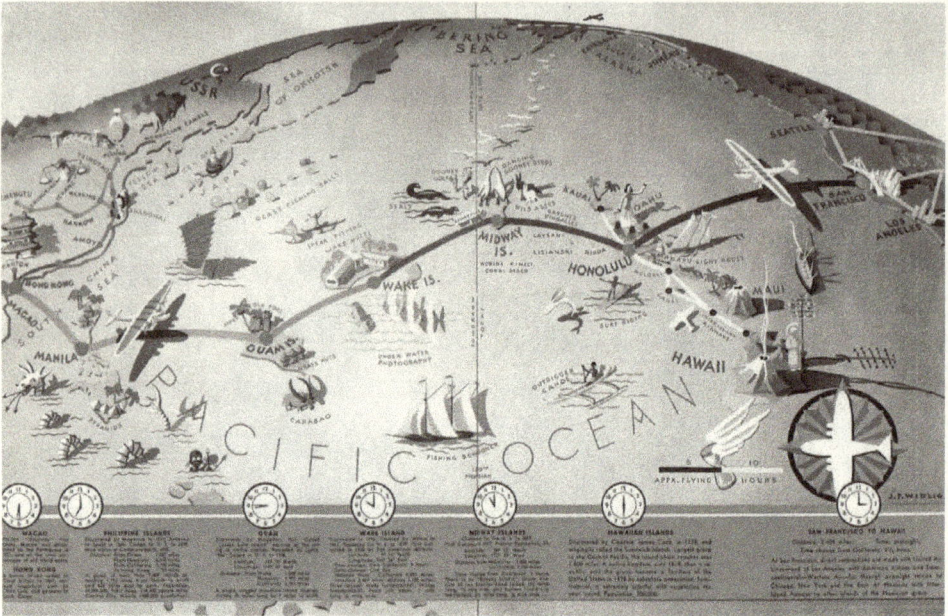

Colorful map of the Pacific Ocean that Halliburton in 1937 sent to future screenwriter James Watson Webb, a prospective member of the *Sea Dragon* crew (author's collection).

Designed and built in 1937 by architect William Alexander (1909–1997), Halliburton's hilltop mansion in Laguna Beach, California, was known as Hangover House (courtesy William Alexander).

"Daring Dick" Halliburton in 1936 during construction of San Francisco's new Golden Gate Bridge. About this time he was inspired to sail a junk across the Pacific from Asia to America (courtesy William Alexander).

As captured by a *Saturday Evening Post* editor, the Golden Gate International and the New York World's Fair ran concurrently (author's collection).

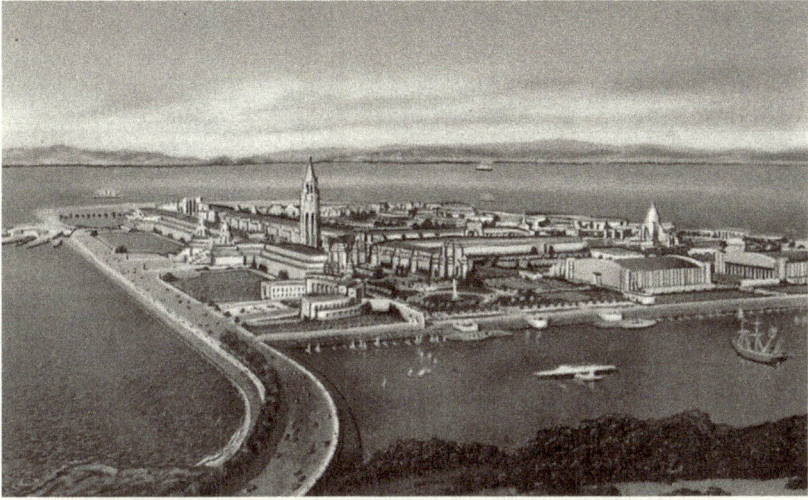

Official view of the 1939 Golden Gate International Exposition. To the immediate left is the causeway leading visitors from the San Francisco–Oakland Bay Bridge to Treasure Island. To the right is Exposition Harbor, where the *Sea Dragon* would dock (author's collection).

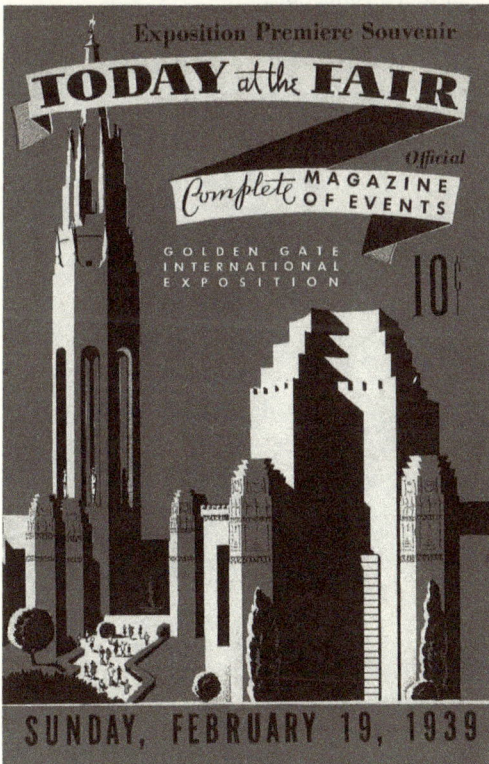

One of many brochures and guides issued to fairgoers (author's collection).

The *Gayway* on the opening day of the Golden Gate International Exposition, February 18, 1939 (Wide World / author's collection).

Record numbers of fairgoers at the Court of the Seven Seas. The eighty-foot statue of Pacifica stands in the background (courtesy Rhodes College's Barret Library).

Nymphet called the "Evening Star" towers over the pools and fountains of the exposition (Wide World Photos, *San Francisco News Bureau,* February 18, 1939).

The Court of the Moon at night was the exposition's Taj Mahal (Joseph Henry Jackson, *A Trip to the San Francisco Exposition with Bobby and Betty,* New York: Robert M. McBride, 1939).

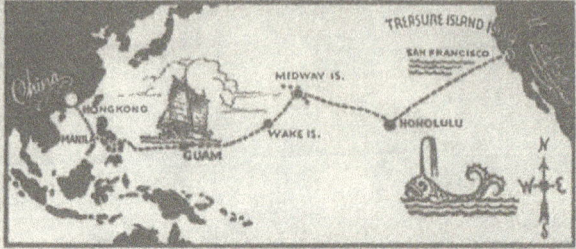

THE HALLIBURTON
CHINESE JUNK EXPEDITION, INC.
OFFERS THIS BEAUTIFUL COMMEMORATIVE SOUVENIR ENVELOPE (COVER)
AUTOGRAPHED BY RICHARD HALLIBURTON

The cachet advertising Halliburton's Chinese Junk Expedition, and the letters about the voyage he intended to send to subscribers. The style of junk, as other matters related to the *Sea Dragon* Expedition, would change over time (author's collection).

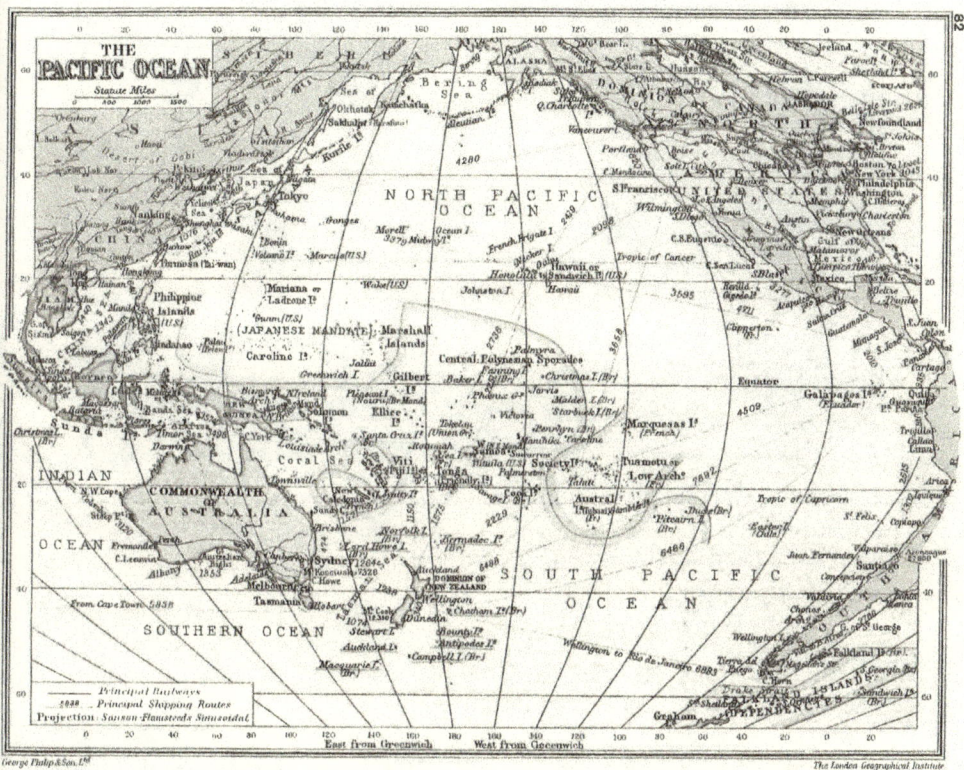

Map of the Pacific Ocean, ca. 1920, showing the major geographic locales associated with Halliburton's final days in the Far East (author's collection).

From left to right aboard the *Coolidge* as the liner is about to depart from San Francisco for Honolulu: Capt. John Welch, Richard Halliburton, and engineer Henry Von Fehren (Acme Pictures Inc. / courtesy Rhodes College's Barret Library).

From left to right in Hawaii are Richard Halliburton, Captain Welch, radioman George Petrich, and engineer Henry Von Fehren (courtesy Princeton University).

George Barstow III (left) and John Rust "Brue" Potter
(originally identified as Robert Chase by photojournalists),
"lads" who would join the crew of the *Sea Dragon*, at the
Mark Hopkins Hotel in San Francisco (courtesy Rhodes
College's Barret Library).

Hong Kong Harbor (author's collection).

A busy street in Hong Kong
(author's collection).

China-Hong Kong–A Native Street   P. 370

Elegant colonnaded buildings along the waterfront at Hong Kong Harbor
(author's collection).

Hong Kong Harbor from Victoria Peak (author's collection).

John Rust "Brue" Potter and Robert Hill Chase (aiming the camera) as they roam about Hong Kong (courtesy Rhodes College's Barret Library).

The initial crew of the *Sea Dragon*. Standing from left to right are Paul Mooney, Henry Von Fehren, James Sligh, Gordon Torrey, John Wenlock Welch, Richard Halliburton, and John Rust "Brue" Potter. Seated from left to right are Richard L. Davis, Robert Hill Chase, George Petrich, and George Barstow III. Except for Potter, Torrey, and Davis, those pictured formed the final crew of the *Sea Dragon*. Not shown are seaman Ralph Granrud, seaman Ben Flagg, world traveler Velman Fitch, boatswain Sun Fook, seaman Kiao Chu, seaman Wang Ching-huo, and the messboy Liu Ah-su (or Ah-shu) (courtesy Princeton University).

A rare photograph of able-bodied seaman Ralph Granrud, who defected from the SS *Coolidge* to become part of the *Sea Dragon* crew. Image dated March 1939 (Acme News Pictures Inc. / San Francisco Bureau / author's collection).

Chinese carpenters construct the *Sea Dragon*, on the right using handsaws to cut the planking ("Log of the Sea Dragon #13," March 2, 1939 / *San Francisco News*).

At the front of the *Sea Dragon*, Halliburton discusses construction points with an unidentified man (possibly Fat Kau or the harbor master). Captain Welch stands to the left with an unidentified man (possibly surveyor or Lloyds of London inspector). Henry Von Fehren, in a white captain's cap, appears in the right foreground. Image dated November 1938 (courtesy William Alexander).

Possibly the only photograph that shows carpenters working on the interior of the *Sea Dragon*. Image dated November 1938 (courtesy William Alexander).

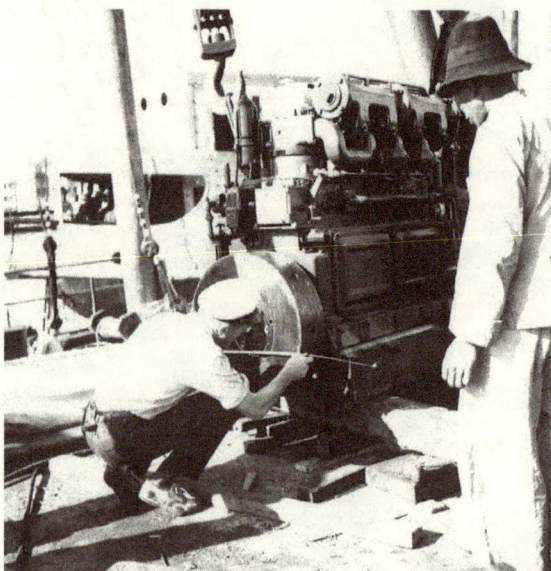

Henry Von Fehren tends to the flywheel of the diesel engine while one of the Chinese carpenters watches (courtesy Rhodes College's Barret Library).

Captain Welch watches as the Chinese carpenters continue their work. Top of photo shows Potter (in white), Chase (to his right), and Mooney (to his left). By the mast stands the ship's surveyor; to his right is Halliburton (courtesy Rhodes College's Barret Library).

The *Sea Dragon* nears completion. Henry Von Fehren is almost certainly the man wearing a captain's hat and sitting aboard the ship. Paul Mooney is either the man wearing a dark shirt and sitting the foreground or the figure standing alongside the pole aboard ship. The match in either case is close. Photograph dated December 1938 (courtesy Rhodes College's Barret Library).

The completed *Sea Dragon* in dry dock
(courtesy Rhodes College's Barret Library).

The *Sea Dragon* afloat (courtesy Rhodes College's Barret Library).

Paul Mooney among children at an orphanage he and Richard Halliburton
often visited while in Hong Kong (courtesy Rhodes College's Barret Library).

Richard Halliburton embraced the children of the world as his own
(courtesy Rhodes College's Barret Library).

One of the last photographs of
Richard Halliburton (Acme News
Pictures Inc. / author's collection).

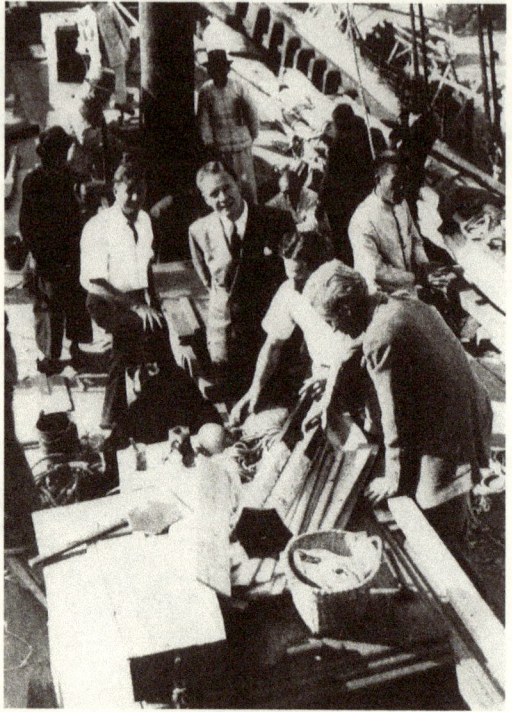

Standing from left to right aboard the *Sea Dragon* are Captain Welch, Halliburton, George Petrich, and John Potter (courtesy Rhodes College's Barret Library).

Paul Mooney and John Potter learn the ropes (courtesy Rhodes College's Barret Library).

An artist on bamboo scaffolding puts the finishing touches on the phoenix painted on *Sea Dragon's* stern (courtesy Rhodes College's Barret Library).

The *Sea Dragon* docked and open to visitors for a guided tour through its corridors (courtesy Rhodes College's Barret Library).

Passersby stop to examine one of the two twenty-foot blue-and-green dragons that climbed each side of the poop (author's collection).

A unique view of the *Sea Dragon* (courtesy Rhodes College's Barret Library).

The *Sea Dragon* on a trial run (courtesy Rhodes College's Barret Library).

The *Sea Dragon* under sail during a trial run
(courtesy Rhodes College's Barret Library).

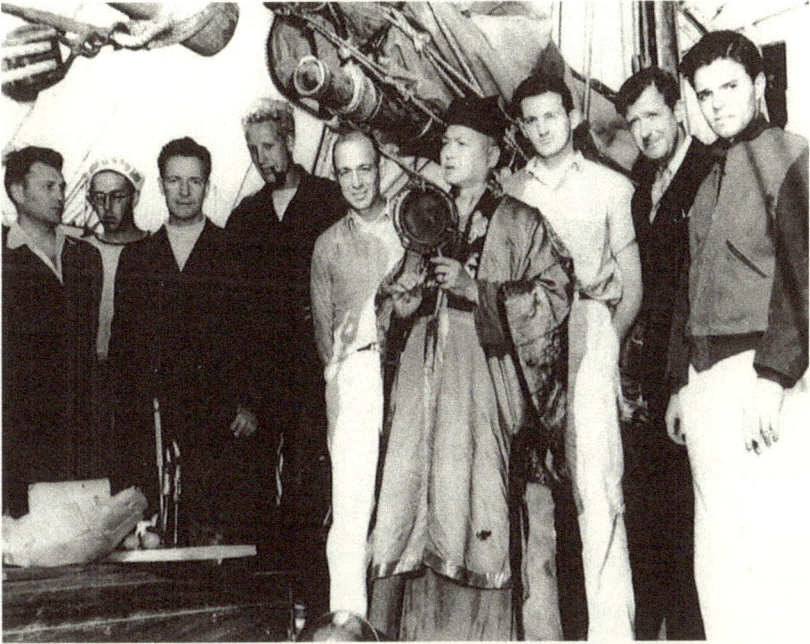

A Chinese priest blesses the *Sea Dragon*'s first effort to cross, ca. February 4, 1939. From left to right are Paul Mooney, George Barstow III, Richard Halliburton, John Potter, priest James Sligh, Captain Welch, and Robert Chase (courtesy Princeton University).

A junk about the size of the *Sea Dragon* alongside the SS *Coolidge* (author's collection).

Artist's rendering based on Halliburton's report of turbulence the *Sea Dragon* experienced during a storm on its first attempt to cross the Pacific. The drawing was published with "Log of the Sea Dragon #13," March 2, 1939 (*San Francisco News*).

Richard Halliburton next to his *Sea Dragon*
(courtesy Rhodes College's Barret Library).

Hong Kong's annual dragon parade roughly coincided with the first sailing
of the *Sea Dragon* in early February 1939 (author's collection).

One of the most famous photographs of the *Sea Dragon*. Although it appears to be Captiain Welch, the figure seen on board has not been positively identified (courtesy Rhodes College's Barret Library).

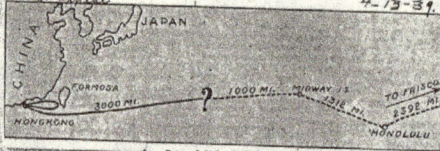

Generalized map of the *Sea Dragon*'s final route published in newspapers (author's collection).

Still Missing

SAN FRANCISCO — (UP) — The fate of Richard Halliburton, world traveler and author, and his crew of 14 aboard a 75-foot Chinese junk remained in doubt today more than a week after the craft reported itself in the path of a typhoon some 1200 miles west of Midway Island.

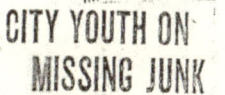

CITY YOUTH ON MISSING JUNK

Navy Searching for Lost Halliburton Ship

The U. S. navy today was searching the mid-Pacific for a Chinese junk reported lost with a crew of 10, including Richard Halliburton, author and adventurer, and a Minneapolis youth.

The youth is Velman Fitch, 21, son of Dr. E. L. Fitch, veterinarian, 5700 W. Lake street.

En route from Hongkong to the San Francisco World's fair, the vessel was reported lost Thursday 2,400 miles out of Hongkong.

* * *

Fitch left Minneapolis last November to work his way around the world. His father received a cablegram from him March 5 saying he had signed with Halliburton for the trip across the Pacific.

Minneapolis Man Among Crew Lost on Chinese Junk

VELMAN FITCH.
Lost on Ocean Trip.

Craft Sailed by Halliburton Is Somewhere on Far Reaches of Pacific.

A Chinese junk sailed by Richard Halliburton and a crew of 10 was lost Thursday on the far reaches of the Pacific ocean. One of the crew is Velman Fitch, 21, son of Dr. E. L. Fitch, veterinarian, 5700 West Lake street.

Halliburton was sailing the junk from Hong Kong to San Francisco's World fair. It left Hong Kong early in March and was expected to reach San Francisco in three months. Radio offices at San Francisco lost contact with the junk last Friday when it gave a position 2,400 miles from Hongkong. The junk was due to reach Midway island April 5.

The navy department gave assurances Thursday that plane, ship and radio facilities would be used in the search for the missing junk.

Dr. Fitch said he had received a cablegram from Velman from Hongkong, March 5, in which Velman said he had signed with Halliburton for the trip. Velman left Minneapolis last November to work his way around the world. He went to Germany and France for a couple of months, then proceeded to Singapore and Hong Kong.

Halliburton's junk has a 75-foot keel and a 20-foot beam. Besides sail it is equipped with a gasoline motor.

Navy Begins Search For Halliburton, Aids

Washington, March 31.—(P)—The Navy ordered a mid-Pacific search today for the Chinese junk in which Richard Halliburton, author and adventurer, set out with a crew of ten from Hongkong to San Francisco.

Navy Orders Search for Halliburton

WASHINGTON, March 31 (AP) —The navy ordered a search yesterday for the Chinese junk in which Richard Halliburton, author and adventurer, set out with a crew of 10 from Hongkong for San Francisco. Commandants at Honolulu and Manila were directed to instruct vessels in their area to keep a sharp lookout for the craft, whose radio had been silent since Friday.

The adventurer reported then that his position was about 2400 miles east of Hongkong.

Hometown newspapers carried the news of a community member lost in the *Sea Dragon* tragedy. These articles focus on Velman Fitch, who, in his own royal road around the world, hitched a ride aboard the *Sea Dragon* (courtesy Rhodes College's Barret Library).

29

History Is Made at Night

That sailors in seaports far from home should go a-roving and con-tract a venereal disease is hardly news. Bawdy limericks and sea chanteys were to be heard in every port in the world. Many were famous. "Who's that knocking at my door?" said the Fair Young Maiden. "It's me and my crew and we've come for a screw," said Barnacle Bill the Sailor. Or as one refrain ran, "Ship me somewhere East of Suez where the best is like the worst / Where there are no Ten Commandments and a man can raise a thirst."

Somewhere, El Dorados promised wealth, fountains of youth con-ferred eternal agelessness, and insular Tahitis offered romantic relief from the hormonal shakes, quivers, and chills. The *Sea Dragon* Expedi-tion featured an almost all-male cast; exceptions included the wives of Fat Kau, and the Scheherazades and Sonias whom Halliburton had seen in a Shanghai nightclub. Sexual intrigue, however, would play as lead a role in the fate of the *Sea Dragon* as it had once played in that of the *HMS Bounty*.[1]

For years, no one said a word about forays by the *Sea Dragon's* crew members into the Colony's hot spots and nightclubs. John Potter came closest to declaring that a scandal or two had occurred, but he spoke of it in a roundabout manner. He said that Gordon Torrey had gotten "some indigenous disease" associated with sailors and peculiar to the Asian continent. Potter failed to mention that he had come down with the affliction as well; he admitted only that he had received an injury when a boom crashed down on him during the junk's first run as he and George Barstow worked the tiller. Although Potter had come to Hong Kong with perfectly righteous motives, upon arriving he shoved them aside to make female conquest his one true aim. With his mother Bertha not around to chaperone him, George Barstow held a similar philosophy, adding whiskey to its wisdom. Little is known of Robert Chase's courting habits, but, he too thought that being beyond the reproving superintendence of the older generation permitted him to swing a hooky hand with impunity. Captain Welch was not so amused. Ashen-faced, he had "lectured [the crew] for hours and hours about what would happen to them if they played about with these Chinese women." Yet they paid no heed. "Even the Russians are all clapped up," he said, "and it's not a pretty picture to see them over at the free clinic every morning." Condoms of vulcanized rubber were available. If Captain Welch dispensed sexual safety information to his "innocents abroad," they clearly didn't receive it.[2]

Often hanging out with Halliburton and crew was the said John Rowland, who resided in the Colony from December 1938 and April 1939. In his opinion, Hong Kong, "one of the most action-packed harbors in the world," was "chaotic" and "very crowded." In the afternoons, Rowland met members of the crew for tea at the Gloucester Hotel as well as spent some evenings with them at the hotel's dinner dances. "[These were] wonderful times," he recalled, "hilarious times." He remarked that "the *Sea Dragon* boys" were instant celebrities who drew "pretty girls" to the dances. Intimacies ensued, and the lads hauled off one or another of these women to their hotel or to their quarters on the *Sea Dragon*.[3]

Captain Welch didn't mind the goings-on until they got in his way. Any sailor who could tie a knot knew about the *clap* or the *drip*, colloquialisms for gonorrhea, which Rowland called "Cupid's Revenge" and Welch called "the old complaint." In those days few people in polite society openly discussed the ailment; it was offensive to mention or even notice it. In dives and on the waterfront, such restriction did not apply. Privy to such things, Captain Welch noted that from one ship alone twenty-eight crew members were admitted to the hospital with clap resulting from unions with Asian prostitutes. The captain suspected that two of the *Sea Dragon* lads—he did not say which two, had contracted the disease and another one was about to contract it. Due to this "final blow," the *Sea Dragon,* expected to sail on January 29, was again detained.[4]

For weeks Welch had seen it coming. A month earlier, John Potter— ruling out an alternative diagnosis of typhus or scarlet fever, outright told him that he had contracted the clap. As he felt he must, Welch advised him to get whatever it was he had treated at once. Potter went to the hospital at once, likely the Bowen Road Hospital on the Peak. The event encouraged Welch to examine closely the articles related to the prerogatives of his captaincy. Concluding that Potter's and Torrey's incapacitation would force him to raise a quarantine flag, Welch judiciously decided that both crew members best be left behind. Apart from its place in naval regulation, contracting a venereal disease brought shame, one to which the families of upper-crust individuals were especially sensitive. Fearful that he might be held responsible, Halliburton tried to persuade Captain Welch to tell the consul that Potter "contracted his disease *before* he signed on the ship." In his refusal, Welch emphasized that the Richard Halliburton Corporation was "legally bound to take care of [Potter] until he . . . returned to the U.S." A sick crew member also meant an idle one.[5]

A week after he reported his condition to Captain Welch, Potter returned from the hospital and said he was "all set to go." Skeptical, Welch told him he needed a doctor's certificate to confirm that he was indeed clear of disease. Potter, unable to produce or acquire such a document, couldn't be signed on. When Halliburton learned about the incident, he "got sore" and "demanded that (Potter) be signed on anyway." Together he and Welch made their way to the consulate and were told that "if (Captain Welch) carried the man to sea and anything happened, (Captain Welch) would be held responsible for it." Further discussions and at least one additional visit to the consulate followed. Just days before the *Sea Dragon's* scheduled departure on February 4, Welch had "an interview with the Public Health doctor at the Consulate." This physician asserted that the captain "might take a chance on Potter" and would not be directly responsible for him. Gordon Torrey, who had also contracted the clap, "made no fuss about being let out." Welch gave him "his passage money back to the States," saying only that he had "paid him off." Torrey lingered in the Colony, went home to the States, then returned to the Far East. Later he worked for CalTex and saw the brutal results of war from the sidelines while stationed in various parts of Asia. Chase, meanwhile, had gotten an ulceration or *chancre* that would be treated later.[6]

When two of the lads (unnamed) approached Captain Welch and told him that they wanted "to get married to a couple of the local tramps," the whole business took a bizarre turn, one which again had Welch running to the "articles" to see what mandate he could enforce. What most flustered him was that these women had come on board the *Sea Dragon* and now wanted to join the crew on their voyage. If they stowed away on the ship, Welch vowed to "lock them up and hand them over to Police." The lads themselves had become "the object of a slightly amused curiosity." "Rather like a bunch of semi-clowns on the loose," Welch said, "the lads all show their excellent breeding in all the best bars and even at the mess; I have been with square-head seamen who had better manners." One square-head, "an old Swede skipper" come to wish Welch luck, did a once-over of the *Sea Dragon*'s "gallant crew" and shook his head. 'With these bunglers,' said the look in his eye, 'you'll need all the luck you can get.'[7]

Instructor Welch's report card for the lads read 'poor.' They had flunked every test. "Not one has done a goddam day's work since they came here." Slackers sat in his classroom, not sailors. One day, while attempting to accustom them to their different roles aboard the ship, he hollered out to them to "set the mainsail." Instead they ran for their cameras. "Some of the things that happen here during the day make me weep," said Welch. Once, while backing the junk "out of the dock to swing compass," he asked the "mate" who was stationed on the poop "to let go the stern line." The mate, "all done up in a fiery red silk jacket," was none other than John Potter. "He let it go alright," Welch fumed, "lifted the 'filthy' thing off the bitt, and threw the whole goddam thing over the side! The dock master was standing there and I thought he was going to faint." At the sight, "all the Chinese on the dock yelped like the devil and I almost hit the dock in my bloody rage."[8]

Crew incompetence became a regular amusement to dockside observers who saw the ship become more of a bordello than a sailing craft. "The whole town takes it as a huge joke now," Welch remarked, "because we have been talking about sailing for weeks." To say good-bye one day and be seen again the next was now pointless. About wasted opportunities, the captain had much to say, but he limited himself to brief remarks about the slackers under his command: "My great discouragement is that this bunch of lugs never attempted to learn a thing about the ship." One lad he credited with knowing how to use a compass, but, he said, "Not one of them can steer" or "take over a watch." The comedy team of Potter, Torrey, Chase and Barstow, knew how to party. To that end, they had turned their "room" into what looked "like a brothel," a "silk bathrobe" belonging to one of them often blocking Welch's way into the hold. Besides enjoyably suffer "love trouble in the groin," the lads liked to eat. Sneering in contempt, cook Sligh told Welch that they went through twelve days' stores in five.[9]

Faced with crew dissension and incompetence, Welch kept his eyes peeled for backup crew, and probably let it be known to fellow customers at the haunts he patronized. Although any post for an unemployed sailor was at the moment seen welcome, an opportunity to serve on the *Sea Dragon* did not suit everyone. Pumped up and ready to go was a "New Zealand aviator." When Welch told him that his quarters would be in the main cabin alongside "the sweet one," the fellow "balked," said Welch, and "backed out cold." Other potential enlistees, one gathers, may have backed out, citing the dangers of the mission.[10]

Crew troubles, meanwhile, took new turns. Sligh, the hotheaded cook, had gotten into a spat with a local merchant and was hauled off by police for disorderly conduct. Halliburton came to his rescue and paid his fifty-dollar fine, an amount roughly equivalent to a week's wages for the average American worker. An infuriated Halliburton, not so sure he wanted to cough it up, asked Captain Welch to fire Sligh, perhaps thinking John Rowland could replace him. Welch refused, saying only that Sligh was "a perfectly swell chap." Neither Welch nor Sligh thought much of Patrick Kelly, a Portuguese American whom Welch called a "mess punk" and "Halliburton's girl." Seeing him every morning lounging around the main cabin in his "robin blue dressing gown of silk," the captain had to chuckle. Told that Kelly was from Macao, he had to wonder "What joint (over there) *she* served her time in." The minor subplot thickened when Sligh, learning that Kelly had spent the night in a hotel with Halliburton, refused to sleep in the same room with a fellow whom he thought a "common whore."[11]

The Potter-Torrey and Sligh-Kelly flare-ups made Welch ill at ease. Unless certain changes to his authority were met, he threatened to resign his captaincy. "It's a lovely ship this," he said. The sights around him got more disturbing to him and put him at his wit's end: "Red silk robes and robin blue jackets or the other way around, mincing walks and lisping voices that shriek with impotency when the going gets tough and they [Halliburton and the lads] find themselves up against the inexorable log book and U.S. law." Next Welch placed at Halliburton's door a copy of the "articles." These in no uncertain terms indicated that Halliburton was only a "seaman" who could not make independent decisions concerning personnel matters. A "showdown" between employer Halliburton and his hired captain occurred around January 28. Welch believed he had told him off: "(I made it clear that) neither he [Halliburton] nor anybody else [was] to have any say in the navigation of the ship, either while the ship was in port or while she was at sea. . . . (So) I told him [Halliburton] that the U.S. Consulate would take no notice of anybody but me in matters of ships business. That I would file charges against him before the Consul if he tried any more to delay my sailing without good and sufficient reason. By the time I was finished she broke down and cried." Welch said he suddenly felt "a bit sorry for" Halliburton, but, he thought, how disastrous it would be if he yielded his command to the incompetent Halliburton. In any case, he won his battle with Halliburton, or so it seemed, as a new contract that resolved the situation "nicely" was ultimately drawn up. There was then the matter of the *cumshaw*, which Welch, with his itching palms, said was "excellent" and "all good clean money." Under Halliburton's nose—or past his blind eye, contraband—not freight—was loaded aboard the ship. The British authorities, once they learned Welch was Australian, glanced elsewhere and, though despicable "limeys," were "decent."[12]

The *Sea Dragon* continued to be a cause of wonder in the Colony. "The old timers laugh at her," said Captain Welch, "and the Chinese say that she is no longer a junk but a coffin." Fearing that the ship might be quarantined, the disgruntled captain paced about. Often he complained that the lads kept dashing up to "the Doctor" and returning with a medical packet. "When Halli gets the news," he noted, "she claps her hands over her forehead and moans, 'Another?' My God, Captain! What are we going to do? Let's sail out of here before we all get it.'" Only the engineer, the radio operator, the cook, and Welch himself hadn't caught the clap. Halliburton hadn't caught it either, and Welch sneered, "There isn't much chance of her getting anything like that wrong with her." Given cause, the captain decided it was best to enforce the articles, "sign them (the sick ones) off," and replace them with a fresh crew—twelve men would be ideal. In his dispatches home, Halliburton, more diplomatic than Welch, said that able seaman Gordon Torrey—to whom he had already given walking papers, at the "very point of an earlier departure was taken *ill* and whisked from the junk to the hospital."[13]

Of all the crew members, Paul Mooney seemed the ablest in Welch's opinion: "The odd part of this," he said, "is that the very chap who has sworn to do no work has turned out to be the most willing and the most decent of all this messy bunch. He is H(alliburton)'s secretary. He (i.e., Mooney) works till all hours at night typing her stuff, then never fails to appear on deck at a reasonable time, get into dungarees and starts to clean brass or sweep the deck. Knows nothing of sailing but is willing as hell. Called me aside and said that in his opinion two of them [the lads] had the wing-up and were using the complaint (clap) as an excuse to back out of the trip." Halliburton called Mooney a "great help" and a man who had "thrown himself heart and soul into the project." Likely Mooney used an energizing drug to increase his work acumen while the lads drank booze. Whatever it had taken, the hardworking Mooney had become a full-blooded character in whatever narrative the soon-to-be retired Welch intended to write about the voyage. The *Sea Dragon* had also become a fully developed character, the unhealthy "child" of a sympathetic captain. Although it "broke (his) heart" to see the ship "running up in the wind" and listing, Welch told his friend, "I wish you could have seen her under sail." The "old timers" told Welch the junk was "lovely," but he knew its flaws only too well.[14]

Most of its trial runs brought to an end, the *Sea Dragon*'s dress rehearsals for the Treasure Island fair began. On Thursday, January 19, the junk, moored by the central waterfront at W. S. Bailey's (or Pier #1) in Kowloon, opened to public view Friday and Saturday. Docked alongside big sea-tested ships, one of them self-assured and even smug, the debutante junk became prey to wondrous and jeering eyes. One photograph shows Halliburton speaking "to a Chinese shipwright," possibly Fat Kau. Behind the bystanders a sign reads, "Storks and Cranes." White buildings with an administrative look appear to one side, Alamo-like buildings to the other. The *Hong Kong Daily News* published photographs of the junk and its crew, noting that the *Sea Dragon* would set sail for the "Lagoon of the Trade Winds at the *San Francisco Fair*" on Sunday, January 22, at six thirty in the morning. While crossing the Pacific, the ship would reach Keelung, Midway, and, at last, San Francisco. Crowds gathered for a last look at the junk—now at Queen's Pier, then at Bailey's.[15]

In Hollywood, the "Scarlett O'Hara business" had been settled, with "British beauty" Vivien Leigh winning the part. On January 26, production began on *Gone With the Wind*. Production in Hong Kong on the *Sea Dragon* Expedition, meanwhile, stalled. January 22 had passed, as had January 29. Welch was anxious to depart, but "the sweet one" kept fiddling while Rome was about to burn. Providing some incentive for the *Sea Dragon* to leave may have been news that the *Pang Jin* in just twelve days had reached Singapore. Homesickness of course was a persistent incentive. When Welch said, "I want to get home so badly," he voiced the sentiments of the entire crew. "I miss the apartment so much and the battles we used to have," he told Wetjen. "I'll fill your shell-like ears with the goddamndest story you ever heard when I once get on that blue davenport, old sock." Now and again, the captain turned to the "two logs" he kept. "They read like a misadventure," he said. "All the truth is there for all to see, and it can't be rubbed out." The logs are now lost.[16]

30

The Awful Truth—Captain Jokstad
Inspects the *Sea Dragon*

For days the *Sea Dragon* sat in the harbor with carpenters and caulkers "tighten(ing) the ship throughout." Captain Welch and Henry Von Fehren "noted every weak point in the junk" and jointly concluded that it would take at least a week to recaulk "some of the seams and refit some of the equipment." Halliburton called these "minor details," but, bitterly disappointed, he lamented that "his first sailing date," January 20, had passed a week before. An earlier sailing date had been January 1, the day he wrote to tell his parents that his crew was "all assembled" and that the jacket he had promised his mother would soon arrive. Halliburton did not discuss mechanics or the operation of the junk, subject areas his civil engineer father might have wanted to hear about. He did send his parents color negatives showing the junk as it had looked just before Christmas. Since then, "dragons" had been added to the ship's stern and these "improved (it) vastly." He offered only one bad tiding: "Some of her peach (colored) paint" had come off when the *Sea Dragon* plunged under the waterline, and this had to be addressed. Troubled that some of the "circus-wagon color" might rub away before the ship and crew reached San Francisco, he seemed little worried that the hull might leak.[1]

By now a fixture in the community, Captain Welch slowly warmed to the sudden attention he received from the crowd. Well-wishers, stopping him in the street, wished him "Bon voyage." On the day the *Sea Dragon* was to sail, "an elderly woman" came that morning to the ship and handed Welch a cross of Saint Christopher for him to clutch onto "in times of stress." Soon he had the statue "working overtime." The captain had brought the big top to town, and the press interviewed him "as a curiosity and a chump." As for Welch, he saw himself not as a captain on a mission but as someone "sent from heaven to amuse the Chinese." Rather than bring China to the fair, Halliburton had brought the fair to China. As supporting cast to the spectacle, the crew now moved from the shipyard to "the public pier where the big ships dock(ed)." Mobs of visitors—tourists as well as resident Chinese, Americans, Europeans, and Eurasians, came to meet them. Welch, who had seen the lads perform little work, now saw them sign autographs as if were their only job. "Some days are almost tragic," he remarked, "but some are so funny that I laugh at them."[2]

As he had conducted tours of Hangover House—he charged admission, Halliburton now conducted tours of the *Sea Dragon*. "The curious came to gape, to wonder and admire the *Sea Dragon* (which) resembled some gay and fabulous toy." Arriving in droves were youngsters and adults, so many that soon Halliburton had to fix "visiting hours." One day twelve Royal Air Force officers came to view the junk and with them workers from the American warships in the harbor come to wish him "a good voyage." Given the junk's "tremendous box office appeal," Halliburton predicted that "thousands" at the fair would flock to see it. Long a lecturer, he was now a ringmaster, beginning his presentation in words akin to the following:

> Our foresail has been dyed yellow, the mizzen-sail, vermilion; as an extraordinary gesture toward simplicity we've left the mainsail white . . . On either side of the poop a Chinese artist has painted a ferocious red and yellow dragon twenty feet long, not counting the curves! . . . A junk lends itself magnificently to color. We emptied the rainbow on hull, stern, and sails. The hull is a brilliant Chinese red, edged at the rail with bands of white and gold. The *glance* of the eyes is black. . . Our Chinese artist used the traditional colors. The crescent-shaped hull is bright vermilion. A twenty-foot blue and green dragon—very ferocious—climbs up each side of the poop. Our sails are dyed a rusty orange. . . . If all this sounds

like a circus wagon, blame not me, but the last hundred generations of Chinese junk builders—for the most careful research proved that this style, and none other, is proper for the *Ningpo type* of junk which we have built for our voyage.

While guests on board might marvel, Captain Welch could only wonder whether the ship's appearance mattered more to Halliburton than its seaworthiness.[3]

Those aspects of junk construction that *Hummel Hummel* captain Dr. Petersen considered superficial were essential to Halliburton who promised his fans a "brilliantly colored oriental craft." The adventurer had conferred with several artists concerning the size and placement of the junk's "eyes." "Rare pictures of famous old Ningpo junks," possibly from postcards, or from Ivon Donnelly's classic *Chinese Junks* (1924), might have inspired the *Sea Dragon*'s "eyes" and other "decorative schemes." Like the blue and green dragons on each side of the poop, the junk's avian iconography might have been, with thematic suggestions from Halliburton, the Chinese artist's own creation. Most eye-catching was Halliburton's "special pride," the stern and its boisterous depiction of a "mythical Phoenix with feathers of brightest hue." Proud emblem of the Golden Gate International Exposition, the phoenix also assured "good luck" (as did other birds classed as *fenghuang*). Its talons extended in readiness to extinguish what vipers the evil forces of nature might send to destroy it, the bird also ensured guidance and protection. "The brightness of his feathers makes up, perhaps, for the fact that the native painter gave him only one leg. Above the phoenix is the ship's name and port—*Sea Dragon*, Hongkong—in great gold Chinese characters. Below the bird is a scene from Chinese mythology, with Oriental angels riding through the clouds on peacocks." Besides wisdom, the brilliantly colored peacock symbolized speed and hasty delivery to one's destination.[4]

January wore on. Welch apologized to Dick Wetjen for not having written for a while but believed his friend would understand. By now, Wetjen knew his friend was desperate to leave, but also knew Halliburton, basking in attention, wanted to linger. There was of course more to it. According to Halliburton, "shameful excuses" explained the departure delays. If a new quarrel did not begin among the crew members, a technical issue arose with the junk. When would *late* become *too late*? Now should the *Sea Dragon* leave Hong Kong on February 1, it would arrive in San Francisco in May 1. One obstacle overcome, another quickly took its place. Even while Welch was writing to Wetjen, "one of the lads" reported that he was suffering from a groin issue. Next, Paul Mooney slipped down the hatchway leading to the main cabin. He was rushed to the hospital with a presumably broken ankle, and his doctor advised him to remain in bed for five days, after which a plaster cast could at last dress his foot. Two thousand gallons of water, meanwhile, were loaded aboard the *Sea Dragon*, along with "enough oil to fuel [the] auxiliary engine for ten days, and a three-months' supply of food for twelve men." As the junk sat in port and the crew satisfied their hunger, food stores soon dwindled. Halliburton said only, "We must replenish." How to cope: "A mental wreck" from "so many shipbuilding battles," Halliburton looked feverish and emaciated, and the "trembles" he had experienced on the lecture tour now returned. What else might recur? In Phnom Penh seventeen years earlier, he was stung by a "poisonous insect" and contracted "a bit of fever" which laid him up for days. Now, as the Hong Kong ordeal seemed near an end, he "was in bed in the Kowloon Hotel," writes Jonathan Root, "with a raging eczema the doctor attributed to nerves." To his parents he kept mum about it, telling them only that he "worked hard" and had "plenty to worry about," but "I love it," he said, "and have never *worried* about anything as exciting as this."[5]

Covering the junk for $10,000, Lloyds certified that the *Sea Dragon* was ready to sail. Among probable other caveats, the underwriting giant recommended caution against the wind with a ship having so high a poop as well as provided information on the safest travel routes. Riders were in place, including for any damages, accidents, or catastrophes that were provable "acts of God." Coming aboard the *Sea Dragon* "before and after its completion," Chief Officer Dale Collins of the SS *Coolidge* believed, as did the Lloyds surveyor, that it "certainly seemed a sturdy craft and well-constructed." Collins, however, thought that "her poop deck (at) about ten feet higher than most Chinese junks of this size" might be too high. He further noted that "her masts, especially the mainmast, appeared very heavy and lofty in proportion to the size of the junk." Six weeks after Collins evaluated the completed *Sea Dragon*, Captain Charles Jokstad, whose SS *Pierce* had "steamed into Hong Kong" on January 26, made his own assessment and found that little had been done to remedy the junk's defects. In his memoirs, Captain Jokstad described Hong Kong's harbor as "one of the most picturesque in the world." Here "ships of all nations (were) tied up amongst the strange craft of China," but "nothing" was more fantastic than the *Sea Dragon*: carved, gilded and painted in dazzling colors, it was the gaudiest thing on the waterfront."[6]

Seventeen years earlier, Halliburton had been a stowaway aboard a freighter bound from Surabaya to Singapore. Charles Jokstad, the youngest person ever to serve in the US Merchant Marine Corps, had been the captain of that ship. Now a world-famous author whose works were found in bookshops in every port, Halliburton was courteously led aboard the *Pierce* and taken to Jokstad's quarters. No longer the young man who had scaled the Matterhorn and swum the length of the Panama Canal, he impressed Jokstad as "personable (and) adventuresome." Touched by his visit, he remarked how much he "enjoyed" reading his books. After a handshake and a bow, Halliburton explained his purpose in Hong Kong and his particular reason for the visit—recruitment. He said his crew was presently composed of "college boys," each of whom "had paid anything from $1,000 to $1,700" to participate in the voyage. They "were to share in the profit" of the book he intended to write about the voyage as well as "share in the profit from admission fees at the San Francisco Fair." Although a good deal, Halliburton, sighing, said "that he had had difficulty getting a full crew together," then asked Captain Jokstad if he could "release one of (his) own men who wanted to join the *expedition*." Already able-bodied seaman Ben Flagg had jumped the SS *Pierce* to join his crew. Although saddened that Gordon Torrey was "ill in the hospital," Halliburton was pleased that an experienced shiphand like Flagg could replace him. Jokstad said he could neither spare a man nor keep one from skipping ship.[7]

The topic turned to geography. Halliburton "hope(d) to keep far enough south to escape the worst of the gales," but he told Captain Jokstad that he abandoned the southern route by way of Manila. Instead, he had chosen a course that would take him "north to the latitude of Yokohama," then "eastward to Honolulu." Jokstad told Halliburton that "at that time of year it was the most dangerous (route) he could take, as he would encounter heavy northwesterly and westerly gales which sometimes would increase to storm force." He suggested "a trial run along the China Coast" as an alternative. As the safest course for the crossing itself, he then recommended that Halliburton "head north and east to the south end of Formosa and then steer in an easterly direction until he got the longitude of Honolulu." If he did this he would have "fine weather all the way." In a kindly gesture, Jokstad mapped the route out for him.[8]

After lunch, Halliburton invited Captain Jokstad to tour the *Sea Dragon*. For the record, the informal inspection that followed became the most important eyewitness testimony of the junk's merits and its chances at sea. Jonathan Root portrays the "salty old sea dog" Jokstad as plainspoken and genial "A man of few words," he was now "at his most terse," letting Halliburton know at once that he wouldn't take the junk "ten feet from the dock." An ill-omen followed. As the two ambled along the wharf toward the *Sea Dragon*, Jokstad saw a rat scurry down the narrow gangplank and disappear into a rubbish heap. At the sight, he told of the belief held by "the white race" that a rat leaving a ship portends doom at sea. Learning that some members of the crew were Chinese, he shared another dread omen: "If a vessel has a Chinese crew," he said, "they will invariably bring cockroaches on board in a bottle, as their superstition is that if a ship has no cockroaches it is bad luck." Halliburton said that he cared little what his crew members brought with them.[9]

Jokstad, said Root , "strode up and down the deck in silence." Captain Welch, "on deck supervising a gang of Chinese workmen who were caulking seams and painting," watched his roving eyes. Jokstad, who "knew" Welch, "greeted him coolly, further undermining Halliburton's confidence in his captain." Within earshot and heedful, several members of the crew stayed quietly in step with the imposing Norwegian, stopping when he stopped, moving forward when he moved forward, all the while listening to every word he said. When Jokstad told Halliburton "that the vessel would never reach the United States under the present conditions," said Root, "a couple of them turned practically white." Jokstad then fired a volley of critical potshots at this or that aspect of the junk. Up and down he looked while attentive student Halliburton followed his line of vision. The sails and masts seemed okay, he remarked, but the shroud lines and stays needed to be bolted, not screwed to the planking: any strong gust of wind could snap them. In his autobiography *The Captain and the Sea*, Jokstad wrote, "If the vessel got into a heavy storm these leg screws would pull out and the mast would go overboard, probably ripping the deck along with it." Root, who contacted Jokstad for this episode of the *Sea Dragon* story, reported as much. Getting the idea, Halliburton replaced the screws with "galvanized iron bolts" and made "minor repairs" to the rigging.[10]

Of greatest interest to Captain Jokstad was the rudder—described by Captain Welch as "seven feet long, planed and iron stropped." Without a strong rudder, one solidly attached to an equally strong tiller, a ship cannot be directed or steered; a rudderless ship is just a floating raft. On the *Sea Dragon*, as was typical of many junks, a tackle, "fastened to the end of the tiller and let out to the rail," adjusted the rudder's height. To help the lads better manage it, Captain Welch rigged an "endless fall" or rope line to the tiller, but, still, its handling eluded them. Deeply curious about this main element of the junk, Jokstad "climbed up to the poop deck," writes Root, "and waggled the tiller" which was secured to the user end of the rudder. In his book Jokstad denounced the rudder as "a very flimsy affair, far from the proper rudder to take a craft of the *Sea Dragon's* size across the Pacific." He also observed that the stock of the rudder was far too close to the hull, maybe an inch or two away, and far too loose as well. Were a good strong wind to come by and boldly strike the rudder, it might snap off or be broken in two. During a storm, in a worst case scenario, it could rip free from its post and cripple the ship. Again getting the idea, Halliburton had "workers from the shipyard fit a wooden collar around the top of the rudder stock which seemed, for the time being, to end its wobble." In a last turn of thought, Jokstad pronounced the rudder not quite right for the *Sea Dragon*. Even if it were, the assembled crew was, in his opinion, no match for its handling.[11]

Next, led by Halliburton, Captain Jokstad stepped down the hatchway. Once in the hold, he "peered into the bilge with disgust." Roaming about, he "poked gingerly at the glass ports in the main cabin with his fingers," then "shook his head balefully." This time he was not so terse in voicing his disapproval. "'Those aren't portholes," he explained, 'They're windows; the sea will smash them before you've gone 10 miles.'" Biting the bullet, the cost-conscious Halliburton ordered "steel sheeting placed over the cabin windows." Next, Jokstad "studied the diesel engine" and did so "with an expression of incredulity." He did not wonder that it should be there, only that it was attached to the ship's ribs with leg screws, not bolts. A little pitching and yawing on the open sea "(will) shake it loose." Again biting the bullet, Halliburton had the engine re-mounted with bolts or, since bolts were not easy to come by, with more leg screws [12]

Captain Jokstad concluded his inspection with random comments about the *Sea Dragon's* general stability. Even as the ship lay idle, he felt far too much motion on and below deck. If he saw the ship on maneuvers in the harbor, he might also have noticed its tendency to list. "The ship rides too high in the water; it'll blow around like a cork," the captain exclaimed, telling Halliburton, "You need at least *another* 10 tons of ballast." *Another* ten tons would imply that the junk *already* had ten tons of ballast—perhaps the fuel tanks and food stores. In a related matter, Root noted that Welch "dumped 10 tons of *concrete* ballast in the hold." "Dumped?" Or poured? Was it in the form of stones, rocks, cinder blocks, bricks, parts of bricks, or loose gravel? Did the ship really need "ballast?" If added, was it properly distributed to prevent listing? Gordon Sinclair said that "many a man told (him) that the junk's engine, her extra water and fuel tanks, her navigation equipment and wireless, made her too heavy for even a calm crossing of the mighty Pacific." Halliburton and Welch also thought the *Sea Dragon* too "heavily built." Why then should it need more weight? John Rowland, who saw the interior of the hull, described it as "one vast space like the inside of a cathedral," but once provisions were loaded aboard, little room could have remained for ten or twenty tons of concrete ballast. True to character, Halliburton wondered if such an expenditure was necessary. He told Jokstad that "they had spent all their money, and much more than what they were supposed to spend." What repairs were done were makeshift, hurried, and the most affordable. During his next docking in Hong Kong, Captain Jokstad was "informed that Halliburton had nailed some boards over the portholes, had made some minor repairs to the rigging, and had taken ballast." Sealing his fate, "he had not changed the rudder, nor had he reinforced the engine."[13]

To his credit, Halliburton *did perform* the trial run that Jokstad rec-ommended—almost certainly the late January "cruise around the harbor" when it was "found that the junk drifted sideways and would not stand up against a stiff breeze." Welch said nothing about any repairs the junk might need. He did mention, however, that Halliburton "got all wet with perspiration" from "the workout" of manning the tiller. "In calm seas, with a steady wind," Halliburton said, "one man can handle this apparatus, but when the waves run high and subject the ship to violent motion, it takes two men to hold the course." Whatever it took to handle the tiller, the captain found any of Halliburton's attempts at 'heterosexual correctness' laughable but congratulated himself for holding back his more sexually explicit free associations. "I was tempted to ask her a slightly smutty question," he said, "but figgered I shouldn't."[14]

Besides highlight Halliburton's incompetence at the tiller, the trial run demonstrated those intrinsic faults of the junk identified by Captain Jokstad. Supporting Jokstad, Captain Welch said that the ship was "very heavy with the engine and fuel and a trifle cranky with it all" Moreover, "She is slow under canvas and can only do six knots with the motor opened up full speed." He also observed, "[The ship] carries too much weather helm, which is very bad and runs up in the wind as soon as the helm is amidships." Alas, "a light breeze" (will cause) the ship to make "two points leeway." Against the "strong monsoons blowing along the Straights [sic] of Formosa," the *Sea Dragon* might list fatally, with the high poop the culprit again. In appearance and performance, the junk looked to him like "a honky-tonk wagon on the loose." 'First Ship of the Admiralty' it was not. He had "fought hard" with Halliburton to remove it, but as he was "not well fixed at the time" and lacked "the cumshaw coming in," he let the matter slide. Still, score one victory for Welch who did persuade Halliburton to remove the two gaudy "5 X 8" figures" mounted on the rigging—almost certainly representations of Tai Toa Fat—by arguing that they might make too alluring a target for the Japanese.[15]

If only by half measures, Halliburton acceded to some of Captain Jokstad's more respected advice. Ultimately, however, he fully ignored the route Jokstad recommended to him. "He had decided to take his chances," wrote the man whom Halliburton called the "most competitive navigator on the Pacific," and "head up north through the Formosa Strait, up to the latitude of Yokohama and then head east to Honolulu." Jokstad thought Halliburton a "fool" on a fool's errand. Many in the shipping community agreed. Regatta stuff at best, the *Sea Dragon* appeared incapable of crossing Lake Michigan, let alone the Pacific Ocean. "In a tone of annoyance," writes Jonathan Root, "Halliburton told Jokstad, 'I've got to (make the voyage). I'm committed. Anyway, adventuring is my business.'"[16]

In 1937 George Gershwin's musical *Shall We Dance* introduced the song "Let's Call the Whole Thing Off." By the end of a very long January, more than one member of the *Sea Dragon* crew, deep green about the gills, shared the sentiment. "Personally," reflected Captain Welch, "I think they have cold feet about the trip! These gallant heroes who sat drinking their rums up in the Mark Hopkins Hotel are no longer the enthusiastic hot shots who landed in a 'blaze of publicity.' Somehow the spark has died. The junk looks woefully small alongside of the *Empress of Japan* and the SS *Pierce* and I am inclined to think that is the reason. Well, here I am and here I stay until it actually blows up and we all come home in complete disgrace and shame." What fire of adventure once burned in the lads now flickered. The fire still raged in Halliburton, but, once fueled by hope, it was now fueled by fear of want. "If Paul and I sink together," he wrote home, "the insurance policy made to him goes to you." Only two months before Halliburton had spoken of his "finances" being "well in-hand"; now he feared bankruptcy. He could only blame himself; after all, he had forked out $500 just to have another dragon painted in his cabin. "Riggers' bills don't mean a thing to *her,*" Welch said, "only whether the snake on the stern is alright. Halliburton made "frantic appeals to America for more capital." The thousand dollar bill had dwindled to a few twenties, and investment capital in the bank had largely vanished. Halliburton owed money— enough to frighten him. "When we get the *Sea Dragon* to the Fair," he said, "[I] expect to clear up all my debts on everything, from the box office profits. Meanwhile, the *Marvels* books will sell and sell." As to recouping any losses, he said, "The first week's income at the Fair may have to go to paying the crew's wages—but at least we (arrive) home—and *safely.*"[17]

31

A Gentleman Knows
When It's Time to Leave

The Dragon Parade had begun. Lasting a month, from January 19 to February 18, its celebration suspended most banking and government operations and brought most businesses and schools to a standstill. A main feature of the event was the colorfully painted, scale-plated dragon float that, to the coordinated steps of the costumed dancers within its massive frame, coiled and writhed down Queen's Road. As more celebrants carrying lanterns attached to the creature's ever-lengthening tail, flares ignited, sticks rattled, cans clanked, and swarming crowds cheered. Symbolically, the pageantry meant that debts were annulled and old friendships rekindled. What was sowed would be plentifully reaped, and in the broad scheme of things, rebirth awaited those who had died. Following a feast would be public exultation. Hatched from the discharge of emotion, the *Sea Dragon* itself seemed an egg. Also it seemed like a sacrificial lamb brought to the slaughter. As they only brought on another "ghastly" delay, Halliburton little noted the Chinese New Year activities. Thanks to them, he could now expect to be only a few miles out to sea when the Golden Gate International Exposition opened to visitors in two weeks, on February 18. From March and then April, arrival time was now moved to May. Good luck and perfect weather would bring the junk and its crew home in two months, or by April 1, at least to Honolulu where they planned a five-day stay.[1]

The repeated delays that rattled Halliburton's nerves also stretched his pocketbook. In a letter home, dated January 23, he reported that he had "borrowed $4,000" on his house in Laguna Beach—perhaps by now even more; the property he assessed at worth $20,000." He also noted that he had "1,500" in an "emergency account" in San Francisco. David Laurance Chambers, friend, fellow Princetonian and his editor at Bobbs-Merrill, advanced him $600 in royalties on "the Orient" book which, like the earlier "the Occident," was selling well. Some (additional) advance money may have come from the *San Francisco News* which published his articles. John Potter evidently threw in $500 dollars, an amount tripled by George Barstow. Bills stacked up. Time and money pressures intensified. Prices for goods, meanwhile, had risen, refugees continued to glut the streets, and bombs blasted near and far. Haste now mattered more than ever. Its moorage fees paid, insurance premiums met, and supplies loaded, the *Sea Dragon* appeared ready to face the dreaded monsoon winds on Friday, February 4. "This was the day I had waited for and worked for, these many months," Halliburton enthused. "The discouragements, the quarrels, the despair that had hung over us for the long period of ship-building and out-fitting vanished. In the battle with bills and Chinese procrastination, the words romance and adventure had faded from our vocabularies. But now, in one exhilarating moment, these words came back again." Perseverance had overcome all obstacles and dissolved every misery, failing, and misstep. The truculent Mr. Halliburton had striven and not yielded. Eleven years before, as the SS *Richard Halliburton,* he dove into the Panama Canal to begin his eight-mile swim from the Atlantic to the Pacific. The din from the crowds that had gathered then matched with the din he heard now. The camera crew's cameras again flashed. Presenting opportunities for "suicide a dozen times," the Panama "adventure had bitten deep into *his* fancy" and made him the very embodiment of 1920s daredevilry. One reporter noted that while Theodore Roosevelt "built" the Panama Canal, Richard Halliburton swam it.[2]

Today as then, camera crews were in place to capture the moment, and, said Halliburton, "half the town crowded unto the docks to wave goodbye." Director and main actor in his own play, Halliburton glanced about as the volleys of applause rippled through the crowd. The crowd was orderly at first, but soon the din became raucous, ominous, and dark. Hands clapped, and fingers jiggled upward to mimic spiraling flames. Brightly colored banners fluttered, rattles clicked, sticks clattered and whistles blew amid boisterous cheers and satiric cackles. Those who had stared in disbelief a week earlier now watched excitedly. Currently docked, the SS *Coolidge* "towered" nearby. Four months earlier the liner, soon to depart again for America, had delivered Halliburton to the Orient. "She was sailing too," Halliburton wrote, "and her crew and passengers joined in the chorus of good-byes." Two British navy planes, notes Root, "roared overhead by sheer coincidence" while firecrackers on the masthead popped. Scores of people lining the rail of the SS *Coolidge* waved toward the junk "without knowing what they were waving at." Many had toured the *Sea Dragon* "during the days preceding," said Dale Collins. These sightseers thought the junk "looked very picturesque in all her red, white and orange paint, complete with Chinese dragons and an American flag, made fast on each side amidships." Running into Halliburton beside his ship, Collins commended him for his bravery. "I am not sure that I am brave," Halliburton responded, "or just really naïve."[3]

If during that "exhilarating moment" Halliburton recalled a singular moment from his past, it was to his first lecture, called simply "life and religion" which he delivered to some kids at the YMCA. Twenty years had passed. Exposed now to bombs and refugees, he was then exposed to boos and hecklers. Famous now, twenty years before he was a nobody. When he addressed them, no one listened; they passed him by, talked over him, or rudely bumped into him. Oblivious, he had persisted. He devised better ways to address indifferent listeners and in time gave accounts of adventures that left audiences spellbound. He had entered the door that led to fame. Lecture agencies wanted to sponsor him, and soon he was filling auditoriums and earning deafening applause. Once, he told "the wildest stories" to a group of underprivileged children whose eyes bulged "big as saucers" with wonderment.[4]

Gear slung over their shoulders, seamen Robert Chase, George Barstow, and John Potter assembled on the deck of the ship. Flush with enthusiasm, they took to their assigned positions and smiled before the cameras. Captain Welch stood at the tiller, engineer Von Fehren at middeck, and mediator Halliburton by the mainmast. Lookout Paul Mooney, who had "refused to let (his injury) prevent his sailing," leaned against the portside rail. He had his camera at the ready, as did Barstow and Chase. Halliburton wore a leathery jacket fitted snugly over a woolen sweater. The lads wore boat-deck shoes and boater's breeches. Barstow wore a white sailor's cap, Von Fehren a skipper's cap without insignia.

Captain Welch donned a captain's cap *with* insignia, dark pants, and beige linen shirt with rolled-up cuffs. Radioman George Petrich waved and wore a black short-sleeve shirt. Dressed in white civvies, cook James Sligh twisted his scowling mouth into the shape of a presentable smile. Two Siamese kittens, two Chow puppies, and "three new men" joined these veteran crew members. Ben Flagg was the most, Patrick Kelly the least experienced of the new seamen. The third party, short-term Hong Kong resident Richard Davis, dropped out at the last minute. Davis, hired as a back-up to Henry Von Fehren, intended to go only as far as Honolulu. The "cherished possessions" of some aboard the *Sea Dragon* had already been sent ahead on "large freighters."[5]

Shouts in English cracked through a cacophony of many languages, hisses, catcalls, and weird sounds. The pitch of celebration soon reached hysteria. "Bon voyage" was often heard, as were garbled equivalents of "Fare-thee-well," or "Good riddance." For others, it was stunned disbelief that the gaudy ship was at last on its way. Told of Halliburton's mission, tourists considered the voyage as history in the making or as an exhibit of the great Fair itself. Cameras flashed, and reporters scribbled down notes. Delegates of the Portuguese community came to send off Patrick Kelly, who was one of their own. Members of the local orphanage Halliburton and Mooney visited might have come to bid their sad adieus. Sailors from other ships watched, and many applauded even as they wondered whether so small a craft could actually make it on the high seas. Eyes shifted from the *Sea Dragon* to its star passenger, who said he had taken extra precautions to ensure the voyage's success. "[I] burned an extra stick of thanksgiving incense before the shrine of our good luck god (*Tai Toa Fat*)," he clarified, "to speed (ship and crew) on our way to the Golden Gate." Hoisted above the tallest mast, an American flag rippled in a listless breeze. Halliburton waved from the gunwales as everyone clapped. Moments later, he stood behind the tiller like a maestro at the podium as a slight breeze rocked the ship. The two-foot plank leading to the deck was pulled away, the mooring lines loosened. "At 4:00 p.m.," wrote the *Coolidge*'s Dale Collins—Halliburton said *at two o'clock*, still plenty of daylight, "the *Sea Dragon* backed away from the pier in Kowloon, gave us three blasts on her small electric horn, and proceeded out of Hong Kong Harbor on her first attempt at conquering the China Seas." John Rowland called the junk "a beautiful sight," especially "when she made her way out of the harbor, with sails set, and the diesel engine making the prow show a bone in her teeth."[6]

A few crew members sent radio messages home and received return messages of congratulations. In his first message, stamped February 4, "4 AM," Halliburton said tersely, "Have Sailed NEXT PORT KEELING FORMOSA ALLS WELL HIGH SPIRITS LOVE." His dispatches more eloquently captured the moment. "The sun shone with unusual ardor on our lacquered hull and painted stern" went one, and " Our new American flag stood out handsomely in the breeze. We could not have asked for a more glorious—or noisier—departure. . . . Down toward the harbor entrance we went, our orange foresail set, engine tugging." The applause tapered off, the crowds at the dock dispersed. Once "outside the harbor entrance," writes Jonathan Root, "Welch ordered the immense white ribbed mainsail raised and, soon afterward, the scarlet mizzen. The ship picked up a brisk starboard breeze and started up the coast of China at better than six knots." Familiar landscapes soon slipped from view; the nine-thousand-mile voyage had begun. Halliburton's plan was to sail the junk northward along the main steamship lane, "then head across the north of Formosa, taking advantage of the Japanese current." Dale Collins, who had just brought the SS *Coolidge* down from Shanghai, told Captain Welch that the weather could suddenly change, and for the worse. "We had experienced strong Northeast Monsoon winds," he said, "but as the wind had moderated considerably that day he (Welch) thought he would try his original plan anyway." It was a good plan. "The first few hours out were splendid," Welch cheerfully reported; "the lads sat upon the whorehouse poop and sang loud paeans of praise at the weather and how splendid it was to be at sea." All was serene. "A poetry ship devoid of weight and substance," Halliburton would put it, a wonder of nature "gliding with bright-hued sails across a silver ocean to a magic land," so was the *Sea Dragon* those first moments when it broke innocently through the waves. "A warm twilight came as a huge moon rose out of the sea," Halliburton waxed poetic, "and (came) a pleasant starboard breeze (as) our foresail caught and drove us steadily up the channel between China and Formosa." "Clearly seen in the luminous night," the mystical "poetry ship" glided along.[7]

As "midnight came," wrote Halliburton, "the watch changed, but nobody, not even the Siamese kittens, would go to bed—not in this tropical night—not on this still and shining ocean." Just as the *Flying Carpet*, the airplane that had taken him around the world, had "dived toward the sea, and skimmed the waves," the *Sea Dragon* tilted toward the sky and touched the heavens. Vessels including passenger liners, freighters, and battleships were in the junk's path, and many of them sent radio messages of good luck to Halliburton and his crew. Through tangled waterways clotted with houseboats, and oily fields of rotting debris, the junk moved northward. Halliburton was "astonished" that no Japanese destroyers or pirate scows blocked his way. Three months ago not a single fishing junk was to be found, as the Japanese sunk any ship venturing offshore. Had the war ended? The sudden appearance of the gaudily painted *Sea Dragon* stunned the many fishermen who from their barges or junks stared wondrously at the unexpected sight that had floated into their dreary world. Some shouted to see whether it was a real ship holding real people. Happy for the applause, the junk's crew members shouted back. Shaking his fist, Captain Welch yelled at the fishermen to get out of the way.[8]

The junk proceeded up the northeasterly path chosen by Captain Welch. The waves freely broke way, the wind sweetly whistled. Traveling at an average speed of 6 knots, the *Sea Dragon* covered 100 miles in 14 hours. If the junk sailed 170 miles in a day and 1,190 miles in a week—"8000 miles in less than seven weeks," Halliburton calculated that they "would be home by the end of March." Their first objective, for now, was the notorious Penghu Islands, a network of landfalls just northwest of Formosa. The seaways to get there were treacherous. Waves bounded high and monsoon-force winds gusted to speeds as high as 40 knots. Gordon Torrey, who had come down this lane to Hong Kong, called it a most daunting corridor. Although the waters the *Sea Dragon* met were "not very rough," said John Potter, "the junk rolled considerably and with the movement there was a tremendous whip of stays and guys, with some chafing." Awakened by "a blast right between the eyes," the "blissful dream voyage" came to a punishing end. "At noon on the second day," said Halliburton, "with incredible suddenness, the sun departed, black clouds raced overhead, the wind swept down, the waves rose." As it headed into a storm, the *Sea Dragon*, prepared well enough for calm waters, was not so well prepared for harsh ones. "Under canvas," Captain Welch said of the junk, "she is slow." In gusting winds of great force (she) also leaned. Blame the masts with their sails—which, as eagle's wings attached to a gull, were large and clumsy. "She was very touchy and not stable at all with all those great, big, heavy wooden masts and all those transverse battens in the sail," recalled Potter: "There was a lot of weight up aloft and she wouldn't go windward at all." The ship was as a heavily armored knight thrown off his horse—useless dead weight.[9]

Storm clouds gathered, and other light craft were hidden behind the heavy waves that assaulted the junk in moments. Captain Welch alertly ordered the sails to be reefed, and he might also have ordered the engine started, as he had done on one of the trial runs. "With the motor opened up full speed," the junk could achieve a speed of six knots, but the strain showed. The now overexerted "auxiliary engine (turned on) full force," said Halliburton, trembled noisily and, overheated, it smoked. The "fumes from the newly-painted tanks and bulkheads, escaping into the main cabin, were almost overpowering" making it necessary to close off the engine room. All the while, the *Sea Dragon* "pitched and rolled," moving leeway when close-hauled. Heavy rains coupled with crashing waves thrashed the hull, drenched the deck, and soaked the crew. Objects and equipment flew through the air. Pots and pans in the galley clanged together or also went for a ride. "Half-dead from seasickness," messboy Patrick Kelly "lay in his bunk." Other crew members, evidently without guide ropes, were dislodged from their work posts. "Seizing the rail with one hand," Halliburton recalled, "we fed ourselves apples and dry bread with the other." The sea never quieted. "One wave pitched us forward so violently," he said, "that the radio aerial, atop the mizzenmast, was torn from its lashings, leaving the wire to thrash about in the wind." As the aerial was set "70 feet from deck to tip," recalled first mate John Potter, it was "impossible to effect repairs." Rescuing the moment, said Halliburton, was Ben Flagg who, hoisted aloft "in a bosun's chair (and) clinging to the gyrating mast, nailed the aerial in place again." Lo and behold, "We held to our course." Tested was the ship's mettle and its "steering system" which was helpless against the "violent motion" the "rough weather" had caused. "Our tiller and compass were on the after deck, unprotected from the weather," and "the steersmen got the full force of spray and rain."[10]

The *Sea Dragon* had again shown its temperament and distinct personality. "Junks are dangerous in a running seaway," Dick Wetjen once told Captain Welch, and so it was with this one. "She didn't groan or cry to like most new ships," reported Welch, "[and] is very quiet even when she has the sea abeam. But what a lively bitch she is. She is the fastest thing I have ever been on in my life. I tried to work sights on the chart table but it was a task. You can hardly stand on her when she gets moving." Unable to adapt to the wind, the ship neither *came about* nor *hove to*. As alarming as its propensity to list was its tendency to nose-dive. About twelve feet across, the ship's prow resembled a flat-bladed shovel and had a castle-like embrasure at its center. While its sleek shape kept it moving steadily, the ship, alternately riding high and low, crashed too forcibly against the fierce, tumbling waves and, rather than punch through them, the ship's bow, after each nose-dive, shoveled huge amounts of water on deck. "She would cock up her tackle," said Welch, "and dip her bow so deep that there was times when I thought she was going straight down, *down* under." During the downpour, the protective oil of their coats, ones "made in China," quickly "washed off." Against such odds, the crew, reported Halliburton, stood firm against an angry sea. Did they? "Most of the slamming about we got," said Welch, "was more due to bad steering than heavy seas. The sea was high and very steep, but she behaves well with it on the beam. Close hauled she makes a lot of leeway. Couldn't put the big mainsail on her because there was nobody left to take it in." Halliburton's belief that comradeship would result from adversity was given its first test.[11]

Its bearings all but lost, the *Sea Dragon* foundered. "At six o'clock on the second afternoon we caught sight of a light-house on the coast of China," recalled Halliburton. This was good news, but "at six o'clock the next morning, the same lighthouse was still in the same place." The junk had "not gained an inch." The storm had shoved it eastward, back to the China coast. The crew could either abandon ship or seek a tow; they chose neither option. The chugging engine pressed, and punched, but the waves kept pushing the ship back. Dale Collins was right: "the northeast monsoon was blowing so strongly that the *Sea Dragon* was unable to sail any further north than Chilang Point," only seventy-six miles north of Hong Kong. Adjusting to circumstances, Captain Welch headed southeast and, according to Halliburton, "tried to sail around the southern tip of Formosa, instead of the northern." Given a good wind, the ship "moved forward again." By the third day out, "despite mountainous seas and howling winds," it had advanced "another hundred miles" and was soon "within 90 miles of *southern* Formosa." Bad news then interrupted this leg of the voyage when John Potter sustained a serious injury. Reported Halliburton: "Potter, our first mate and the most experienced sailboat-man aboard, in struggling with the mainsail boom, ruptured himself." Wisely, he was "put to bed, and watched anxiously (by one or another crew member) to see if 'rest'—the 'rest' of a roller-coaster—would cure him." The junk lumbered along for two days. After consulting with Captain Welch, Halliburton judged it best to return to Hong Kong. Welch was miffed and blamed Potter alone for the turnaround. Years later, when no one was left to dispute him, Potter blamed what happened squarely on the tiller and fellow crew member George Barstow. Barstow's mishandling of the tiller had caused Potter to end up with "a busted pancreas and some ribs." Potter also blamed "the big, heavy wooden masts, and all those transverse battens" which had made the ship far too "weighty" and prevented it from going "windward." Heading into calm waters, fewer crew members now worked the ship. Halliburton said, "With Potter prostrate on his back in the forward cabin, and Mooney able only to act as a lookout, we were short two men at the tiller. Consequently Flagg, Barstow, Chase and I had to stand watch at the tiller, four hours on, four hours off, day and night, in pairs." Captain Welch now manned the tiller. The newly recruited Ben Flagg, whom Welch called "a damned good man," relieved him; as a reward, he would be promoted to second mate.[12]

Nasty weather returned. In the midst of crashing waves and whip-lashing gusts of wind, the ship rocked, its planking creaked, and its masts tilted irregularly. A few crew members struggled with seasickness and puked. First mate Potter's moaning got louder. "A high fever had come upon him," Halliburton reported. "For two days and nights he was unable to take food or water," he elaborated. "Obviously he was not getting better, but much worse. Welch and I held a council of war. The next possible port was Keelung—300 miles away—at Formosa's *northern* tip. We might try to take Potter here, but there was no guarantee we could make it." The "next port of call was Midway Island—4000 miles beyond." Midway was halfway home. "For half an hour we suffered from an anguish of indecision. Should we push on and hope to make Keelung, or should we take no chances, and retreat to Hong Kong?" Hong Kong was three hundred miles away in one direction, Keelung three hundred miles in another. "One more look at Potter's flushed face and fevered eyes," Halliburton later wrote, "and we voted to turn round." Sighs of relief followed: "A life—and a very fine life—might be at stake," said Halliburton, "so we swung the tiller with a mighty swing, spun the ship about, and set our course back to China." Immediate medical attention was not to be had. "The Japs wouldn't let us land in Keelung," John Potter much later explained, "because it was shortly after the *President Hoover* had run ashore down the coast." The beaching of the SS *Hoover* on December 11, 1937, had sparked Japanese suspicions of a covert spy operation in the making. "(The Japanese) distrusted Halliburton," said Potter, "and were under the impression that his desire to put me ashore was simply another hoax to land an agent." A retreat to Hong Kong was the more prudent option. Fifty miles from the southern tip of Formosa and three hundred miles east of Hong Kong, Halliburton believed a strong northerly current would accelerate the junk's speed and bring the sailors faster to safety.[13]

Four storm-filled days had passed since the crew saw the "lighthouse on the China coast." Heavy rains and rough waves had pummeled the *Sea Dragon*. Meeting favorable winds, Captain Welch, after adjusting the rigging, aimed the ship southward and guided it down the main channel toward Hong Kong. No report is given of the diesel engine; evidently it had proved itself of little use or it had malfunctioned. At some point, Welch hove to. For thirty miles the junk purposefully "drifted." With only an "inner-harbor chart" to assist him, and two land "sights" to mark the way, the junk "crept through a treacherous confusion of reefs and islands" and through "Lyman Pass." Early afternoon of the sixth day out of port, Gap Rock Lighthouse and next the lights of Victoria Peak came into view. Aboard the *Sea Dragon* there were cries of elation. Hours later, at about "11:30 that night," the tattered craft, without mishap, "was brought" into Bailey's Harbor. Welch's cool-headedness and seamanship had saved the day; truly a failed mission had become a successful retreat. "Before going ashore," Halliburton reported sadly, "I glanced at our good-luck god (Tai Toa Fat), still sitting and smiling in his little temple," (and) I decided he was a fake and a hoax." Maybe one or two dockworkers were on hand to receive the beaten crew. Maybe only a few carters and midnight strollers stopped to gaze as, one by one, the men tiredly trundled off the ship. According to Welch, first mate John Potter was "*sneaked* ashore." Halliburton worded it differently, saying that "an ambulance rushed (him) off to the hospital (on the Peak)" where "the doctors assured him he could be mended without an operation, but (they) would not consider allowing him to sail with the *Sea Dragon* again."[14]

The first crossing was thus aborted. Why? The immediate accounts differed.

32

A Priest, a Whore, and a
Siamese Cat Walk Into a Bar . . .

The *San Francisco News* reported that a storm and an injured crew member brought the *Sea Dragon* back to Hong Kong. In the fourth and last of his "Letters from the *Sea Dragon*," first issued on February 16, 1939, Halliburton explained, "A series of misadventures upset our well-laid plans and delivered us back in this British colony." He then apologized to his subscribers for mailing the letter from Hong Kong rather than from Formosa, which represented "the first lap of (the *Sea Dragon*'s) voyage across the Pacific." Halliburton's account of events reached a wider audience nearly six weeks later, March 19, when it appeared in the *Boston Sunday Globe*. The byline duplicated the earliest newspaper reports of the mishap: "Halliburton's Junk Battles Storm at Sea—Adventurer and His Crew Put Back to Hong Kong to Rush Injured First Mate to Hospital." Much about the return to Hong Kong was prudently covered up. "Perplexities" remained, however. Halliburton told one side of the story. In his last known letter, dated February 13, 1939, Welch told another, sharing with Dick Wetjen "the story behind the story of the turnabout of the great body of adventurers" and emphasizing that "*all* the news from the NEWS" was not really "*all* the news." To his audience of one, Welch stepped up to the podium and proclaimed,

"Clap has whipped many a good army and navy, so what chance did a junk have in the face of an almost universal complaint."[1]

Welch agreed with Halliburton that the "first few hours out were splendid." Seated "upon the whorehouse poop," the lads rejoiced over the fine day and being at last on their way. The captain steered the ship along the coast to Swatou, then toward Amoy with sailing going smoothly into the fourth day. About nine o'clock the next evening, when the *Sea Dragon* was "well north (and) bucking a fresh breeze with a nasty steep sea," Welch turned the steering over to the (unnamed) second mate, with instructions to keep it at "70" and to summon him when "he had the Breakers Point light on the four points." Believing all was well, he retired to his cabin. While he slept, and with the junk pointed east, first mate John Potter, already quite ill, relieved the second mate and took over the steering. Hours later a terrible pounding awakened Welch. Running out to see what was the matter, he saw that the steering was at 100," not "70." Enraged, he looked at the man he had assigned to the post and was told by him that Potter, now "lying under the tiller as sick as a dog," had changed course at midnight and pointed the ship *due east* rather than *northeast*, the direction originally charted. Welch chewed Potter out while Potter laid there, groaning. Thinking he had "the sulks and seasickness," the shaken captain had him taken to his quarters, and, putting himself at the helm, swung the ship around.[2]

Daylight spread out across the waves. Realizing he was not progressing northward, Captain Welch made "a pretty run" toward Formosa. At 5:00 p.m. the next afternoon, Sligh told him that Potter was "really sick." Halliburton later said that Potter "had been suffering torments." In point of fact, "a high fever had come upon him," and "for two days and nights he was unable to take food or water." Checking in on Potter, Welch listened to him complain of "swollen nuts and the whole bagful of tricks." But the captain knew his condition was "just plain wind-up" or gas—and the clap. By the time the *Sea Dragon* was within fifty miles of Formosa, Potter's condition had worsened, but soon his fever lifted and he demanded food. Up and about, he boasted to Sligh that he had recovered his strength. Learning this, Welch ordered him back to bed, threatening to proceed to Midway should the first mate leave his bunk for any other reason than using the toilet. Potter returned to his bunk. Welch wanted to continue to Midway anyway; Halliburton was not so sure he should. The two "held a council," one where tempers flared and one which Halliburton described as "a council of war." Halliburton insisted they take Potter to the nearest medical facility at "some port on the south end of Formosa," and Welch insisted they sail onward. For all the captain cared, Potter "could stay there and rot before [he] put in to any port to land him." Welch further contended that there was no need to worry because the "medicine chest" held "plenty of clap medicine." Halliburton walked off, then returned to tell Welch that "if anything happened to Potter while he was on the junk it would mean social ruin for him when the junk got back." Halliburton, said Welch, "begged and cried,"and reluctantly he "gave in" to him.[3]

As the ship neared Hong Kong, a worried George Barstow asked Welch to examine him. Grudgingly, Welch acceded, and spotting" a nice little rose" or "chancre," he nearly threw a fit. "So, there I had Potter and Barstow with that packet," he told Wetjen, "Mooney laid up with an infected foot and the messboy, Halliburton's punk (Patrick Kelly), seasick for four days." Details were kept from the press, and nothing was said about Barstow, Chase, or Torrey (who hadn't been aboard the junk). The press, thanks to Halliburton's likely intervention, exhibited little curiosity about Potter's "fever." Of his *other* "injury," however, the duped press made a fuss. "First thing we knew," the amused Welch said, "was that New York had headlined the turn-about and was saying that Potter was seriously ill. Presidents of Corporations in New York, his mother and sisters and all his friends cabled the consulates, even one from the Secretary of States [*sic*] office and no news or information went back as to the true story. The consulate told me that H(alliburton) had better do something about it or else they would. So, now it has become a rupture. A nice clean rupture too. Caused by hard work when the bastard was the most useless man on the ship even when she was tied up to the dock." Question was, had Potter suffered a rupture or injured his appendix? Over the years, Potter himself contributed to the uncertainty, showing off his scar and invariably remarking that he received it when a boom on deck struck him. He failed to mention, however, when this accident had happened. In 1946, after the war in the Pacific ended, "first mate" Potter, who incidentally had served in the US Navy, reported that he had "sustained a smashed rib and a few strained muscles when the boom of the stay-sail struck (him) while the ship was yawing." The *China*

Morning Post offered much the same account in a news flash appearing *after* the *Sea Dragon's second* sailing: "A young American, Mr. Potter," the feature read, "was in the junk when she left Hong Kong on her first departure for San Francisco, but strained himself and had to go into hospital when the junk put back." The whitewashed version of the story became the official one: an injury unrelated to the sexual activities of those aboard the *Sea Dragon* had brought the ship back to Hong Kong. Potter perpetuated the story, or myth, from the start. To the end of his days, he maintained that the injury from the swinging boom—and that injury alone—was the reason the *Sea Dragon* returned to Hong Kong. True enough, he had gotten hurt physically, but about contracting any "indigenous disease" he said nothing.[4]

Immediately upon his return to Hong Kong on February 9, Halliburton radio-grammed his parents about what had happened: "Potter Very Ill—Three Days Out—Forced Return (to) Hong Kong—Junk Okay." The next day he wrote them a letter, enclosing an explanatory clipping and remarking, "The six days at sea were tough but good sport." At least no one had lost a limb or died. The *Sea Dragon* itself had returned to port mauled but intact. Halliburton and Captain Welch imputed the "beating outside" to poor handling by an inexperienced crew. Once the junk was in dry dock, only a fin keel was needed to keep it from "rolling and rocking in heavy seas." Still, while the ship "held together" and the engine "worked perfectly," Potter was in the hospital with a rupture, Mooney had an aggravated ankle injury that might have gotten worse, and Barstow was "threatened with illness." Miffed by his crew's incompetence, Halliburton decided to replace several of them with professionals, whom he identified as "two or three Scandinavian seamen who know what to do." His faith in the enterprise unaltered, he reset his arrival time at the fair from the beginning to, now, the end of May.[5]

Jonathan Root contended that Halliburton was "embarrassed by his return to Hong Kong," and explained that "the loss of time put him in critical financial difficulty." Halliburton admitted to being "tired and embarrassed, but not seriously." Money matters continued to eat at him. For weeks his coffers remained close to empty—even with the slight cash infusion from his publisher. In a bind, he could barely come up with the money needed to repair the storm-battered *Sea Dragon* and do it well. He could not even pay the crew the small amounts promised to them once the ship reached San Francisco. Root notes that Halliburton's "fellow shareholders were afraid the crew would be angry enough to levy an attachment on the junk." They could do so because, "under maritime law, a crew is entitled to half its wages when its ship docks at any port." When the junk "stopped at Midway, the crisis would hit that much sooner and (Halliburton) cabled for money to his publisher and his friends." The $10,000 he expected to clear at the fair, however, was offset by his debts—notably by an equivalent $10,000 owed to Central Hanover Securities. What advance royalty, and crew participation money he received, maybe $2,000 total, had dwindled away. He had some reserve funds, notably $1,500 in a savings account. He could also turn to his caring parents or, in a pinch, to his cousin Erle. Looking to the future, he talked with Potter about selling the junk once it had reached the fair and had made its fortune in excursion fees. He might also have talked with Potter about racing another junk there for an $8,000 prize offered to the winner. Days passed. What little money he had in hand now trickled away, as did the prospect of good weather. Halliburton's race to the Golden Gate International Exposition now seemed a rush to the bank. And always was heard the old refrain of "We sail tomorrow."[6]

In a letter sent to Midway, Wesley Halliburton replied to his son's February 10 letter and told him he must be the strong one—the bulwark "to shoulder the responsibilities, be the leader, and carry on regardless of all difficulties." While the elder Halliburton had learned some details concerning the turnabout, he was not so sure that he had gotten the whole story. "Maybe," he wondered, "you do not tell us if you break a leg or suffer from that oriental dysentery or whatever else that may happen to you." Wesley's source was the *Memphis Commercial Appeal* which had mistakenly identified James Sligh as the crew member stricken with "appendicitis." As the cook was "a very important personage indeed," the news, Wesley said, was "distressing." Wesley added, of uncertain meaning, that, in an emergency, he trusted his son's know-how "with a case of obstetrics" more than his knowledge concerning "a case of appendicitis." Writing on, he "wondered as a last resort what (son Richard) would do (next)." Wesley thought it "a tragedy" for Richard to leave without Torrey and Potter, and he hoped they might board a Clipper, join the *Sea Dragon* "at Midway or Honolulu, and complete the trip and have the thrill of sailing into the Golden

Gate and up to Treasure Island." He wanted Mooney to "be careful and let his foot get well," but he admitted, "It is difficult to resist activities when nearly recovered."[7]

Presumably the "lads" checked into one or another of the Colony's better hotels, tidied up and returned to their carefree ways. Saturday, February 11, two days after their return, Captain Welch learned of their activities when, "with one of the consuls and the manager of the Dollar Line," he stopped for a "nightcap" at a drinking establishment—perhaps the Peninsula Hotel, one frequented by American sailors and tourists. His report of the incident that followed is partly cryptic. "Somebody had come in to the bar and whispered that it would be less embarrassing if I stayed where I was for a while," he recounted. "It appears that the only pure one of the lads had been boasting of having laid one of Kowloon's belles and she heard about it. So, she ups and heaves a glass of beer in his face. Before the elite of Hongkong. No less than the Flying Corps of the Army and the Navy of His Majesty. Nice beginning." Continuing to fuss, the "belle" widened her search for those whom she believed had harmed her. "Somebody told her that I was in the other Bar," said Welch, "so she came flying in and came over to me. Says she, 'What I think of your lousy rotten crew and your lousy junk, Captain Welch!' Then she left. Nice girl that. Lots of spirit. Nice bitch too. I didn't blame her though because I heard that after (Robert) Chase thumbed his nose at her while she was sitting with another chap who had his back to Chase." That same night a "scuffle" broke out at the Peninsula Hotel's bar, which "finished up in a knock down drag out fight in the street." Said Welch: "All the limey officers were properly shocked at the primitive display. I was in a corner with two flight chaps whom I have known for about a month and they went out to report for me. The news was all sad. There was a nice little gathering of about a hundred coolies watching it and it will probably leak into the press tomorrow."[8]

"Nightcaps" were forcing Welch to dig deeper into his pockets. Predictably, when Halliburton next day got aboard the *Sea Dragon,* he "nailed her for some money" and, from what had to have been a reserve 'rainy day' account, promptly received "twenty gold." Sick and looking pale, Halliburton put up little resistance. "She looks like that wrath of god," Welch said. "Now she has the *scabies* or Chinese itch. Says he doesn't know where he got it, but there is a story going about Hong

Kong that on the night before we sailed from here he engaged a room in the biggest hotel across the harbor and had two, I said two, Chinese boys spend the night with him. A newspaper reporter told the yarn a few days ago. Said that all they were waiting for was something to really get on him and they were going to break loose with the complete story." "Yarn" or unvarnished truth, the "complete story," never reached print. One must wonder if it was much of a story anyway. The "slight skin itch" Halliburton experienced in February could have been related to a much earlier medical condition, or been the "eczema" noted by Root. What ailments Halliburton brought to the Colony could also have mingled with those already there. During the writing of *The Royal Road to Romance*, Root noted that Halliburton "suffered a crippling recurrence of trembling, nausea, rapid heart-beat, and weakness and was told by a Nantucket doctor that he had a goiter and that it would have to be removed. He went into New York to a specialist instead and received a diagnosis of hyperthyroidism." Over-activity as under-activity of the thyroid gland can afflict the skin, even cause itching. Before going off to China, Halliburton had complained of "a fierce attack of hay fever": decreased energy and watery, itchy eyes resulted. For Welch, *scabies* was *scabies* and bed mites were the cause. Transmitted by skin-to-skin contact, "scabies" generally, but not exclusively, shows up in the hip and midriff section of the body. Halliburton confided ito his parents that he had a "slight skin itch" and seemed befuddled as to its origin. The ailment—and slander—might have prompted him to go "on holiday to Macao" for a few days. Such a meaningless trip left Welch baffled. "I have an idea I will be coming home as a prisoner on a cruiser if this keeps up," the captain said. "Gone nuts. Anyway between the clap and chancres and the itch and a broken foot and an infected toe I am having a swell time." Captain indeed, he found himself in command of damaged goods aboard a broken ship.[9]

The *Sea Dragon* languished in port and could have done so until it rotted. Limited funds and time prohibited a complete overhaul, but over the coming weeks its leaks were plugged, its hull was tarred, its rigging was adjusted, and its diesel engine was tuned and tightened more strongly to its mounting. The tiller and the rudder seemed in working order, but a new keel was ordered. Captain Welch said it was "twenty five feet long, four feet deep and (ran) from the engine room to the stern post" and installed "to stop the leeway she makes when we are close hauled (when turned to the wind)." Following Halliburton's account, Root notes that the eighteen-foot-long keel was fastened to "reduce (the junk's) frightening tendency to roll" and, as S. L. Kahn understood it, "to improve (its) sea riding qualities." The Coolidge's Dale Collins later reported that the new keel was never installed. Were that the case, likely it was decided that the existing keel— especially if weighted with lead, would do.[10]

33

Proudly We Hail

~~~~~~

Three weeks separated the *Sea Dragon*'s return to Hong Kong on February 9 and its next departure on March 4. Welch stayed committed to his business. As to events in the outside world, he admitted he was "kind of out of touch here at the yard." These days he and Sligh were the only ones living aboard the craft. "The rest of them," Welch said, "are *scattered* all over the place. . . . I sleep on the junk and the others are at the hotel." He didn't mind the arrangement: "I am comfortable here and it is very quiet at night." The lads showed up during the day to do *some* work, but, by now proven slackers, went "ashore again right after lunch." Of camaraderie, they had learned something; of navigational skills, they had learned nothing. "One of the lads, Chase, went up in a bos'ns' chair this morning," said Welch, "and treated it like the first great adventure" of his life. Grandstand, that Chase could do, but could he work the sails or scrub the deck? Forget it. When he wasn't snapping pictures or stumbling about the deck, 'Peck's Bad Boy' George Barstow was "ashore getting treatment for his chancre." Incorrigible, he afterward brought a guest to his den to share a bottle of scotch. Loud rumbling ensued. Patrick Kelly, another "mess," proved an even more useless asset. Nearly as useless, Halliburton at least signed the checks. With his bad ankle, Mooney was questionable. Except for Ben Flagg, who in relief had earlier made things "a bit easier" for him, none could be trusted to watch the tiller. Miffed, he realized that on most shifts he alone would be manning the tiller.[1]

Needed were recruits with maritime skills. One such prospect, whom Halliburton discovered, was a "Norwegian mate" named Johannsen, an ornery man "as big as a house and plenty tough." Welch found him made to order. "If I get Johannsen this week," he said, "I shan't have to worry any more, I won't even go on deck when she is in port. He is all sailor and he can handle this thing alright." Johannsen was strong, and, what was most important to Welch, he swore allegiance to his captain and no one else. The Norwegian also boasted he could whip raw recruits of any temper into shape. With one look at the lads, he knew what work needed to be done. "Joost give them to me, Captain," he snarled, "and I'll make everything alright for you in a week." Laughing, Captain Welch told him that "he could have them forever—the sky is the limit." Johannsen, looking murderous, gave Barstow the once-over. When the captain told him that "new mate (Johannsen) didn't think much of him," Barstow exploded: "If he strikes me," he warned, "I'll sue him." Johannsen probably didn't want the job anyway. Besides, his asking price of $500 a month was steep. He also wanted return passage money. Hindsight once again argues that Welch should have hired several Johanssens, but, after interviewing a couple, decided he didn't want to captain a "gang of hooligans."[2]

In 1939, tensions again rose in China, and the Japanese moved closer toward Hong Kong. Welch set the Sea Dragon's new sailing date as Monday, February 20; Halliburton scheduled the departure for Saturday, February 25; Paul Mooney, who called Hong Kong "a depressing place to be," hoped he could leave "today." Lately, Hong Kong was near to being bombed and thrown into mayhem. Nazi spies working for the Japanese were everywhere. Each week ships filled with refugees from Germany arrived in Hong Kong; refused entry to the city, the refugees headed north for Shanghai. What followed reads like an 'ms' found in a bottle: "Halliburton and a group of us came to China to buy a junk to sail it to the San Francisco Fair," Mooney wrote. "But we found we had to build one instead and it took longer than we planned. A fortnight ago we started home in the crazy cockleshell, and after a week had to return to Hong Kong to keep a member of the party from dying. Tomorrow we start once more for Midway, Honolulu, and San Francisco. It will take perhaps three months to make the voyage. There are 13 in the party, and our *Sea Dragon* is a gaudily-painted, 50-ton sailboat with an auxiliary motor. You can read about us in the papers." In his last known letter home—postmarked February 27, Mooney told his mother that "with all the interest the trip held for him, he would be glad to see the coast of California again." Halliburton hadn't written home in a couple weeks, and his worried parents had cabled him for "some sign of life." In a letter dated Thursday, February 22, he promised to answer them from the "junk radio" because it was "so expensive otherwise." He also blamed the latest delay on the Chinese New Year—"every(thing) was closed." Of changes to the junk, Halliburton said only that "the fin-keel" had "been" put on" and that he and the crew were "trying to find a leak in the hull." Again all boded well. "On Sunday surely, we depart again," he wrote, "and there will be no turning back."[3]

Repairs to the junk were rushed. What upgrades were made is largely unknown—though a gyro-stabilizer could have been installed to counteract its sideways motion and copper plating put on the existing keel for added protection. The surveyor came by and *again* gave it clearance, reiterating statements from his earlier pre–February 4 report that "the hull (was) well-constructed (and) of good, sound wood," and that its "frames (were) hardwood bolted through with galvanized iron bolts." A cause for concern remained the high poop. "[The] *utmost vigilance is necessary*," he cautioned, "when handling the vessel under sail." Jonathan Root notes that Halliburton subsequently "insured the ship for $10,000." Dale Collins of the SS *Coolidge*, after he read the report, met with Captain Welch to discuss, besides safety issues, a mid-ocean rendezvous. The liner made its round-trip crossing of the Pacific, with stopovers, every six weeks or thereabouts; in its two-week crossing, it ran from San Francisco to Honolulu, to Kobe or Yokohama, Shanghai, Manila, and, lastly, to Hong Kong. Departing from Hong Kong on February 10, it would be leaving for the port city from San Francisco on March 13.[4]

March 1: all was in readiness for the *Sea Dragon*'s second trans-Pacific crossing attempt. "The weather is moderating here," Halliburton wrote home, "and now we'll have March, April, May, for our crossing—better than January, February, March. With our professional crew we may cut down the sailing time 2 or 3 weeks—and arrive about the middle of May." This time Potter would stay behind. Although his cast would not be off for three more weeks, Mooney would make the trip. Halliburton noted

that he had "two white sailors," probably Ralph Granrud formerly of the
SS *Coolidge* and Johannsen. Granrud would make the trip, but Johannsen
would not. The unnamed part-Burmese interpreter who had assisted
Welch during his negotiations with shipwright Fat Kau might have been
one of the "four Chinese sailors"; the others were boatswain Sun Fook,
seaman Kiao Chu, seaman Wang Ching-huo, and messboy Liu Ah-su
(or Ah-shu). A recently married Chinese couple—perhaps learning that
Halliburton once intended to bring an Asian couple, also requested passage on the junk. Root writes that Halliburton "wanted the wife pregnant
so the baby could be delivered in mid-Pacific." Said Halliburton, supposedly: "I've plenty of experience as a midwife." Had the birth happened in
American waters, it would have been the *first recorded* instance of *birth
tourism*; later, the child could sponsor his or her parents as legal US
citizens. Perhaps other persons asked Halliburton for passage; paid the
right fee, he might have considered it. Second thoughts were not given
to the pandas or to the Siamese kittens. Two Chow puppies, however,
were admitted aboard the *Sea Dragon*. Guardians of temples, they now
served as guardians of the ship.[5]

Just before he departed, Halliburton received two cables from his
parents wishing him well. Besides tell them he would make a speedy voyage home, he told them him that he was keeping costs down. "Each day
I planned to be gone on the junk where I could radio you for 10c a word
instead of $1 a word cable," he said, his concern for money and fear of
want now becoming an obsession. Money drifted easily from his pocket,
but he said every investment "would be made back with profit at the
Fair." Dated March 3, his last letter to his parents was meant to placate
them. "One more—one last—good-bye letter," he began timorously and
uncertainly. As the drums in his head rolled, Halliburton picked himself
up: "We sail, again in a few hours—(the ship) far more seaworthy than
before." The crew of course was "far more expert now"—two good sail-
men—and four Chinese—(these last four) "old hands at junk sailing," and
two of whom would return to Hong Kong once they reached Honolulu.
Regretful that he had left Torrey behind, Halliburton was even more
regretful that he had left Potter behind. Bored, Mooney remained eager
to leave. "High spirits," said Halliburton, had returned. Blessings had at
last taken off their disguises. "The delay has been heart-breaking," he
said, " but worth it in *added safety*." Food and water for three months
and "oil for twenty days running" assured that no one would starve.

"Bickering and feuds among the crew" persisted, but, said Halliburton, these "will soon pass." Shared strife and faith in a common cause would win the day. His was a "swell idea," he still believed: "Everything is coming out as I dreamed." Once at sea, he intended to radio his parents every few days so that they could follow his adventure as it unfolded. "Think of it as wonderful sport, and not as something hazardous and foolish," he told them. He also hoped his parents, and maybe a few of his friends, would be at the fair to greet him. "I embrace you all," he said. To his "sweet Mother," whose birthday had come, he offered "an extra hug." She remembered his many good-byes, especially the time eighteen years earlier when he and fellow Princetonian Mick Hockaway boarded the freighter *Ipswich* bound for Europe, he told him, "Just about a month from today I'm set adrift, with a diploma for a sail and lots of nerve for oars." The metaphor still applied.[6]

In a send-off letter of his own, Wesley Halliburton told son Richard that his letters "perked up" Nelle Nance and himself. He was "glad" his son was "getting nearer and communication (could) become normal again." Wesley also addressed Richard as "Bully Boy," associating him with "Bull-Moose" Party leader Teddy Roosevelt, who came close to losing his life on an adventure on the ominously named "River of Doubt" in South America. During the Great War, Roosevelt had lost his beloved son Quentin when the plane he piloted was shot down behind enemy lines. Wesley filled his longish letter not with harrowing news but with the sort of topical things Richard liked to hear. Nelle Nance amused herself by playing cards and mah-jongg with friends. Her interest in her yard had increased, her duties as president of the garden club kept her on the go, and she occasionally attended a movie. Wesley owned large tracts of land in the Memphis area and expressed concern for the well-being of the tenant farmers who worked it. He admitted to being "restless," but land deals and investments kept him busy. He called the Pacific crossing "the most gripping" of Richard's adventures but said nothing about his and Nelle Nance's worries about the voyage. "I knew you would meet all obstacles and overcome them," he told Richard. "But I also knew how trying the job would be." Wesley understood his son's sorrow at having "underestimate(d) the cost of the enterprise," but he thought the price "very secondary to the ultimate consummation of it." In the end, he hoped "that the ledger may break even: if it does, I am content."[7]

Wesley then had "an inspiration": to rendezvous with his son at Honolulu and with him "experience storms, sea-sickness and the grumblings of the crew (drafted into) the adventure." Nearing seventy, and no longer spry and adventurous, he at once had second thoughts, concluding that it was better that he and Nelle Nance meet up with their son at the fair. They would be holding "a banner . . . waving vigorously to the Vikings of the Pacific." Richard asked his parents about their recent trip to Mobile, Savannah, and Charleston. In a response Richard never heard, they said that they had suffered one "inferior hotel" on the way, and the discomforts of "touring," but "the gardens were all beautiful and every day was sunny." Wesley said he could "chat on," but promised that "lots of mail" would await Richard at Midway. He signed off, "Remember me to the Crew." No longer a parent but rather a personal friend, he said, "Love, Wesley," not, "love, Dad."[8]

Few people in America talked about the *Sea Dragon* Expedition, certainly not in the same way they would talk about the first moon landing thirty years later. Americans were more likely to talk about atomic power and what *wonders* it might bring. They also talked about superhighways one day crisscrossing the country, and worried about mounting tensions in Europe. Many commented on the latest Hollywood film releases. They also marked passings. Notables in 1939's obituary columns included actor Douglas Fairbanks Sr., 56; war correspondent Floyd Gibbons, 52; discoverer of King Tut's tomb Howard Carter, 54; and psychoanalyst Sigmund Freud, 83. News broadcasts reported an attempt on the life of Italian dictator Benito Mussolini. The front page carried a statement by Adolf Hitler declaring Great Britain Germany's chief enemy, and one Germany must defeat. Editorials on such government programs as the *WPA* and *CCC* regularly appeared in newspapers. Some economists predicted "hard times" were ending. More Americans were joining the ranks of the middle class, and, although stocks often drifted irregularly, they were as often steady and mixed.

Whatever the dark moods, the bright lights of a bleak decade never dimmed. The news of the day was of every kind, as reality and celluloid reality, advertising and policy-making, fact and fiction cooked up a new stew of information for the public to ingest. Actor Wallace Beery was wearing his first mustache in years, and footballer Douglas "Wrong Way" Corrigan was about to make his film debut. Hits at the box office included *The Hound of the Baskervilles*, with Basil Rathbone as Sherlock

Holmes, Nigel Bruce as Dr. Watson, and newcomer Richard Greene as Sir Henry Baskerville; *Made for Each Other*, with Carole Lombard and James Stewart; *Idiot's Delight* with Clark Gable and Norma Shearer; and *Ninotchka* with Greta Garbo and Melvyn Douglas. All drew record crowds. Released on February 17 was *Gunga Din*, a blockbuster epic like the *Sea Dragon* Expedition but set in a British outpost in India. A bombshell of a book, John Steinbeck's *The Grapes of Wrath* was soon adapted into film. Songwriter Hoagy Carmichael published the "Hong Kong Blues." Walt Disney's *Snow White* (1937) was a smash that lost money, as was *Pinocchio* (1940). Off in another realm, Cardinal Eugenio Pacelli, elected the new pontiff, assumed the name Pius XII.

In San Francisco, the fair was still the main focus of civic life. February 22, Washington's Birthday, would bring tens of thousands of people to the event. The "New" Folies Bergere troupe would soon perform, and Banjo-eyed vaudevillian Eddie Cantor had already opened a big show. Ernie Pyle was traveling across America and would soon arrive at the fair. His column usually appeared below Halliburton's "Log of the *Sea Dragon*" articles. Travel Congress Delegates were also slated to arrive, so were the President and First Lady and the King and Queen of England. Seemingly oblivious to the war in China, pageant sponsors would soon select a beauty queen for "Japan Day" while the swans in the pond at China Village gracefully glided about. Boy Scouts would soon set up camp at Golden Gate Park, and the tent city neared full operation. Upcoming events at the fair were regularly broadcast. The PTA requested that the already small admittance fee for young people be dropped to ten cents, and others asked that parking areas be expanded. Apart from stirring events everywhere else, Halliburton could only scream—he was still in Hong Kong and had no income to show for it.[9]

34

Don't Give Up the Trip

The night before the *Sea Dragon* sailed, "a great party," wrote attendee John Rowland, was thrown for the crew "at the best café in Victoria." At the gathering, the freelance journalist got up the nerve to ask Halliburton if he wanted to join him on a trek down the Yukon River into Alaska. Amused, Halliburton scotched the idea. The *Sea Dragon* adventure, he said, would be his "last." Earlier, when Rowland had asked if he could join the crew, Halliburton had consented. Rowland had some practical experience serving on a ship and seemed eager, but he also wondered about the ship's chances. Calling it "a floating eggshell on the sea," the Hong Kong harbor master told him that he wouldn't "sail past the harbor entrance in her." Halliburton heard similar forecasts; these and the frightening specter of endless ocean hardly comforted him. The romantic readiness once enlivening him had dulled; the value he once placed in the mission had wilted.[1]

Crew members who by now knew the creak and crack of every plank that ran across the deck also had misgivings. Resident skeptic Gordon Torrey reproached Halliburton as "irresponsible," as a man recruiting and leading individuals "who couldn't know the real dangers in sailing the Pacific in such a vessel." At the "great party" of champagne and clanked glasses, Rowland saw a dismal atmosphere of wan smiles, dispirited talk, and grim outlook. Ever optimistic, Halliburton yet believed the crossing could be done. Like so many of those heroes he had emulated, he would do a brave thing. He would walk down the plank to the ship as confidently as Caesar had crossed the Rubicon, Sidney Carton in *The Tale of Two Cities* had mounted the scaffold, or the Pied Piper had whisked off the children from Hamlin.[2]

On Saturday, March 4, 1939, the ropes connecting the *Sea Dragon* to its moorings were unwound from their posts. Docked nearby was a British battleship. Smaller craft, mostly sampans and fishing huts, floated in the harbor. The day had opened with temperatures in the high sixties; by 2:00 p.m., temperatures were in the low seventies. The spring fog had lifted, and easterly winds, moderate clouding, and early morning mist and drizzle were predicted for the region. Clearer skies would prevail by late afternoon. "The sun," writes Cathryn Prince, "was a hazy smear in the sky above." No rains or dense fog threatened. Visibility was estimated to be five miles. In San Francisco, it was fifty-six degrees and partly cloudy. After they stowed their gear, the crew again took to their stations and (maybe) cheered as Paul Mooney, perhaps on crutches, made his way down the gangplank. Just as evacuees begged Dr. Petersen to take them aboard the *Hummel Hummel* as it fled war-torn China, so evacuees from a city about to crumble probably begged Halliburton to take them aboard the *Sea Dragon* for a fee. Only Velman Fitch, a university dropout, caught the fever of wanderlust by Halliburton's *Royal Road,* is known to have been admitted.[3]

Why March 4 was chosen as the day of departure remains unclear. That FDR, exactly six years earlier, had made his famous speech exhorting Americans to fear nothing but "fear itself" was only coincidental. Not so coincidentally, some of the dangers lurking in the waterways going north had been lifted. Japanese general Akira Tanaka, chief of staff to the Japanese military commander in South China, had arrived in Hong Kong two days earlier to make amends for his country's shelling of British territory. In actions emphasizing that "British complaisance" was "no longer to be expected," the governor of the Colony was compensated with $20,000 in Hong Kong currency and given sanction that light shipping could move freely up the coast (and the Pearl River) without being stopped by Japanese warships. The lifting of restrictions was good news to most mariners, who could count on some protection from the British navy.[4]

Halliburton's "skin itch" lingered, but, "if somewhat weary," he was otherwise "in perfect health." After the *Flying Carpet* Expedition, he reported that "he would be glad when this trip was over and he would be his own master once more." The current trip over, he would again be glad. Seeing the extent to which Japan was pushing the region to the brink of war, he knew he was leaving a city about to collapse. As bombing raids on the Colony were no longer just threats, escape by any means seemed better than almost certain maiming or death. "It is always the case," Halliburton once said, "that residence in a new country is happiest just before departure." According to Jonathan Root, the adventurer "spent the first hours of every voyage in an agony of apprehension." That agony miserably returned, but crossing the Pacific in a junk still excited him. From Wilfred Crowell Halliburton had learned that a "big barge anchored in the small ships basin at the Fair (held) a big sign (that read) 'Reserved anchorage for Halliburton's *Sea Dragon*.'" Maybe the risks were worth the prize after all.[5]

Through Madame Chung and pilots stationed in Hong Kong, Halliburton had made some effort to obtain backup support from the Clippers flying across the Pacific. Help could also come from the *Coolidge,* sailing from Honolulu directly to Yokohama "via the Southern Track" and on a direct line to meet up with the *Sea Dragon.* That was it. Should the *Sea Dragon* run into trouble, no escort or trailing ship was there to rescue it. No helicopter flew overhead to monitor its progress; no satellite pinpointed its whereabouts. A radio alone connected Halliburton to the outside world. *Watchful-eye technology* did not exist in 1939. Halliburton could only hope that in a mid-ocean emergency a friendly commercial or cruise liner crossed his path. The unlikelihood of that happening must have troubled him, and his crew.

New Year's celebrations in Hong Kong had ended. The *Sea Dragon,* which seemed a part of those celebrations, had left port and then returned, hobbled. Little in the way of fanfare accompanied this next departure: firecrackers didn't pop, drums didn't roll, and no blissful exultations from the crew. While a salvo of cheers had bade the ship farewell a month before, jeers and muted roars would see it off now. "The old timers laugh at her [the *Sea Dragon*]," Captain Welch had written, "and the Chinese say that she is no longer a junk but a coffin." Forebodings were several and pointed: cockroaches were seen coming aboard the junk, and a rat was seen leaving it. Chinese myth held that the dragon was a land animal and therefore not adapted to the sea. Drawn to the quiet spectacle as to a public execution, a small crowd gathered to watch with idle curiosity and bewilderment as the ropes holding the odd tuna-eyed vessel to the dock were loosed. Some waved good-bye to the junk as they would to any ship leaving port. Sightseers, gazing down into the harbor from Victoria Peak and seeing many nameless vessels enter and leave, made idle comment.[6]

Standing at the dock, journalist Gordon Sinclair wished Halliburton the best of luck. "Sometimes he seemed glad to see me," Sinclair reported, "[but] at others he was apathetic and so unresponsive that I wondered if he knew I was talking to him." Halliburton had "no time to care about people." Alienated by occasional choice from others, he at times seemed as willfully alienated from present reality. On that "last day," his wearing "such non-nautical clothes as a white sweater under a top coat" perplexed Sinclair. Halliburton had slept poorly and had evidently developed an indifference toward eating. Sinclair also remarked that his friend's "face was lined and gaunt" and that he seemed "pretty close to a nervous breakdown." Just as Halliburton turned from the dock to board the *Sea Dragon*, Sinclair said that they "shook hands a bit grimly and said something of no consequence." The journalist attempted humor: "I said, 'Don't get your feet wet Dick.' He said, 'Don't take any *wooden nickels*,' and we parted. It was anti-climax." Sinclair also took several photographs of the ship, including what he believed was the last photograph of the "glorious adventurer," who "stood on the high stern and waved." Concealing the pitiable state of nerves racking within him, Halliburton appeared worn but carefully composed.[7]

Winds were still and waters calm, boding that the Pacific might remain true to its name. Boyishly bold, the *Sea Dragon* looked ready for play. The diesel engine kicked in and roared. Once the junk was out of port, its sails were hoisted and flapped. The prow lifted its nose, and the ship, aimed northeast, sailed by the guiding buoys that led to the sea. "Swept by a following breeze out into the South China Sea," the *Sea Dragon* skimmed across the waves. Fishing boats bobbed indifferently on the tranquil waves. Maintaining a steady speed through now-familiar waterways, the junk covered "better than 100 miles a day along a somewhat zigzag course". As on the first attempt, it rode the waves splendidly and "in ladylike fashion [slid] gently through moderate swells." A radiograph sent from the junk on March 5 expressed confidence: "Radio March 5, 1939 JUNK SEA DRAGON VIA SAN FRANCISCO SAILED AGAIN TODAY SOUNDER SHIP BETTER CREW FINE WEATHER RADIO SEADRAGON SAN FRANCISCO POSTAL TELEGRAPH HURRYING HOME LOVE." Another radiogram, dated the same day, from Halliburton to lecture manager "Monica" of the Alber-Wickes agency, noted simply, "Have sailed."[8]

The *Sea Dragon* sailed northeast from Hong Kong toward the southern tip of Formosa. Short of Keelung, it headed due east on the course Captain Welch had attempted during the first sailing. Skies were clear, waters calm. Past drifting or resting boats, and maybe a honking steamer or two, the junk made its way. With its mainsails taken in to sail closer to the moderate wind, the *Sea Dragon* then entered the vast Pacific Ocean, a wide-open stage whose audience was water, endless water. Halliburton and the crew found no villagers or fishing boats in the wings, certainly no orchestra in the pit. To their right and left and many fathoms beneath them was nothing but water; it stretched without end to where it met the sky. "It is difficult to sense the immensity of the ocean," wrote naturalist and adventurer Gifford Pinchot of an earlier cruise into the South Seas. "The ship is like a beetle crawling under a bowl constantly pushed forward. Day after day the voyager remains in the exact center of the same circumference. Occasionally a rain squall or a bank of clouds may seem to place him nearer one edge than the other, but the sun comes out and the clouds disappear, and once more the distance to every part of the horizon is just the same." The *Sea Dragon* had sailed beyond every still object that suggested a human presence and now entered into the sea's enchanting illimitability. As evening came, and as he often did when at sea, Halliburton "looked (overhead) for the Great Bear, the Lion, and the Unicorn, and all the other animals an imaginative mind can find in the stars."[9]

A radio message dated March 9 reported that "all was going well." The next radio message was transmitted on March 13, 1939: "JUNK SEA DRAGON VIA SAN FRANCISCO 1200 MILES AT SEA ALLS WELL." Learning that the *Sea Dragon* was "well out," Wesley Halliburton cheered from the sidelines: "Come on old Junk, ride the waves and toss them about if you will, but keep coming. A happy landing at Midway. A quick start again and a well-organized and happy crew for this last lap." He supposed that the voyage was "bound to be exhilarating." The little ship traveled "miles nearer each sunrise to its goal" through calm waters, toward warmer weather and gentler breezes. In time, it passed the Ryukyu Islands, which included Okinawa, and other of the smaller islands of Japan. Skies were fair, waters unruffled. Halliburton calculated the span from Hong Kong to San Francisco to be nine thousand miles—or the distance, west to east, from San Francisco to Baghdad. In nine days the *Sea Dragon* had sailed eastward some 1,200 miles, or the distance from San Francisco to Denver, Colorado, a reckoning of which Halliburton was likely aware. Now only 7,800 miles separated the ship from its final destination; with good luck, the *Sea Dragon* could reach San Francisco by the end of May. For two weeks the junk cruised east-northeast on a course nearly parallel to the Tropic of Cancer and presumably along well-charted shipping lanes. For days no word came from the *Sea Dragon,* a period of silence broken on March 16 when Halliburton sent another radiogram to Monica of the Alber-Wickes agency: "Can accept October lecture dates." Like Jules Verne's around-the-world traveler Phileas Fogg, should he fail, he did not want to clerk in a bank—or join bread lines or eat at soup kitchens.[10]

The *Sea Dragon* moved steadily towards Midway Island. On March 19, nearing the point of no return, Halliburton sent his parents a much-awaited radiogram: "Radio March 19, 1939 JUNK SEA DRAGON VIA SAN FRANCISCO HALFWAY MIDWAY ARRIVING THERE APRIL FIFTH SKIPPING HONOLULU WRITE CARE PANAMERICAN MIDWAY AIRMAIL LOVE." At the time, the *Sea Dragon* was about 900 miles southeast of Yokohama and about 1,500 miles due west of Midway. John Potter, whose skill at the helm Captain Welch questioned, later suggested that Halliburton should have bypassed Midway and, taking advantage of the trade winds to the south, gone directly to Honolulu. Aeronautical engineer Amos Wood calculated that the *Sea Dragon,* just before it entered the storm system that brought it down, was "on a true course of 100 degrees and at a position of 31 degrees north and 155 degrees east," or "about 150 miles east of a small island named Ganges and about 1,600 miles west of Midway island." The reading indicated that the junk was "heading for the Hawaiian Islands (anyway) rather than directly for San Francisco."[11]

The SS *Coolidge*, on its return trip to the Orient, left Honolulu March 18 and stuck to the southern route. Plans were still in place for a rendezvous with the *Sea Dragon* "to supply the junk with fresh fruits and vegetables." Dale Collins, still second-in-command, received the following communication from Captain Welch on March 19: "JUNK SEA DRAGON,

MARCH 19, 1939 1148 GCT. DALE COLLINS, PRESIDENT COOLIDGE. LAT. 29.50 N. LONG 144.20 E NOON 19TH, 8PM 1925 FROM HONG KONG BOUND MIDWAY ALL WELL, LEFT FOURTH (MARCH) APPRECIATE LATTER [*SIC*] PLEASE KEEP CONTACT OUTWARD. REGARDS ALL." This was Welch's first message to the *Coolidge*.[12]

March 20, Captain Welch indicated that his noon position to be at latitude 30.35 north and at longitude 146.54 north. "By plotting his positions and approximate estimated course and speed," noted Collins, "we could see that we should pass fairly close to the SEA DRAGON in about four days' time." On March 21, Welch stated his noon position as "30 degrees—52" W and 149 degrees—34"E." Collins deduced from the change of position that the *Sea Dragon's* "average speed" was about six knots. At noon on March 22, Welch indicated his position at latitude 30.24 and longitude 152.23 east. Responding quickly, the *Coolidge's* Captain Ahlin assured Welch that "we would endeavor to steer directly for the *Sea Dragon*, and, weather permitting, we would stop, lower a boat and give them a few fresh supplies."[13]

Weather conditions during these pivotal moments were recorded in the *Coolidge's* log. Basically a paper black box, the chief virtue of a ship's log is its immediacy, as it is a pilot's record of events entered within moments of their occurrence. Little in the way of drama or suspense attaches to a log, which may explain why Captain Welch kept two accounts of the *Sea Dragon* Expedition, one a *log*, the other a *narrative*. In recounting a disaster, a log can read like an accident report written under duress. The log of the SS *Coolidge*, however, shows presence of mind throughout and notes with an unhurried hand both the customary data and information about the increasingly enlarging and fast-moving waves the liner encountered. The lookouts initialed these notations, and the bottom of each page was signed by "Master K. A. Ahlin and Chief Officer D. E. Collins." As an aspiring singer might make music from a menu, Collins in his later "The Royal Road Across the Pacific" arranged the log's unconnected, random series of meteorological events into readable copy.[14]

According to the *Coolidge's* log, the southwesterly winds the liner encountered on March 21 (pre-meridian time) were little more than gentle breezes, traveling at four to six miles per hour. Skies were partly cloudy but sufficiently clear, and surface water temperatures were likely in the sixties Fahrenheit. Collins gave the *Coolidge's* position as "27

degrees-08' N and 171 degrees-11' E, course 283 degrees and speed
18.74." Captain Welch, he said, was "maintaining an average speed of
about 6 knots," and gave his own position at noon "March 22" (pre-
meridian) as "30 degrees-24' N, 152 degrees-23' E." The evening of
March 21, the *Coolidge* crossed the 180th meridian, or International
Date Line where its "next calendar was automatically advanced (from
March 22) to March 23." Changing wind conditions were now entered
into the log with frequency and given a rating on the Beaufort Scale
based largely on observation. Low wind is classified as Force 3—the
rating recorded in the *Coolidge*'s log on March 21, and as Force 4. The
rating for high winds—those with speeds exceeding twenty-five miles
per hour, begins at Force 6. Readings of oceanic behavior, as those relied
upon by the crews of the *Sea Dragon* and *Coolidge*, were subject to human
error; while useful, they were not a perfect measure. Dangerously high,
a Force 10 indicates winds of over fifty miles per hour, accompanied by
waves reaching upward of thirty feet. Treacherously high, a Force 11
indicates winds exceeding seventy miles per hour; now a gale becomes
a whole gale (or near hurricane). Alarmingly high, a Force 12 indicates
winds that soar to over one hundred miles per hour; now a whole gale
becomes a typhoon. In all these high Force levels, which it seems both
the *Coolidge* and *Sea Dragon* endured, hard water strikes with ferocity
as surges of white foam and spray limit visibility.[15]

The entry in the *Coolidge*'s log for March 23, a new day in a new
hemisphere, reported northeasterly winds at Force 7 and Force 6. The
gentle breezes had increased from a moderate to strong rating and from
a strong to a near gale-force rating, or from thirty-one to thirty-eight
miles per hour. Clouds converged, darkening the sky. Swirling breezes
developed into harsh winds, and, with the water level rising, waves
crashed against the hull and lashing rains swamped every deck and corridor. "Wind increased to gale force," the log reads. "Very rough seas,
shipping water over foredecks continuous, rain with frequent dense
squalls." Wind speed soon decreased, "then built up (as indicated by
the) rough northwesterly seas." The "vessel (was now) pitching easily." Said Collins, "From midnight to noon, we had experienced a strong
southwesterly gale and a very rough westerly sea but by 4 p.m. the wind
had decreased to Force 4–5 (on the) Beaufort Scale, and the sea had
moderated considerably." The *Coolidge* had so far only *courted,* not *wed*
disaster.[16]

But what was happening to the smaller ship if the bigger one was struggling? The SS *Coolidge* weighed 21,936 tons, the *Sea Dragon* about 40 to 50 tons. The typhoon's battlefront might well have extended for hundreds of miles. With swells reaching forty feet or more, the *Coolidge* slowed to six knots. The liner stood several stories high, and thirty or forty feet separated the top of its hull from its waterline. By comparison, only twelve to fifteen feet separated the top of the junk's hull from its waterline. Into its third week at sea, the *Sea Dragon* rode smoothly. March 23, traveling at a steady five knots through the most desolate waters on earth, it headed into a storm. Five days after the first message, Captain Ahlin received a second message: "NOON POSITION (LATITUDE) 30.50 N (LONGITUDE) 154.30 E WIND SE 4 O'CAST MOD SEA HEAVY NW SWELL SPEED 4 KNOTS COURSE 75 DEGREES TRUE WOULD BE HAPPY TO COOPERATE WITH YOU 0123 GCT REGARDS WELCH AND HALLIBURTON." Forty-mile-per-hour winds reached gale force. Winds of this strength can derail a moving or sitting train. Waves of thirty feet or more can bury a village.[17]

Within the same time frame, the "heavy seas" that threatened the *Coolidge* also threatened the *Sea Dragon.* At precisely 12:40 GMT, Welch sent a message to the *Coolidge:* "SOUTHERLY GALES WIND SQUALLS LEE RAIL UNDERWATER WET BUNKS HARDTACK BULLY BEEF HAVING WONDERFUL TIME WISH YOU WERE HERE INSTEAD OF ME. REGARDS, WELCH." Clearly ship and crew were now in the clutches of far more wicked weather than the "touch of rough weather" they encountered during the first attempted crossing. What with the "lee rail" submerged and the bunks soaked, the junk was filling up with enough water to sink it. Forty-five years later, in 1984, radio operator Charles Dunn of the SS *President Cleveland* claimed that he had received "two ominous messages about food being gone, the ship filling with water and in danger of breaking up." The *Cleveland* was making its regular run "along the sunshine belt to the Orient"—near the *Sea Dragon* but farther from it than was the *Coolidge*. If Dunn was correct, the *Sea Dragon* still moved forward with the bow scooping up water into the ship faster than the crew or gunwale scuppers could bail it out.[18]

On March 24, the "rough seas" and "dense squalls" returned.

35

Vanishing Point

A pall of gloom now filled the air. Reads the *Coolidge's* log for March 24 1:00 a. m.: "Mainly overcast," meaning 'dark clouds obscured the stars.' Temperatures were in the high fifties. "Pitching in heavy northwesterly head swell," soon the "vessel (was) laboring and shipping water along port side." By noon, a "steady westerly gale" accompanied "very rough confused seas and high heavy westerly head swells." Below the still "overcast to cloudy" sky, the ship pitched and yawed. As evening approached and skies grew darker, the "vessel (was) running under reduced speed in high breaking westerly sea and swells. (It was) pitching and laboring heavily," heaving to the port, then to the starboard side.

Within hours the wind had climbed from a Force 4 on the Beaufort Scale to a Force 6, then increased to Forces 7 and 8 before falling to Force 7 again. Wind velocity subsequently increased to Forces 8-10, roughly 40 to 65 miles per hour, then hit the dreaded Force "11 plus." Called still a "violent storm" with wind speeds ranging from 64 to 72 miles per hour, a storm of "hurricane" ferocity loomed. By noon the wind returned to Force 8, then rose again to Force 9, indicating a "strong gale" with a wind speed of 47 to 54 miles per hour. By six o'clock, the winds were again at Force 10; these "precipitous proportions" signaled an attack by a "storm" with a wind-force of 55 to 63 miles per hour. In rapid succession, Beaufort Scale estimates of Forces 10–8, 8, 9, and 7 meant that the weather had become perilously volatile. Passengers reported wave heights of 50 to 60 feet. If these observations were accurate, the winds that created these waves had reached an intensity of 80 to 90 miles per hour. A witness called the storm "next to the most fearful ever recorded."[1]

The storm, or network of storms, assailing the *Sea Dragon* was probably as fearful. The wind velocity rocking the *Coolidge* rose steadily, reported Dale Collins. Soon it reached "hurricane force" or "10 plus" on the Beaufort Scale. "Laboring and founding," the ship was soon "pitching deeply in (a) mountainous westerly head swell, high very rough sea." With the sea "increasing in size" and "green water" about to pour "over the forward deck," said Collins, "we hove to for one hour in order to unship all ventilators on and fore deck." Secured were "all booms (and) hatches, etc." Next, "Captain Ahlin (reduced) speed to 60 rpm (about 6 knots) as the seas were crashing over the foredeck in a never-ending bombardment. This reduction in speed relieved the excessive pounding but we still dove, shuddered, and scooped tons of water over the fore deck the remainder of the night. He estimated the seas to (be) between 40 and 45 feet in height (and 400 feet apart)." By "seas" Collins meant "swells." Swells of "40 and 45 feet" are as high as a four-story building. Swells of twenty to twenty-five feet in height can endanger a battleship. Despite all its precautionary measures, the *Coolidge* nearly capsized. Had they known of the *Sea Dragon*'s plight, those aboard the giant ship could only wonder what a storm believed to have reached Force 12 strength could do to the very small seafaring craft.[2]

Between Welch's "noon position" and the "southerly gales 12:40" transmission, forty minutes elapsed. During that time emergency steps were likely taken to keep the ship afloat. Although the diesel engine needed to be water-cooled, it could conk out if immersed and deprived of air. The ship's speed of four knots was, in any event, far short of its five-and-a-half knot potential. As water gushed into the *Sea Dragon*'s hull, crew members took turns at the bilge pump. The bunks were reported "wet," an indication that the ship was flooding. As the water rose, the pump-men, submerged to their waists, and engaged in Sisyphean drudgery, could not eject water faster than it entered. Sick with horror, the crew had little time *to react* and no stamina left *to act*. Once disaster struck, most may have been trapped below while only a few stood on deck to work the ship.[3]

By that third week, the crew began to tire. Stomachs groaned and muscles ached. Before leaving San Francisco, Halliburton said he would be "living on salt beef and hard-tack cooked by a Chinaman." At the time he believed food would be the least of his worries. But supposing that most of the food on the *Sea Dragon* had been consumed—evidence of the hearty appetites of hardworking men, or of overindulgence or theft—crew morale had to have been low. If food was rationed, altercations as to its equal distribution might have ensued. In late January, cook Sligh reported to Welch that certain crew members "had eaten twelve days [*sic*] sea stores up in five [days]." In reply, Welch told the cook that "he could starve them all he liked at sea." Except for a couple short messages sent from the junk, the period from March 16 to March 23 is the "lost week" in the *Sea Dragon*'s three-week voyage. The ship's food stores exhausted, crew members were reduced to eating "hard-tack." During those last days when the ship dawdled in port, much food could have been consumed without being restocked. If those stores were restocked, a food shortage after only twenty days at sea makes one wonder whether last-minute evacuees from the Colony contributed to its consumption. That the food stores were contaminated and had to be chucked is a more likely scenario.[4]

The *Coolidge* did not respond immediately to Welch's messages. Lookouts were again posted and given two-hour shifts. High winds and heavy swells were reported for March 24, and so were rough seas. This last phenomenon meant that the choppy waves created swells that *did not break*. Welch communicated increasingly unsettled weather conditions. His first message noted "moderate sea" and "heavy NW swell," and his second one noted "southerly gales" and "wind squalls." What size waves these winds produced— from crest to crest, trough to trough, can only be guessed. High, howling winds can turn into squalls. Strong, spiraling, drenching rains capable of disorienting the hardiest seamen often accompany these winds. Strong winds create big waves. Big waves move faster than small ones and brings greater destructive force to what lies in their path. With speed and a sturdier bow, the *Sea Dragon* might have punched through these waves, but only at great risk to its structural integrity.[5]

What forecasting instruments Captain Welch had on hand to complement his own instincts is imperfectly known. Hazard a guess, however, that headed his way was a slow-moving warm-front storm over which charged a fast-moving cold-front storm, one bringing with it steady phalanxes of high-towering waves. Although the exact nature of the storm the *Sea Dragon* met is little known, Dale Collins gauged its diameter at "between 600 and 800 miles." Thinking the storm much smaller, Welch could have steered the junk into the storm's eye—and into the breeding grounds of even vaster storm fronts. If so, the *Sea Dragon* stood little chance. If, in his hopeless efforts at escape, he supposed that the storm front was, say, half a mile in width—and that he could outflank it or run horizontal to it, he would have fared no better. In the event that he was unable to punch through the waves, he would have been smart to swing the ship around—that is, if it still had its rudder. By this time the sails could have collapsed and the hull been breached. As heavy rains in the pitch-black sky pummeled the tiny craft, the crew, thrown into confusion, ran for cover or got tossed overboard.[6]

In history, it is fun to ponder what might have been. As he had on the first sailing, Captain Welch might have assigned two crew members to the tiller as well as tossed an anchor into the water to steady the ship. On March 23, the *Sea Dragon* was 900 miles east of Yokohama and 1,500 miles west of Midway. Welch deemed the "little bitch" fast, even sleek, but the junk was neither sturdy nor tough. Now, moving at a reported speed of five knots through lashing rains, curtains of waves as high as thirty or forty feet, and winds howling at forty miles an hour or more, the *Sea Dragon*—-supposing all its best attributes were in place, struggled forward with its engine running and its sails partly raised. By March 24, the storm had grown into a typhoon, or Pacific hurricane. Four minutes after the "bully beef" message, Welch transmitted a graver one: "JUNK SEA DRAGON 1244 GCT MARCH 24 [TO] CAPTAIN AHLIN PRESIDENT COOLIDGE FROM 0200 TO 1030 G[REENWICH] C[IVIL] T[IME] HAD S[OUTHER]LY GALE HEAVY RAIN SQUALLS. HIGH SEAS BAROMETER 29.46. RAN WITH DOUBLE REEFED FORESAIL. 1100 GCT WIND CHANGED TO WEST FORCE 6, BAROMETER 29.54 RISING. TRUE COURSE 100 5.5 KNOTS POSITION 1200 GCT 31.10 NORTH 155.00 EAST ALL WELL WHEN CLOSER MAY WE AVAIL OURSELVES OF YOUR DIRECTION FINDER REGARDS WELCH." Rather than hundreds of miles, the *Coolidge*, in Welch's mind, may have been only ten or so miles distant. If so, he might have been steering the *Sea Dragon* in the known path of the *Coolidge* to intercept it. No one knows. The *Coolidge* had a

50-watt radio, which was standard for ships of its time and of its size, but its range was limited and during a storm messages transmitted and received could be static-filled or wholly unclear. Captain Welch's and Captain Ahlin's calculations, at any rate, indicated that the two ships would be "near" one another on March 25. The "heavy rain squalls" and "high seas" forced Captain Welch to *double-reef* the foresail to keep the junk from listing. Although the junk was apparently in grave danger, no distress signal went out. Welch sent a clearly articulated message, so the radio still worked. The barometer was at "29.46," then "29.54" and *rising*. Ordinarily when the barometer rises a storm is clearing; when it falls, one is likely but not certain. If the barometric readings didn't sound an alarm, the gale-force winds from the south changing to "west force winds"—Force 6 on the Beaufort Scale—should have done so. At Force 6, winds accelerate from twenty to thirty miles per hour and fifteen to twenty-foot waves with white foam crests form in long procession. Heavy rains add to the torment, water spray reduces visibility, and seawater, streaming over the deck, rushes into the hold. Thus afflicted, the reeling *Sea Dragon* staggered forward, then sank.[7]

Chief Officer Collins maintained that any action he took to address Welch's "cries for help" would have been an overreaction. Said he, the "bully beef" and "direction finder" communications "were the last messages that we received from the *Sea Dragon* and they gave no indication that the *Sea Dragon* was in any particular danger at the time sent nor did the messages indicate any anxiety on the part of Welch or Halliburton even though it was evident that the *Sea Dragon* had been experiencing some of the same rough weather we had met during the early morning of the 23rd." Wisely Captain Welch reefed the foresail—this is known. Not known is if he simultaneously "ran" the diesel engine; of course water, bubbling and gurgling, could have infiltrated the gaskets and gone into the fuel line, soon halting the engine. Propulsion lost, the ship, now flotsam, could then drift indefinitely. "Heavily over-engined," the *Sea Dragon*'s "bottom," as was reported in Hong Kong, "might have been ripped out in a storm." True or not, Captain Welch's indication that the "lee rail" was "underwater" should have signaled to any ship within range that "the junk was in dire distress."[8]

Able to determine his position, Welch was unable to determine his direction. If he carried an emergency navigation chart, he either couldn't get to it or, amid catastrophe, couldn't use it effectively. In the event, he might have relied upon celestial navigation, recognizing the stars overhead as familiar "moving bands of light" and identifying what landmass or ocean current lay directly below whatever star the *Sea Dragon* passed. Unfortunately, the blinding storm took away those visual cues. Without fixed points of reference to guide him, he had to resort to dead reckoning. Steadfast, he stayed on course, but, his bearings lost, he asked for a "direction finder" from the *Coolidge*. Had that request not been made, the rest of his message might read as a progress report.[9]

By 2:30 a.m. on Saturday, March 25, the swells had flattened and lengthened and decreased in speed and intensity, but sea levels remained high. As dawn neared, it was still cloudy with "very high, rough north-westerly and heavy swells flatten(ing)." The "vessel (is) pitching heavily at time(s) and shipping heavy sea over forward." Captain Ahlin maintained "strict and double lookouts" and offered a bottle of champagne to the first man who spotted the *Sea Dragon*. By noon, March 25, the storm had abated to a Force 4 rating before returning to a Force 5 or 6; by midnight, the rating rose again to Force 7. Throughout the day, it was still "overcast," then the "wind dropped to (a) breeze." The *Coolidge* still labored and continued to "pitch deeply and (receive) seas over (the) bow in a long northwesterly head swell." Designated lookouts remained at their stations, "keeping sharp watch for junk '*Sea Dragon*.'" The "heavy seas" broke "(the) starboard ladder leading from (first) deck to Promenade deck . . . (and) boat covers for numbers 7-9-13-15-6-8-14 were torn by wind." If the storm menaced the *Coolidge* that harshly, it must have menaced the *Sea Dragon* even more so.[10]

Like other ocean liners of its day, the SS *Coolidge* would have carried a Radio Direction Finder for triangulating signals from two or more stationary radio transmitters to pinpoint a targeted location. The ship's fifty-watt radio was considered standard equipment. It lacked, however, the long-distance tracking power of the Generalized Estimating Equation (GEE) used by the British Royal Navy, or Long Range Navigation (LORAN) used by the United States Navy. Both technologies were developed during World War II. In electrical storms, clearly transmitted messages are not so clearly received. John Potter had spoken to the pilot (presumably) of a Boeing Clipper when the plane arrived in Hong Kong. He learned that at Guam, "the Clipper pilots were told that the radio operators there had been in contact with the *Sea Dragon*, but owing to static it was difficult to hear her."[11]

Friday, March 23, at 8:00 a.m., shortly after Welch gave as his last position as 31.10 degrees north and 155 degrees east—about one thousand miles due west of Midway Island, radio operators aboard the *Coolidge* noted that the junk's radio faded out. Drawing within "three hundred miles" they "were unable to hear a sound" from the craft. Collins did not indicate what radio frequencies were used. He said only that broken batteries or a lost antenna caused the sudden lapse in communication. According to the US Weather Bureau, winds on March 24, a Saturday, attained velocities of forty and fifty miles per hour, creating the "same mountainous seas" the *Coolidge* encountered. Operators at Mackay Radio subsequently alerted all ships believed in the region to keep watch for the junk and listen for its radio signals. Lifted up by the wind, tons of water—salt water weighs sixty-four pounds per cubic foot, had by now fallen with crashing force on the tiny craft. Moments after Captain Welch asked for directional assistance from the *Coolidge,* transmissions from the *Sea Dragon's* shortwave radio then ceased. Potter noted that the junk's aerial, "strung between (the) main and mizzen masts" some seventy feet high, was vulnerably positioned. When the junk rolled, "there was a tremendous whip of stays and guys, with some chafing." Fallen rigging lay about the deck in a crumpled heap. Perhaps as happened during the junk's first attempted crossing, the aerial was sent "adrift."[12]

Checking his options, Collins decided that he could not rescue *the Sea Dragon*. He later revised his earlier reading of Welch's message. "Being a personal friend," the captain, for the sake of a good joke, could exaggerate, he said. Today it seems odd that Collins would think the *Sea Dragon* was not in danger, or that the last message Welch sent showed no anxiety. Both he and Welch had seen the 'wolf,' but only Welch cried 'wolf.' A rescue mission at sea can be costly, however, and with its own safety at risk, the *Coolidge* either would not or could not come to the junk's rescue. Confronting the same turbulent waters, the liner suffered its own torments as passengers became seasick, tumbled down stairs, or tripped over debris. In such a melee, a collision with another craft sent off course could and did happen—the *Coolidge* struck the *Frank M. Buck* on March 6, 1937, near the Golden Gate Bridge. When the *Sea Dragon* sent its last (known) message, the one asking the *Coolidge* to assist it with her direction finder, the *Coolidge* was itself In the grip of a catastrophic storm. Waves were maybe fifty feet high and crashing heavily against the struggling vessel. Furniture tipped over, objects flew, and lights flickered on and off. Crew and passengers clung to guardrails. Seasick victims stuck to their cabins; some were "flung across their staterooms and down companionway steps."[13]

While it is uncertain how near the *Coolidge* got to the *Sea Dragon*, Collins, as noted, supposed the liner was "approximately 300 miles from (the junk) on the night of March 24." By comparison, when the *Titanic* radioed for rescue assistance in the North Atlantic in April 1912, the *Carpathia* was 38 miles away. Although the rescuing liner seldom achieved a speed of 14 knots, Captain Arthur Rostron said he drove it hard at nearly 17 knots to reach the site in about 3 1/2 hours. If the *Coolidge* was 400 miles away, even traveling at a speed of 20 knots, it would have taken the liner 20 or 30 hours to reach the approximate position of the *Sea Dragon's* last radio point. Twenty knots was overtaxing the vessel; in the ensuing world war the top speed for battleships designed for rapid assault and breakaway was 20–30 knots. Had the *Coolidge* been within 100 miles of the *Sea Dragon*, it would have taken 4 to 5 hours to arrive on the scene, sufficiently close for Collins to report that he didn't even see "a stick floating." Savage winds and waves had, in his opinion, dismembered the junk. Had they sailed in "normal weather conditions," Collins was "reasonably sure that Halliburton and Welch would have sailed in through the Golden Gate not later than May 15." Wouldn't it have been marvelous if it had? Mysteriously, on March 25, when the *Coolidge* neared where it believed the "*Sea Dragon* should have been (allowing for her last known course and speed)," the "only thing . . . sighted," said Collins, "was a large glass ball commonly used by Japanese fishermen" to keep their nets and droplines afloat.[14]

36

Live a Little, Die a Lot

The storm cleared, but no *Sea Dragon* emerged triumphant into the light of day. Kept anxiously awake by the silence, Halliburton's publicist Wilfred Crowell telephoned Richard's parents on the evening of March 25. He told them "only that radio contact with the junk had been lost," and "that there was probably nothing to worry about." A couple days passed with no word from the ship. "Fear Felt for Halliburton's Junk," read a headline from the *San Francisco News* dated Monday, March 27, three days after all communication with the vessel had ceased. The article featured a picture of Halliburton, captioned "famous author and explorer, and his junk, the *Sea Dragon*, while she was under construction at a Hong Kong shipyard." The byline read, "Explorer's Ship Radio Is Silent," and "Adventurer's Junk Unreported on Pacific Voyage." Known by now was that the *Coolidge*'s radio operator had received the junk's position at some 1,200 miles west of Midway Island, and that appeals had been sent out "to all ships on the trans-Pacific sea lane between Honolulu and Yokohama to attempt to pick up signals from the junk."[1]

Those early days hopes were strong that radio communication with the junk would be reestablished. Queried by reporters about the matter, John Potter thought that, even with the inventive skills of radio operator George Petrich, the junk hadn't the resources aboard to address radio failure. To the best of his knowledge the junk had "no electronics, no fridge, no winches, no generator, and (had) radio batteries (that) were charged by running up the main diesel engine." Responding to Crowell's urgent appeals, the Navy used its powerful radio at Honolulu to contact the *Sea Dragon* but heard nothing. With assistance from Senators Kenneth McKellar of Tennessee, Charles Linza McNary of Oregon, and Hiram Warren Johnson of California, Crowell next enjoined the *Navy* to send ships to the rescue. The Naval Department, with its reply that that they "had no vessels in that area," compelled Crowell "to rely on the merchant marine and the Clipper planes in the search."[2]

The articles Halliburton wrote about the *Sea Dragon* Expedition that had appeared exclusively in the *San Francisco News* suddenly stopped. Although newspaper articles about the unfolding mystery were always dated a day or two later than the first wire-service reports, they kept the public minimally informed and often made it seem as though large flotillas of ships were hunting for the junk. The *Sea Dragon*'s radio was still "silent" on Tuesday, March 28, or "was unheard for several days." Some basic specifics were clear: "The sturdy 75-foot craft when it left Hong Kong March 4 [was carrying] Mr. Halliburton and a crew of 10 Americans and four Chinese." Several photographs of *Sea Dragon* crew members appeared in newspapers on March 28; Acme News published a photograph of George Barstow and John Potter; also published was a photograph of Ralph Granrud of Tacoma, Washington. Followers of the news started glancing past the customary details for new information. New was an item in the March 28 *San Francisco News* which reported that at 8:00 p.m. (PST) "today" the westbound liner *Empress of Canada* would "pass within a few miles" of the *Sea Dragon*'s last position. 'Old news,' no trace of the junk was found. New news? On Wednesday, March 29, the Associated Press noted with emphatic finality, "Author's Chinese Junk Vanishes After Gale."[3]

Readers were still left to wonder. Perhaps the junk was drifting about on its side. Perhaps it had bottomed up. Maybe it was just a little lame and chugging along. "The *Sea Dragon* had 30 days' fuel for its auxiliary motor," it was learned, "and sufficient food to reach Midway." Accounts of the ship's fate shared similar details: the date of departure, number of crew members believed aboard, and supposed location of the junk when it was last heard from. These accounts quickly grew old. Halliburton's die-hard fans turned their radio dials clockwise and counterclockwise to find the station carrying the latest news. Some even wrote directly to Halliburton's publisher, Bobbs-Merrill, to get the full scoop, but company managers were as much in the dark as anyone. For a time, the story's ending, not so much written and rewritten as embroidered and re-embroidered, had some in the general public wondering whether the *Sea Dragon*'s disappearance was all a big publicity stunt by an author who liked having his name in the paper. Never before had Halliburton's adventures received such riveting and ongoing attention. Maybe the news bulletins were all fabrications. Of recent memory was the "Panic Broadcast" of October 30, 1938, when the Mercury Players under the direction of Orson Welles had aired made-up *War of the Worlds* so plausibly that listeners everywhere believed Martians had emerged from a UFO and were terrorizing the countryside. Conceding that a disappearing act was, theoretically, a clever ruse, John Potter nevertheless thought the junk too far away from San Francisco for it to be a likely explanation.[4]

Ships of several callings continued to comb the area of the last radio signal, noted roughly as "about 1000 miles west of Midway Island," or as "1500 miles from Midway." By March 30, a Thursday, the sixth day after the last message was sent from the junk, concern for Halliburton's safety amplified. That same day, the *China Morning Post* reported, "Fears are now entertained that the Hong Kong trans-Pacific junk *Sea Dragon* has foundered in mid-ocean." Crowell, meanwhile, insisted that "the Navy Department search for the missing vessel." Already he had issued statements that the *Sea Dragon* had merely wandered off its charted course, yet his fears for the worst were deeper than were his hopes for the best. His and others' pleas for an all-out rescue mission increased in tempo if not in volume. Crew members' families and friends, as well as Halliburton's representatives in San Francisco, now leaned more heavily on the government, Coast Guard, and Navy to launch a major search. Seen as too costly and time-consuming—and believing it better to adopt a wait-and-see attitude while commercial vessels diverted their own time and resources to the task, a search by those departments was put off.[5]

Although John Potter thought the "failure in radio communication" proved nothing, he admitted that the junk could be "in serious difficulty." However, "I doubt it," he gave the matter further thought: "She is pretty well found and seaworthy; I have sent messages to relatives of the crew in America, telling them that." Still, absence of a radio signal, while it proved nothing, implied a good deal. Queried by reporters, the Mackay Radio Company issued a March 31 statement that it was "probable a typhoon had engulfed the craft." A storm of typhoon-like fury had "at least disabled the radio." Marine superintendent S. W. Fenton said that the company "had been operating on a regular contact schedule with the Halliburton party, and that the last contact was made Friday morning when the junk was 2400 miles from Hong Kong bound for Midway Island." He added that a "severe typhoon" had "swept the area" and that, with the loss of radio contact, "ships in the vicinity" had been asked to "watch for the junk."[6]

Details slowly emerged. United Press and its affiliates now released the names of those feared lost: Captain John W. Welch; first mate John B. Flagg; engineer Henry Von Fehren; licensed radio operator George I. Petrick [sic]; chief cook James N. Sligh; travel writer Richard Halliburton; able-bodied seaman Ralph Granrud; Paul Mooney; Robert R. [sic] Chase; George E. Barstow; Velman E. Fitch; and four Chinese, Sun Fook, Kiao Chu, Wan Ching-huo [sic], and Liu Ah-shu. "Those who were originally signed on included John Rust Potter, Gordon E. Torrey, and Richard L. Davis, of the United States, and Patrick Kelly, of Hong Kong, who remained in Hong Kong." Hometown newspapers concurrently devoted space to crew members from their respective communities, notably the *Washington Star* for Paul Mooney, the *Minneapolis Star* for Velman Fitch, and the *Boston Globe* for Robert Chase.[7]

March 31. Nearly a week had passed since the *Sea Dragon's* radio signal failed. The papers for that day reported the junk "Still Missing." Admirers of the author were gratified to learn that an air search accompanied the sea search; indeed, Pan American Airway officials rerouted trans-Pacific Clippers westbound from Honolulu, putting them "slightly off their regular courses" so that their crews could scan the area where the *Sea Dragon* had sent its last radio messages. The Navy Department, meanwhile, assured Senator McKellar that it would send vessels to help in the search. Ocean liners were also rerouted to lend a hand. By April 1, the search had clearly widened. Hope remained that the *Sea Dragon*, its radio disabled and diesel engine damaged, was slowly making progress "mostly under sail." Fans of the author could draw a sigh of relief: a violent storm had injured the ship but not sunk it. Still, efforts by sea and air "reported 'no trace' (of craft or crew)." While independent search teams were eager to lend assistance, the Navy, except for attempts at radio contact, was "making no official search." Sunday, April 2, beneath the more chilling headline "Richard Halliburton's Chinese Junk in Area Swept by Typhoon Last Week," the *Boston Globe* noted that "fears (were) felt for [the] safety of [the] adventurer and his crew when a storm cut off communication." By Monday, April 3, devout followers of the story supposed that the *Sea Dragon* had vanished, but without a body, how could they be certain?[8]

From the flood of uncertainty, none other than noted author Albert "Dick" Wetjen surfaced. Identified as a "writer of sea stories" by the *San Francisco Examiner* reporter who interviewed him, "Wetjen, (believed) to have known Captain Welch for years, said the *Sea Dragon* was in capable hands." The feature had Welch's photograph. Next to it was one of Wetjen with his ear bent to a marine radio transceiver as he awaited news about the junk. "If anyone can bring Richard Halliburton's *Sea Dragon* into port, Captain Welch will do it," the writer winked: "Anything can happen at sea, especially in a typhoon. I'm not giving up hope until the middle of the week. But if they haven't reached Midway by then . . ." Committed skeptics like Wetjen who were not so ready to accept the verdict of death listened intently for that faint bell ringing from the grave. April 4, the *San Francisco News* reported that the craft could still "manage to reach the naval base at Midway Island." But it also stated, "The maximum deadline allotted (for such an arrival) by marine experts, who have plotted the craft's progress from her last radio reported position [is nearing]." Eleven days had passed with navigators agreeing that, even traveling under sail, the *Sea Dragon* "should near the island today or tomorrow or early Thursday (her fourteenth day missing), at the latest."[9]

Cheerier alternatives to the grimmest news added to this faint glimmer of hope. On April 5, 1939, a certain Ralph Lundquist, currently a mimeograph operator in the US Customs Building and formerly a crew member of the *Ning-Po*, offered the most compelling of these alternatives. He recalled how the *Ning-Po* was thought lost on its historical 1915 voyage to the Panama-Pacific Exposition, likening that picturesque junk's unreported fifty days to the *Sea Dragon*'s unreported twelve days:

> In mid-winter of 1912, four days before Christmas, we left the Japanese port of Shimidzu (near Yokohama). The second week out we struck a storm of hurricane proportions. At noon, wind and seas increased. By 1

o'clock there was no time for log books. At 8 p.m. the log might have read: Sails lost, main boom broken, wooden water tanks ditto, tightened leaks, 26 inches fresh water left in starboard tank, hull leaking, soundings: 8 p.m., main hold, one foot (water level), manned (bilge) pumps. Gale at height, 3 p.m.; Seas, cross, seep, and breaking, wheel rope carried away. 3:07, new wheel rope rove; all oil used. Rudder post damaged, 18-inch lengthwise crack above water line, unable to repair until storm abates. All hands safe. Cook's arms injured by scalding water. Fire out in galley. 8 p.m.: Great seas, riding easily to improvised sea anchor.

So close is the resemblance of these recollections to what might have happened aboard the *Sea Dragon* that they read as a page taken from Welch's intended book about the adventure. "At midnight," Lundquist continued, "the most mountainous seas that I have ever seen rolled the Chinese junk along. It was stupendous. One minute on top of the world, the next in a canyon of the sea, then up again, all through the night into the morning and the next day, and a new storm, and more days and more storms. They never ceased. Reefing and double-reefing, pumpship [*sic*] and patching leaks, having repaired our rudder, boom and sundry things, neglecting to call at Hawaii, we sailed stubbornly on." Like the *Sea Dragon*, the *Ning-Po*—off course by two hundred miles, had sent a message to a larger vessel, here the *Honolulan,* for help to establish its position. Once put on a proper course, the junk sailed down the coast of California and "anchored inside the San Pedro breakwater" at midnight on February 19, 1913. Although the *Ning-Po*'s plight was not a duplicate of the *Sea Dragon*'s, both ships were headed for a world's fair, both encountered similar hardships—leaking, difficult rudder control, "sundry things," and both had communicated their position to another vessel. "It's my belief," concluded Lundquist, "(that) the Halliburton ship was damaged in a storm, that they have repaired the damage and are now making slow progress." He concluded, "Anything can happen in sail and storms." Just a few years before the *Ning-Po* rescue, Jack London, following mounting fears that he with the *Snark* had vanished, reappeared. Missing vessels, concluded Lundquist, were more likely to be found than lost.[10]

April 5, the SS *President Adams* passed the supposed spot where the *Sea Dragon*'s last communication was sent. Nothing. Faint word was of course better than no word at all. Reported one news service: "The latest 'No sign of Halliburton junk' message was received by Mackay Radio last night from the freighters *Jefferson Davis*, bound from Japan to New York, and the *Torak* (or *Toorak*), out of Los Angeles. Both ships kept a watch as they passed through the area where the junk was last heard from." The glimmer of hope that the *Sea Dragon* would reach Midway under sail had all but dissipated. Still, maybe it was foundering—somewhere? With social unrest at home and the threat of war in Europe and the Pacific, certain California senators and President Roosevelt, himself a former Assistant Secretary of the Navy, received heated-up pressure to extend the search for the missing junk. Holding to its line that a full-scale search would be impractical, the Navy limited its search efforts to sightings by one or two of its ships diverted from their regular course to the junk's last reported position. April 6, 1939, readers learned that the Navy Department had before it "a request from Mr. Halliburton's business representatives (and Senator McKellar), asking that the cruiser *Astoria*, enroute to Japan, be diverted to search the area west of Midway Island from where the *Sea Dragon* last reported 13 days ago." And, none other than Albert Wetjen inspired the call to arms.[11]

Attempts at radio contact continued. The route of the Pan American Airway planes was 1,200 miles south of the junk's last reported position—too far away for visual contact, but with radio stations at Honolulu, Midway, Wake, and Guam, the airline tried to make radio contact with the *Sea Dragon*. The coast guard cutter *Taney*, some 2,500 miles to the south, repeated the effort. Halliburton's "local representatives" asked the Navy to chart the junk's *probable drift* and tell ships in the vicinity to remain on the alert. Except for a reported last sighting weeks before of the junk "400 miles north of the Marcos Islands," not an atom of substantive 'new' news about its whereabouts surfaced. Days passed. Inquiries to newspaper editors were made. Subscribers to the *San Francisco News* in particular wondered if the matter of the missing craft had been laid to rest. Monday, April 10, news reports contained the now-familiar refrain that no ships had spotted the junk: "Navigators who (like the naval experts) plotted the *Sea Dragon's* course from her last radio-given position of '1200 miles west of Midway' were not pessimistic, but saw chances lessen that she might make the naval base under sail." The *Sea Dragon* was now overdue at Midway by almost a week. Optimists like Dr. E. Allen Petersen of the *Hummel Hummel* put little trust in the conclusion this dreaded news strongly implied: "There are calms (in the Pacific) lasting weeks," he said, implying that the ship could drift for days under sail. "The engine," he noted, without elaboration of the point, "would prove of little aid in a calm." Although "clumsy to handle," junks, he said at last, "are extremely seaworthy." Despite the eighteen-day silence from the junk and the failure of search teams to find it, to his mind the *Sea Dragon* and its crew were safe.[12]

Amid the hoopla, the US Coast Guard studied requests for a "cutter to be sent to the area west of Midway Island to search for the *Sea Dragon*." These pleas were put on permanent hold. "Sending a cutter from San Francisco or Seattle would require 12 or 13 days," it was said, "and use up all the fuel required to conduct a preliminary, let alone a thorough search." The decision to sit still was reached on April 12, the twentieth consecutive day the junk went unreported. By April 13, newspapers spoke of the "fate" of the *Sea Dragon*. The *Memphis Commercial Appeal* even inserted a map showing the route of "the missing Halliburton craft"; a question mark indicated the general location—some "1000 miles (from Midway)," where the junk's radio "went dead." That same day, pressured by California congressman and onetime San Francisco Harbor Master Richard Joseph Welch and by Tennessee native and Secretary of State Cordell Hull, President Roosevelt assured the public that "all naval vessels in the Western Pacific and all trans-Pacific passenger vessels have been notified to watch for the *Sea Dragon*." Naval officials said that it was doing "everything possible," but, as "no Navy ships or planes (were) in the immediate vicinity where the *Sea Dragon* was last reported," a quick response was unrealistic. Also on April 13, Wesley Halliburton received a letter—unpublished by the newspapers and unsigned. In it a Lieutenant Reed was quoted as saying that the "port authorities at Hong Kong had told him that it (the *Sea Dragon*) was the finest junk of its kind that had ever left the harbor." Based on that consoling news, Wesley supposed all on board were "safe." Not everyone agreed. The April 14 issue of the *Hong Kong Telegraph*, for instance, reported that the ship was "believed lost in the Pacific." The *Telegraph* also announced that the wedding of original crew member Richard Leslie Davis and Eileen M. Clewer had been solemnized, with fellow crew member John Rust Potter among those on the registry. The reception was held at the Peninsula Hotel; the honeymoon would be in Honolulu.[13]

By mid-April, three weeks after the *Sea Dragon*'s last reported message, news about the junk's whereabouts had gotten briefer. After several days on the front pages, word of the ship's fate retreated to the interior pages of most newspapers. Readers were reminded that Halliburton's was a trans-Pacific cruise from Hong Kong to the Golden Gate International Exposition, and that radio operators of the Mackay Radio Company, Pan American Airways, the Navy, and the coast guard had failed to make contact with the junk. In addition, the *Sea Dragon*'s last reported position placed it in the center of a storm of cyclonic proportions. Ardent fans still kept watch. A letter to the editor of the *San Francisco News* from one Sarah Pratt, published on April 18, reminded readers of the sort of heroism Halliburton stood for in a world gone mad with technology. In her view men of daring like him put "safety" last in their attempts to conquer nature or reach a star, but over what Halliburton called "adversity" they had ultimately triumphed. Bare bones news reports were not as positive. The front page of the April 20 edition of the *San Francisco News* carried the headline "Fears for Safety of Halliburton Rise." These fears grew. The general feeling was—if a large craft had little chance in a typhoon, what chance had a small craft? Richard, however, had led his mother Nelle Nance to believe that nothing "could possibly go wrong." When she contacted Gordon Torrey about the junk's seaworthiness, he likely expressed doubt. Still, both she and Wesley kept faith that their son and his crew would be found, but by April 24 that faith wilted. For weeks, they had received numerous letters of hope; now they received numerous letters of condolence.[14]

"Not a clue" about the junk's whereabouts emerged. Diehards continued to wonder, and read into the slightest hint of one the deepest significance. Passing steamers for years kept a lookout, thinking they just might run into the junk. For months at least, both the Navy, which did not make an actual search until much later, and the Coast Guard, which too remained idle, only made efforts to contact the junk by radio, as faith in the resourcefulness of radio operator George Petrich persisted well into the fourth week of April. Reliance upon the skills of Petrich held some merit. Commented an operator in the San Francisco office of Mackay, Petrich was "one of the best in the business." Indeed, "Even if high seas had swept away his antenna and damaged other equipment, (George) probably could have sent out signals with makeshift materials," namely "with an old receiver or with 'a tube and a couple dry cells.'" That not a peep was heard from him indicated that the junk "must have gone down." April 29, a report came that the junk had "foundered," having been "pounded" by the same "fifty foot waves" that had struck the *President Coolidge*. The report also mentioned that Chief Officer Collins had spoken with Captain Welch who told him that he was "dubious about the *expedition* and had *added* 18 inches to the junk's keel. because 'it was rolling too much.'" Welch also told him that he had little "confidence in his crew" and "had signed on four or six junk sailors." If sailors were added—a tantalizing element in the mystery—it explains the food shortages as well as explains why Welch, now with experienced sailors working the ship, reckoned it safe to undertake a voyage he had long thought "dubious."[15]

May 3, 1939, the *San Francisco News* published a cartoon of the *Sea Dragon* by "Rodger." A banner attached to the ship's mizzenmast bore the name "RICHARD HALLIBURTON." The cartoon's caption read, "His last adventure?" Although pointed questions remained, reports of the *Sea Dragon* were soon lost to other news. On April 22, the *San Francisco News* reported that the *Tai Ping*, knowing little or nothing about the fate of the *Sea Dragon*, was journeying on to Treasure Island. Three weeks later, on May 13, word came that the junk *Tai Ping* had been rescued. News reports pictured owners "Mr. and Mrs. John Anderson" alongside *their* junk, not Halliburton alongside *his* junk. Following "the old trade route of early American sailing vessel," the Andersons' goal was to reach the Golden Gate International Exposition—perhaps for the $8,000 prize awarded to the first such craft to reach the fair. Their voyage had been "interrupted by storms," and their disabled vessel had been "towed into safety by Japanese fishermen." In Yokohama the craft was repaired and incompetent sailors replaced with experienced ones. Undaunted, the Andersons, and the *Tai Ping,* recommenced their trans-Pacific voyage.[16]

37

Mistaken Identity

*F*ate implied *doom*. But was Halliburton really lost? Proof was needed, or, as Paul Mooney's mother, Ione, would tearfully insist, final peace needed its obsequies. Nearly sixty days had passed since the junk's last signal when Secretary of the Navy Claude A. Swanson, following a conference with chief of naval operations Admiral William D. Leahy, Tennessee Democrat Senator Kenneth McKellar, and Wesley Halliburton, at last ordered the *USS Astoria* under the command of Richmond K. Turner to begin a search for the junk. The quest would begin "from a point (400 miles) north of Marcos Island and following the small sail-craft's probable course."[1]

Wednesday, May 17, the following headline appeared: "Navy Planes to Hunt Halliburton, D.C. Aide." Exactly a month before, the cruiser had arrived in Yokohama with the ashes of Ambassador Hiroshi Saito. Now as then, the ship had permission from the Japanese government to search the area in a route that had them going from latitude 38-10 north, longitude 155 to fifteen miles from Midway Island. Equipped with floatation gear, four biplanes (technically Curtiss SOC-1 Sea-Gulls) carrying two men each were called into action. Exact coordinates of the search were not reported, nor was it reported whether the radius of the search was narrowed or expanded. Aiding them in their search were only the naked eye and binoculars. "Sunlight streaming into the cockpit, warming it," writes John Alt, "the men slid open the canopy to let wind rush in and to better see any specks on the water." Often the pilots flew their planes so low they could see themselves reflected in the water.[2]

While the press made it sound as though numerous planes and ships, like cavalry to the rescue, had converged on the spot where the junk last reported its position. This action was neither concerted nor timely, as the Navy conducted its initial search quite late, long after many other ships had combed the region and the junk was presumed lost. Although they conformed to established protocols, the thankless search efforts were at best the belated obsequies which Ione Mooney wished for her departed son. Consider the very size of the task—the Pacific Ocean is about sixty-three million square miles in extent. Wrote *You Can't Go Home Again* novelist Thomas Wolfe: "The sea makes a city." Not one but hundreds of cities compose the largest body of water in the world, and in that context the *Sea Dragon* was just a paper clip in a desk drawer, a trinket in a midtown boutique, a flea crossing an eight-lane highway.[3]

The search began May 21 and ended, according to Admiral Turner, May 26. Late to start, this quest raised Ione Mooney's hopes that her son Paul was still alive and fueled suspicion that its real mission was to mark Japanese activity in the region. In length about 1,480 miles were covered, in width an average of 103 miles—in all, nearly 152,000 square miles, a figure Admiral Turner later disputed but did not correct. The search proved "fruitless." Admiral William D. Leahy concluded that "no trace was found of the *Sea Dragon*" while Admiral Turner concluded that the results were "negative." This was depressing news certainly; however, it was noted that the ship might still "be drifting helpless" somewhere.[4]

By June 1939, the belief had become universal that the *Sea Dragon* and its crew were lost. On the eighth of that month, the Department of Navy, following a brief inquest, declared that the *Sea Dragon* and its crew had perished. Photographs of the famed author appeared with a summary of his storied life in the June 11 edition of the *Memphis Commercial Appeal*; read the byline, "Life's Odyssey Ends for Richard Halliburton." In an interview published the next day, Nelle Nance Halliburton declared stoically that her son had died on the sea he loved. "Somewhere," she said, whipping away a tear, "Richard is finding new worlds to conquer." Wesley remained stoic. He was two when his thirty-two-year-old father,

John, died and barely thirteen when his forty-four-year-old mother, Juliet, died. Now seventy years old, he had outlived both his sons.[5]

Watching events further unfold, Admiral Leahy assured Wesley Halliburton that a lookout for the junk would continue. Recently, on June 16, dire news reached the public that the *Pang Jin*, another junk, had foundered in the Red Sea, but was intact. Cordell Hull, meanwhile, informed Navy Secretary Swanson that he had gotten word from the Foreign Office of Japan that their shipping companies would be on the alert for the *Sea Dragon*.[6]

Briefly front-page news, the *Sea Dragon* story within days began drifting toward the lower columns of the back pages. From the start, news about the junk appeared alongside movie reviews, society features, Hollywood gossip, and want ads. Interest faded. Hopes of the *Sea Dragon*'s miraculous reappearance now toddled into the realm of make-believe and wishful thinking. From time to time tantalizing news did surface. Naturally drawn to it were the families of those who had participated in the expedition. On January 6, 1940, the *Boston Globe* noted, "Believe Wreck Sighted Is Junk *Seadragon* [sic]," but the identification was mistaken. Particular mention was made of "a Milton young man, Robert Chase," given up as lost, whose will was filed for probate." The singular irony was that Chase, had he listened to Gordon Torrey and not gone on that second sailing, would have returned home and received a nearly $1 million inheritance from his grandmother in a matter of months. That sum of money could have financed many *Sea Dragon* adventures. Call the incident an entry in Ripley's *Believe It or Not*.[7]

In July 1939, besides the *worst* bad news, news arrived home that the *Sea Dragon* had at last materialized and Robert Chase and company were saved. To be exact, on July 13, the SS *Coolidge* sighted the *Tai Ping* about 600 miles due east of Yokohama. This was the very same junk Captain John Anderson and his wife, Nellie, had started to sail from Kobe three months earlier. That the craft was first mistaken for the *Sea Dragon* indicates that some news followers and search parties believed Halliburton and crew might still be at sea. Making its regular round-trip run, the *Coolidge* was seven days out from Honolulu when it spotted the *Tai Ping*. Both ships were roughly 250 miles from the last known position of the *Sea Dragon*. As did other passengers (including indomitable Chief Officer Dale Collins), journalist Violet Sweet Haven at first thought that the re-emergent *Tai Ping* was the *Sea Dragon*.[8]

Captain William O. Kohlmeister (who on May 4 had replaced the retiring Captain K. A. Ahlin) knew the risks of getting too close to an unidentified ship, as well as the cost in time and money of a mid-ocean stop. With Chief Officer Collins at his side, Kohlmeister wished only to wave to the craft's occupants as he passed it by. Then, recognizing Anderson from a past captaincy role, Kohlmeister had the *Coolidge* draw close to the unknown vessel. Violet Sweet Haven wrote, "Laughter ran along the spacious decks (of the SS *Coolidge*)" as novelty-starved passengers and crew thrilled "at the prospect of a giant six-hundred-and-fifty foot vessel moving toward a forty-one foot junk." In a loud voice, Kohlmeister, or one of his officers, asked if he could in any way assist the amusing smaller craft, but got no response. "The little junk and its crew," said Haven, "seemed almost eager to proceed into the dreary leagues of uninhabited ocean unaided." Captain Anderson then shouted back that he could use some fresh water. The *Tai Ping* had seven full water tanks, and now only half of one was left. Captain Kohlmeister also gave Anderson "fresh eggs, soap, maps, a quarter of beef, a ham, ice cream, bread and two flashlights (and batteries)." In his own report Collins added to the list of shared provisions "cigarettes, distress flares, magazines, soap, matches, etc."[9]

The craft hurried forward, then lingered. Some passengers aboard the *Coolidge* found it thrilling to be at a full stop in the middle of the ocean, and they hung over the guardrails to watch. A few, "touched by the sight of the girl (Nellie) and the drifty craft, had showered armfuls of American magazines and San Francisco newspapers, cartons of cigarettes, gum, candy and jars of cold cream" on the party. Spotting the "tiny gray kitten" and "mischievous puppy," both of whom, noted Haven, were "chas(ing) each other around the narrow deck," a couple passengers even tossed snacks toward them. Insistent, Haven persuaded Dale Collins to let her interview Captain Anderson, but the two exchanged few words. Haven fared better in her interview with Anderson's wife, Nellie, whom she called "a petite and extremely pretty girl" and whom she identified as the chronicler of the voyage. "Among five husky bearded men," wrote Haven, this able young lady, decked out in "a bright red sweater, khaki shorts, blue anklets and a colorful beret," leaned against one of the masts, "making hasty jottings in her log book." Eager to learn more, the journalist asked her about her days at sea. The woman graciously responded, "Crossing the ocean in a Chinese junk gives you lots of exercise, cooking and washing, and pulling on ropes." Asked what she did to pass her time—-particularly, if she did any "embroidering," she replied that she did "some sewing." Through some laughter, Nellie then exclaimed that she was "forever making coffee." At eight in the morning, "porridge and coffee" were served; at noon, "soup" and what she could only describe as "hash." Asked about "tea," she said that she and the crew had cut down on tea. She further informed Haven that the men were "well-fed." Medicine? "We have a bottle of mixture," Captain Anderson or one of the crew added a voice, "but it isn't any good. We could use some stomach medicine." Soon it was delivered. Also delivered, "just in case he (Anderson) might want to signal a steamer in the darkness," were several "emergency rockets."[10]

The liner and the junk parted ways in July. On October 3, news bulletins reported that five people were rescued on a junk whose journey from Shanghai to San Francisco had been interrupted by storms. Was it the *Sea Dragon*? Sadly, no, it was the *Tai Ping*—now 105 days out of Shanghai (or, rather, Kobe) and in the Puget Sound area, far off its course. Haven said that the "Chinese diesel engine" had conked out, and that the ship "could not sail by the wind, only before it." Mangled from top to bottom, its rudder loose and its radio lost or broken, the junk was towed to safety in Vancouver. Those on board, all gravely ill and haggard in appearance, were identified as a Russian, a German, two Norwegians, Captain Anderson, and his wife, Nellie. For days the junk had just drifted along. To shelter themselves from the sun, those on board lived under canvas sheeting and for cooking they used a kerosene stove. Over the course of many sunrises and sunsets, their food supply had dwindled down to a small quantity of rice and ten gallons of water. Half-starved and close-quartered, the crew had to have been reduced to a not-so-noble savagery. What happened to the gray kitten and mischievous puppy will never be known. When he saw seaman Burton Antill from the rescuing ship produce a loaf of bread, one of the crew grabbed it and "bit at it ravenously." All were "almost hysterical," said G. C. Jones of the *Discoverer*. Quipped Haven, "Romance ends at the docks."[11]

At sea five times longer than the number of days the *Sea Dragon* had been, the *Tai Ping* gives some idea as to the ordeals its more illustrious counterpart might have faced had it survived the storm. Enduring the worst and uncertain of survival, would those aboard the *Sea Dragon* have warred or worked together? All memory of the hardships the crew of the *Tai Ping* endured was released by the exhilaration they felt at their salvation. Haven concluded, "Pirates . . . typhoons . . . famine . . . rain storms . . . illness . . . Broken engine . . . Lack of water . . . lost at sea . . . dangers, yes, but the brave sea-goers were contented. They had staked their lives on adventure—and won!" The *Tai Ping* would voyage onward to San Francisco. Although no mention was made of a reward given to the first junk to reach Treasure Island, or of a race with Halliburton, the indomitable Captain Anderson, even after his purely lucky landfall in Vancouver, remained determined to reach the fair. He proceeded then to Panama, upward to New York, and finally to Boston.[12]

38

The Summing Up

The bell from the grave of the sea did not ring. What remained for Wesley and Nelle Nance Halliburton was the business of closure—addressing "the necessary things." Both were mentioned in their son's will as his principal beneficiaries and as the recipients of proceeds from the no fewer than three insurance policies issued in his name. A life insurance policy valued at $30,000— one without an accident clause, had been fully paid. Also fully paid was a $3,750 insurance premium to Lloyds; what settlement, if any, was paid to the beneficiaries of accidental death insurers is unclear. Also named as a beneficiary was Richard's own Princeton class of 1921. Paul Mooney was the last person named, a status he would retain only if he and Halliburton did *not* die together. In January 1939, just before the *Sea Dragon*'s first attempt to cross the Pacific, Halliburton returned a "change of policy" form requested from Gay Beaman, his financial advisor in Laguna Beach. Careful wording clarified the chain of custody: "If Paul & I sink together, the insurance policy made to him goes to you," he said. "I do not want it to go to his family." Less entangled was the matter of the party's cash assets; these included, as of January 1939, a $1,500 "emergency fund" in a simple savings account in a San Francisco bank. Royalties (as much as "9,000) from the two *Books of Marvels* brought the total value of the estate to over $50,000, or between $500,000 and $1,000,000 in today's money. "Hangover House," meanwhile, was assessed at about $20,000; today, just the land it sits on would be worth several million. After an offer by actor Eric Linden to buy the house for $12,000, and an attempt to sell it at public auction, it sold in 1941 to US Marine Corps pilot Wallace T. Scott and his wife, Zolite, for $7,500, roughly a fourth of what it had cost to build. Likely Halliburton would have sold the property, which he thought an investment, just as he intended to sell the *Sea Dragon*, but not for so dismal an amount. The fate of the used Dodge coupe is uncertain.[1]

Clearly it was Halliburton's wish to travel through life as lightly as possible. Apart from Hangover House and a used automobile, he had surprisingly few tangible assets. Furnishings in the house were sparse, including a few mementos from his travels, as well as some artwork. His library consisted of about two hundred books—mainly humanities texts and multiple copies of his own works. For his part, Halliburton thought a bequest to a library the perfect gift to posterity. In a last will and testament made out on August 10, 1937, signed before witnesses, and filed on October 12, 1939, he named Princeton University as the curator of "The Richard Halliburton Geographical Library." Once she learned of the Halliburton bequest, Bertha Barstow offered funds to Berry College in Rome, Georgia, to build a library bearing her only son's name. What was called the Barstow Memorial Library would become the Berry College Elementary and Middle School. The benefactor's wishes, in Halliburton's case, suffered a similar misreading. While his private papers, letters, and manuscripts formed the Richard Halliburton Collection, what books Princeton received from the Halliburton library did not form an adjunct Halliburton archive, but were absorbed into the school's general collection.[2]

For probate purposes and to hasten the transfer of any and all assets belonging to their son, Wesley and Nelle Nance Halliburton, with the assistance of the First National Bank of Memphis, filed a petition in the *Chancery Court* to have *both* Richard and Paul Mooney declared legally dead. Chancellor in the proceedings was Tennessean John E. Swepson. The *Memphis Commercial Appeal* reported, "The suit was brought to establish death to collect a $10,000 Federal War Risk Insurance Policy and obtain title to the Halliburton estate in California. Insurance companies holding policies for approximately $50,000 waived (their right of) defense and signified (their) willingness to pay after court certification of death." During the two-day proceeding, Wesley Halliburton, resident of 2275 Court Street, "real estate man, and father of the lost author," was asked to take the stand. In his testimony he remarked that "the *Sea Dragon* had provisions for 90 days, enough engine fuel for a little less than 90 days, and water for about 110 days." Those aboard the ship, he said, "expected to make the Midway Islands in 30 days, and had planned to contact the SS *President Coolidge* for fresh food supplies enroute." Introduced into evidence were the maps and the log of the liner. The Navy Department's efforts were duly noted; the agency had ordered "all boats in the vicinity to keep a lookout for the junk." Also introduced into evidence was notice of the United States cruiser *Astoria's* search over 152,000 square miles of territory. Impressed, Wesley Halliburton remarked, "That is four times the area of Tennessee."[3]

Courtroom discussions added nothing new to what was already known about how the *Sea Dragon* met its end, but they remained *on point*: "The last radio message, 'our lee rail is awash,' came hours after the weather had turned 'nasty' and sails had given way to the auxiliary motor. From reports of other craft in the typhoon area and United States Naval Weather Stations, tremendous waves must have tossed the *Sea Dragon* before it either sprang a leak or foundered." That said, the jury concluded that the *Sea Dragon* "sank with all on board 'on or about March 24' between Hong Kong and the Midway Islands." Case closed. Recapped Jonathan Root: "The court's decree served two functions; it permitted the payment of life insurance benefits to Halliburton's now pressed parents, and it punctuated all the eulogies by people who had known him intimately and slightly, and who were trying to define the loss." On October 5, 1939, Richard Halliburton was officially declared dead.[4]

In the latest news for overseas travelers, LaGuardia Airport had opened in New York on October 15. Other headlines reported that baseball great Lou Gehrig had been stricken with "polio"; though he was not quite thirty-eight years old, he would die two years after delivering his famous farewell address at Yankee Stadium. Elsewhere, news sources reported that the British battleship *Thetis* was attacked, mortally wounded, then sunk. Subsequently, a French ship was sunk, then another, the *USS Squalus*, with families of the victims waiting in hope and with fear. In other naval news, the pride of the German naval fleet the *Bismarck* had been launched on February 14. In movie houses *Topper Takes a Trip* starring Roland Young and Billie Burke entered into its third successful week. Box-office hits such as *The Dawn Patrol* with Errol Flynn and David Niven, *Stagecoach* with John Wayne and Claire

Trevor, *The Oklahoma Kid* with James Cagney in his first Western, and soon *Gone with the Wind* would premier. *Good-Bye, Mr. Chips* sentimentalized England, the private boys' school, and the war effort. In August, "Mexican Spitfire" Lupe Velez and Johnny "Tarzan of the Apes" Weissmuller, after several futile attempts to reconcile, finalized their divorce. Other news included King George VI and Queen Mary of England's tour of Canada and America. Word was out that the Dionne quintuplets of North Bay, Ontario (Annette, Cecile, Yvonne, Marie, and Emilie), were getting ready to meet British royalty on March 13. A blockbuster news story about this same time reported that child star Jackie Coogan and his mother were to split his fortune. Only $250,000 remained of "The Kid's" cash, but Jackie was to get $125,000 in real estate and own the rights to his many silent films. March 23, the day the *Sea Dragon* disappeared, First Lady Eleanor Roosevelt visited the Bay Area. She described San Francisco as "one of the cities of the United States which has real charm-," and added, "I always enjoy coming back to it." At the fair she visited the *Federal Building* and remarked that the Pillars of the Forty-Eight States were "striking but not beautiful in the daytime." Commenting on Bay Area weather, she recommended that one wear a "very warm coat." The First Lady evidently had no idea about the *Sea Dragon*, that it was headed for America, much less that it was lost. Nor did Hitler, whose armies now occupied Bohemia and Moravia. By this time Czechoslovakia had been annexed to the expanding Nazi empire.[5]

The likelihood of an overseas war with Japan was not yet driven home to the American public. Before the attack by Japanese warplanes on Pearl Harbor on December 7, 1941, and remand of Japanese Americans to detention centers, the Japanese were at the top of America's minority list. Even as war waged between the Chinese and Japanese ten thousand miles away, in America Japan and the United States remained studiously cordial. As a Japanese exhibit had been in the Panama Pacific International Exposition of 1915, so too had a Japanese Pavilion returned as a main attraction of the 1939 Golden Gate International Exposition. Most popular was a pond where visitors could watch the trained cormorants catch some odd fish called "ayu." In proud obedience, the birds then 'delivered' the fish to their masters, who were standing by in tiny boats. On May 2, 1939, Japan Day was celebrated on pavilion grounds with both Japanese ambassador Kensube Horinocchi and army major Maurice Stubbs saluting the Thirtieth Infantry Exposition Company during the ceremonies. If a political statement on behalf of greater Chinese and

American relations, the Halliburton Chinese Junk Expedition was now a long-retired circus act.[6]

While Halliburton's fans sat glued to their radios for any news about the *Sea Dragon*, most Americans were focused on events closer to home or in Europe. Attendance figures at the fair continued to peak; at one event alone, crooner Bing Crosby sang to a crowd of eighteen thousand as visitors continued to pour into the fairgrounds. Newspapers regularly mentioned events at the Golden Gate International Exposition. Noted as often were the Clippers that flew across the Pacific and connected Asia with America. Of course the fair and ocean flight were but two of the many topics the media addressed: US military buildup ("Uncle Sam Chooses His Weapons"); government spending ("The Balance Sheet on *WPA*" and "The New Deal an Inventory"); the failure of the Versailles Treaty; the end of the civil war in Spain; the papal appointment of American Francis Spellman as archbishop of New York; ongoing kidnapping, rape, and murder cases; and vacation ideas, movie debuts, Hollywood gossip, and society news. So much news. One could drown—in news.

Zoltan Korda's technicolor *The Four Feathers* was due for American release in August. Its subject was civil unrest in late nineteenth-century British-occupied Egypt and Sudan, and its exotic locales conveyed something of the Halliburton touch. Called "the radio with pictures," television now winked its invasive eye. Ostensibly serving their purpose, the Works Progress Administration and Civilian Conservation Corps often fell short of funds. Markets often dipped or were dull, and silver was sent to America in trade-balance agreements. Like the morning and the evening star, Eleanor Roosevelt's column "My Day" *daily* appeared, Ernie Pyle's travel commentaries, the comics (*Nancy* was already a favorite), Hollywood gossip, society notes, want ads, and a spot on the daily doings of the Dionne quintuplets. Periodicals regularly featured editorials. These ranged from state-of-the-world editorials such as "Behind the News" by Arthur Caylor to "Fair Enough" by Westbrook Pegler, a notably acerbic columnist famed for calling FDR a "feeble-minded fuehrer. Amid such tabloid features as "What Hitler's Lightning War Will Do to England," or "Is America Sufficiently Armed?" matters in Asia seemed irrelevant, for the moment. Celebrating the Pacific rim nations, the Golden Gate International Exposition itself would close late in October 1939 and reopen in May 1940.[7]

Almost overnight wartime needs transformed Treasure Island, with its convenient airstrip, from an amusement park into a US Navy base.

Many ships of the *President Lines*, including the SS *Coolidge*, were subsequently converted into convoys for troops, supplies, and munitions. Ships were sunk; when they did it garnered news, but their sinking was not a rare occurrence. The *RMS Lancastria*, once a passenger and cargo ship and now a troopship, was torpedoed by a German submarine off Saint Navaire in France on June 17, 1940. It had four thousand people aboard and sank in fewer than twenty minutes. By comparison, the fate of smaller civilian ships like the *Sea Dragon* seemed irrelevant. What with the war and the postwar boom, Richard Halliburton seemed an old travel bag filled with dated suits and ties and thrown into the attic. His books seemed at best living fossils from a distant past, with descriptions of lovely landscapes turned into battlefields, and accounts of daring deeds that seemed scripted. Halliburton's daydreams came across as inapplicable to the hard-core global outlooks of the coming atomic age. Jonathan Root saw Halliburton, at last, as the "great romancer" people of every generation need: "Richard," he wrote, "had never been a part of worldly reality in the sense that statesmen and corporations, or even movie stars, exist in a space of time and have an impact on the tangibles of life. Richard was the embodiment of day dreams, the public manifestation of private fancies. He could have frolicked across the tapestry of history at any point, and mankind would have recognized him. His was what any of us like to believe we might have been." In Antiquity, he would have been Herodotus, in the Middle Ages a knight errant, in Modern Times—Richard Halliburton.[8]

Months after the Chancery Court made its ruling, the Bobbs-Merrill Company sent a circular to newspaper editors which noted the "sincere and persistent interest of readers all over the country in the last adventure of Richard Halliburton." The document, offering nothing new, mainly expressed uncertainty as to the "actual fate of the *Sea Dragon* and its venturesome crew." While what transpired "must remain forever one of the unsolved mysteries of the sea," it was stated, "for Richard Halliburton it must have been an end set to the tempo of a gallant life." The idea of sailing a junk across the Pacific still thrilled and inspired. Shortly after Halliburton's declared death, another man proposed completing the Chinese junk trip, and indicated his intention in an article submitted to the *Memphis Commercial Appeal*. The editors at Bobbs-Merrill thought the plan not "uncomplimentary" to the famed traveler's memory. Richard's father, Wesley, thought it "a very original idea."[9]

39

Neptune's Realm

D uring a sojourn in China that lasted from the fall of 1938 through the winter of 1939, significantly a longer time than was planned, it is generally agreed that Halliburton built, equipped and attempted to sail a Chinese junk, the *Sea Dragon*, across the Pacific Ocean from Hong Kong to San Francisco. What happened in Hong Kong during those nearly six months, and once the junk left port, is still debated among historians. No one now doubts that the *Sea Dragon*, becoming one of the mysteries of modern maritime history, vanished without a trace. Press coverage of the ship's fate, blending the probable with the possible, was initially ample—and perplexing. Slowly accepted was the plausible certainty that the *Sea Dragon* with all on board had sunk. Said Oscar Wilde, "The truth is rarely pure and never simple." Often it eludes. Did rushing waves and furious head winds break the junk apart? Did all on board know the ship had been struck a blow from which it could not recover? What panic, if any, ensued? Did the *Sea Dragon* sink in the blink of an eye or float about for days while its crew, dazed and confused, clung to a plank or fallen mast? Did the ship list to starboard right, then to port left, and in moments begin to roll end over end? Did it just fill with water and go under?

Were storm surges, underwater volcanoes, or giant whirlpools responsible for the *Sea Dragon*'s demise? Every conceivable theory has been put forth—lightning, fire, lunar eclipse, even the angry hands of Neptune breaking the ship to pieces or a vicious current coiling about it python-like and crushing it. Did the junk collide with another ship? Apocryphal and biblical theories of the *Flying Dutchman* or Jonah and the Whale sort have been advanced. As everyone on board could hear the loudening roar of an approaching storm front, the *Sea Dragon* more likely succumbed to a rogue wave or a charging column of wind-agitated waves that sent it to the ocean floor. But why blame the elements? Blame pirates—or Japanese warships. Large- and small-scale Japanese submarines ranged shark-like as far as Midway Island in the mid-Pacific and as far as the eastern shores of Australia to the south. "My own idea," wrote travel writer Carveth Wells, "is that while passing through the innumerable small Japanese islands that lie between the Philippines and Hawaii, Halliburton unintentionally—or otherwise—landed on or came near within reach of one of the secret Japanese naval bases. Taking him for a spy, the Japs probably shot him out of hand and buried his body at sea." From a submarine asked to trail the *Sea Dragon*, Japanese intelligence would have had no difficulty intercepting Halliburton's radio transmissions or randomly using the junk for target practice. Even were it certain, however, that a Japanese torpedo brought the *Sea Dragon* down, lessons in risk management, human nature and navigation could still be learned from the ship's time in port and at sea.[1]

No one survived the *Sea Dragon*'s probable violent end. There were no witnesses to tell the story. Help arrived late, too late. Search teams loitered; once on the scene, no heat detectors aided them in their search. No black box, like those routinely installed in jet liners to record flight data and cockpit voices, were around to reconstruct the ship's last dire moments. Today, racing boats even in the most treacherous seas or remotest parts of the globe are rarely lost. Tracking devices are far too sophisticated, and rescue teams can reach an imperiled ship in moments. In Halliburton's day radio transmissions alone reported a ship's whereabouts. The disappearance in July, 1937, of Amelia Earhart and her Lockheed Electra could have reminded the adventurer and his contemporaries that human beings safely enclosed in modern metal capsules could still become irretrievably lost.[2]

Captain Welch called the *Sea Dragon* "a stout bitch." Against weaker odds the junk showed moxie; against graver ones it faltered. The *Sea*

Dragon entered a big body of water, which was terrifying enough, but then it entered a big storm. Typhoons with sixty- to seventy- mile wind-force that hit a shoreline community can lift semi-trucks, tear roofs from buildings, knock down telephone poles, and uproot trees. On the sea, they can sink riverboats, upend oil barges, and cripple ocean liners. In their path, the *Sea Dragon* was a teacup in a tempest, a fly struggling in an Olympic-size swimming pool; odds are it had as much chance of survival as a tepee in a twister. As an experiment, float a tiny box in a washing machine as the agitator begins its cycle—slow, fast, spin, rinse. The machine suddenly starts shaking and seems about to burst. Once adjusted, it settles down, continues its cycle, then stops. Remove the tiny box and see what's left of it.[3]

In "Ocean Tow," Captain Welch described what could happen to a ship—in this case, the tug the *Coringa*—under the worst conditions at sea. His narrative is a portentous witness to what might have happened aboard the *Sea Dragon*. The tug, like the junk, entered a storm field, encountering "drizzling rain" and a "wind . . . blowing with increasing violence." Then the sea erupted. "At every dive (the) *Coringa* made," Welch writes, "we were treated to a salt-water shower." What came next was apocalyptic: "The sun went down in a gray murk, and as the darkness came over the sea, it blew a gale. There was no sky, no longer the comforting horizon. It was as though we had been suddenly cast adrift in a universe of our own with nothing but foam-sheeted sea that leapt at us from the Stygian darkness." Welch held watch while the ship's skipper named Manning stood at his side. To hide from "the stinging, blinding rain (that was) beating a hellish tattoo on [their] faces," both "crouched behind the canvas dodger." Their fears mounted. "We hung on bitterly," wrote Welch, "as she would throw her bow high in the air, roll her weather rail deep as the thundering sea would go clean over her. . . . The squalls increased in violence as the night wore on and the early dawn showed us a North Atlantic that was a gray desolation of flying scud and vicious white-fanged sea. The horizon had disappeared behind a roll of gray-green walls of water that marched toward us in formidable array, their tops a-tumbling, roaring cascades driving before the implacable fury of the gale." The "onslaught of the sea" had damaged the surface of the tug. "The wind (meanwhile) had increased to whole gale force and the sea, milk white with spindrift, [was] stinging our eyes and piercing our faces with a thousand needles. The smoke from the funnel, close abaft the bridge, streamed to leeward, mixing the acid bitterness with the spray." The engineer was "battened down amid the stench of hot oil and stifling, sooty heat" and of "the escaping steam and the pounding engine . . . the water sometimes rose knee-deep on the floor of the engine-room and they [the crew] clung desperately to the iron ladders as the *Coringa* dove headlong into the sea." Continued Welch: "When (the engine room crew) came topside through the fiddly gratings, they had the starring, bloodshot eyes of madmen. Nobody talked anymore. There seemed nothing to say. We forgot the day and the date. We seemed eternally damned to be imprisoned in this new desolate world of screaming wind and racing seas. (The) *Coringa* tossed madly, her propeller thrashing the sea at every dive, making the ship tremble like a stricken thing. We touched the nadir of human misery." That the crew of the *Sea Dragon* had reached a similar "nadir" even before the storm came is likely; then they and the ship disappeared. But how?[4]

The *Coolidge*'s Dale Collins noted that while "many things could have happened (to the *Sea Dragon*), three possible causes of her loss seem most probable":

> 1. She may have run before the wind and sea for too long a period instead of being brought about and having a sea anchor put out;
> 2. She may have been dismasted, which in turn would probably have opened up her deck planking allowing water to get below decks; or
> 3. She may have fallen off her course enough so that she was broadsided in the trough of the seas and thereby capsized.

As to *precisely* what happened, Collins was uncertain. His given causes tantalize but hardly solve. The first one reads not as a cause really, but rather as a preventive measure not taken. The second and third causes are general and conform without argument to accepted views. "From reports of other craft in the typhoon area and United States Naval Weather Stations," the *Memphis Commercial Appeal* reported on October 5, 1939, "tremendous waves must have tossed the *Sea Dragon* before it either sprang a leak or foundered." When the Chancery Court met to decide whether Richard Halliburton should be officially declared dead, none of these "possible causes" were brought into court.[5]

John Rowland saw the Sea *Dragon* as "a mixture of Chinese junk, Spanish galleon, and Halliburton: these last two additions to her design, no doubt, were the root cause of her demise, though stouter ships than the *Sea Dragon* have failed to weather a typhoon." He called the ship a "floating eggshell." Unlike Collins, who focused on the external forces that caused the junk to sink, Gordon Torrey, like Rowland, chose to focus on Halliburton's "artistic and creature-comfort fantasies," those incorporated into the *Sea Dragon's* design which "robbed it of any semblance of reasonable sea-worthiness":

> **First,** to accommodate a great diesel engine and elaborate living and office space below deck, she (the *Sea Dragon*) was left wide open below from stem to stern, depriving her of water-tight bulkheads at frequent intervals. These are found in all sea-going Chinese junks and account for their extraordinary survival characteristics under the typhoon conditions which are in the western Pacific.
>
> **Second,** to have it pictorially Chinese, as an exhibit item, it was given a Soochow river-type junk spoon bow and a great, towering poop section aft to accommodate the painted dragons.
>
> **Third,** to have nice working light during the days in the below deck area, a long trunk-type raised section was built on deck, to accommodate glass skylights . . . in the kind of seas I had seen in the Formosa Straits. . . . Solid water could come right over that low spoon bow, wipe out the sky-light trunk and fill the lower deck section with enough water to enable that great heavy diesel engine to take her to the bottom.

Based upon "the kind of seas" he had witnessed in the Formosa Straits—a zone representing the first leg of the *Sea Dragon's* voyage, Torrey believed that the junk's chances of sailing the entire length of the Pacific from Hong Kong to San Francisco were quite slim. Even a voyage lengthwise across Lake Michigan might have challenged its 'long-range' capabilities.[6]

Still, the *Sea Dragon* had gotten far; why didn't it get farther? While Torrey implied that the junk was to its detriment *overbuilt*, the *Sea Dragon*, lacking the bulkheads that governed safety, a strong rudder, and solider frame, was more than likely *underbuilt*. Mariner Charles A. Borden remarks that "the general feeling of shipyard men in Hong Kong at the time was that the loss of the vessel was partly due to interior changes that greatly weakened the hull, including the removal of several orthodox compartments to make deluxe living quarters." As to the organic nature of junk design in general, Halliburton and Fat Kau may have been of one mind, but the reconfiguration of the Sea Dragon into a sort of Chinese caravel not only deformed it, but threatened its performance.[7]

Views regarding the junk's tragic end changed. By the time World War II ended in 1945, Collins, who offered one view, had had ample time to reflect. He remarked simply that the Sea Dragon was "pooped," and "unexpectedly." He explained that a huge wave, stirred up by the force of the wind, had rear-ended the vessel, at once incapacitating it. Boxers would call this a haymaker delivered just behind the right or left ear. If rear-ended, the Sea Dragon might have been retreating from the storm. If it "keeled over," it was doomed. Drenched with water fore and aft, the junk would have sunk within moments. But if the junk was not in retreat, one must suppose that a violent wave of the size that Collins supposed struck it hard and instantly disabled it. Wave behavior can be tricky. A wave above the water, for instance, is known to have a ripple effect below the water, a subtle aquatic phenomenon that crew members aboard a ship can neither see nor feel. In any case, a ship rising to the height of an oncoming wave or waves—in an effort to ride them, will more often than not overturn or be thrown backwards. If a strong keel is in place, the ship can upright itself. According to artist James Garrett, who for Click Magazine drew a picture explicitly depicting the Sea Dragon's last moments, the junk, amid total darkness, pitch-poled. Tossed upside down by a giant wave, it somersaults high in the air with crew members, like shrapnel from a dynamite explosion, literally blasted from their posts. From the summit of "mountainous waves" and falling down into deep trenches of water, the ship—as its trial run with Fat Kau had somewhat foreshadowed, was violently lifted up and slammed down deck-first on the mat of the sea, again and again.[8]

Collins did not use the terms pitch-pole or breaking wave. When he spoke of the fatal wave, he probably referred to a knockdown wave. Named for a term borrowed from the boxing ring, knockdown waves punch what's before them full force. Volleys of them just clean up the mess the first one left. In Left For Dead—Surviving the Deadliest Storm in Modern Sailing History, Nick Ward, who sailed on the Nicolson thirty-foot yacht Grimalkin that competed on the Irish Sea in the Fastnet Race of 1979, describes his firsthand experience with these waves. "Each knockdown sent all . . . of us hurtling through the air," he said of the constant bombardment, "then crash-landing on the boat or, worse, in the water—whichever way, we were either awash or completely immersed in bitterly cold water." Ward survived the ordeal, but twenty-one of his fellow mates perished.[9]

According to early *Sea Dragon* researcher Edward Howell, Collins suspected that the *Sea Dragon* was struck not by a knockdown wave but by a "monster (or killer) wave," a naturally occurring phenomenon whose very existence was doubted in the 1930s. Monster waves can tower upward of 90 feet, and some of them may climb as high as 120 feet. Few who have borne witness to them have lived to tell the tale. Antarctic explorer Ernest Shackleton, a man not known to fib, reported his seeing such a wave near the South Georgia Islands in 1916, but naysayers believed he had seen just an unusually tall wave. In July 1958, the tallest wave ever recorded, measuring 1,720 feet, formed when an earthquake struck Lituya Bay in Alaska and caused a tremendous landslide. The mega-tsunami that followed brought additional massive destruction to the region. In 2000, satellite imagery from the British vessel *Discovery* confirmed that waves of appalling size do exist, and that monster waves must account for the loss of the many ships that have perished without a trace over the years.[10]

Today waves are better explained by meteorological science than by mythographers who attribute them to the temper tantrums of Neptune. Waves are of many kinds; while most are normal, some are quite abnormal. Four times the size of those surrounding them are rogue or killer waves which owe their formation largely to brutally high winds and strong opposing crossing currents. No ship in their path is safe. Against such waves mounting in height and moving violently forward, the *Sea Dragon*, with no time to wave a flag of surrender, would have perished in seconds. Numerous known parallels exist. In recent memory, on October 1, 2015, the 790-foot cargo ship *El Faro*, while making its way in the Bahamas just off the coast of Florida, ran into Hurricane Joaquin. Winds accelerated to seventy miles per hour, waves rose to fifty feet. The ship disappeared beneath the waves, and lost were the thirty-three crew members on board.

Possibly an immense, thunderous wave—perhaps many of them in succession, struck the junk. When suspended between two enormous waves, ships' hulls can crack and disassemble in the blink of an eye. Front-heavy and back-heavy, a craft could split in two. If unable to navigate over or plow directly through a cresting wave, a ship could otherwise find itself crushed by the force of water. "(Captain Welch) might have waited too long to heave to," Dale Collins remarked, "and thereby found himself running before such a heavy sea (as the *Coolidge* itself was experiencing) that it was impossible to bring her about." 'Pooped' by a wave or waves, the junk was "left helpless in the trough of the sea." Visibility blocked and steering upset, the ship pitched and yawed recklessly. Frontally attacked and double-enveloped by lashing winds and pounding rain, water swamped the deck and rushed into the hull. Uneven weight distribution in the ship's hold caused it to lean further. "Perhaps she just capsized," an ultimately baffled Collins said of Halliburton's ship.[11]

If detached from its mounting, the diesel engine could have slammed into the hull and breached it. Scarcely a foot thick and pounded repeatedly by heavy waves, the *Sea Dragon*'s sides might have been battered and mortally so. Without watertight compartments to slow it, the ensuing rush of hard water would have been so quick that the sailors trapped inside would have perished. And by now they were either thinking or crying out, 'We should have listened to Torrey, We should have listened to Torrey.' A climactic roundhouse "poop" could have sent the dazed ship reeling. At that point, little could be done. Ship and crew were as raw meat to a hungry lion. But suppose, just suppose. With the wind at his back and the ship beginning to list, Welch, blinded by the heavy rain, may have attempted to steer the craft out of danger, but, as he might have said himself, he was only *spitting in the wind*. Likely the masts had split or fallen, so what use were they? A standing target for direct hits by wind and rain, the high poop, just a wooden turret or siege tower, couldn't be reefed like a sail. For the *Sea Dragon* to maintain a steady course, the diesel engine had to operate at a high level of efficiency. A clattering racket to those not used to its sound, to engineer Von Fehren that same racket was a pleasant purr. The first attempt at a crossing, however, showed that, against an onslaught of high waves, the engine, once revved up—and purring, could drone uselessly in neutral; if so on the second attempt at a crossing, it was dead weight on a dying ship.

Of the different causes offered for the junk's demise, one may add *fire*. The *Sea Dragon* was a tinderbox on water—it carried two thousand gallons of combustible fuel, or nearly fifty barrels of it, and it was also supplied with an unspecified kind or make of stove. Halliburton noted, "We've outfitted our kitchen with a fine stove—we've food & water for

90 days—heavy clothes, some rum—4 guns against pirates." A lightning bolt could have struck the junk. Alternatively, there could have been a short in the wiring system. On this floating tinderbox, combustion was ever possible. A small fire stove, besides a hot griddle, was located in the galley. Skylights were located below deck. Several crew members, including Halliburton and Mooney, smoked cigarettes. Von Fehren smoked a pipe, and Captain Welch smoked cigars. A fire could have started when the stove malfunctioned or when a casually tossed cigarette butt landed in the fuel tanks, forcing the unprotected crew to breathe smoke. Overdriven, the diesel engine could have blown up, tearing the ship apart as surely as any rogue wave. The engine reportedly had a poor exhaust system. Oblivious to its faults, Halliburton noted only the "overpowering" fumes that "escaped (from the engine) into the main cabin from the newly-painted tanks and bulkheads." Diesel fuel, thicker and oilier than gasoline—-though denser than gasoline in energy, emits less carbon monoxide than lighter "auto" fuel. In the early days of its use diesel fuel released a black smoke and may have done so in the model sold second-hand to Halliburton. Needless to say, inhaling the smoke could cause dizziness.[12]

After the first aborted sailing, the junk's defects, rather than capabilities, were assessed. Among other things, it was supposed that the addition of a fin keel would solve the junk's reckless motion. If such a keel was added, one may then wonder if it was properly installed. While a fin keel may not have been necessary for the *Sea Dragon* to brave the open seas, a strong rudder was. Captain Jokstad blamed the loss of the rudder for the ship's doom; he suspected it might come loose in heavy battle with a storm. In July 1940, while the SS *Pierce* steamed ahead from Honolulu to the Orient, he sighted what he was certain was that very rudder entangled with seaweed and floating among some other wreckage. Jokstad figured that the barnacles attached to this rudder had taken over a year to form. The rudder's location, about two thousand miles east of Yokohama, seemed to him correct for the amount of time it would have had to drift from the last reported position of the *Sea Dragon*.[13]

In 1945, a thirty-foot skeleton of a ship with part of the keel reached the shores of Pacific Beach, California, some 450 miles due south of San Francisco. Several ribs were still attached to the washed-up remains. Like the rudder, the wreckage was identified as belonging to the *Sea Dragon*, whose disassembled parts might have left a debris field that eventually extended over hundreds of miles. Marine surveyor and Halliburton enthusiast K. M. Walker, suspecting the debris was from the *Sea Dragon*, theorized that it "might have been carried by Japanese currents to North America and down the coastline." While the "large-headed all bronze bolts" on the frame indicated the craft's Asian origin, the conjectured length of the ship, 150 feet, made it an unlikely candidate for the Halliburton junk. Timbers belonging to wooden boats are found often enough along the California coast, but their identification as remnants of the *Sea Dragon* is at best only tempting. The sightings of the rudder and then the keel reopened speculation that parts of the junk had been found, but, during the war, the Japanese launched deadly torpedo attacks against a high number of Chinese junks and some junks just went on daring missions or drifted off-course, never to return. Could the *Sea Dragon* reside among their remains? Little short of recovering the panel from the junk's stern bearing the words "*Sea Dragon*—Hong Kong" could prove its provenance.[14]

Attractive theories, most conjectural, have arisen from the little that is presumed. Might they all be *jettisoned*? Captain Jokstad said that he had been contacted by Halliburton's attorney about the possibility that his client was still "alive on some island." After carefully examining the spot where the junk presumably disappeared, Edward Howell, theorized that a landfall could have been made on the Los Jardines Islands ("The Gardens"). Located close to the equator just to the east of the Marshall Islands and not far from the International Date Line, these "disappearing (or phantom) islands" were sighted by earlier explorers and mapped out in the eighteenth century by Captain James Cook. They are not shallow coral reefs, said Howell, but volcanic *seamounts* whose sudden rise and fall is determined by the activity of the earth's tectonic plates. If the *Sea Dragon* collided with such a *seamount*, considerable damage would have been done to the ship's hull; if the ship had successfully landed on one, it would only have been a matter of time before it went under water with the *seamount* itself.[15]

Paintings by Joseph Turner of ships in crisis during a storm offer visual clues to the *Sea Dragon*'s fate. So do films. Audiences in Halliburton's time watched *China Seas* (1935), a romantic adventure aboard a steam-tramper bound for Singapore that shows the ensuing havoc when a storm loosens a heavy piece of equipment and lighter items on deck. Both *White Shadows in the South Seas* (1928), a silent, and the *The Caine Mutiny* (1954) dramatize the catastrophe a typhoon can create. Storm sequences (set supposedly on the Isle of Manikoora) at the climax of John Ford's *The Hurricane* (1937) provide additional pictures. Boats and buildings, including a church on high ground, are demolished; trees are uprooted; and people are tossed into the rushing waves. No quarter is given to man or the man-made. Consider then Hurricane Michael, which struck the Florida Panhandle in 2018 , whose crippling destructiveness is recorded in countless film takes. The television documentary *Underwater Universe*, a comprehensive look at ocean behavior, gives multiple explanations of what can happen to small craft on ruffled seas and waterways. Depictions of other ships locked in misery——in art and books, in movies and newsreels, can at best only suggest the *Sea Dragon's* final throes.

More reliable are the reports of the junk's behavior under sail. Waves thrash the ship and climb over its sides onto the deck. Deluged by waves, the masts collapse, the planks rip loose, the rudder breaks off, the poop splinters; simultaneously or in quick succession, glass shatters, rigging tangles, and what is not tied down is flung overboard. Shaken, the diesel engine is dislodged from its bed and thrown against the hull, producing a gaping hole. No bulkheads are in place to resist the subsequent flooding. What can one do? Battling the water spray from wind and rain, the steersman, looks up to a pitch-black sky and sees little. Nautical equipment no longer of use to him, dead reckoning alone is his guide. Then comes a punch delivered with such force that the ship and all on board recoil, senseless as to what struck. Broadsided or struck from the front or rear, the *Sea Dragon* leans to its side, fills with water, and sinks. Fractured syntax and broken chords better than well-wrought sentences and melodic lines convey the horror of those final moments.

Hastily built, and not to specifications that ensured its safety, the ship was condemned from the start. Through errors of omission and coherence, it had flunked most of its tests. Numerous mishaps occurring all at once or in sequence might have sealed its fate, sending a medley of alarms to a crew unprepared to address them. The only facts agreed upon are that the junk entered a storm, its captain called for assistance, and the junk sank. Disaster struck well into the night. While digital sculpting can recreate the craft, wreck simulation using computers, which has yielded breathtaking results when applied to the

Titanic's sinking, might brighten our picture of the *Sea Dragon's* tragic end. Graphs showing failures of mechanical function as well as wave height, wave direction, and wave interval might offer visual clues; but, without correct data input, the simulation cannot hold up as proof of the junk's actual final moments. Some envisionings, however, continue to intrigue. Captain Welch, as Dale Collins suggested, may have attempted to steer the craft out of on-coming traffic, but, with water filling the ship, his efforts were futile. The *Sea Dragon* almost certainly entered the eye of the storm. Water from pounding rain washed across its deck while violent waves smashed into the hull. Already the prow, with its tendency to dip, had nosed beneath the surface. Junks had an Achilles heel; they leaked, and the *Sea Dragon* leaked constantly. If the deck ("lee rail") was "underwater," the hull filled with seawater. Its masts slanted or broken, rigging in a jumble, the *Sea Dragon* was now a raft adrift. The combined forces of wind, rain, currents, and waves had already torn loose the junk's radio antenna and shattered its skylights. Its rudder and engine now useless, the vessel was already half-submerged when large waves blanketed it. If its hull was breached, the storm-lashed vessel—heavily weighted with a diesel engine and added ballast, listed, rolled over, then sank without a gurgle.[16]

40

The Long Dusk

So much for the sailing ship the *Sea Dragon*. What of the sailors who lost their lives? Once the junk drifted from the well-traveled shipping lanes into unknown waters, a cold fear might have paralyzed the crew—but under the worst stress, maybe they stood cool and collected. From interviews conducted fifty years later, John Potter's descriptions of the calamities the crew underwent on the trial run must be judged recollections in tranquility. They tell little about the plight of the crew at the moment of supreme crisis. Nor do they articulate fear in action. Persists a main likelihood: for twenty days a group of human beings, of the most diverse temper one could gather into one place, lived on a floating barge called the *Sea Dragon*. Out in the middle of the ocean, with their food and water near empty, they focused only on survival. In such circumstances strength can wilt. Crew members squirm, panic, or set heatedly to what tasks their captain orders them to do. With "the starry bloodshot eyes of madmen," they might otherwise crawl into some nook or cling to something they believe might float. Amid all the frenzy, the chow puppies are thrown overboard because they consume; or, because their caretakers are desperately hungry, the puppies are themselves consumed. Say some, the junk's final scene was acted out in seconds. Captain Jokstad believed the end came so fast that the crew didn't even have time to put on life preservers. Captain Welch's last two messages, sent minutes apart, suggest that the crew had less time to pray than it had to cuss.[1]

As far as is known, neither Captain Welch nor anyone else aboard the *Sea Dragon* launched flares, color rockets, emergency beacons, or Roman candles. Weakened by hunger, the flegling crew, now bleary-eyed, had performed poorly during the trial runs and first attempted crossing. The professionals whom Halliburton thought to replace them with might have better looked the part of sailors, but against so daunting a wind and rain opponent would have been outmatched. A quick-acting Ralph Granrud *might* have grabbed the tiller but in vain. Useless at the tiller, Halliburton, goes one theory, remained or slept in his quarters while "look-out" Paul Mooney struggled to free himself from fallen rigging. Unassisted, Welch had everything to do at once. George Petrich, his radio dead, lost focus. Seeing the diesel engine loosened from its mounting, Henry Von Fehren, drenched with water, agonized, wrench in hand without a nut to twist.[2]

Giant sheets of water weighing tons suddenly poured down on the craft. No longer able to battle the elements, it heeled, rolled, flipped, and tumbled. In what Amelia Earhart called the "livid loneliness of fear" in her poem "Courage," crew members on deck and below were likely tossed about as well. If Halliburton wasn't knocked out by a swinging beam or boom, crushed by rigging, impaled by a splinter or nail, thrown hard against the rail, or flung by the force of wind and rain into the water, he surely drowned, dying of exposure in seconds or minutes. He might have ducked for cover or, as he had done during a stormy trial run, spread-eagled on the deck. Still, he could have floated for days, even weeks, nearly or quite dead.[3]

During the Pacific crossing of the *Free China*, Paul Chow said that the moment he encountered foul weather, he "looked out to the water . . . as dark as the navel of the Mother Earth . . . (and could see (only) the foam pushed out by the stalled junk while rolling with waves." Not knowing the storm's full width, Welch might have tried to outrun or skirt it. This presumes he could still steer the ship. The crew of the *Free China* had thought to hoist up their ship's rudder so it wouldn't break off. Hysterical and soaked in water, the crew of the *Sea Dragon* had little time to address any sudden malfunctioning of their ship. Likely the rudder broke loose within seconds of the last message made to the *Coolidge*, making the tiller as useless as a butter churn without a crank. If Captain Welch was shouting commands, the crew might not have heard them. If the storm arrived unannounced, many of the crew could still have been in bed—"the heavy rolling at night (making) the ship a cradle that rocked (them) to sleep." If all hands were on deck—and each knew his role, the question becomes did they act resourcefully or panic.[4]

In *The Perfect Storm*, author Sebastian Junger notes "the zero-moment point" when a ship pressured by the forces of wind and sea finds itself in an irreversible position: the ensuing "transition (among the crew) from crisis to catastrophe is fast, probably under a minute." Lights on board go out, and each member of the crew works his assigned position in total darkness. This effort quiets the internal poundings of mounting fear. The noise of the storm stifles voiced orders from the captain or first mate. While one embattled crew member might attempt some spectacular feat of daring—maybe lifting a fallen mast off a comrade, another, shivering in fright, might take refuge behind some coiled pile of rope. Fortunately, the "zero-moment point" is brief. Still, how brief?[5]

Now and again, someone will ask what Halliburton's last moments were like. Was his a "last stand" like that of General Custer? Unlike Errol Flynn's portrayal of the Indian fighter in *They Died With Their Boots On*, some say he was not the last to be killed at Little Big Horn, but among the first. Was Halliburton then the first member of the *Sea Dragon* crew to die, the last one, or one somewhere in the middle? Was his death speedy and painless? If he saw it coming, did he accept it gracefully, stoically, or was he hauled away kicking and screaming? If mortally wounded by flying debris or tossed overboard, did he languish and suffer, cry out to the Almighty or thank the powers that be that his debts were paid? If he threw himself prone on the deck as he had during a trial run, did a falling beam finish him off? By comparison, closure of a kind was achieved when the intact bodies of Will Rogers and his pilot, Wiley Post, were carted off after their Lockheed Explorer crashed near Point Barrow, Alaska. Scores of the stringed eye patches belonging to Post were strewn about the cockpit like some species of long-tailed mice. In the typewriter Rogers used to write his feature stories was one he had barely started and had tentatively entitled "Death." Both Post and Rogers died instantly. *Merciful death people* might hope that

Amelia Earhart—rather than captured, humiliated and beheaded by the Japanese, fatally struck her head on the instrument panel of her plane the moment it hit the water.[6]

In truth, the *Sea Dragon* only shared with Earhart's Lockheed Electra the fate that it disappeared. Witnessed was Earhart flying off from a runway in New Guinea, and Halliburton sailing off from a port in Hong Kong. The *Sea Dragon* crew, if as tired as Earhart, otherwise seemed cheerful. They sailed toward the open sea while the sight of the Hong Kong skyline and hills grew smaller. Soon the harbor, the big ships, and all places on shore could be only dimly seen. Inlets and islands, barges and boats now appeared to be illusions. Crew members no longer sighted or sniffed weeds and fetid waste debris; the air moistened by the dewy breeze was fresh. It became quieter, as the only sounds were the wind and the creaking of the ship's sinews and joints as the waves of the South China Sea lapped against its sides. The curtains to the Pacific then opened. All at once the men on the *Sea Dragon* were surrounded by a never-ending sea. The vessel seemed small, those on deck even smaller. As even the Hawaiian Islands are hard to spot from the air, what of a diminutive junk carrying no fewer than a dozen sailors separated from deep-water eternity only by a plank of wood? Whether exhilarated with joy or paralyzed by fear, the sailors likely yearned to tread once again on solid ground.

The *Sea Dragon*'s end was probably *brutal* and *quick*. As noted, one mast, then another collapsed. Pressure from the pounding waves and shifting cargo opened a gap in the hull. The steering mechanism wouldn't respond, the planking across the deck split, and water flooded the hull and the deck. About to go under with the ship, Halliburton froze or sought cover. The mostly sea-weary crew were numb with fear. Their food, stained with rodent urine, maybe riddled with cockroaches, was almost gone; their water, probably close to putrid, had run low. That being so, they were reduced to eating hardtack and to lopping in the sea-water that dripped from the rain into their mouths. One of them might have been struck by debris. Another might have gotten tangled in the rigging. Still others might have been swept overboard. A healthy adult of average strength can tread ocean water for an hour, maybe two. Under water, he can hold his breath for a minute, maybe two, before yielding to the natural impulse to breathe. Even if he is only a few feet beneath the water, he might wonder in his dazed state whether he was returning to the surface or sinking. During those chaotic moments, that "100-yard stare that surfers get after a two-wave hold-down or near death experience" might show on his face. In a drowning scenario, the heart, when normal breathing is interrupted, goes into panic mode. When breathing is attempted, salt water enters and, once in the stomach, creates a burning sensation; death rapidly occurs. Asphyxia by submersion and water-immersion injuries are common and often recorded. Deprived of oxygen even for moments, a sailor going under can suffer severe and irreversible damage to his internal organs. Suffocation follows quickly.[7]

The only joy that comes from sinking with one's crew is the shared participation in a joint venture. Offering added joy was "Fiddler's Green," the Shangri-La for sailors lost to the ol' Devil Sea. "The very heaven of such men had to be a ship," Dick Wetjen wrote, "a vessel so large as to be immune to wind and tide, bearing on board everything that a sailorman could wish for. Taverns and girls and music; good food, warm bunks, and old shipmates who had gone before." On the sea floor were other "strange ships . . . ghostly ships . . . with giant serpents, outlandish gods, and cool-skinned maidens with seaweed for their hair." Wetjen joined that subterranean crew in 1947; he was forty-seven years old. Pallbearers at his funeral included friends Herb Caan and Jacland Marmur—and maybe the ghost of John Wenlock Welch.[8]

Today, the diesel engine rests three or more miles below sea level in a debris field littered with blackened concrete shards, a crumpled heap of encrusted timbers (the bones of the ship), nails, coins, swivel cannons, tin cans, assorted mementos, personal effects, and maybe a safe. For their discovery these await further advances in sonar imaging and visualization technology—and a financial risk deemed worth the prize. High-definition cameras, 3-D mapping systems and remotely operated underwater vehicles—deep-sea robots, underwater drones and crawlers, might spot wood-planking from the junk hanging from a ledge, or its high poop asleep on the sea-bed, but offer no proof about what really happened. The image of the *phoenix* on the *Sea Dragon*'s bow, no doubt, would no longer be there. The Greeks associated this miraculous creature with Hera, consort of Zeus and mother of the gods, and consequently with renewal. Through its association with the solar bird Benu in Egyptian mythology, the phoenix also symbolized redemption and travel in the afterlife. Familiar with the mythologies of both East and West, it was a lovely story. Although he wondered how he would arrive there, Halliburton had no delusions about death. Over the years, he saw the ruins of great civilizations and learned that everything becomes nothing sometime. Gibraltar will crumble, the temples of the world will come to ruin, and tin will oxidize after forty-five hundred years. Henry Van Dyke, Halliburton's teacher at Princeton, wrote, "No human progress is unbroken and continuous. No human resting-place is permanent. Where are Pharaoh's Palace, and Solomon's Temple, and the House of Caesar, and Cicero's Tusculum, and Horace's Sabine farm?"

Halliburton knew by heart many of the poems of Rupert Brooke, among them "The Great Lover" and its concluding lines:

> The best I've known,
> Stays here, and changes, breaks, grows old, is blown
> About the winds of the world, and fades from the brains
> Of living men, and dies.
> Nothing remains.

The French have a philosophy of decay whose emphasis in part is the body's postmortem ride from carcass to total extinction, atoms to atoms, molecules to molecules. Hard-core in its realism, the philosophy, which holds that matter can neither be created nor destroyed, is also *pantheistic* in its spiritualism: as Ariel proclaims in *The Tempest*, eyes become pearls, bones become coral—"five fathoms deep," all turns "into something rich and strange." The course leading to those coral bones and pearly eyes is a pathetic sight. Enlarge the picture of tiny organisms feasting on corpses to that of a whole school of piranhas devouring a hapless donkey whose misfortune it was to have fallen off a tottering wicker bridge. Falling into a vat of liquid oxygen and getting it over at once is better than falling into the carnivorous sea. The bodies of the *Sea Dragon's* crew could have floated for days, prey for macro- and microscopic marine predators.[9]

Paul Mooney liked clear-cut endings and conceptually satisfying conclusions. He had read Djuna Barnes's surrealist novel *Nightwood* with interest. Troubled by its conclusion, he asked a friend who knew its author, "[Did the heroine] apologize (for a miscue she had committed), commit suicide, or sleep with a dog, or go crazy, or only faint?" Barnes enjoyed alternate endings, while Paul preferred certain ones. "The rest of the book may amuse you," he went on to say to his friend, "if you've missed it so far; but personally I like my mystery at the beginning and not at the end." In "My First Night," a story about camping in the woods that Paul wrote when he was fifteen, he spoke of being prepared for darkness when, as "the test of the true camper," the "goblin Fear, banished until nightfall, comes to the front and dances in the fantastic shadows thrown by the firelight, and gives voice to his shadowy self in the moaning of the wind." The night revealed what the day concealed.[10]

With Halliburton's death, the wind moaned. Thousands of letters from grieving and bewildered fans arrived at his parents' home in Memphis. Some offered condolences, others still doubted that the worst was so. Even while his books continued to sell, images of the man, once so familiar from press releases, dimmed. "Fellow adventurer, author and lecturer Lowell Thomas," who had created many of the myths surrounding Lawrence of Arabia, delivered Halliburton's eulogy to a "jammed auditorium" at the Second Congregational Church of Waterbury, Connecticut, "where Richard had frequently lectured." Nelle Nance, meanwhile, kept the torch lit; whenever her phone rang, she hoped it was her son calling.

A wing of the most dedicated fans of Richard Halliburton, those with a taste for the supernatural, continued to believe that his spirit wandered about—in what what metaphysical poet John Donne called "the progress of the deathless soul." Accomplished artists, like magicians, are masters of deception—even after death. Some people believed that Halliburton, an odd concoction of artist and magician, had faked his death only to reappear suddenly on the horizon reincarnated in another headline. Rumors spread that his ghost haunted Hangover House. Residents in the area even insisted they saw "strange, flashing lights" through its windows and silhouetted figures along its walls. William Alexander, the house's designer and builder, found the stories amusing. Ultimately he moved from New York City and resettled in the Hollywood Hills. There, from the perch of a house he built for himself, he watched daily to see if the *Sea Dragon* was sailing towards shore. Others, too, have glimpsed far out into the sea for such a sighting.[11]

Discovery of the fisherman's glass ball—it a sort of talisman, might have added to the drama of ghostly returns. As it was, the ghost did materialize, sort of. Nearing the end of life, Wesley Halliburton put up $450,000 to build a tower at Rhodes College as a cenotaph honoring his son's memory. Called the "Memorial Tower," it now serves as the headquarters of the college's president and staff. Built "of brick and concrete," the majestic stone structure resembles the four-hundred-foot Tower of the Sun that welcomed visitors from all over the world to the Golden Gate International Exposition. The bronze plaque at its base reads in part, "To Richard Halliburton—Traveler-Author-Lecturer. This Tower is dedicated to memorialize a wonderful life of action, romance and courage. Erected by his parents Wesley and Nelle Nance Halliburton. 1961–62." Words on the Tower alongside an image of Icarus hail Richard Halliburton as "a daring modern Icarus (who) flew too near the sun." Although he designed his own version of the waxen wings that brought Icarus to his doom, Halliburton did not think much of that other mythic dreamer. As a public figure, he seemed to revel in acclaim, but, even with fame, he suffered from a deep inner loneliness; while its symptoms often found momentary relief—through public acclaim, devotion to work, and the love of family and friends, its cause never found a cure. In the sea's shifting moods and ultimate unknowability, he found an analogy to something in himself and in the sirens' enchanting song some purpose in his final mission.[12]

Halliburton's death had to it a tragic grandeur. He could have choked on a pea, succumbed to a mosquito bite, suffered a stroke while bending down to tie his shoe, or strangled himself when the scarf he was wearing got tangled up in the hub of a car wheel in motion. He did not die in Davenport, Iowa, as would film idol Cary Grant; on the pot as had King George II of England; or in a freak auto accident as would cowboy star Tom Mix. A knight in linen armor who wielded a pen instead of a lance, Halliburton was, as he once called himself, "Don Quixote Jr." He did not return to his village, however, and, scorning all books of heroic adventure, die, like the throwback romantic, peacefully in bed. Nor would he suffer the torments of aging he often foresaw—blindness, madness, incontinence, and immobility. He might have wished for a sublime death like that of Nikos Kazantzakis's "modern" Odysseus who, endless wanderer that he was, at last becomes a flame. He did plan, once, to live with friends for a time on a Greek island, but he thought little then of cosmic endings.[13]

Halliburton's death marked the end of an era of bold undertakings into an unknown world. The Great Depression ended with America's entry into another Great War, and soon American troops waged war on two fronts. For people back home, hearing some well-tailored dandy offer lectures on the perils of freelance roving seemed passé amid head-lines of soldiers truly risking their lives overseas. After World War II, vast changes in weapons and communication technology transformed the world. Places Halliburton had visited, many of them destroyed, no longer seemed quaint. Perspectives had changed. Wesley Halliburton didn't think his son, who disliked "modern progress," would have liked the world that emerged after 1939. Maybe, maybe not. A new book based on a new adventure was always on his mind. He of course wrote more than two books, but, after the second, he wrote: "I want to do a third travel book while youth predominates, and I must hurry, for (my) spirit and outlook are changing so terribly fast." May the foregoing pages have served as that *third or* next unwritten book, and conveyed that Richard Halliburton and the crew of the *Sea Dragon* will be missed.[14]

Notes

Letters

The "Letters from the *Sea Dragon*" (hereafter cited as Halliburton, "Letters from the *SD* #1," November 20, 1938, etc.) appeared in the following sequence:

#1, November 20, 1938 (Shanghai).

#2, January 18, 1939 (Hong Kong).

#3, January 27, 1939 (Canton).

#4, February 16, 1939 (Hong Kong).

The "Log of the *Sea Dragon*" (hereafter cited as Halliburton, "Log of the *SD* #1," December 12, 1938, etc.), at first published exclusively in the *San Francisco News*, appeared in the following sequence:

#1, December 12, 1938 (Monday).

#2, December 13, 1938 (Tuesday).

#3, December 26, 1938 (Monday).

#4, December 27, 1938 (Tuesday).

#5, January 23, 1939 (Monday).

#6, January 24, 1939 (Tuesday).

#7, January 25, 1939 (Wednesday).

#8, January 30, 1939 (Monday).

#9, January 31, 1939 (Tuesday).

#10, February 1, 1939 (Wednesday).

#11, February 28, 1939 (Tuesday).

#12, March 1, 1939 (Wednesday).

#13, March 2, 1939 (Thursday).

#14, March 3, 1939 (Friday). (Except for its three opening paragraphs, nearly identical to Halliburton, "Letters from the *SD* #3," January 27, 1939.)

#15, March 4, 1939 (Saturday). (Except for a few stylistic changes and the omission of the last three paragraphs, identical to final paragraphs of Halliburton, "Letters from the *SD* #3," January 27, 1939.)

#16, March 19, 1939, *Boston Sunday Globe*. (Nearly identical to Halliburton, "Letters from the *SD* #4," February 16, 1939. And, except for two added paragraphs ending the article—about avoiding storms and noting the southern route of the *Sea Dragon*—the version published in the *Buffalo New York News*, date-stamped April 1, 1939, is identical to the *Boston Sunday Globe* version.)

The *Letters from Hong Kong*, the correspondence from Captain John Wenlock Welch to Richard Wetjen, are dated as follows, and their place of origin is indicated. Transcribed by Louella Kenan Sawyer, the envelopes and holographic originals have not survived. (Hereafter, these letters are cited thus: Welch, *LHK* #1, October 24, 1938, etc.)

#1, October 24, 1938, Kowloon Hotel, Hong Kong.

#2, October 29, 1938 (Saturday), Kowloon Hotel, Hong Kong.

#3, November 2, 1938, no stated origin.

#4, November 2, 1938, no stated origin (probably added to letter #3 above).

#5, November 16, 1938, Kowloon Hotel, Hong Kong.

#6, November 24, 1938, Kowloon Hotel, Hong Kong.

#7, November 25, 1938, Kowloon Hotel, Hong Kong (possibly affixed to letter #6 above).

#8, November 30, 1938 (fewer than fifty words and probably added to letters #6 and #7 above).

#9, December 5, 1938, no given place of origin.

#10, December 6, 1938 (appears to be an independently sent letter).

#11, December 16, 1938, Kowloon Hotel, Hong Kong.

#12, December 21, 1938, Kowloon Hotel, Hong Kong.

#13, January 29, 1939, Hong Kong Island.

#14, February 13, 1939, Hong Kong Island.

Abbreviations

Sources most often cited appear below:

John H. Alt, *Don't Die in Bed—The Brief, Intense Life of Richard Halliburton* (Atlanta: Quincunx, 2013); hereafter, Alt, *DDIB*.

Michael E. Blankenship Archive, Richard Halliburton Collection, Barret Library, Rhodes College, Memphis, Tennessee; hereafter, MEB Archive.

James Cortese, *Richard Halliburton's Royal Road* (Memphis: White Rose Press, 1989); hereafter, Cortese, *RHRR*.

Richard Halliburton, *Richard Halliburton—His Story of His Life's Adventure as Told in Letters to His Mother and Father* (Indianapolis: Bobbs-Merrill, 1940). Hereafter, this source is identified as *RHL* and *His Story*. Omitted parts of the published letters, when quoted, are noted as *RHL-P* (*RHL*-Princeton).

Gerry Max, *Horizon Chasers—The Lives and Adventures of Richard Halliburton and Paul Mooney* (Jefferson, NC: McFarland, 2007); hereafter, Max, *HC*.

Paul Mooney, *Letters of Paul Mooney*; hereafter, *PML*.

Richard Halliburton Collection, Barret Library, Rhodes College, Memphis, Tennessee; hereafter, RHC, Rhodes.

Jonathan Root, *Halliburton—The Magnificent Myth* (New York: Coward-McCann, 1965); hereafter, Root, *HTMM*.

San Francisco News; hereafter, *SFN*.

William R. Taylor, *A Shooting Star Meets the Well of Death* (Abbeville, SC: Moonshine Cove, 2013); hereafter, Taylor, *SSMTWOD.*

Preface

1. For "fairy queen" usage, see Welch, *LHK* #13, January 29, 1939. If a reference to J. M. Barrie's Peter Pan and Queen Mab, the fairy queen who taught Peter Pan how to fly, those familiar with the story must suppose Welch to be the villainous Captain James Hook whose hand Peter cut off and whose fear of crocodiles is obsessive. One may infer, at any rate, that Welch thought Halliburton 'flighty.' Other pejoratives for Halliburton occur throughout the letter. For "bachelor" reference, see Richard Halliburton, *The Book of Marvels—The Occident* (Indianapolis: Bobbs-Merrill, 1938), sleeve (quoted). For Halliburton's fear of exposure and maintenance of a façade, compare Joseph Morella and George Mazzei, *Genius and Lust: The Creativity and Sexuality of Cole Porter and Noel Coward* (New York: Carroll and Graf, 1995), 5 passim. For Halliburton's public persona, see, for instance, Root, *HTMM*, 126 et seq. For the Hollywood rumor mill and damage limitation to negative press releases, see Kieron Connolly, *Dark History of Hollywood—A Century of Greed, Corruption, and Scandal behind the Movies* (London: Amber Books, 2014), 151–52 (Cary Grant), 60 (William Haines), 58–60 (Rock Hudson and Mamie van Doren). Welch's *Letters from Hong Kong* may sit alongside Thomas Wolfe's *The Starwick Episodes* as examples of 1930s homophobic literature. See *The Starwick Episodes*, ed. with an introduction by Richard S. Kennedy (Baton Rouge: Louisiana State University Press, 1989). Halliburton's first biographer Jonathan Root skirts mention of Halliburton's sexuality, but he cannot resist clever innuendo. From the multiple options available to him, he chooses Radclyffe Hall's novel *The Well of Loneliness* "about the tortures of lesbian love" as exemplifying the "moral revolution" of the 1920s. Implied are Halliburton's own "tortures" as a gay man in a presumed anti-gay world. See Root, *HTMM*, 149–50 (quoted). For Welch on Paul Mooney, see *LHK* #12, December 21, 1938.

2. For members of the press corps (e.g., Halliburton and Wolf Blitzer) glimpsing scenes of war from vantage points, see John B. Powell, *My Twenty-five Years in China* (New York: The MacMillan Company, 1945), pp. 301–2. For Davis, see John Seelye, *War Games—Richard Harding Davis and the New Imperialism* (University of Massachusetts Press, c2003), p. xiii, and p. xiv (quoted).

3. See "Notes" above for main sources. See also "A Note on Sources" in Max, *HC*, 243–44. Quoted is Richard Halliburton, *The Royal Road to Romance* (Indianapolis: Bobbs-Merrill, 1925), 284. For the "autobiography," see *RHL-P,* June 28, 1931. Barstow's literary interests are noted in letter from Bertha Barstow to Martha Berry (1865–1942), founder of Christian-based Berry College, Berry College archives, November 25, 1939. Also consulted was "the William Barstow Family genealogy (1635–1966)," prepared by Arthur Radasch, at Hartford Historical Library, Hartford, Connecticut; mentioned therein is the library at Berry College created to honor the memory of George Barstow III.

4. Compare Richard Davis, "The Mysterious and Strange End of Richard Halliburton," subtitled "What Happened to the Most Famous Man in America?" (Internet). In England, even after he was well known in America, Halliburton seems to have been quite unknown. See *Letters*, August 8, 1927, 275. For Rupert Brooke poem, see *The Collected Poems, with a Memoir by Sir Edward Marsh* (London: Sidgwick and Jackson, 1918, 3rd

Edition Revised, 1942, Reprinted 1953), 134-36. See Gordon Sinclair, *Bright Paths to Adventure*, with illustrations by Stanley Turner (Toronto: McClelland and Stewart, 1945), 85-99, at 87 and 90 (quoted). For Potter's comments, see Cortese, *RHRR,* 181 (quoted). The "little boy playing Indian" remark ends Jonathan Root's biography of Halliburton. See *HTMM,* 278. Quoted is Carveth Wells, *My Candle at Both Ends—The Autobiography of John Carveth Wells* (London: Jarrolds, 1950), 108.

1. Glorious Adventurer

1. Compare Davis, "Mysterious and Strange End of Richard Halliburton." Even after he was well known in America, Halliburton seems to have been quite unknown in England; see *Letters,* August 8, 1927, 275. For Halliburton's fear of the public losing interest in him, see Taylor, *SSMTWOD,* 271.

2. See Paul Fussell, introduction to *The Great War and Modern Memory* (Oxford: Oxford University Press, 1975). For the "trench sensibility," see Paul Fussell, *Abroad— British Literary Traveling between the Wars* (New York: Oxford University Press, 1980), 4-8 passim.

3. See *Time,* July 8, 1940, 69-71. See George Weller, "The Passing of the Last Playboy," *Esquire,* April 1940, 58, 111-12 (quoted).

4. For H. L. Mencken, see Root, *HTMM,* 99 (quoted). For it being a wicked world, see Halliburton, *Royal Road,* 49 (quoted). For "opinions," see Root, *HTMM,* 216 (quoting Halliburton). Root believed Halliburton "had no sense of politics."

5. For changing pictures of America, see Jon Meacham, "The American Dream: A Biography," *Time,* July 2, 2012, 26-39. For "the old Europe," see Frank G. Carpenter, *Carpenter's World Travels—France to Scandinavia,* 3 (quoted). The multivolume series is invitingly entitled *Carpenter's World Travels: Familiar Talks about Countries and Peoples.* We are assured that "reading Carpenter is seeing the world"; he is "on the spot and the reader [is] in his home." For a cultural crossing bringing together Christianity and Islam, see David Boyle, *Toward the Setting Sun—Columbus, Cabot, Vespucci and The Race for America* (New York: Walker, 2008), 85, 128. For the hopes inspired by the Golden Gate International Exposition, see *SFN,* "World's Fair Edition," Wednesday, July 6, 1938, 15 (quoted) passim.

6. For "embodiment," quoted is reporter for the *Review of Niagara Falls,* April 13, 1938; see Alt, *DDIB,* xviii. For credo, see Halliburton, *Royal Road to Romance,* supra, 3-4 (quoted). For estimate of Rupert Brooke, see *RHL,* August 19, 1927, *His Story,* 275 (quoted). See also Nigel Jones, *Rupert Brooke—Life, Death and Myth* (London: Head of Zeus, 1999; 2014); for Jones, 165 (quoted) and for Saki, 47 (quoted). For Halliburton's aims, see Root, *HTMM,* 116 (quoted); for Halliburton thinking "the world of the past had been a mistake," see 56-57. For Halliburton's philosophy, and his debt to Oscar Wilde, whom he quotes, see *Royal Road to Romance,* supra, 3. For "liberty or death," see *RHL,* editorial note, *His Story,* 19; for shaking off the "paternal yoke" and the thrill of "being released from every tie" (both collegiate and religious), see 72-73. For "shoulds," see Karen Horney, *Neurosis and Human Growth—The Struggle Toward Self-Realization* (New York: W. W. Norton, c950), 17, 19 passim. For Halliburton's values, compare Root, *HTMM,* 56-57. For espousing bachelorhood and expressing disdain for the settled life, compare Root, *HTMM,* 59. For another "bachelor" reference, see Halliburton, *Book of Marvels—The Occident,* sleeve. For opposing an "even tenor" existence, see *RHL,* December 5, 1919, at

51; also see *RHL,* October 15, 1923, 220. For "quest for riches," see Halliburton, *Royal Road to Romance*, 4; for "prosaic," see 5. Among the writings Nelle Nance Halliburton, a teacher, might have consulted in rearing her sons, Richard and Wesley Jr., are John Dewey's *My Pedagogic Creed* (1897), *The Child and Society* (1900), and *The Child and the Curriculum* (1902). For Nelle Nance's role in her son's education, as well as that of teacher Mary G. Hutchison, see Root, *HTMM*, 39-40. For context, see in general *The New Century 1900-1914—A Changing World,* History of the 20th Century (Andromeda Oxford Ltd.: Chancellor, 1993). For age issues, see Howard P. Chudacoff, *How Old Are You? Age Consciousness in American Culture* (Princeton, NJ: Princeton University Press, 1989), 119 passim. Also see, in general, John R. Gillis, *Youth and History*, Studies in Social Discontinuity (New York: Academic Press, 1974).

7. For the (social) value of romanticism, see Halliburton, *RHL*, July 16, 1927, in *His Story*, 273 (quoted). For "boors" see Joris-Karl Huysmans (a source for Oscar Wilde), *Against the Grain (A Rebours)*, with an introduction by Havelock Ellis (New York: Dover, 1969), 25 (quoted). For "search," see *RHL*, sleeve (quoted). For his books as an appreciation of the arts, see Root, *HTMM*, 126 (quoted). For youth on the move, see Errol Lincoln Uys, *Riding the Rails*, 19-20, 185, 186. For the tawdry side of the royal road, see Paul J. Bauer and Mark Dawidziak, *Jim Tully—American Writer, Irish Rover, and Hollywood Brawler*, with a foreword by Ken Burns (Kent, OH: Kent State University Press, 2011). For hobo world, see, in general, Richard Wormser, *Hoboes—Wandering in America, 1870-1940* (New York: Walker, 1994). For a comparative look, from the 1950s, of what dangers lurked on our nation's highways, see Ginger Strand, *Killer on the Road—Violence and the American Interstate* (Austin: University of Texas Press, 2012).

8. For Oscar Wilde on education, see the 1891 essay "The Critic As Artist." For pictures of fellow students, see Halliburton, *Royal Road to Romance*, supra, 2-3. See Ted Conover, *The Routes of Man—How Roads Are Changing the World and the Way We Live Today* (New York: Alfred A. Knopf, 2010), 7 (quoted). For "ultra-modern generation," see *Letters*, January 9, 1921, *His Story*, 70 (quoted). For "routine," see *RHL*, December 6, 1920, *His Story*, 68 (quoted). For a "monotonous confined respectable life" as "horrible" to Halliburton, see *RHL*, October 5, 1918, *His Story*, 10-11. For historical trends in education, see was Charles J. Sykes, *The Hollow Men: Politics and Corruption in Higher Education* (Washington, DC: Regnery. Company, 1990).

9. Halliburton's long association with the Alber-Wickes lecture agency argues for his commitment, as they, to moral development and spiritual growth. The tag line of the lecture service from one of its brochures, read "the richest heritage of the great is the influence for the leader of Yesterday on the Youth of Today," might be his own: "There are boys and girls in this town—in every town—who will become good useful citizens because of (the platform's) splendid Christian influence." For his antipathy toward military regimes, see Halliburton, "Log of the *Sea Dragon #5*," January 23, 1939. For Carveth Wells, see his unpublished letter to Elizabeth Cleveland, September 28, 1933, author's collection, quoted.

10. Quoted is Richard Halliburton, "Out of Russia," in *Seven League Boots* (Indianapolis: Bobbs-Merrill, 1935), 227-28. For Halliburton's comments about Russia, see Robert Richards, "Halliburton Letters Depict the Man Himself," *Press-Scimitar*, July 19, 1940. For short essay on Russia, see *RHL*, November 7, 1934, 358 et seq. For Halliburton's "anglomania," see *RHL*, September 28, 1919, 31. For claims by London reviewers that Halliburton's *Royal Road to Romance* was "anti-British," see Root, *HTMM*,

130. For mixed views of the Englishman in America, see Wyndham Lewis, "If I Were a British Agent," in *America, I Presume* (New York: Howell, Soskin, 1940), 203-28. For communism in America, see "Red Revolution—In America and in Russia," with (I) Malcolm Logan, "These Terrible Reds," and (II) William C. White, "Home Office of the Revolution," *Scribner's Magazine*, June 1930, 649-65. For elitism, see William A. Henry III, *In Defense of Elitism* (New York: Doubleday, 1994).

11. For "John Jones," see *RHL*, November 27, 1936, *His Story*, 386. For lack of friends as a "vital part of (his) life," see October 15, 1936, 383. He was not blind to human misery: "Chicago swarms with bums and beggars," he wrote. "There are 400,000 men out of work here." See *RHL*, September 30, 1930, *His Story*, 308. For "Rough Road to Reality," see *RHL*, August 19, 1927, 275 (quoted).

2. Great Wall of China

1. Relevant online features include "United States Freezes Japanese Assets—July 26, 1941" and "How US Economic Warfare Provoked Japanese Attack on Pearl Harbor." See George Ashmore Fitch, *My Eighty Years in China* (Mei Ya International Edition, Copyright in the Republic of China, 1967), notably 130 passim. For a journalist's life in China before Halliburton's last arrival there, see H. G. W. Woodhead, *Adventures in Far Eastern Journalism—A Record of Thirty-Three Years' Experience* (Tokyo: Hokuseido, 1935), notably informative in its presentation of foreign courts, consulates, opium trafficking, and the results in China of the World War I peace settlements. For earlier Far Eastern ventures, see *RHL*, September 3, 1922, 172-77. For US financial assistance to develop Manchuria, see *New York Times*, Sunday edition, January 16, 1938. For America extending itself in the Pacific, see Christine Negroni, *The Crash Detectives—Investigating the World's Most Mysterious Air Disasters* (New York: Penguin Books, 2016), 66-67. For US policy towards China, see, for instance, *Time,* December 11, 1939, XXXIV 24, 17-18.

2. See PML, copy of letter to Harriette Janssen sent to Gerstle Mack, November 24, 1938. For worsening Anglo-Japanese relations, see Hallett Abend, *My Life in China—1926-1941* (New York: Harcourt Brace and Company, 1943), 294-300.

3. In general, see Upton Close, *In the Land of the Laughing Buddha—The Adventures of an American Barbarian in China* (New York: G. P. Putnam's Sons, 1924). For narcotics trafficking, see Hendrik DeLeeuw, *Flower of Joy* (New York: Lee Furman, 1939), 136 (quoted). Also see Elsie McCormick, *Audacious Angles on China* (Shanghai: Chinese American Publishing Company, 1922). US Foreign Services representative Addie Viola Smith owned a copy of the original publication, which was later published by the Appleton-Century Company in 1923, and again in 1934. For degree of interest by Americans in China and the Far East, compare Hallett Abend, *My Life in China*, op. cit., 5.

4. Daughter of two Presbyterian missionaries, Pearl Buck (1892-1973) taught classical Chinese; she later won the Nobel Prize for Literature. For Walter Futter, who produced *India Speaks*, see Alt, *DDIB*, 201-5. Quoted is Edward Alexander Powell, *Asia at the Crossroads Japan, Korea, China, Philippine Islands* (New York: Century, 1922), 37-38. Also see Lucian Swift Kirkland, "What Japan Thinks of Us," *Travel*, August 1923, 10-15, 37-38.

5. For American interests in China, see John B. Powell, *My Twenty-Five Years in China* (New York: MacMillan, 1945), supra, 37-38 passim. For the land and its people (one view), see Peter Fleming, *News From Tartary—A Journey from Peking to Kashmir* (New York: Charles Scribner's Sons, 1936).

6. Powell, *My Twenty-Five Years in China*, 37–38 passim.

7. For Halliburton's attire during his early romp across Asia, see *RHL*, April 23, 1922, *His Story*, 144; and for Mandalay and Bangkok, see September 3, 1922, 172 (quoted).

8. For Chinese as "reckless," Halliburton, *Royal Road to Romance*, op. cit., 361; for Chinese as making peculiar sounds, see 339–40. For Chinese, also see Halliburton, "Log of the *SD #1*," December 12, 1938. For Japanese, see Halliburton, "Letters from the *SD #3*," January 27, 1939; and "Log of the *SD #3, 4,* and 5." For differences between Indian and Chinese shopkeepers, see *RHL*, November 21, 1922, *His Story*, 194. For the rickshaw race, see Halliburton, *Royal Road to Romance*, op. cit., 144–45; for "wop culture," see 152; for "Chinks" (in a quote by a Russian named Demitri), 359. For Chinese as "most interesting," see *RHL*, December 7, 1922, 196. Work habits identified ethnic groups for Halliburton. As a young man working briefly in New Orleans, he wrote his parents, "I make the best longshoreman you ever saw, and can handle a two-wheel truck like a nigger." See *RHL*, August 5, 1919, *His Story*, 24. Later, in 1920, on a hike with friends in the mountains of Montana, Halliburton said of his Indian guides: "These Indians are worth the trip to see. Lots of them speak English and have the funniest names." He added, "They are like our niggers, entirely irresponsible, and we've been delayed by their breaking faith." See *RHL*, July 20, 1920, *His Story*, 59 (quoted). Cf. Root, *HTMM*, 55. For other racial references see *RHL*, July 20, 1920, 59 and September 17, 1921, 92–93. As a young man, he placed people of color on a lower rung than whites without hesitation. An exception to this prejudice is his reverential account of Henry Christophe, "the Black King of Haiti" who was once "a dishwasher in a public bar" and later "a leader in (a) black rebellion." For Henry Christophe, see Halliburton, *Book of Marvels—The Occident*, op. cit., 119–27, quoted at 119. Ethnocentric, Halliburton thought it "extraordinary" that "such a *primitive race*" as the Seri Indians living in Lower California "should live so close to (presumably civilized) Hollywood." See *RHL*, July 13, 1934, *His Story*, 351–52 (quoted). Apologists for Halliburton will argue that his racism was casual and not uncommon for his time and place. The social conditions of the South in early 19th-century America might offer him a degree of pardon, as he grew up in a segregated, racially divided culture. But even though he intended no malice, he had, besides decided human sympathies, equally decided racist turns.

9. For miscegenation, see *RHL*, 187–88 (quoted); also see December 29, 1919, quoted at 54. For "the Asiatic's resentment of the Caucasian," see Root, *HTMM*, 92. For some modern perspectives on ethnicity, see Jack Weatherford, *Savages and Civilization—Who Will Survive?* (New York: Fawcett Columbine, 1994), 236 (quoted). On balance, see Ivan Van Sertima, *They Came Before Columbus—The African Presence in Ancient America* (New York: Random House, 1976). It is probably true that the blending of races and cultures alarmed Halliburton. However, when writing about the "slim Sonias" of Blood Alley in Shanghai, he seemed amused by interracial mingling: "When Chinese blood and foreign blood are mixed, especially if the foreign blood is Russian, Portuguese or French, the devastating result is something to write home about." Halliburton's ideas of miscegenation are traceable partly to racialist notions trending in his day. One source was Lothrop Stoddard's *The Rising Tide of Color Against White World Supremacy* (New York: Charles Scribner's Sons, 1920). In F. Scott Fitzgerald's *The Great Gatsby*, this work is retitled "The Rise of the Colored Empires" by "Goddard," and Tom Buchanan notes its putative theme of the "white race (being) utterly submerged." Halliburton stayed for a time as a patient at the *Battle Creek Sanitarium* in Michigan. It is not evident from his writings whether he shared founder John Harvey Kellogg's theory of eugenics as well

as his dietary theories. For Kellogg, see Tristram Stuart, *The Bloodless Revolution—A Cultural History of Vegetarianism from 1600 to Modern Times* (New York: W. W. Norton, 2006), 436. In an early letter, Halliburton notes "a hundred shades of negroes (in Marseilles)—Chinese—Indians—a drop out of the east" and, implicitly, a similar blending at the fair. See *RHL*, December 19, 1919, at 54. Over his twenty-year career, Halliburton crossed barriers, and learned that there were multiple points of view. See, for instance, "*Allah's Children*," in *Second Book of Marvels—The Orient* (Indianapolis: Bobbs-Merrill, 1938), 83. Rejected in 1935 for its outspokenness by the Bell Syndicate, "The Wickedest City in the World," also entitled "Cairo—the Capital of Sin," was #38 of the fifty articles, appearing from 1934 to 1935, the editors commissioned him to write. A manuscript copy is now in the *Richard Halliburton Collection* at the Princeton University Library. For commission, see Root, *HTMM*, 236. For the *Bell Syndicate* stories, see *RHL*, March 1, 1934, *His Story*, 349. Showing Halliburton at his best as a writer—tackling leprosy, the inmates of Devil's Island, the French Foreign Legion, and the king of Arabia and the *Koran*—these fifty stories should be published in a single volume. He thought the leprosy story "a terror," stating, "[It] may be thrown out by some of the editors." See *RHL*, February 12, 1935, *His Story*, 364. Halliburton spent ten days among the Dyaks. See *RHL*, May 10, 1932, *His Story*, 338. For "kid book," see Max, *HC*, 172.

10. See Michelle Cottle, Notebook, "Race Baiting: All the Rage," *Newsweek*, February 20, 2012, 5, col. 1 (quoted). See Halliburton, *Royal Road to Romance*, 361, 339-40. For history, see, in general, Ric Burns and Li-Shin Yu, dir., *American Experience*, "The Chinese Exclusion Act," (May 2018), DVD. For contemporary views of the Chinese in America, see Greenberry G. Rupert, *The Yellow Peril, or Orient vs. Occident as Viewed by Modern Statesmen and Ancient Prophets* (1911; on-demand repr. of orig., Nabu Press, 2012). For reversal of views as more Chinese students came to study in America and global markets were expanding, see Jonathan D. Spence, *The Chan's Great Continent—China in Western Minds* (New York: W. W. Norton, 1998), 166. Historian and futurologist H. G. Wells noted the different world that would one day emerge once turmoil in China settled. See his lectures in *The Way the World Is Going—Guesses and Forecasts of the Next Few Years* (London: Ernest Benn, 1928). For physician Dr. Margaret Chung, see Max, *HC*, 178. Dr. Chung was "mother" of the squadron that assisted in establishing air and naval supremacy over the Japanese during the Saipan invasion in June 1944. A milestone in Allied strategy, the capture of Saipan removed a major link in the Japanese inner-island defense chain. Born in Stockton, California, Addie Viola Smith (1893-1975), once assistant trade commissioner in Shanghai and practitioner at the US Court of Law in China, became in 1939 the American consul in China.

11. See, in general, John Keay, *China: A History* (New York: Basic Books, 2009), 499 et seq. Also see John Keay, *Empire's End—A History of the Far East from High Colonialism to Hong Kong* (New York: Scribner, 1997). In general, see Powell, *My Twenty-Five Years in China,* supra. Especially for views of Chinese civic life, see Paul Samuel Reinsch, *An American Diplomat in China* (London: George Allen and Unwin, 1922). Reinsch was an American minister based in China from 1913 to 1919.

12. For "China scene," the unpublished advertising circular to the "Letters from the *Sea Dragon*" is quoted. *Time* magazine publisher Henry Luce admired Chiang Kai-Shek. See "Chiang Kai-shek's 10 *Time* Magazine Covers," online. Also see Milestones: 1937-1945, "Henry Luce and 20th Century U.S. Internationalism," US Department of State, Office of the Historian, online.

13. For background information, see Keay, *China: A History*, op. cit. For a brief treatment of the Sino-Japanese War and its coverage, see Arthur Hacker, *China Illustrated—Western Views of the Middle Kingdom*, with a foreword by Frederic Wakeman (London: Tuttle, 2004), 274–82. For contrasting and parallel views, see Harry Alverson Franck, *Glimpses of Japan and Formosa* (1924; repr., New York: D. Appleton-Century, 1939). For Chiang Kai-shek's rise to power, see Abend, *My Life in China*, supra, 28.

14. See Woodhead, *Adventures in Far Eastern Journalism*, 73 (for Washington treaties, quoted), 75 (for drug trafficking, quoted), 85 (for Chiang Kai-Shek, quoted). For main events of the day, including Ernie Pyle's travels, see the *New York Times* and *SFN*, 1938–39. For Immigration Act of 1924, and its impact, see "Coolidge Signs Jap Bill," *Omaha Evening Bee*, Omaha, Nebraska, May 26, 1924.

15. See H. C. Thomson, *The Case for China* (New York: Charles Scribner's Sons, 1933), 101 (quoted).

16. Quoted is George Washington's 1796 "Farewell Address." Hobnobbing with notables was an aspect of the Halliburton persona that others imitated, including admirer Ellery Walter. See his *The World on One Leg* (New York: G. P. Putnam's Sons, 1928). For Roosevelt's imperialism, and jingoism, see John Seelye, *War Games—Richard Harding Davis and the New Imperialism* (University of Massachusetts Press, 2003), 5. For Captain Alfred Thayer Mahan's "anticipation of increasing trade in the trans-Pacific zone, chiefly China," see 2. Mahan was the author of *The Influence of Sea Power Upon History, 1660–1783* (1890).

17. Quoted is (Halliburton acquaintance) Charles Caldwell Dobie, *San Francisco's Chinatown*, with illustrations by E. H. Suyham (New York: D. Appleton-Century, 1936), 1, 2. For the Harte expression used in an article, see C. Whitney Carpenter Jr., "The Peculiar Heathen Chinee," *Travel*, June 1919, 23–28, 48. Also see "The Hate Merchants," *Bonanza*, 1960. Set in the 1880s, this episode reflects an attitude toward Chinese still quite common in the America of the 1930s. As an introduction to China, see episodes 1–10 of the documentary *The Pacific Century* documentary. Specifically, see *The Pacific Century*, episode 1, "The Two Coasts of China: Asia and the Challenge of the West" (Annenberg/CPB Collection, 1992), DVD.

18. For China's world market potential, see Carl Crow, *400 Million Customers* (Harper and Brothers, 1937). For "show of American solidarity," see Don Skemer (curator of manuscripts at Princeton University), "A New View of Richard Halliburton's *Sea Dragon*," March 17, 2014, online, quoted. Beneath the headline "Little Hope Remains for Earhart's Safety," the *Harrisburg Telegraph* carried the feature "Chinese Demand War with Japan," subheaded "Japan's Invasion Arouses Ire of Nation's Leaders," on July 10, 1937. For China's failure to Westernize, see Halliburton, "Log of the *SD #9*," January 31, 1939. For Halliburton's earlier thoughts on China, see *RHL*, November 15, 1922, *His Story*, 192 et seq.

19. For Stilwell in China, see Barbara W. Tuchman, *Stilwell and the American Experience in China, 1911–1945* (New York: MacMillan, 1970, 1971).

20. Those "big blank spaces in the map" are noted in Arthur Conan Doyle's *The Lost World*. The specific quote reads, "[They] are all being filled in, and there's no room for romance anywhere." See Paul Zweig, *The Adventurer—The Fate of Adventure in the Western World* (Princeton, NJ: Princeton University Press, 1974), 226. Zweig quotes Doyle's *The Lost World* (Berkley Medallion edition, 1965), 12. For "map over which he (Halliburton) wove the web of his travels rapidly changing," see *His Story*, xi.

3. Why a Junk?

1. For junks as "ramshackle," see *His Story*, 401–402. For later remarks about junks, also see Halliburton, "Log of the *SD #1*," December 12, 1938, quoted. For "several years (spent) in wandering by sea and land," see Halliburton, "Letters from the *SD #1*," November 20, 1938 (quoted). For early views of Canton (Guangzhow), see *RHL*, November 21, 1922, *His Story*, 194 (quoted). For junks, see Max, *HC*, 198 (quoted). Also see Hans Konrad Van Tilburg, *Chinese Junks on the Pacific—Views from a Different Deck*, New Perspectives on Maritime History and Nautical Archaeology (Gainesville: University Press of Florida, 2013), 23–24. Also see Ivon A. Donnelly's classic *Chinese Junks and Other Nautical Craft* (Shanghai: Kelly and Walsh, 1924; 1930). For the word "junk," see Douglas Phillips-Birt, *Fore & Aft Sailing Craft and the Development of the Modern Yacht* (London: Seely, Service, 1962), 61. Compare G. R. G. Worcester, *The Junks and Sampans of the Yangtze* (Annapolis, MD: United States Naval Institute, 1971), 28 *et seq*. For functional definition of a junk, see Jean De La Varende, *Cherish the Sea: A History of Sail*, trans. from the French by Marvin Saville (New York: Viking, 1958), 322.

2. For "schooner" and "small boy game," see Halliburton, "Log of the *SD #1*," December 12, 1938, (quoted). For Halliburton's childhood interest in junks and China, see *RHL*, editorial note, 1. Also see *RHL*, August 13–August 21, 1912, 3–6. Cf. Root, 39–41. Letters written at age 12 or 13 that might have reported the event are presumably lost. For lost correspondence, see *RHL*, 3. For dim recollection, see *RHL*, July 18, 1912. A mnemonic object, the "real sea-going model" compares irresistibly to the sled named "Rosebud" in Orson Welles' *Citizen Kane*. For "sport," see Halliburton, "Log of the *SD #1*," December 12, 1938, (quoted). For "fun of it," credit Amelia Earhart, *The Fun of It— Random Records of My Own Flying and of Women in Aviation* (New York: Brewer Warren and Putnam, 1932). For initial decision to sail a Foochow-style junk—"the best and most picturesque"—see *Honolulu Star-Bulletin*, September 28, 1938. For the Dartmouth lads, see James Zug, "Sea of Dreams," *Dartmouth Alumni Magazine*, July-August, 2014, online.

3. By the mid-1920s, Erle Halliburton (1892–1957) had moved his business operations from Duncan, Oklahoma, to Los Angeles, where he resided in a mansion in Berkeley Square. A pioneer in the oil service industry, Erle also played a key role in the expansion of aviation technology, founding *Southern Air Fast Express* in 1929. These and other offshore business enterprises made him among the richest men in the world. See Max, *HC*, 74–76. Erle and Vida had three daughters and two sons. See "Erle Palmer Halliburton," *The Encyclopedia of Oklahoma History*, online.

4. For the "Trans-Pacific Yacht Races," see Wikipedia and Wikiwand. Also see Albert Soiland (Honorary Commodore), *Trans-Pacific Ocean Races and Trans-Pacific Yacht Club: Facts, Fancies and Some Gossip about One of the Most Unique and Interesting Yacht Clubs in the World, and the Races It Sponsors* (Los Angeles: privately printed, 1937). For the "Marine Exposition," see 60. Jack London's "Sea Dragon" was a schooner named the *Snark*, which he, wife Charmian, and his crew sailed beyond the Hawaiian Islands in 1907. London later recounted his adventures in *The Cruise of the Snark*.

5. For "ordinary sailing craft," see Halliburton, "Log of the *SD #1*," December 12, 1938, quoted. For the history of junk sailings, see "Log of the *SD #2*," December 13, 1938. Once he had chosen a topic for an adventure, it was customary for Halliburton to "read every book in the library on the field," take "lots of notes" for "a clear vision of" his purpose and destination," and become "properly soaked" in the material. See *RHL*, July 4,

1925, *His Story*, 244. One source he surely heard about was "Classic of Mountains and Seas," a geographical text from the second millennium BC. Lost now, but preserved in part from reports by a fifth-century BC Buddhist missionary, the work suggested that well-established links between Asia and the Americas reached back to the beginning of recorded time. In 1761, French scholar Joseph de Guignes published *Recherches sur les Navigations des Chinois du Cote de l'Amerique, et sur quelques Peuples situes a l'Extremite Orientale de l'Asia*. The work's premise, that the Chinese had often sailed to the Americas, spawned noted controversy and heated debate well into Halliburton's time. For Robert Ripley and the *Mon Lei*, see Neal Thompson, *A Curious Man—The Strange and Brilliant Life of Robert "Believe It or Not" Ripley* (New York: Crown Archetype, 2013), 315 et seq.

6. For the vessel's name, see Halliburton, "Log of the *SD* #1," December 12, 1938. For Sir Francis Drake, see Derek Wilson, *The World Encompassed: Drake's Great Voyage, 1577-80* (New York: Harper and Row, 1977). For dragon lore, compare Marie Schubert, *Minute Myths and Legends* (New York: Grosset and Dunlap, 1934), 76-78.

7. For Columbus, see Halliburton, "Log of the *SD* #1," December 12, 1938, *Cincinnati Enquirer*, January 29, 1939 (quoted). For "most seaworthy," see "Log of the *SD* #1," December 12, 1938, quoted. For "cultural exchange between America and Asia in ancient times," notably by Sinologist Joseph Needham (1900-1995), see Tim Severin, *The China Voyage—Across the Pacific by Bamboo Raft* (Reading, MA: Wesley-Addison, 1994), 5. Told of Severin's plans to sail across the Pacific on a raft, Needham remarked, "The voyage is extremely important, not only in the study of exploration but also in the study of civilization in general," 6. An earlier voice (than Halliburton) in marking trans-Pacific contacts between Asia and America was *National Geographic* contributor Harriet Chalmers Adams (1875-1937). See Durlynn Anema, *Harriet Chalmers Adams— Adventurer and Explorer*. Greensboro, North Carolina, 1997, p.76.

8. For whereabouts in July 1937, see *His Story*, 390-91. For "junk plans," see *RHL-P*, August 30, 1937, *His Story*, 391 (quoted). For Wilfred Crowell (1891-1945), see *RHL*, January 24, 1938, 294 (quoted). For Swanson's encouragement, see Root, *HTMM*, 253, see 248-249.

9. For "Fair people," see Halliburton, letter of February 21, 1938, in *His Story*, 394 (quoted). Cf. Root, *HTMM*, 258. For publicity agents including Art Linkletter, see Max, *HC*, 274, chapter 15, note 9. Clyde M. Vandeburg was director of publicity and promotion for the exposition. He worked closely with the manager of the Radio Division, Arthur "Art" Linkletter, who later told me that both he and Vandeburg regretted sponsoring the ill-fated *Sea Dragon Expedition*. Thanks are extended to Art Linkletter; letter to author, January 26, 1998. For being "furiously busy," see *RHL*, February 21, 1938, *His Story*, 391 (quoted). For the Chinese merchants, see *RHL*, February 21, 1938, 394. For background of Chinese investors, see Charles Caldwell Dobie, *San Francisco's Chinatown*, op. cit., 119-37. For getting affairs in order, see *RHL-P*, June 25, 1938.

10. For "wonderful sport," see *RHL*, March 3, 1939, 432 (quoted). For obsession with a project, Richard Halliburton letter to Noel Sullivan, December 15, 1930 (quoted), Noel Sullivan Papers, Bancroft Library, University of California—Berkeley.

4. Lecture Circuit Drives Halliburton to the Sea

1. For Oklahoma City engagement, see *RHL*, November 15, 1936, *His Story*, 374. For detective in Oklahoma City, see Root, *HTMM*, 245-46 (quoted). For his "not reliving

anybody's life," see *RHL*, February 26, 1928, *His Story*, 281 (quoted). For being "wearied by the company of others," see November 27, 1936, 385 (quoted).

2. For reluctance to meet people, see *RHL*, October 15, 1936, *His Story*, 383. For "gland life," letter to Noel Sullivan, date unclear, likely 1927 or 1928. See Charles E. Morris III, "Richard Halliburton's Bearded Tales," *Quarterly Journal of Speech* 95, no. 2 (May 2009): 123–47, 129, 144n32. For warm weather preference, see *RHL*, February 13, 1936, *His Story*, 376. For being "unsociable," and "wearied by the company of others," see *RHL*, November 27, 1936, *His Story*, 385 (quoted). For being "unsociable" and "wearied by the company of others," see *RHL*, November 27, 1936, *His Story*, 385 (quoted). For Halliburton's wish to "fall in love," see *RHL*, July 20, 1926, *His Story*, 262 (quoted).

3. For Halliburton's reaction to the recorded sound of his voice, which he found "strange," see *RHL*, March 10, 1933, *His Story*, 345. For health issues, see, for instance, *RHL*, 227, 234, and 235. It has been estimated that Halliburton spoke before five thousand audiences in his career; this figure seems high. Consulted was "Halliburton's First Lecture—In Insane Asylum," January 4, 1938 (unidentified newspaper clipping), MEB Archive. When Halliburton was once called "boy," he felt flattered. To himself, he felt "79." He was slight of stature—perhaps 5 feet, 8 inches tall, and seldom weighed more than 160 pounds. For need to exercise, see *RHL*, *His Story*, 203 (quoted); for golf as exhausting, see August 15, 1925, 247–48, quoted at 247; for canceling engagements, see January 20, 1936, 376 and January/February 1936, 376; for the strain of schedule and "hoarseness," see January 20, and January 26, 1936, 374–75 and March 6, 1937, 387; for "grind" and "hoarse voice," see May 1, 1936, 378. Reportedly, Halliburton spoke to no one all day—"just rested (his) throat." According to the editorial note, he went into the hospital to learn if there was additional "trouble" besides that "with his voice." For the tour producing anxiety, see, for instance, March 1, 1936, 376. For "sinus drainage," see *RHL*, March 1928, *His Story*, 282 (quoted). For goiter and "trembles," see August 10, August 13, and August 16, 1924, *His Story*, 237–38 (quoted); for "trembles" also see editor's comment, 239; for strain from motoring and boating, see August 10, 1924, 236; for canceling "two-a-days," especially later in his career, see 376. For goiter specialist in New York, see Cortese, *RHRR*, 102, who quotes Halliburton's report of his symptoms (also quoted here). For throat problems, see, for instance, *RHL*, May 1, 1936, 378. For medicines, see *RHL*, January 15, 1936, 375. For being tense, see *RHL*, March 1936, 377 (quoted); also see October 1926, 264.

4. For the Battle Creek Sanitarium, consulted was M. V. O'Shea and J. H. Kellogg, *The Body in Health*, The Health Series of Physiology and Hygiene (New York: MacMillan, 1916). See in general Andrew Scull, *Madness in Civilization—From the Bible to Freud, From the Madhouse to Modern Medicine*. London: Thames and Hudson, 2015, 272. Also see Root, *HTMM*, 41, and T. Coraghessan Boyle, *The Road to Wellville: Story of John Harvey Kellogg, Inventor of Cornflakes; A Novel* (New York: Viking, 1993). Cf. Alt, *DDIB*, 37–38 and 186–89. For the administering of "nerve exercises" by a Dr. Seabury, see 188.

5. For losing control of his life, see *RHL*, February, 1928, *His Story*, 281 (quoted). For Riley's view of the circuit (which differed little from Mark Twain's), see Elizabeth J. Van Allen, *James Whitcomb Riley—A Life* (Bloomington: Indiana University Press, 1999), 201 (quoted). For receptiveness to criticism about his writing, see *RHL*, September 29, 1926, 263. For becoming cynical, see *RHL*, November 27, 1936, 386. Cf. Root, *HTMM*, 77. Quoted is Shattuck, from R. Scott Williams, *The Forgotten Adventures of Richard Halliburton—A High-Flying Life from Tennessee to Timbuktu* (History Press, 2014), 97–98.

For lecturing as "tempestuous," see *RHL*, November 18, 1923, *His Story*, 222; for circuit as a "grind" and "a mad gallop," see March 2, 1927, 268; for strain of tour, see February 1928, 278–81, quoted at 278; for succession of eight appearances, see February 6, 1928, 280; for lecturing each day, see January 19, 1927, 267; for increase of bookings, see Root, *HTMM*, 245; for lecture season, roughly October to May, see *RHL*, 209, also 233. For free time to write, see *RHL*, April 1927, (top) 269.

6. See Arnold Henry Savage-Landor, *Everywhere—The Memoirs of an Explorer* (New York: Frederick A. Stokes, 1924), 288–89 (quoted). See *RHL*, *His Story*, 276 (quoted). For Halliburton's repertoire of stories, see Root, *HTMM*, 102. For "old reliables," see *RHL*, January 2, 1934, *His Story*, 348–49; for new stories, see editorial note, 259, 397; for effect of replacing old with new stories, see November 10, 1927, 278; for Halliburton *letting go*, see June 25, 1923, 217. For Halliburton being out of the ordinary and startling, see *RHL*, March 1926, 260 (quoted). For self-appraisal following engagement before packed house at *Orchestra Hall* in Chicago, see November 10, 1927, 278. For attention to the "reactions" of his audience, see *RHL*, editorial note, 233. For "furious motion," see *RHL*, November 5, 1927, 277–78 (quoted).

7. See Robert Byron, *The Byzantine Achievement* (London: George Routledge, 1929), author's note, quoted.

8. For Paris stay, see Root, *HTMM*, 245, op. cit. (quoted); for Birmingham, 245 (quoted). For sudden exhaustion, see *RHL*, August 10, 1924, *His Story*, 236; for need for rest, see February 6, 1928, 280.

9. For "respite," see *RHL*, February 1936, *His Story*, 376. For "the routine of a fixed existence," see Halliburton, *The Flying Carpet* (Indianapolis: Bobbs-Merrill, 1932), 13. For attitude toward regular employment, see *RHL*, December 6, 1920, *His Story*, 69. For Roxy Theatre personal appearances, see *RHL*, May 7, 1933, *His Story*, 346. For the tour aging him, see *RHL*, *His Story*, 268.

10. For "chucking" the grind, see *RHL*, February 1928, *His Story*, 281. For heavy bookings, see *RHL*, February 1928, 280. For sleeplessness, see *RHL*, March 1927, *His Story*, 269 (quoted). For last tour, see Root, *HTMM*, 261. For cancellations, see *RHL-P*, June 25, 1938. Also consulted was note to "Monica" at the Alber-Wickes agency, in Lowell Thomas Collection at Marist College's Archives and Special Collections, Poughkeepsie, NY. For "visiting and revisiting cities, towns and schools from the coast of Maine to the Continental Divide," and, by 1927, crossing "over to the Pacific side," see editorial note, *RHL*, *His Story*, 266–67.

5. East Is East, and West Is San Francisco

1. For "money thrown around," see *RHL*, May 27, 1930, *His Story*, 306 (quoted). For California, compare Nathanael West, *The Day of the Locusts* (1939).

2. For this period in Halliburton's life, see "India Speaks" in *His Story*, 340 et seq. Also see Max, *HC*, 63 et seq. For other friendships, see Alt, *DDIB*, 199–200.

3. For Halliburton with sailor from Maine, a photograph with an explanatory description on the reverse was consulted. Author's collection.

4. For Bohemian Grove, see, for instance, *RHL*, July 28, 1936, *His Story*, 382. For Erle Palmer Halliburton (1892–1957), see "Historic Los Angeles—Berkeley Square Resurrecting a West Adams Street Lost to the Freeway," online. Richard spells his first name "Earl," *RHL-P*, May 24, 1938. For Gertrude Atherton (1857–1948), see Emily Wortis

Leider, *California's Daughter: Gertrude Atherton and Her Times* (Stanford, CA: Stanford University Press, 1991). For early life and connections, see "Gertrude Atherton and Ambrose Bierce," *California History—The Magazine of the California Historical Society*, Winter 1981. For Atherton and Halliburton, see Max, *HC*, op. cit., 177–79. For Halliburton meeting Atherton, see *RHL*, February 1928, *His Story*, 280. For Florence "Pancho" Barnes and Halliburton, see Max, *HC*, 70–73.

5. Atherton's *Resanov* appeared in 1906; Resanov, born in 1764, died in 1806. For a contemporary reassessment of the man, see Owen Matthews, *Glorious Misadventures—Nikolai Rezanov and the Dream of a Russian America* (New York: Bloomsbury, 2013). Another notable example of cooperation between Russia and America occurred in the nineteenth century: "In the Tsarist years between 1860 and 1870," wrote John B. Powell, Russia's leaders "were anxious to obtain the assistance of Americans in the development of Siberia." See *My Twenty-Five Years in China*, 225 (quoted).

6. See Nathan Miller, *New World Coming* (New York, 2003), 1 (quoted). For art and architecture lost or destroyed, see Henry La Farge, ed., *Lost Treasures of Europe—A Pictorial Record*, with 427 photographs (New York: Pantheon Books, 1946).

7. For Norton see Allen Stanley Lane, *Emperor Norton—Mad Monarch of America* (Caldwell, ID: Caxton Printers, 1939). For Beachey, see Frank Marrero, *Lincoln Beachey—The Man Who Owned the Sky* (San Francisco: Scollwall, 1997). For beginnings of aviation, see Gerrie Schipske, *Early Aviation in Long Beach* (Arcadia, 2009); and Gavin Mortimer, *Chasing Icarus—The Seventeen Days in 1910 That Forever Changed Aviation* (Walker, 2010).

8. For early San Francisco see Gertrude Atherton's *My San Francisco—A Wayward Biography* (Indianapolis: Bobbs-Merrill, 1946); Curt Gentry's *San Francisco and the Bay Area—Present and Past*, with maps and photographs (Garden City, NY: Doubleday, 1962); and Barnaby Conrad's *San Francisco—A Profile with Pictures* (New York: Viking, 1959). A fine modern treatment is Michael F. Crowe and Robert W. Bowen, "Night Life—High Jinx and High Balls," in *San Francisco Art Deco* (Chicago: Arcadia, 2007), 35–42. See also George Robinson Fardon's *San Francisco in the 1850s: 33 Photographic Views*, with an introduction by Robert A. Sobieszek (Rochester, NY: Dover, 1977); and Gertrude Atherton's *An Intimate History of California* (New York: Blue Ribbon Books, 1936). Halliburton's secretary Paul Mooney read books by Herbert Asbury (1889–1963), author of *The Barbary Coast—An Informal History of the San Francisco Underworld* (New York: Alfred A. Knopf, 1933). Captain Welch of the *Sea Dragon* and Dick Wetjen hung out at *Izzy Gomez's Cafe*, a well-known haunt of writers and reporters (including *Examiner* reporter Herb Caen). For Izzy's Café, at 848 Pacific Street, see James R. Smith, *San Francisco's Lost Landmarks* (Sanger, CA: Word Dancer, 2005), 82–83. Smith describes Izzy's as "the gathering place for aspiring writers." A regular, writer William Saroyan, "immortalized the place" in his play *The Time of Your Life*.

9. For a full account of the Gaines-Halliburton meeting, see Alt, *DDIB*, 318, 321–25. Cf. Root, *HTMM*, 249 et seq. For specifics of the crossing, see *San Francisco Examiner*, June 6 (Saturday), 1936. Cf. Williams, *Forgotten Adventures of Richard Halliburton, op. cit.*, 144. For Golden Gate Bridge, see *Life* magazine, Golden Gate Bridge (cover) issue, May 31, 1937. For the history of the bridge, see Kevin Starr, *Golden Gate: The Life and Times of America's Greatest Bridge* (New York: Bloomsbury, 2010). Also see Michael F. Crowe and Robert W. Bowen, "Marvels of Engineering, in *San Francisco Art Deco*, op. cit., 83–100.

10. Ibid.; Alt, *DDIB,* 321–25.

11. For the New York World's Fair, see, in general, David Gelernter, *1939—The Lost World of the Fair* (New York: Free Press, 1995). Also see Alt, *DDIB,* 319 et seq. For view of America, see Max, *HC,* 102.

12. Within a year of the Oakland Bay Bridge's opening, the groundwork for the fair that changed Halliburton's life was being laid. For details on the Golden Gate International Exposition, familiarly called "the magic city floating on San Francisco Bay," see bibliography. Information on the fair is from *SFN,* Wednesday, July 6, 1938, 17. Other sources consulted include the many brochures issued for the fair, as well as numerous scrapbooks devoted specifically to the event. Newspaper clippings, postcards, photographs, and ticket stubs often were affixed to scrapbook pages. In addition to the sources in the bibliography, see Richard Reinhardt, *Treasure Island—San Francisco's Exposition Years* (San Francisco: Scrimshaw, 1973). For Halliburton and the fair, compare Alt, *DDIB,* 319–20 et seq. For the reclamation project, see Stuart O. Blythe, "The Fair Is Ready To Open," *Golden Gate International Exposition* Premieres February 10 and 19," *California—Magazine of the Pacific,* February 1939, 5–9.

13. The "new inspiration" came about the time the *Eleventh Olympiad,* a two-week series of competitive athletic events, commenced in Berlin. For more information on the Berlin games, see *The Nazi Games: Berlin 1936* (PBS, 2016), video. Also see Richard D. Mandell, *The Nazi Olympics* (New York: MacMillan, 1971).

14. For the announcement of the junk plan, see *RHL,* July 14, 1936, *His Story,* 381 (quoted). For parents' trip, see *RHL,* July 28, 1936, *His Story,* 382. For discussion of the house, see *RHL,* October 15, 1936, *His Story,* 383–84, description at 384. Also see *RHL,* July 6, 1938. Halliburton did suggest that his parents visit Olympia, birthplace of the Olympics. See *RHL-P,* June 25, 1938. For "1407 Montgomery" see Max, *HC,* 129. See *RHL,* July 1 and July 14, 1936, *His Story,* 381.

6. Halliburtonland

1. For "floating city, see *Official Guide - 1940 - Golden Gate International Exposition on San Francisco Bay,* quoted at 21. The full name of the *Paris Exposition,* which lasted from May 25 to November 25, 1937, was *Exposition* Internationale des Arts et Techniques dans la Vie Moderne. See Wikipedia. For the Golden Gate International Exposition in its time, see, for instance, "*Golden Gate Exposition* Opens With a Wild West Wallop," *Life,* March 6, 1939, 11–15; "Jeweled Radiance," *Magic of Night,* in *San Francisco Examiner,* Wednesday, February 15, 1939, section 3, 1–12; An important first-hand rundown is "Today at the Fair," *Complete Magazine of Events, Exposition* Premiere Souvenir, Sunday, *San Francisco Examiner,* February 19, 1939, 16 pages; "Treasure Island of 1939—Coloroto," *Popular Mechanics Magazine,* May 1938, 649–56, 128A–29A. Views of the fair are readily available. See, for instance, Gail Hynes Shea, "Treasure Island Fair: *Golden Gate International Exposition,*" historical essay, online. A main source for the fair is the "*Exposition* Edition," *SFN,* July 6, 1938. Another main source was Alexander Gross, F.R.G.S., ed., *Famous Guide to San Francisco and the World's Fair—Pictorial and Descripti*ve, with 9 maps and 78 illustrations (San Francisco News Company, 1939). Also see *Official Daily Program—World Championship Rodeo,* Livermore Day (noted), May 21, 1939; and *Official Guide Book (of) Golden Gate International Exposition on San Francisco Bay,* includes foldout map, 25 cents (rev. ed., 1939), 116 pages. For Fair personnel and

368 NOTES TO PAGES 44–45

committees, *SFN*, supra, 16 was consulted, as well as various exposition guides. The fairgrounds were called "Exposition City," or the "Pageant of the Pacific." The bibliography contains various sources providing additional information on the Golden Gate International Exposition and its fundamental aims. For quick reference to dates, see "List of World Fairs" in Wikipedia. For a contemporary treatment, see Jim Marshall, "How the West Throws A Party," (Golden Gate International Exposition), *Colliers—The National Weekly*, February 18, 1939, 21–23, 64–65. For a modern perspective, see James R. Smith, *"Golden Gate International Exposition—1939,"* in *San Francisco's Lost Landmarks*, 147–74. See also Crowe and Bowen, *San Francisco Art Deco*, op. cit., 101–25, quoted ("last great Civic event") at 101. It is hard to duplicate for readers today the visual impact of the fair. See, however, "Magic in the Night," above, for evening views in color of the Court of Reflections, Elephant Towers, the Court of the Flowers, the Tower of the Sun, and the Court of the Seven Seas; for similar views from Yerba Buena Island, see "Magic City Afloat." "Magic in the Night," *Official Souvenir—The Golden Gate International Exposition* (San Francisco: Crocker, 1939), unpaginated. For "color achieve(d)," see "Jeweled Radiance," supra. Among the best photographs were those taken by photographer Gabriel Moulin (1872–1945). See Anna Burrows, "The *San Francisco Golden Gate Exposition*," online. For A. F. Dickerson, see "Lighting the Treasure Island World's Fair—1939," online. Dickerson is also noted in "High Times, High Visions: The *Golden Gate International Exposition*," online. For "Modern Aladdin," see "Magic in the Night," op. cit. For Walter D'Arcy Ryan (1870–1934), the electrical wizard who created the light at the *Panama-Pacific Exposition*, see Wikipedia. A. F. Dickerson's debt to Ryan is noted by E. T. Buck Harris in "Magic in the Night." For color experiments, see *SFN*, July 6, 1938, 18.

2. Consulted (and used throughout) are brochures and maps of the fair, notably a thirty-by-twenty-inch foldout map entitled the *Golden Gate Exposition*, a John Fix Feature presented by the Gilmore Oil Company and an insert in the *San Francisco Examiner*. Outside the color-drawn fairgrounds, numbered squares on the map identify exhibits and buildings; inside the fairgrounds, corresponding numbers in circles indicate exact locations. Other maps were available to fairgoers. A pocket-sized map "Your Guide to Treasure Island" was issued "Compliments of (the) *Oakland Tribune*." Handy four-page maps were available in local hotel lobbies or at the fair gate. In addition, the San Francisco Chamber of Commerce issued maps of both the city and the fair, and the exposition's official guides contained maps. On February 17, 1939, the *San Francisco Chronicle* issued its *"Golden Gate International Exposition"* extra; one of the supplement's seven sections included a map. See also the helpful map in Crowe and Bowen, *San Francisco Art Deco,* supra, 106–7. For hopes of a better tomorrow brought about by the Golden Gate International Exposition, see "World's Fair Edition," *SFN*, Wednesday, July 6, 1938, 15 (quoted) passim. For the phoenix, see Joseph Henry Jackson, *A Trip to the San Francisco Exposition with Bobby and Betty* (New York: Robert M. McBride, 1939), 13. For *Tower of the Sun*, see 88 (quoted). For Transparent Man, see 30 (quoted).

3. See, in general, Laura P. Ackley, *San Francisco's Jewel City—The Panama-Pacific International Exposition of 1915* (Berkeley, CA: Heyday, 2014). Also see Official Publication: *The Panama-Pacific International Exposition at San Francisco 1915*, the Albertype Edition. Consulted was "Lagoon Reflects Beauties of Nippon—Colorful Display from Land of the Rising Sun," in "Jeweled Radiance," 1–12; *San Francisco Examiner*, Wednesday, February 15, 1939, Vol. CLXX. No. 46, Section III, page 2. Specifically, the 1915 Exposition also announced that San Francisco had emerged triumphant from the ashes of the great

earthquake and fire of 1906. See "Treasure Island of 1939—Coloroto," supra (quoted). A main source for the Temple of Religion and Tower of Peace is Stanley Armstrong Hunter, Temple of Religion and Tower of Peace at the Golden Gate International Exposition (San Francisco, 1940), 96 pages with photographs.

4. Cf. "World's Fairs," World History in Context—Gale, online. For references to buildings and exhibits, see, besides earlier mentioned sources, "High Times, High Visions: the Golden Gate International Exposition," online. The source also notes A. F. Dickerson. A "vast, animated diorama covering the end wall of the Electricity and Communications Building" (identified as "The Story of the West") portrayed "three phases of California's development." Exposition postcard. For Hall of Science, see 26-27; the La Brea Tar Pits is the noted spectacle.

5. References are to postcards, brochures, and advertisements from the "fair."

6. For Gayway, see Jackson, Trip to the San Francisco Exposition, supra, 84-85 (quoted). For "native villages," see Popular Mechanics, supra (quoted). Those who did not see the Mata Hari of fan dancing here could catch her solo act at the Music Box at 859 O'Farrell not far from Omar Khayyam's celebrated dinner club. Consulted was a program for the Music Box at 859 O'Farrell, author's collection. Rand also appeared at the Golden Gate Theatre. For Sally Rand causing a commotion at the fair, see SFN, May 2, 1939. For exhibits, see Official Guide Book, supra.

7. The Golden Gate Exposition map from the Gilmore Oil Company, supra (see note 2 above), was consulted, as well as other maps of the fair.

8. For the Japanese and other pavilions, see the postcard exhibit—for instance, "Golden Gate International Exposition—Alameda Info.com." Numerous brochures were also consulted. The Chinese Pavilion replicated idyllic aspects of Chinese village life. False portrayals of Chinese culture given at the Panama-Pacific International Exposition of 1915 were less noted at the later fair. See "Underground Chinatown: Racism at the Fair; the 1915 Panama-Pacific International Exposition," Chinese History Society of America Museum, April 2, 2015–April 2, 2016, online. The Palace of Mines, Metals and Machinery might have had a (privately sponsored) German exhibit.

9. Halliburton did not see the fair in its finished form. In a letter dated June 13, 1938, he mentioned how quickly the "Fair Grounds" were rising (RHL-P), and in one dated August 31, 1938, he mentioned a visit to his "anchorage" (RHL, His Story, 399). In a "Confidential Prospectus" he sent to James Watson Webb, Halliburton included a "booklet" entitled "TRANSPACIFIC" and containing views of the fair, a map of the Pacific, a picture (not a photo) of the junk he would sail, and an envelope with the cachet of "the Sea Dragon." For "spirit," a mailer entitled "1939 World's Fair Facts" is quoted. For Halliburton's 'Taj Mahal' experience, compare Root, HTMM, 81-82.

10. For "most exciting concession," see RHL, August 31, 1938, His Story, 399.

11. Various brochures were consulted concerning facilities; see those cited in note 1 above. For parking, see "Treasure Island of 1939—Coloroto," op. cit., 128A. SFN and the San Francisco Examiner regularly noted the fair's celebrity attendees.

12. See Jackson, Trip to the San Francisco Exposition, op. cit., 65-66.

7. High Cost of Daring

1. Salaries and profit margins were topics he certainly believed were of interest to his father. For "big plans and fat fees," see RHL, March 16, 1928, His Story, 282; for

money to be made in Hollywood, see May 27, 1930, *His Story*, 306. For financial rewards from junk expedition, see *RHL*, September 10, 1938. For money fueling optimism, see *RHL*, June 25, 1938, 398. For *Hangover House* and its history, see Max, *HC*, 139–71. Cf. Alt, *DDIB*, op. cit., 312–17. Costs and financial expectations naturally varied. Author F. Scott Fitzgerald, a heavy spender, kept income ledgers indicating about $450,000 in lifetime earnings and $30,000 in withheld income. "If we accept a 20-times measure, the modern equivalent of Fitzgerald's income would be roughly $500,000 (annually). But a person earning $500,000 today does not live as well as Fitzgerald did." An "elegant villa" he rented near St. Raphael" in France cost $79 a month. See William J. Quick, "Reading F. Scott Fitzgerald's Tax Records," *American Scholar*, Autumn 2009, 96–101, quoted at 97 and 101.

2. For Chinese merchants, see *RHL*, April 12, 1938, *His Story*, 396; also see March 26, 1938, 395. For "junk prospects" looking "black," see *RHL-P*, May 24, 1938. The October 7, 1938, letter to "Monica" at the Alber-Wickes agency is in the Lowell Thomas Collection, Marist College Archives and Special Collections. Also see Taylor, *SSMTWOD*, 193. For small investments, see *RHL-P*, July 14, 1938.

3. For project taking "firm hold," an editorial note, see *RHL*, *His Story*, 398. For "first sight of foreign land," see July 6, 1938, 398. For "forming a corporation," see *RHL-P*, June 18, 1938.

4. The budget was in the "Confidential Prospectus" Halliburton sent to James Watson Webb III (1916–2000), who became a noted film editor for such productions as *State Fair* (1945), *The Razor's Edge* (1946), *Letter to Three Wives* (1949), and *Broken Arrow* (1950). A twelve-page handwritten letter accompanied the prospectus, along with a "trans-Pacific" itinerary, route information, and maps. Letter sent from *Chancellor Hotel*, August 30, 1938. Author's collection. Evidently others, besides Webb, received a prospectus. See *RHL*, June 17, 1938, *His Story*, 398. To his parents, he wrote that he followed an established budget "carefully," and had raised more money than he reckoned he would need—"with (on his return home) several thousand dollars unspent." See *RHL-P*, November 21, 1938 (quoted).

5. Ibid., "Confidential Prospectus," quoted.

6. Cf. Root, *HTMM*, 262. Schwabacher-Frey's most famous publication was B. K. Beckwith's *Seabiscuit—The Saga of a Great Champion*, "Copyright 1940 by Wilfred Crowell." The book's first chapter is entitled "It Can't Be Done," its last "It Can Be Done." For identifying business and personal addresses, *The San Francisco and Bay Counties Telephone Directory* for May 1939 has proven useful.

7. For the interviews, compare Root, *HTMM*, 24. For applications, some five thousand for Shackleton's *Antarctic Exposition* of 1913, see Margery Fisher and James Fisher, *Shackleton* (Cambridge, MA: Riverside, 1958), 308 et seq. For "three sporty girls" applying to Shackleton *expedition*, see Fisher 308. Always the gentleman, Shackleton regretted that there were "no vacancies for the opposite sex on the (*Antarctic*) *Expedition*" (Fisher 308). For both the Shackleton and Halliburton expeditions, courage and romantic temperament counted as much as technical skill. For Shackleton's interviewing style, see Fisher, 314–15. For outdoor camps for teenage girls who might have applied, see Susie Seefelt Lesieutre, "Joy Camps: the Camp Craft Camps for Girls," *A Northern Wisconsin Adventure, Wisconsin Magazine of History,* Summer 2015, 36–49. Girls learned emergency medical techniques, tent-pitching, paddling, cooking, and other skills needed to survive in the wilderness. Almost certainly some Boy Scouts applied, as did the trio who

won an opportunity to go on an African safari with Martin and Osa Johnson ten years earlier. For Putnam-sponsored enterprise, see *Three Boy Scouts in Africa* (New York: G. P. Putnam's Sons, 1928); the authors, the scouts themselves, were Douglas L. Oliver, David R. Martin Jr., and Robert Dick Douglas Jr. Scouting had become increasingly popular in America, with the Scouts claiming well over a million members by the time this photo was taken. FDR encouraged scouts to lend a helping hand to relief agencies during the Great Depression, and scouting leaders also provided training to members of the *Civilian Conservation Corps,* a creation of the Roosevelt administration. Like Halliburton, Admiral Richard Byrd and Charles Lindbergh were associated with the movement.

8. For labor history, see R. Conrad Stein, *The Story of the Child Labor Laws* (Chicago: Children's Press, 1984). For Shackleton's qualities, see Neville Peat, *Shackleton's Whiskey* (London: Preface—The Random House Group, 2012), op. cit., 35–36, quoted at 36. Also see F. A. Worsley, *Shackleton's Boat Journey*, with a narrative introduction by Sir Edmund Hillary (New York: W. W. Norton, 1977). For Shackleton's interviewing style, see Fisher, *Shackleton*, supra, 314–15. For the popularity of the expedition and the faith given to Halliburton as its organizer, see Root, *HTMM*, 24. Although Halliburton flunked a naval exam at Princeton, a difficult one on "range-finding" that required math skills, he fared well in "Ordinance and Gunnery, Navigation, [and] Seamanship," courses taught by a Professor Eisenstadt. Still, his maritime skills were limited. For the naval examination, see *RHL*, September 10, 1925, *His Story*, 250; see also October 5, 1918, 10. For Eisenstadt courses, see *RHL,* October 5, 1918, 10.

9. For the Chancellor, advertising brochures from the period were consulted. Halliburton noted the hotel as his headquarters in a letter to Mrs. Vida Halliburton dated June 28, 1938, RHC, Rhodes. For cable cars, see Curt Gentry, *San Francisco and the Bay Area,* supra, 15 et seq. Thomas Cook and Son was at 689 Market. For addresses, consulted was the 1939 *San Francisco and Bay Counties Telephone Directory,* op. cit.

10. Quoted is *RHL-P*, September 10, 1938. "Longer intervals" would be required between the mailings of the last four letters. See Halliburton, "Letters from the *SD #2,*" January 18, 1939. A circular for the *Sea Dragon Expedition* is quoted. For subscriber information, see Halliburton, "Letters from the *SD #1*," November 20, 1938. Quoted is the circular advertising the "Letters from the *Sea Dragon.*"

11. See *RHL-P*, July 14, 1938 (quoted). For radio opportunities, see *RHL*, July 20, 1926, 262; also see February 21, 1938, 394. Various brochures were consulted concerning radio as a vehicle for exposure and income; these include Carveth Wells, *Exploring America with Conoco and Carveth Wells* ("the Man They Call 'Radio's Truthful Liar'"), Station WEAF, New York, 1933 (through April 15) (Conoco Travel Bureau—America's Foremost Free Travel Service), 16-panel brochure. Author's collection. For royalties, see *RHL-P,* January 23, 1939. For the $2,500 junk, see *RHL-P,* July 6, 1938. For the *Flying Carpet's* return, see Halliburton, *Flying Carpet,* op. cit., 352 (quoted).

12. See *RHL-P,* July 21, 1938 (quoted). Omitted from *His Story* are letters focusing chiefly on investment issues. These are dated June 18, July 6, July 14, and July 21, 1938.

8. "The Lads"

1. For Myrtle Crummer, see Gaylord A. Beaman, *A Doctor's Odyssey—A Sentimental Record of Le Roy Crummer: Physician, Author, Bibliophile, Artist in Living* (Baltimore: John Hopkins Press, 1935); copy in author's collection inscribed by wife Myrtle

(Crummer), 20 et seq. and, in particular, 24. For the Crummer-Ingram divorce, see the *Los Angeles Times*, May 2, 1938. Consulted was *RHL-P,* August 1, 1938. Also see Root, *HTMM*, 31. For "plugging away," radio broadcasts, and "nothing ventured," see *RHL-P,* July 21, 1938 (quoted). For the *Books of Marvels*, see *RHL-P,* June 17, and January 23, 1939.

2. For promoter(s) and other financial dealings, see *RHL-P,* June 18, 1938; the letter is omitted from *His Story*. For asking for money from his father, see James Cortese, "Richard Halliburton—He Called the Road Royal," *Delta Review* 4, no. 5 (September 1967): 50. For "wanderlust," noted by Welch in newspaper article, see *LHK* #7, November 25, 1938. Also see Root, *HTMM*, 262 et seq.

3. For Hollywood friends, see Max, *HC*, 70 et seq. For Erle Jr.'s participation, see *RHL*, May 28, 1938. For information about the finances and Erle's eye operation, see *RHL-P,* July 14, 1938. For "dismal flop" in conversation with Erle Sr., see *RHL-P,* June 18. For Potter's participation and financial contribution, see *RHL-P,* July 21, 1938. For genealogical information, see RHC, Rhodes. Also see Root, *HTMM*, 263. Also see James Zug, "Sea of Dreams," op. cit., online. Information on the Potter family can be found online. For Colonel Wilson Potter, see "Colonel Wilson Potter" (obituary), *New York Times*, Thursday, June 13, 1946. Potter was interviewed by investigative reporter Alan Landsburg. See Landsburg, "Richard Halliburton," *In Search of Missing Persons* (New York: Bantam Books, 1978), 9–29. For investment figures, including pledge from Vida, see *RHL-P,* August 1, 1938. Privacy concerns dictated that investment details be omitted from *His Story*. Vida, born in 1894, died in 1951, Erle in 1957; their five children, Erle II, Zola, Vida, David and Ruth, far outlived them. For sum requested for Erle Jr.'s admission, see *RHL-P,* June 25, 1938.

4. My thanks to Rauner Special Collections Library, Dartmouth College, Hanover, New Hampshire, for use of its archival materials, including alumni magazines, related to John Potter (1914–1996), Gordon Torrey (1914–1992), and Robert Chase. For substantial photo record of the lads, see RHC, Rhodes. Halliburton first mentioned Torrey by name in a letter dated November 2, 1938, *RHL-P.*

5. Ibid., Dartmouth. For "distrust," see Root, *HTMM*, 263 (quoted). For "over the top," see *RHL-P,* August 1, 1938 (quoted). In his letter to J. Watson Webb Halliburton notes that, besides his own pledge of $5,000, he had "raised" $5,000 from Potter, $5,000 from Torrey, and $5,000 from Chase. He told Webb he needed $5,000 more. Richard Halliburton letter to J. Watson Webb, August 30, 1938.

6. Quoted is *RHL-P,* August 1, 1938. For Halliburton's communications with Crummer by phone, see *RHL-P,* September 4, 1938 (quoted). For his faith in the project, see *RHL-P,* October 7, 1938; also see *RHL*, August 1, 1938, 399. Companies with investment promise were also companies sponsoring the fair. Consulted were brochures which listed donors.

7. For *Standard Oil* and other companies, see Powell, *My Twenty-Five Years in China*, op. cit., 126. For Shell Oil, see *RHL-P,* August 1, 1938. For the diesel engine, see *RHL-P,* July 14 and August 1, 1938. Dale Collins noted that the engine was "for emergency use only." See "The Royal Road Across the Pacific," April 1940, at 502. Available online: *Proceedings Archive* U.S. Naval Institute.

8. See *RHL-P,* July 14, 1938, quoted. For solicitations of money and for the Linkletter recollections, see Root, *HTMM*, 258–59 (quoted). Root says that Halliburton customarily appeared "at meetings and conferences decked out in his spats, homburg and cane, with (again) a lace handkerchief stuffed up his coat sleeve." For fair management and its commitments, see Art Linkletter, *Confessions of a Happy Man*, noted in Greg Daugherty,

"The Last Adventure of Richard Halliburton—the Forgotten Hero of 1930s America," *Smithsonian*, March 25, 1914. Also see Max, *HC*, 174 and 274n9. For the "$35,000," see Root, *HTMM*, 259 (quoted). Root notes that Halliburton "was as broke as ever, but it never showed" (260).

9. For "Confidential Prospectus" and letter, see chapter 7 ("High Cost of Daring"), note 5 above. For "Kweilin Incident," see *New York Times*, August 25, 1938. The pilot, Hugh L. Wood, and four Chinese survived. There were eighteen fatalities.

10. For George Barstow III's background, see Root, *HTMM*, 22, 31. For George Barstow I and II, see Wikipedia. For obituary of George Barstow II, see *New York Times*, November 18, 1932. George III also had two sisters, one of whom resided in Memphis. The will of George Barstow II was consulted. George III was born on August 22, 1917, and his parents' house at 237 Upper Montclair Avenue still stands. Before 1930, Bertha resided with her husband at 876 Park Avenue; after 1930, she lived at the Surrey Hotel at 20 East Seventy-Sixth Street. For records of the Barstow family, contact the Providence Historical Society in Rhode Island; the Barstow name dates to the 1600s. George Barstow II and Bertha Kellogg Barstow (1878–1959; daughter of James Crane Kellogg of Elizabeth, New Jersey, and sister of Morris Kellogg of New York City) are both buried in a cemetery in Sharon, Connecticut. My thanks to Barstow family historian Patricia Suprenant for this information. For Mrs. Jerome P(illow) Long, see *Commercial Appeal* (Memphis, TN), April 18, 1939 (quoted). George Eames Barstow I, grandfather of the *Sea Dragon* crew member, published a monograph entitled *The Effect of Psychology on Americanism* (1920) opining his views on capitalism and labor, civic duty and commitment. Halliburton mentioned a "Mrs. Barstow," perhaps the sister who lived in Memphis and knew Nelle Nance Halliburton. *RHL-P*, September 4, 1938. He also noted that Barstow would "probably fly with Chase." *RHL-P*, September 10, 1938.

11. For the "check" from "super-cargo" Barstow, see *RHL-P*, September 10, 1938. For Bertha Barstow's contribution, see a September 23, 1938, letter written to Vida "one hour" before sailing, MEB Archive (quoted). The letter reads, "Mr. George Barstow III is investing $4,000 in the corporation," but clearly Bertha controlled the purse strings.

9. O, Captain, My Captain

1. For Thompson and Faucon, captains in literature about whom Halliburton had knowledge, see Richard Henry Dana Jr., *Two Years Before the Mast* (New York: Signet, 2006), with a new introduction by John Seelye, 82, 148, 172 (quoted). For Halliburton and Dana, see *RHL* July 30, 1921, *His Story*, 82. Real-life captains included the SS *Madison* skipper, whom Halliburton thought tyrannical. But to tame a wicked and wild crew, captains had to be stern. See full text of Halliburton's letter to his parents, February 13, 1923, RHC, Rhodes. Also see *RHL*, February 13, 1923, *His Story*, 207. For the debased "bo's'n of the *Gold Shell*, who acted as though he were in charge, see Halliburton, *Royal Road to Romance*, supra, 138. Note also the American freighter captain, "a kind-hearted old piece of salt" with whom he shared meals and sleeping quarters, keeping him segregated from "the vulgar crew." Halliburton, *Royal Road to Romance*, op. cit., 309. For the captain who, shouting, asked Halliburton "if he had a first-class passage," see *RHL*, October 21, 1922, *His Story*, 190. A long-tenured sea captain was Charles Jokstad; he will appear later. See Root, *HTMM*, 26–28.

2. For the SS *Hoover* incident and the last days of the *Dollar Steamship Line*, see

"The Takeo Club: The Wreck of *the S.S. Hoover*," Part II, on-line. Records of Yardley's career are to be found at *San Francisco Maritime Historical Park* at Fort Mason in San Francisco. My thanks to the *San Francisco Maritime NHP* for sharing these and other pertinent documents. For Halliburton's impressions of Richard Yardley, see *RHL-P*, May 28, 1938 (quoted). For "first port," see *His Story*, 421. Today on Green Island, Pi-tou-chiao Lighthouse stands as a memorial to the *Hoover* wreck.

3. *RHL-P*, May 28, 1938 (quoted). For Yardley obituary, see *Oakland Tribune*, May 31, 1938. For investment capital from Richard Yardley, see *RHL-P*, June 25, 1938, in which Halliburton stated, "Yardley has $2500."

4. For hiring captain in Hong Kong, see *RHL-P*, July 14, 1938 (quoted). Muscular language appears in Welch's "Farewell to Sail" and *Letters from Hong Kong* (*LHK*). Quoted is *LHK* #13, January 29, 1939. Halliburton noted the "salty" language. See "Log of the *SD* #1," December 12, 1938. For "yarns," see *Hong Kong Daily News*, January 19, 1939 (quoted). When his "Farewell to Sail" (*Man Magazine*) was published, Welch's middle name was rendered "Wenloch." In Halliburton, "Letters from the *SD* #1," November 20, 1938, the spelling is "Wenlock." For quotations concerning Welch's experience with sailing ships, see Halliburton's letter to J. Watson Webb, August 30, 1938. Author's collection. Also see Root, *HTMM*, 24–25. For being a "Scotchman," see *RHL-P*, August 7, 1938 (quoted).

5. See *RHL-P*, August 1, 1938 (quoted).

6. For "sail(ing) in his sleep," see Halliburton, "Letters from the *SD* #3," January 27, 1939. Welch's name appears in the *Register of Commissioned Officers, Cadets, Midshipmen and Warrant Officers of the United States Naval Reserve*, July 1941, online. For Welch's lack of junk experience, see *LHK* #5, November 16, 1938. Experience sharpened Welch's seaman skills. In the 1930s, the minimum requirements for successful ocean navigation included a barometer, compass, chronometer, sextant, and pelorus, as well as ephemeris tables and naval charts (in use from the eighteenth century in the West and by the 1930s used worldwide). The barometer, compass, and charts are the only navigational instruments Halliburton and Welch mentioned. No reference was made to a sextant, proficient use of which requires months of training.

7. For marriage information, consulted was the *Berkeley Gazette*, March 1930 (which has list of those applying for marriage licenses), and the *Oakland Alameda Directory*, 1935. From Washington State, Bridgeford graduated from the University of California–Berkeley in 1923. The *Los Angeles City Directory* for 1935 shows Welch and Barbara E. Bridgeford residing at the same address. Her father George E. Bridgeford was from Kentucky. In an August 30, 1938, letter in author's collection, Halliburton told J. Watson Webb that Captain Welch was married. Also consulted was *Naturalization Petition for the Southern District of California* 188 (1927).

8. For Jokstad's acquaintance with Welch, see Root, *HTMM*, 27. For Jokstad's own recollection, see *The Captain and the Sea* (New York: Vantage, 1967), 189. Jokstad owned and operated the Sea Captain's Motel at 2322 Lombard Street in San Francisco, so he was close by for interviews with *Chronicle* reporter Jonathan Root. For John Wenlock Welch, "Ocean Tow," see *Proceedings Magazine*, US Naval Institute, vol. 63/2/408 (February 1937): 206–14, quoted at 214, online. Welch's "Signaling and the *U.S. Merchant Marine*" appeared in *Proceedings Magazine*, US Naval Institute, vol. 62/1/395 (January 1936): 79–81, quoted at 80, online. The *Coringa* trip, which suggests the broad sweep of Welch's naval experience, began in Antwerp, then continued to Gibraltar and Malta, then took him through the Suez to Port Said, to Colombo, and finally to Singapore. For

Torrey's opinions, see Landsburg, "Richard Halliburton," op. cit., 23. For compass use, see Halliburton, "Letters from the *SD* #4," February 16, 1939; and Welch, *LHK* #13, January 29, 1939.

9. For the swimming, see Welch, *LHK* #2, October 29, 1938. For Welch's general appearance, compare Root, *HTMM*, 24. For Cortese's comments, see *RHRR*, 160.

10. "Ocean Tow," 206–214, quoted at 214; "Signaling," 79–81, quoted at 80.

11. For Wetjen knowing Welch, see *SFN*, April 3, 1938. The *Sea Dragon Expedition* might have given Welch his first command post. "Funny, but until last week," he told Wetjen, "I haven't given much thought of what it will be like to have the command for the first time really." *LHK* #9, December 5, 1938. For Welch having "master's papers," see Alt, *DDIB*, op. cit., 327 et seq. For doubt in the matter, see Max, *HC*, op. cit., 182, 276n2. For Welch's credentials, I have relied on the research of Edward T. Howell, who on January 30, 1972, requested background information about Captain Welch from the superintendent of the Merchant Marine Academy in King's Point, New York. Referred ultimately to the US Coast Guard Reserve, Howell was told by Captain T. McDonald, "We are unable to identify Captain John Welch as ever having been a licensed U.S. merchant seaman." McDonald added, "We have no records available on either the attempted voyage or any persons involved." Indeed, the ship might not have been registered. On record, in any event, is a Certificate of Competence for Second Mate for Foreign-going Steamships Only issued to Welch on February, 1919 by the Merchant Service. Also note his name on *Register of Commissioned Officers*, note 6 above. "Captain," incidentally, was a rank of utmost respect, but in the world of private boating, it had looser application. See, for instance, Errol Flynn, *Beam Ends* (New York: Longmans, Green, 1937), 83–84, 97. While the requirements for a "certificate" varied, those for a "master's certificate" might include three years of active service. Compare Captain C. Bradley, O.B.E., "A Master's Memories," *Sea Breezes*, July 1950, 32–34, at 33. Bradley, who began his sea career as an apprentice at age fourteen in 1922, received the master's certificate at age twenty-three in 1930. See the American Sailing Association for standards and qualifications, online. Also see USCG National Maritime Center, official center for Merchant Marine Credential information. For "Shark Gotch," see Albert Richard Wetjen, *The Chronicles of Shark Gotch*, The World's Work (1913. Ltd., *The Master Thriller* Library (London: Kingswood, 1913), frontispiece (quoted). For his lack of humor, see p. 1 (quoted); for his "iron hand," see p. 71 (quoted).

12. For this book I viewed hundreds of sea-related films from the period to see the Welch name or hand. Early talkie *Way for a Sailor* (1930) with John Gilbert and Leila Hyams seemed offhand a likelier guess than most. The movie concerns a rollicking group of merchant marines let loose in port; Wallace Beery appears in the film as a grumbly shipmate, and writer Jim Tully assumes the role of a curly-haired deckhand. For the "sailing fraternity," see *Hong Kong Daily News*, January 19, 1939 (quoted). For Nelle Nance meeting Captain Welch, see Max, *HC*, 182. Also see *RHL-P*, August 7, 1938. Cf. Root, *HTMM*, 24. For Nelle Nance's opinion of Welch, compare Cortese, *RHRR*, 159. For Welch's friendship with Collins, see unpublished letter from Dale E. Collins to Edward T. Howell, May 6, 1946, above. For "old man Ahlin," see Welch, *LHK* #5, November 16, 1938.

13. Welch's actual words: "Keeping notes as I go along. Halliburton told me tonight that *she* has put me in her first newspaper article. I am an old seaman, it seems, and the others are just playing at it." *LHK* #6, November 24, 1938. For Cortese comment, see *RHRR*, 160. For possible "beer" preference, see *LHK* #6, November 24, 1938. For

another characterization of Welch, see Alt, who calls Welch "mainly an expert at promoting himself." *DDIB*, 327 et seq. For Potter's initial and Halliburton's ultimate opinion of Welch, see Root, *HTMM*, 32 (quoted). For "veteran sea-dog," see Halliburton, "Letters from the *SD #1*," November 20, 1938. For "veteran seaman," see "Log of the *SD #1*," *SFN*, December 12, 1938; and *Cincinnati Enquirer*, January 29, 1939. *Hong Kong Daily News*, January 19, 1939 (quoted). For "master," see "Log of the *SD #1*," December 12, 1938.

10. The Black Magic of Machinery
and the Wizardry of Radio Communication

1. For Von Fehren, see *Hong Kong Daily News*, January 19, 1939 (quoted). A number of photographs exist of him. For Hannah, born in 1905, see *1940 Census*. For "black magic of machinery," see Halliburton, "Log of the *SD #1*," *SFN*, December 12, 1938. For Potter's assessment, see Cortese, *RHRR*, 180 (quoted). See "Log of the *SD #1*" (as it appeared in the *Cincinnati Enquirer*, January 29, 1939), quoted.

2. For Von Fehren's address and telephone number, see *The Pacific Telephone and Telegraph Company* for May 1939, op. cit., 435. For vitals, see *US Declaration of Intention 63251* (245, US Department of Labor, 1925) for Heinrich Johannes Von Fehren (born January 5, 1904), online. For Halliburton's high regard for Von Fehren, see *RHL*, February 23, 1939, 431 (quoted). Confirming evidence of Von Fehren's employment aboard the *Zaca* has not been found. Garth A. Basford is noted as the *Zaca's* engineer in Templeton Crocker, *The Cruise of the Zaca* (New York: Harper and Brothers, 1933), 5. A press photo was released of the *Zaca* docked in San Francisco. Author's collection. A "Garth X. Basford" lived on 146 South Charles Avenue in San Francisco. Consulted was the 1939 *San Francisco and Bay Counties Telephone Directory*, op. cit., 26. Crew member John Potter noted that Von Fehren worked aboard the "*Zaka,* a 90 foot schooner that belonged to a Dr. Crocker in Los Angeles." See Cortese, *RHRR*, 180. Crocker's unpublished logs, held at the *California Academy of Science* in San Francisco, do not show that a Henry Von Fehren worked aboard the *Zaca*. For Crocker (1884–1948) and his membership in the *Trans-Pacific Yacht Club*, see Soiland, *Trans-Pacific Ocean Races and Trans-Pacific Yacht Club*, op. cit., 126. Crocker appeared in the Bohemian Grove play *Ivanhoe*, which Halliburton attended. My copy of the play (with napkin tucked inside) is signed by Richard Halliburton and dated "August 1, 1936." Within the Halliburton orbit of contacts, Crocker lived near Montgomery Street at 945 Green Street in North Beach. He also had an office nearby at the Shreve Building. See *San Francisco and Bay Counties Telephone Directory*, 93. For Welch on Von Fehren, see *LHK #5*, November 16, 1938. For Halliburton's "diplomacy," see *RHL*, November 21, 1938 (quoted).

3. For Earhart and the *Itasca,* see W. C. Jameson, *Amelia Earhart—Beyond the Grave* (Lanham, MD: Taylor Trade, 2016), 67–68; see 67 for Bendix direction finder and 63–67 for interaction with radio equipment (including interferences).

4. For "radio communications," see *RHL-P,* July 14, 1938 (quoted). For importance to Halliburton of radio communication, see, in addition, *RHL-P ,* July 21, August 1 and November 21. For the hiring of a radio operator (presumably Petrich) said to have participated in the search for Amelia Earhart, see *RHL-P*, August 1, 1938 (quoted). For Petrich, see Halliburton, "Letters from the *SD #2*," January 18, 1939. For his position with the *Matson* line, see *South China Paper,* January 13, 1939. For radio being as good as

a telephone, see *RHL-P*, January 1, 1939 (quoted). Petrich (born in 1904) had a brother named Jesse (1907–2002), a noted seaman and sailor's union activist or "wobbly" who operated mainly in the coastal communities of the Pacific Northwest. The brothers came from a family that had made its fortune in boatbuilding and lumbering. On their own from an early age, they both committed to a life at sea; however, the course of their lives "diverged," and, once adults, they seldom crossed paths. For information about the Petrich family, I am indebted to Jesse's son Captain Wil Petrich. Halliburton's letter to J. Watson Webb in author's collection notes that Petrich was married. "All three of these men," Halliburton told Webb, "have wives and children—and will take no chances." George was married no fewer than five times, and he had at least one child, George Jr., who was killed in a traffic accident in the 1940s. His older brother Jesse, *was* married five times. "Frequent divorce was a common malady of the (sailing) profession." Information on George and Jesse Petrich, courtesy of Jesse's son navigation expert Wil Petrich, March 16, 2016. Records do not confirm that Petrich was the same radio operator who had searched for Amelia Earhart for the Matson Navigation Company or another service. W. C. Jameson notes that the radioman aboard the *Itasca* was Leo G. Bellarts, but George Petrich could have been a member of the radio team. See *Amelia Earhart*, 94. For the *Itasca* search, see Lt. Stephanie Young, "*Itasca* and the Search for Amelia Earhart," online, posted July 2, 2012.

5. Quoted is *RHL-P,* September 10, 1938. About operating the mimeograph machine, or stencil duplicator, manufacturer unknown, Halliburton said the "mechanical problem(s)" involved in cranking at least "70,000 stamped, addressed, and autographed" letters off a cylinder were "terrific." Thinking mastery of the device represented an acquired skill, he told his parents that Mooney initally struggled but ultimately operated the machine "expertly." *RHL-P*, September 28, 1938 (quoted). Often, however, the results were "inexpert." See Halliburton, "Letters from the *SD #2*," January 18, 1939. Familiar with office equipment, Paul had worked as an ad writer for a travel agency where he might have been exposed to printing machines. For the Davises, see Max, *HC*, 168–71. For Paul's comments, see *PML*, to Gerstle Mack, September 12, 1934 (quoted). For Beaman contact, see *RHL-P*, July 6, 1938. The "flying" book was perhaps a reprint edition of the late Martin Johnson's ghosted *Over African Jungles*. Little is known of Mooney's maritime background. For the record, he once filled out an "Application of Seaman's Certificate of American Citizenship," June 3–7, 1924, #65850; the document includes a photograph and thumbprint, and it identified him as five feet, seven inches tall and having a scar on his right hand. For Mooney's unemployment, see *PML*, letter to Gerstle Mack, September 9, 1937 (quoted). Mooney and Gerstle Mack (1894–1983), who was originally from San Francisco, shared an interest in architecture. For the appeal of New Mexico for freelance writers and artists, see Marta Weigle and Kyle Fiore, *Santa Fe and Taos—The Writer's Era, 1916–1941* (Santa Fe, NM: Ancient City, 1982). For Mexican connections, see Max, *HC*, 61–62. Paul was himself a friend to Jaime Martinez del Rio, an ardent Roman Catholic and a leading member of a patrician Mexican family that had lost its wealth during the Mexican Revolution. Del Rio was the first husband of actress Dolores del Rio and the second cousin of film star Ramon Novarro. Del Rio was also an aviation enthusiast and friend of Florence "Pancho" Barnes, herself a friend of Richard Halliburton and Paul Mooney. In December 1928, Del Rio died of blood poisoning in Berlin, Germany; Paul Mooney was among those at his bedside. See the *Washington*

Post, December 8, 1928. See Max, *HC*, 71 passim. Halliburton traveled through Mexico in 1928. See *RHL*, May 1928, 285-86. For "tourist class," see *RHL-P*, September 10, 1938. For Mooney as ghostwriter of Kurt Ludecke's *I Knew Hitler*, see Max, *HC*, 119-28.

6. For occupants of *Hangover House*, see Max, *HC,* 168 passim. For Mooney's comments, see letter to Harriette Janssen ("Hattie"), November 24, 1938 (quoted).

7. For crew "fixed," see *RHL-P*, August 1, 1938 (quoted).

11. The Royal Road to Romance in America

1. For Dr. Chung, see Max, *HC*, 129, 178, 191.

2. For the "Six," see Dobie, *San Francisco's Chinatown*, op. cit., 124 et seq. For "trading instinct" see Halliburton, "Letters from the *SD #3*," January 27, 1939 (quoted). For Dr. Margaret Chung, see Max, *HC*, 178. Along with Halliburton, China Clippers pilot Edwin Musick was "a white celebrity known for his involvement promoting trans-Pacific transportation," and he "helped to legitimate (Chung's) social position with the San Francisco Chinatown." See Judy Tzu-Chun Wu, *Doctor Mom Chung of the Fair-Haired Bastards—The Life of a Wartime Celebrity* (Berkeley: University of California Press, 2005), 132 (quoted). For "connecting the world," see Ben Wilson, *Heyday—The 1850s and the Dawn of the Global Age* (New York: Basic Books, 2016), xi-xii. In 1935, Bobbs-Merrill published Edward Morley Barrows's *The Great Commodore*, an account of Matthew Calbraith Perry, who had opened isolated Japan to world trade through gunboat diplomacy. For "community of aims," see 361 (quoted). Besides reading about Perry, Halliburton read about acclaimed trans-Pacific voyager Nakahama "John" Manjiro (1827-1898), who knew Perry. Manjiro helped the Japanese improve their navy, and he also translated Nathaniel Bowditch's 1802 *American Practical Navigation* into Japanese. Bowditch's work remains a key reference for mariners to this day.

3. A refreshing walk through Union Square took Halliburton into Maiden Lane, where at a favorite diner, the Blue Lagoon Restaurant at address 153, he met with Atherton, Margaret Chung, or Home Insurance Company president Harold V. Smith. Evidently, Halliburton also dined with Atherton at her residence on 2101 California Street. See *RHL-P*, July 6, 1938, and July 21, 1938. For Atherton and "Young Intellectuals," see Emily Wortis Leider, *California's Daughter: Gertrude Atherton and Her Times* (Stanford, CA: Stanford University Press, 1991), 290. For references to Halliburton, see 310 and 316. For Atherton, see Wikipedia. For dining with Atherton, see *RHL-P*, August 1, 1938. For Atherton's activities, see Max, *HC*, 177-79. For Resanov, see Owen Matthews, *Glorious Misadventures—Nikolai Rezanov and the Dream of a Russian America* (New York: Bloomsbury, 2013), and for Wilde, see 322 (quoted). One Harold V. Smith, a possible friend of Halliburton's, lived at 59 Maiden Lane; Hong Kong physician John McElney sent him a photograph of the *Sea Dragon*. The photograph, with its provenance, appeared in the *New York Herald Tribune*, following the legal declaration of Halliburton's death on October 5, 1939. Also on Maiden Lane were Fred Solari's at 19, the Tony Pandy at 51, and La Buvette at 134. In a Western Union telegram dated September 15, 1938, and sent to the Bohemian Grove Club from the Chancellor Hotel, Halliburton invited author Charles Caldwell Dobie to join him at the Blue Lagoon Restaurant "for a cocktail party next Sunday afternoon from 5 to 7 . . . to say goodbye before sailing for China." Charles Caldwell Dobie Papers, Bancroft Library, University of California–Berkeley. Sources

consulted include brochures, letters, and the *San Francisco and Bay Counties Telephone Directory* for May 1939.

4. For *new* book material, see *RHL-P*, August 1, 1938. The "America book," which will be discussed later in greater detail, is noted in *RHL*, June 25, 1938, *His Story*, 398. Halliburton intended to write an adventure story about the *Sea Dragon* Expedition similar to the one he had written about the *Flying Carpet* Expedition. For "book-in-the-making," see *RHL*, June 25, 1938, *His Story*, 398. Also see *RHL*, May 24, 1937, 390 (quoted).

5. For an itinerary of *Royal Road to Romance in America*, see *RHL*, 397–98.

6. When a student at Princeton, Halliburton said he had adopted the philosophy of utilitarian expediency. See *RHL*, November 2, 1920, *His Story*, 65. As early as 1933, Halliburton was talking about what he called "my American book." See *RHL*, November 24, 1933, *His Story*, 348. For "American book," also see *RHL*, March 19, 1936, *His Story*, 377. This "America" or "American" book formed the core of the *Book of Marvels—The Occident*. For the "Royal Road to Romance in the United States" specifically, see editorial notes, *RHL*, 397–98. For Thomas Wolfe, see *The Four Lost Men*, ed. Arlyn Bruccoli and Matthew J. Bruccoli (Columbia: University of South Carolina Press, 2008), xx (quoted).

7. For Halliburton's views, see "Straight Talk from Russia," *Seven League Boots*, op. cit., 189–90. The investigative bodies were the Overman Committee (1919), Fish Committee (1930), the McCormack-Dickstein Committee (1934–1937), and (Martin) Dies Committee (1938). See "House Un-American Activities Committee," in Wikipedia. For Wesley on *U. S. A.* book, see Cortese, *RHRR*, 157 (quoted).

8. For *Cavalcade*, see A. G. Linkletter, *America! Cavalcade of a Nation* (A. L. Vollmann, 1940), 16 pages. Also see *Official Guide Book*, 101 (quoted). Also see Gross, *Famous Guide to San Francisco and the World's Fair*, supra; and Earle Weller and Jack James, *Treasure Island: The Magic City; The Story of the Golden Gate International Exposition, 1939–1940* (San Francisco: Pisani, 1941). For Thomas Hart Benton mural, see Carol Vogel, "Thomas Hart Benton Masterwork Goes to Met," *New York Times*, The Arts, December 12, 2012. Both fairs struck positive notes. Book illustrator and ethnographer Miguel Covarrubias (1904–57) created six murals entitled "The Pageant of the Pacific" that eventually went on tour.

9. For Dale Collins's opinion, see "Royal Road Across the Pacific," 501–12, quoted at 501. For Richard Henry Dana, see *RHL*, July 30, 1921, *His Story*, 82. Halliburton's style recalls not Dana, but Twain. See *RHL*, February 28, 1922, *His Story*, 137.

12. *You Never Die in Your Dreams*

1. For trip, see *RHL-P*, September 28, 1938. For description of Charleston, the source consulted was a letter from Wesley Halliburton to Richard Halliburton, March 24, 1939, RHC, Rhodes. "The azaleas and camellias looked more beautiful than the days before," he wrote, adding, "Even the desolate piney plains looked radiant, and the subject ceased to be 'tabu.'"

2. See *RHL-P*, October 7, 1938 (quoted). According to the January 19, 1939, issue of the *Hong Kong Daily News*, Halliburton said that it would be a "tough crossing." For the rhapsody on junks, see Halliburton, "Letters from the *SD* #1," November 20, 1938. Also see the "first of his Bell stories" ("Log of the *SD* #1," December 12, 1938), in *His Story*, 401 et seq. Much has been written about the SS *Coolidge* online. When war with

380   NOTES TO PAGES 82–86

Japan broke out, the ship helped evacuate American citizens from Hong Kong and other parts of Asia. Converted into a troop convey, it sunk on October 26, 1942, shortly after striking a mine off Espiritu Santo, an island in the remote Pacific region of Melanesia; rescue was timely and no fatalities resulted. For the layout of the SS *Coolidge*, see the *Dollar Steamship Lines* Cabin Plan for the SS *Hoover* and SS *Coolidge*, with photographs and foldout of schematics, 1933. For boats as "new world," see *RHL*, November 21, 1938, *His Story*, 411 (quoted).

3. Another "last trip" was the *Flying Carpet* Expedition. A January 28, 1987, letter from Juliet Halliburton to William Short, curator of Richard Halliburton Archive at Barret Library, Rhodes College, was consulted. Richard had visited Juliet at her home in Greensboro, North Carolina. For the "first breath of tranquility" coming over a full year after that trip, see *RHL-P*, November 12, 1933. For the earlier "last trip," see Christine Sadler, "Halliburton Thinks 'Flying Carpet' Is His Last Romantic Travel Story: Memphian Wants to Stop While Books Have Distinct Flavor; Brooke's Life Next," February 15, 1933, Nashville newspaper, RHC, Rhodes. For assurances, see *RHL-P*, September 10, 1938. Compare edited version in *RHL, His Story*, 400.

4. See *RHL-P*, August 31, 1938 (quoted). For funding, compare Root, *HTMM*, 263–64 (quoted). Also see Alt, *DDIB*, 324 et seq., and 331 et seq. For "27,000" raised, see *RHL-P*, September 10, 1938. For Azores, see *RHL*, July 6, 1938, *My Story*, 398 (quoted). For trip, see *RHL-P*, September 10; also see August 1, August 31, September 4, and September 4. He rejoiced over his parents' stays in Istanbul, Athens, Budapest, Vienna, and Prague, but "their Odessa-Moscow trip" drew his concern. Nazi activities also drew his concern, but Wesley and Nelle Nance did not alter their itinerary. Richard wrote them, and asked, "Was Salzburg all you hoped? Everybody has blacklisted it except the Nazis. Hope you found my letter waiting in Vienna." See *RHL-P*, August 1, 1938. For Halliburton's happiness about their return home, see *RHL-P*, August 31, 1938 (quoted).

5. For exercise routine, see *RHL*, July 21, 1938.

6. See *RHL-P*, August 31, 1938 (quoted). He probably knew Dobie through friend Noel Sullivan, the nephew of late California Senator James Phelan, patriarch of the Villa Montalvo estate. For gatherings ("dinner parties") of twelve or more, including one at a Mrs. Harvey's where they all might have met, see *RHL-P*, June 25, 1938. Captain Welch evidently attended a separate "farewell party." See *LHK*, November 2, 1938. When Nelle Nance formed her impressions of Welch is uncertain. For delays, see *RHL*, August 31, 1938, in *His Story*, 399. For being "swamped," see September 10, 1938; for "rushing around day and night," see July 6, 1938, 399. Also see Root, *HTMM*, 24.

7. See *RHL-P*, July 14, 1938 (quoted).

13. *Columbia, the Gem of the Ocean*

1. For last "10 days" and "mature" captain and engineer, see *RHL-P*, September 28, 1938. For "safe as I can be," see *RHL-P*, August 1, 1938 (quoted). For "last trip," the source consulted was letter from Juliet Halliburton to William Short, January 28, 1987.

2. For developing unrest in Europe, see Cabell Phillips, *From the Crash to the Blitz, 1929–1939—The New York Times Chronicle of American Life* (New York: MacMillan, 1939), 551 et seq. For Chamberlain, see 549. For departure, see *RHL-P*, September 28, 1938. Newspaper clippings of the event are in RHC, Rhodes.

3. See Violet Sweet Haven, *Many Ports of Call* (New York: Longmans, Green, 1940), 204, 206 (quoted).

4. Ibid., 206 (quoted).

5. For departure information, see *RHL-P*, September 28, 1938.

6. For Crowell, see Root, *HTMM*, 22. For letters, stories, etc., see *RHL-P*, September 28, 1938 (quoted). For "Wishing," see Haven, *Many Ports of Call*, 208 (quoted).

7. See Collins, "Royal Road Across the Pacific," op. cit., 501–12, on Welch at 501 (quoted). Collins's later remarks about Welch also appeared in the newspaper; quoted here is a clipping (source cut) from "My Scrapbook, 1939–1940," kept by Halliburton fan Ralph B. Vawter and now in the author's collection. As is so often the case with clippings in scrapbooks, the article's text appears without the date. For Collins knowing Welch, see unpublished letter from Dale E. Collins to Edward T. Howell, May 6, 1946.

8. For Midway Island, see Wikipedia. For distance, see *RHL-P*, September 28, 1938. Dr. E. A. Petersen noted that estimates of the time the crossing would take, by "both Japanese and foreigners," ranged from three to six months. "My guess as to sailing time was under ninety days," he later indicated. See Petersen, *Hummel Hummel* (New York: Vantage, 1952), 56. "Phenomenal runs" by "the tall ships (Yankee clippers) of America and England" from Hong Kong to Boston, New York, and London were noted from the 1844s. See Barrows, *Great Commodore*, op. cit., 199–200; Halliburton was familiar with Barrows' recent book. Halliburton was also conversant with the details of Ferdinand Magellan's circumnavigation. The leg of the journey across the South Pacific from the tip of South America to the Philippines, where Magellan met his death, required ninety-eight days. See Antonio Pigafetta, *Magellan's Voyage—A Narrative Account of the First Circumnavigation*, trans. and ed. R. A. Skelton (New York: Dover), 1; also see map, 32–33. Halliburton figured at this time that ninety days or fewer would be required for the crossing. Incidentally, the *Hummel Hummel* was thirty-three days out from Shanghai before it reached Yokohama. See Petersen, *Hummel Hummel*, op. cit., 44. On July 12, 1938, Dr. Petersen sailed out from Yokohama, aware that five thousand miles separated him from the California coast. See *Hummel Hummel*, 48–49. Some navigators estimated three to six months for the crossing, but Petersen believed that he could accomplish it within ninety days with "the advantage of the favorable Japanese current and the prevailing winds." He ultimately made the voyage in eighty-five days. Halliburton learned of the feat by November, if not October.

9. Disparaging names for Halliburton litter Welch's correspondence. For Halliburton's public persona, see Root, *HTMM*, 126 et seq. For gorgeous design and color scheme, see Welch, *LHK*, October 21 and October 29, 1938. For "a picnic," see *LHK* #2, October 29, 1938.

10. For getting along "beautifully," see *RHL*, November 2, 1938, *His Story*, 410. For friction between Welch and Von Fehren, see *RHL-P*, November 21, 1938. Compare Root, who says Von Fehren was the only one who "did not argue with Welch." Calling him "good-natured and somewhat phlegmatic," Root said that Von Fehren "busied himself with the below-decks installation of the engine whenever Welch went on a rampage." See *HTMM*, 25. Alt suggests that Welch and Von Fehren worked in tandem, Welch "bark(ing)" out orders to Von Fehren to start the engine, and Von Fehren heeding those orders. See *DDIB*, 346–47. As early as October 21, Welch wrote, "I have had to stand the engineer on his feet through interfering, but he is alright now. I did it very nicely so

that there would be no hard feelings." *LHK*, October 21, 1938. For engine and fuel tanks, see Halliburton, "Log of the *SD #12*," March 1, 1939; also in *His Story*, 420 (quoted). Halliburton commented, "In this department (of diesel operation and maintenance), we put our whole trust in Henry Von Fehren." Shipping and storage costs for the diesel engine were probably high. For Paul's remarks, see *PML*, letter to Harriette Janssen, November 24, 1938 (quoted). Halliburton liked the fuel efficiency of the diesel engine, and he preferred the cost of diesel fuel over that of gasoline for a regular gasoline-powered engine. For diesel engines, see Wikipedia. Also see *Engines Network*, "The History of Diesel Engines," online. For the engine aboard the *Snark*, see Jack London, *The Cruise of the Snark* (1911; repr., London: Mills and Boon, n.d.), 28, also 23. For sailors' comments, see 44 (quoted).

11. For "harassments," see Halliburton's letter to "Monica" at the Alber-Wickes agency, October 7, 1938, op. cit. (quoted), Lowell Thomas Collection, Marist College Archives and Special Collections. For reading Atherton autobiography, see *RHL-P*, October 7. For "watchman" episode, see Collins, "Royal Road Across the Pacific," op. cit., 501. For another telling of the story, see Cathryn J. Prince, *American Daredevil: The Extraordinary Life of Richard Halliburton, the World's First Celebrity Travel Writer* (Chicago: Chicago Review Press, 2016), 212.

12. For the *Coolidge's* comforts, see Haven, *Many Ports of Call*, op. cit., 198–207. For Halliburton stargazing, see *Royal Road to Romance*, 395.

13. Haven wrote, "A thousand people would be aboard, meals would be served each day, forty-five thousand pounds of prime meat would be used as the total of a hundred and forty thousand meals which go to make up a round trip to the Orient." A crane brought hundreds of crates of fruits and vegetables, amounting to tons of food, aboard the vessel. See *Many Ports of Call*, op. cit., 204–5, quoted at 204. For the particulars, a number of menus from the SS *Coolidge* for the years 1938 through 1940 were consulted. Author's collection.

14. For "hot sun and calm seas," see *RHL-P*, October 7, 1938. For *The Summing Up*, a book Halliburton recommended that his mother read and that he himself intended to read on his trip to China, see *RHL-P*, June 25, 1938. For being in Hawaii, see *Honolulu Star-Bulletin*, September 28, 1938 (Wednesday). For Kanaka boys, see Haven, *Many Ports of Call*, supra, 210 (quoted). For the Brooke poem "Waikiki", see Rupert Brooke, *The Collected Poems, with a Memoir by Sir Edward Marsh*, 3rd rev. ed. (London: Sidgwick and Jackson, 1942), 126 (quoted). For interviewing, see Haven, *Many Ports of Call*, supra, 211. Halliburton might have met up with island poet laureate Don Blanding and renowned surfboarder Duke Kahanamoku, both friends of his. For Duke Kahanamoku (1890–1968), see Williams, *Forgotten Adventures of Richard Halliburton*, op. cit., 120–21. For Kahanamoku as a member of the Trans-Pacific Yacht Club, see Soiland, *Trans-Pacific Ocean Races and Trans-Pacific Yacht Club*, op. cit., 203. For Don Blanding, see Root, *HTMM*, 137. For "perfectly dull," see *RHL-P*, October 7, 1938. Photograph of team with Petrich is in RHC, Rhodes.

14. *"Japanese, If You Please!"*

1. Quoted is letter to "Monica" at Alber-Wickes agency, October 7, 1938, Lowell Thomas Collection, Marist College Archives and Special Collections.

2. See Nellie Bly, *Around the World in Seventy-Two Days and Other Writings*, reprinted with a foreword by Maureen Corrigan (New York: Penguin Books, 2014), 260 (quoted), 261 (quoted), 263 (quoted).

3. See Haven, *Many Ports of Call*, op. cit., 222–23 (quoted). Also see *RHL-P*, October 7, 1938 (quoted). Cf. *RHL, His Story*, 408. For "stay," see Halliburton, "Log of the *SD* #3," December 26, 1938 (quoted). Halliburton, *Royal Road to Romance* (1925) recounts the 1923 Japan trip. Also see letters of January 4, January 9, January 23, January 29, and February 13, 1923, *His Story*, 201–7. The third log reappeared in a syndicated version. According to the editor of the *Utica (NY) Observer Dispatch*, the text of a later edition of the piece comes from the second of a "series of articles prepared by Richard Halliburton," and it appeared in the *Observer Dispatch* on April 9, 1939, two weeks after Halliburton's last message from the *Sea Dragon*. The *Utica* version combined "Log of the *SD* #3" (above) and "Log of the *SD* #4," December 27, 1938.

4. For the Japanese sojourn, see "Author of 'Glorious Adventure' Here (Yokohama) to Make Pacific Crossing," *Japan Times & Mail*, October 7 or (Saturday), 1938, (quoted). Other noted passengers cited were George Brant, C. F. Cress, S. A. Stolaroff, and Henry F. Kay. Brant might have been Irish-born actor George Brent (1904–1979), who before coming to Hollywood and making a number of appearances as a leading man opposite Bette Davis, was a member of the *IRA Active Service*. Brent boasted that he had acted as a runner for service founder Michael Collins. Copies of documents are in the RHC, Rhodes.

5. For information given to reporters, see *Japan Times*. For arrival times (as Halliburton understood them), see *RHL-P*, October 7, 1938. For the group dividing, see *RHL-P*, October 7, 1938 (quoted). For plan to spend just five days in Japan, see *RHL-P*, September 10, 1938.

6. For assurances from Japanese, see *RHL-P*, October 7, 1938. At the time, Ambassador Grew (1880–1965) figured regularly in news pertinent to America's relations with Japan. Clairvoyant, Grew predicted the attack on Pearl Harbor months before it happened. What he might not have known is that he would be interned in Tokyo for nearly a year before his release back to the US. Unable to rescue himself, he had come came to Halliburton's rescue earlier in 1938; either he or his office enabled the writer to procure the needed documents. For news given to reporters, see *Japan Times*, supra. For Anita Grew, see Root, *HTMM*, 157 (quoted).

7. Popular when it hit bookstores, and available to Halliburton (and Mooney), Irving Stone's *Sailor on Horseback—The Biography of Jack London* covered the adventures of the *Snark* (New York: Houghton-Mifflin, 1938). See bibliography. For Jack London in Japan, see Earle Labor, *Jack London—An American Life* (New York: Farrar, Straus and Giroux, 2013), 196–99. For Haven's outlook on Japan, see Violet Sweet Haven, *Gentlemen of Japan—A Study of Rapist Diplomacy* (Chicago: Ziff-Davis, 1944), sleeve (quoted).

8. For a portrait of Japan (and its vast network of islands and island dependencies) and the Japanese, "said to be warlike, cultured and wealthy," and "an enigma," see Barrows, *Great Commodore*, op. cit., 230 et seq. For atrocities, see James Yin, Ron Dorman, and Young Shi, *The Rape of Nanking—An Undeniable History in Photographs* (Chicago: Innovative, 1996). Also see Don Tow's website, "Massacre and Atrocities in Hong Kong During World War II," April 2007, online; "Why Does No One Ever Mention the Atrocities Committed by the British Empire?" online; George Monbiot, "How Britain Denies Its Holocausts," April 23, 2012, online. Cf. Frank Welch, *A Borrowed Place—The History of*

Hong Kong (Kodansha America, 1993), 407. For "little evidence" of Japan in "great war" by its people, etc., and for initial reactions to the Japanese, see Halliburton, "Letters from the *SD #1*," November 20, 1938. Compare Haven, op. cit., 222–28, quoted at 222.

9. For "little evidence," see Halliburton, "Letters from the *SD #1*," November 20, 1938.

10. See Halliburton, "Log of the *SD #3*," December 26, 1938 (quoted). For contrast of rural and metropolitan Japan, compare Harold Butcher, "The Pittsburgh of Japan (Osaka)," *Travel*, August 1928, 18–24, at 19. For earlier visit, see *RHL*, May 30, 1932, *His Story*, 338–39. For changes in Japan since the 1932 visit, compare Prince, *American Daredevil*, 213 et seq.

11. For "little man with sword" and for writers, see Halliburton, "Log of the *SD #3*," December 26, 1938 (quoted).

12. Ibid. (quoted).

13. Ibid. (quoted). For the Gibraltar incident, see Root, *HTMM*, 74–76. Says Root, "He was a journalist," Richard told authorities, "and he had taken pictures merely to illustrate a magazine article. He knew photographs were forbidden, but because there were so many camera shops near the fortress he assumed the law was not enforced" (76). On September 10, 1938, Halliburton told his parents that he was "seeing the Jap Navy"; this statement is not found elsewhere. See *RHL-P*. For German flag, see Petersen, *Hummel Hummel*, op. cit., 30.

14. See *PML*, copy of letter to Harriette Janssen sent to Gerstle Mack, November 24, 1938 (quoted). For "bottom of the China Sea," see Halliburton, "Log of the *SD #3*," December 26, 1938.

15. Pictures of a Floating World

1. For reactions to the Japanese, including reaction to "rumors," see Halliburton, "Letters from the *SD #1*," November 20, 1938 (quoted). Compare Haven, op. cit., 222–28, quoted at 222. For "dazzling display," "neon signs," and "Tokyo after nightfall," see "Letters from the *SD #3*," January 27, 1939 (quoted). (6. See Welch, *LHK #2*, October 29, 1938 (quoted). See *RHL*, November 2, 1938, *His Story*, 411 (quoted). Also see Halliburton, "Log of the *SD #2*," December 13, 1938, *Utica (NY) Observer Dispatch*, April 9, 1939 (quoted).

2. For "September crisis," see Halliburton, "Log of the *SD #9*," January 31, 1939, *SFN*, (quoted). For prohibitions, see "Log of the *SD #3*," December 26, 1938, quoted. For "national shrines" permission, see *PML*, letter to Alice Padgett, December 31, 1938 (quoted). For Alice M. Padgett, see Max, *HC*, 24–225. Also see Cortese, *RHRR*, 167–68. For cinematized views of the air strikes, see *Unbreakable Spirit* (released August 2018), about the Japanese bombing raids on Chingqing from 1938 to 1942. Directed by Xiao Feng, the film features Bruce Willis and Adrien Brody.

3. Ibid., Halliburton, "Log of the *SD #3*," December 26, 1938 (quoted).

4. Ibid., Halliburton, "Log of the *SD #3*," December 26, 1938 (quoted). "Log of the *SD #4*," December 27, 1938, *SFN* (quoted). For an earlier run-in for spying (at Gibraltar), see *His Story*, January 9, 1922, 119–20.

5. Ibid., Halliburton, "Log of the *SD #4*," December 27, 1938 (quoted).

6. Quote is from George Washington's 1796 Farewell Address. For some persons of note Halliburton met, see Root, *HTMM*, 73–74, 153, 232, 236.

7. See *PML*, December 31, 1938, quoted. For Halliburton liking his "few days" in

Japan, see *RHL*, November 2, 1938. For thirty-day prediction, see *RHL-P*, October 7, 1938 (quoted). For "last thing he saw," see Halliburton, "Log of the *SD* #4," December 27, 1938 (quoted).

16. Shanghai Pen Pal

1. For "friends" accompanying him, see Halliburton, "Log of the *SD* #1," December 12, 1938, *SFN*. For accompanying parties *"from* Shanghai," see Halliburton, "Letters from the *SD* #2," January 18, 1939.

2. Lieutenant (also Chief Officer) Dale Collins of the SS *Coolidge* reported that Halliburton and Welch intended to buy supplies in Japan, get the necessary transit papers, and proceed directly to Hong Kong on a British ship. In Hong Kong, Collins found them in "high spirits and eager to start negotiations for a junk that would meet their requirements." Changes of plan often occurred, and it is not unlikely that Halliburton and Mooney reunited with Welch at Kobe or a port on the China coast, maybe Shanghai or Swatou. Welch said nothing about the junk search Halliburton and Mooney conducted. Nor did he mention his own arrival in Hong Kong. Halliburton noted that he arrived in Hong Kong on a *passenger* ship; the *Suiang* was both a passenger and a cargo ship, but it was not German operated. For "bout of dysentery," see Welch, *LHK* #1, October 24, 1938.

3. For dangers, see Powell, *My Twenty-Five Years in China*, 294 et seq. For USS *Augusta* and "settlements," see Halliburton, "Log of the *SD* #6," January 24, 1939, *SFN*. For political situation in Shanghai, see Abend, *My Life in China,* op. cit., 286–293.

4. For print culture in Shanghai, just before Halliburton's arrival, see Paul Hutchinson, "New China and the Printed Page," *National Geographic Magazine,* June 1927, 687–722. For Ruth Harkness, see, initially, Wikipedia. Harkness' journey prefaced a series of events in the Halliburton story that will be discussed later. Halliburton's own written reports of a devastated China complement those of Dutch businessman and journalist Karel Frederik Mulder (1901-1978), whose photographs and letters record the same sights and events. Like Mulder and others in that war zone, Halliburton risked his life. For Mulder, see "Timeline" (for Karel Frederik Mulder [1901–1978]), *International Institute for Asian Studies* (58 Autumn 2011). For political forces, see Betty Peh-T'i Wei, *Old Shanghai* (Hong Kong: Oxford University Press, 1993), 27–28; for banks, see 28–29. For Shanghai as a "wasteland," see *RHL-P,* November 2, 1938.

5. Twenty years earlier, Halliburton had written comparably about the ravages of war. For his touring ruined cities in 1919, see *RHL*, November 3, 1919, *His Story*, 46–48. In Shanghai, a place he found "as Chinese as Pittsburgh," he was once offered, curiously, an editorial position for a *semi-Chinese* newspaper. See *RHL*, November 27, 1922, *His Story*, 195–96.

6. For one view of Shanghai at this time, see Stella Dong, *Shanghai—The Rise and Fall of a Decadent City* (New York: William Morrow and Sons, 2000), 1–2 (quoted). For maps of the districts, see pages after xi; also see photograph of Bund, title page. Compare Powell, *My Twenty-Five Years in China*, op. cit., 299–300. For "Paris of the Orient," Petersen, *Hummel Hummel,* op. cit., 15 (quoted). For *Paris*, see Halliburton, *Book of Marvels—The Occident*, op. cit., 58 (quoted). For "fine new road" and other comments, see Halliburton, "Log of the *SD* #5," January 23, 1939 (Monday), *SFN* (quoted). For recurrent hostilities, consulted was "Chinese Fall Back After Savage Battle," *Oregon Daily Journal*, September 13, 1937, front page. For Shanghai several years before

Halliburton's arrival, see George F. Pierrot, *The Vagabond Trail: Around the World in 100 Days*, with a foreword by Lowell Thomas (New York: D. Appleton–Century Company, 1935), 161–87. For Halliburton's earlier visit to Shanghai, where he stayed at the YMCA ("as usual"), see *RHL*, November 27, 1922, *His Story*, 195–96.

7. For buildings, see Halliburton, "Log of the *SD* #5," January 23, 1939 (quoted). For Shanghai impressions, also see "Log of the *SD* #6," January 24, 1939. For French Concession and International Settlement, see Peh-T'i Wei, *Old Shanghai*, supra, 11. For maps of the districts, see Dong, *Shanghai*, op. cit., after xi; also see photograph of Bund, title page. Compare Powell, *My Twenty-five Years in China,* op. cit., 299–300. For Butterfield and Swire, and companies in the district, see Peh-T'i Weh, *Old Shanghai*, 22–23.

8. For Ruan, see Richard J. Meyer, *Ruan Ling-Yu—The Goddess of Shanghai*, with DVD (Hong Kong: Hong Kong University Press, 2005). For her death, see 1 et seq. For the Russians, compare Powell, *My Twenty-Five Years in China*, op. cit., 51 et seq. For discovery of American cinema, see Peh-T'i Weh, *Old Shanghai*, 42 (quoted). For Astor House Hotel, see Powell, supra, 7 passim. For Majestic Theatre, see Peh-T'i Weh, *Old Shanghai*, 43. For Dame Margot Fonteyn and others, see 32–33.

9. For Jewish community and religious organizations, see Peh-T'i Weh, *Old Shanghai*, supra, 45–52. Halliburton's views are expressed in "Log of the *SD* #5," January 23, 1939; and "Log of the *SD* #6," January 24, 1939.

10. For "foreigners" and "midpoint," see caption to Halliburton, "Log of the *SD* #7," January 25, 1939, supra (quoted). Japanese troops took control of the International Settlement on December 8, 1941. For McGovern, see *RHL*, December 4, 1931, *His Story*, 330. For details of settlement, see Powell, *My Twenty-Five Years in China*, 325–29. Chapei, later Zhabei, merged with the Jing'an District in 2015, and its forty-plus square miles now constitute the principal interior hub of modern Shanghai.

11. For bleakness of the city, see Halliburton, "Log of the *SD* #6," January 24, 1939 (Tuesday), *SFN* (quoted). Cf. Powell, *My Twenty-Five Years in China*, 325–26. For US Marines' presence, consulted was "Japan Grab in Shanghai Halted By U.S. Troops," *Los Angeles Evening Herald Express*, front page, December 3, 1937.

12. Ibid., "Log of the *SD* #6," January 24, 1939, *SFN* (quoted).

13. For Chinese terrorists, see *SFN*, May 11, 1939.

14. For nightlife, see Halliburton, "Log of the *SD* #7," January 25, 1939, *SFN* (quoted). US-British film *The White Countess* (A Merchant-Ivory Production, 2005) suggests the Shanghai that Halliburton experienced. The movie's Russian countess Sofia Belinskya (Natasha Richardson) and former US State Department official Todd Jackson (Ralph Fiennes) are united as the result of tragedies and the war. In *India Speaks*, the cantina scenes of Halliburton seated with friends suggest his mild-mannered demeanor at this and other nightclubs. For dance halls, compare Wells, *My Candle at Both Ends*, op. cit., 109.

15. For "die now" attitude, see *PML*, letter to Alice Padgett, December 31, 1938 (quoted). For brawl, see Halliburton, "Log of the *SD* #7," January 25, 1939 (quoted).

17. Toward the South China Sea

1. For the junk search, see Halliburton, "Log of the *SD* #4," December 27, 1938; and "Log of the *SD* #5," January 23, 1939.

2. See Halliburton, "Log of the *SD* #8," January 30, 1939 (Monday), *SFN* (quoted). For full detail of purchases, see *RHL-P*, November 11, 1938.

3. For Nicholas and the second junk, see Halliburton, "Log of the *SD* #12," March 1, 1939 (Wednesday), *SFN* (quoted). For "finest junks," see "Log of the *SD* #4," December 27, 1938 (quoted). For Ningpo style of junk, see Robert F. Fitch, "Life Afloat In China—Tens of Thousands of Chinese in Congested Ports Spend Their Entire Existence on Boats," *National Geographic Magazine*, June 1927, 665–86, at 683 (pictured). Consulted for Ningpo and Ningpo-style junks was Wolfgang Asbach, *"The Hangzhou Trade—A 1/30 Scale Model of a Trading Junk from the Shanghai Area, Model Shipbuilder,"* No. 81–84, 1992-1993 (Mariner's Museum Library), 3–11.

4. For costs, and size of junk required for a diesel engine, see Collins, "Royal Road Across the Pacific," op. cit., quoted at 502. Appreciated would be a better documented account of the junk search down the Chinese coast. Besides Halliburton's general remarks, there are the equally general remarks of Collins and Welch. Welch evidently wrote several letters to Wetjen from Yokohama, at least one from Shanghai, and another from Swatow; this correspondence is presumed lost. "The skipper of the 'Suiang' that I came south from Shanghai on (and who thought a trans-Pacific junk voyage foolish) told me that I could ship out any time on one of their ships but with only two mates; I can't see it at any price." On October 21 or October 24 (the dates of his first surviving letter to Wetjen), Welch said he had been in Hong Kong a week, which would confirm Halliburton's report that his captain had gone directly to the Colony (without first stopping on the Chinese mainland at Shanghai and Swatou). "Came here on a limey ship called the 'Suiang,'" Welch himself wrote. "Good skipper: said I was a goddam fool (to sail a junk across the Pacific) but gave me a drink or two every night on the way." But did he sail directly to Hong Kong? Welch mentioned his stop in Shanghai (*LHK* #1, October 24, 1938), as well as his stop in Swatow (November 2, 1938). For the junk price increasing to $5,000, see *RHL-P*, November 21, 1938.

5. For the purchase of a junk in Amoy, the *Japan Times* (evidently) for October 8, 1938, was consulted. The newspaper noted Halliburton's intentions: "[He will] purchase a junk (there) which will be equipped with a radio and an auxiliary engine, the latter shall be used only in time of calm weather or going in and out of port." Copy of article in RHC, Rhodes. For the story of the *Amoy*, see Halliburton, "Log of the *SD* #2," December 13, 1938, *SFN* (quoted). For Foochow junks, see Van Tilburg, *Chinese Junks on the Pacific* op. cit., 82–90; for other junks and their characteristics, see this work's index. For "taste of the war," see Halliburton, "Letters from the *SD* #1," November 20, 1938.

6. For beauties of the region, see Graham S. P. Heywood, *Rambles in Hong Kong*, with a new introduction and commentary by Richard Gee (Hong Kong: Oxford University Press, 1992), 40 et passim. For likeness to Hebrides, see 4 (quoted); for Mirs Bay, 92 (quoted); for wild region, 84; for birds, 92 passim; for Hang Hou, 70 and 79; for crowded beach, 70; for flora, 86.

7. For "strangling blockade" and "most interesting cities," see Halliburton, "Log of the *SD* #4," December 27, 1938. For "back door home" and "goal," see *RHL*, November 15, 1922, *His Story*, 192 (quoted).

18. See Hong Kong, the Riviera of the Orient, and Die

1. See George Ashmore Fitch (1883–after 1966), *My Eighty Years in China* (Taipei, Tawan: Mei Ya, 1967, privately printed by the author. Also see Fitch, "Life Afloat In China," op. cit. While Peking (Beijing) at this time had no appreciable skyline, Shanghai had a metropolitan skyline that resembled that of New York or Chicago. Hong Kong had the peaks as its skyline. Also see "Hong Kong—Britain's Far-Flung Outpost in China," with 16 illustrations, *National Geographic Magazine,* March 1938, 349–60. For slum clearance, see Nigel Cameron, *Illustrated History of Hong Kong* (Oxford: Oxford University Press, 1991), 234. The street and harbor scenes of postwar movies *Soldier of Fortune* (1955) and *The World of Suzie Wong* (1960) little predict contemporary Hong Kong with its glitter of neon signs, glass steel-framed high-rises, shopping malls, moving walkways, and world-renowned escalator system, but recall rather the Hong Kong of 1938.

2. See Elizabeth Bisland, *In Seven Stages: A Flying Trip around the World* (1891; repr., United Kingdom: Dodo, n.d.). For Bisland's description, see 42; for streets and walks, 45; for climate, 48. See Bly, *Around the World*, op. cit., 260 (quoted); for Japan, 261 (quoted); for "Hong Kong," 260 (quoted).

3. For the look of the Colony in the 1930s, period business cards, postcards, photographs, travel brochures, and maps were consulted. Also see "Hong Kong—Britain's Far-Flung Outpost." For "ambulatory merchants," see *Travel*, August 1928, quoted at 6 (with photograph of street traffic).

4. Business cards, postcards, photographs, and maps in the author's collection and dating to the 1930s supplement the literary descriptions of the Colony. Also see "Hong Kong—Britain's Far-Flung Outpost." For "ambulatory merchants," see *Travel*, August 1928, quoted at 6 (with photograph of street traffic). For Japan's "mixed feeling," see Halliburton, "Log of the *SD* #4," December 27, 1938 (quoted).

5. For spying, see Philip Snow, *The Fall of Hong Kong—Britain, China and the Japanese Occupation* (New Haven: Yale University Press, 2003), especially at 36 passim. For espionage and its history, see David Kahn, *The Reader of Gentlemen's Mail: Herbert O. Yardley and the Birth of American Intelligence* (New Haven: Yale University Press, 2004). Halliburton knew that Oxford-educated diplomat, Arabist, and famed travel writer Gertrude Bell (1868–1926) worked as a spy for British intelligence in the Middle East. For Nazis in the Colony, compare John Rowland, *Slipping the Lines—Adventures around the World in Peace and War* (North Battleford, SK, Canada: Turner-Warwick, 1993), 245. Also see *The Many Lives of Herbert O(sborn) Yardley*, Unclassified, online. Besides *The American Black Chamber*, Yardley (1889–1958) wrote *The Chinese Black Chamber—An Adventure in Espionage* (London: New English Library, 1983), covering his years in Chungking. He also wrote a series of espionage thrillers. These included *The Blonde Countess*, a work about German intelligence prying open US secret codes during World War I that was later made into the film *Rendezvous* (1935) starring Rosalind Russell. For the Office of Strategic Services (wartime intelligence agency), see Wikipedia.

6. A main source consulted here is *Information for Travellers Landing at Hong Kong* (Thomas Cook and Son, 1919); the guide includes steamship and railway schedules, consulate and hotel addresses, site descriptions, itineraries, maps, and photographs of streets. The guide also contains information on Kowloon, Canton, Macao, Swatou, Amoy, Foochow, etc. For China and its culture, see, for instance, Frank Johnson Goodnow, "The Geography of China—The Influence of Physical Environment on the History and Char-

acter of the Chinese People," *National Geographic Magazine*, June 1927, 651–64. Guides
were available to servicemen stationed here. Among the best extended introductions
is S. H. Peplow and M. Barker's now-rare *Hong Kong About and Around*, 2nd ed. with
map (Hong Kong: Commercial, 1931), featuring a convenient foldout map and covering every topic from reckless motorists to fire brigades, religious rites and adoptions,
pirates, and markets. For the history of Hong Kong, Halliburton possibly consulted
Geoffrey Robley Sayer's *Hong Kong, 1841–1867: Birth, Adolescence and Coming of Age*
(Hong Kong: Hong Kong University Press, 1937). For earlier visit to Macao, see *RHL*,
November 21, 1922, *His Story*, 195.

7. Ibid., *Information for Travelers*. See Halliburton, "Log of the *SD* #8," January 30,
1939, *SFN* (quoted). For "thumbing their noses," see "Log of the *SD* #4," December 27,
1938, supra. The *Sui An* might have been the *Suiang*, likely overhauled, which had taken
Captain Welch down China's coast.

8. Ibid., Halliburton, "Log of the *SD* #4," December 27, 1938 (quoted). See Welch,
LHK #1, October 24, 1938 (quoted).

9. For "magnificent junk," see Halliburton, "Log of the *SD* #3," December 26, 1938;
for "carved joss shrine," see "Log of the *SD* #4," December 27, 1938 (quoted); for "better
luck," see "Log of the *SD* #8," January 30, 1939. For "nine thousand max," see Welch,
LHK #1, October 24, 1938.

10. Identification of this Kowloon Hotel as the Peninsula "Kowloon" Hotel, or the
seven-story cube-shaped Kowloon Hotel on 2 Hankow Road, is likely incorrect. A bill
of sale on parchment pasted inside a twenty-four-dollar teak- and camphorwood chest
Paul Mooney purchased in December 1938 from one Kwan Tung Cheong bears the
Kowloon Hotel address, 508 Canton Road. See Max, *HC*, 229. For the Oriental arts-and-
crafts shops, see *Another Hong Kong* (Hong Kong: Emphasis Limited, 1999), after 255.
Cat Street was a mecca for curio hunters like Paul; see *Another Hong Kong*, 257. At the
Jade Market, Mooney purchased figurines of a fisherman, a philosopher, and a dragon,
and he sent these and other items to his mother in Washington, DC, probably in the
chest. Richard himself made a number of purchases for his parents and friends: little
black lacquer tables, china horses, four elephants, a "Mai Jjong" game, two ivory heads
with ebony stands, six lacquer cocktail glasses, and set of blue horses. All these items,
including the five junk models, cost thirty dollars, and they might also have filled the
chest. See Max, *HC*, 229–30. Mooney's friend William Alexander owned the chest until
his death in 1997. A descendant of Alexander's now owns it. Near the wharves, the
hotel was a short walk to the ferry that took Halliburton and company across Victoria
Harbor to Hong Kong Island and to more wharves. For the terminal station, see *National
Geographic Magazine*, March 1938, op. cit., pictured opp. 360. Long rows of commuter
vehicles, storehouses, wharves, and docks then fronted the harbor; today a ferry from
the same station crosses the harbor from Kowloon to Hong Kong. While parking lots
and storehouses are now gone from the area, rebuilt, reconditioned wharves and docks
remain. For a description of hotel, see Sinclair, *Bright Paths to Adventure*, 93 (quoted).
Compare *RHL-P*, November 11, 1938 (quoted). For Welch's description of the hotel, see
LHK #1, October 24, 1938 (quoted). For Mooney's remarks, see *PML*, letter to Harriette
Janssen, November 24, 1938 (quoted).

11. "For "colonial imitation," see Halliburton, "Log of the *SD* #4," December 27,
1938 (quoted). For "commercial transportation companies," see Van Tilburg, *Chinese
Junks on the Pacific*, op. cit., quoted, 172. See "Log of the *SD* #10," February 1, 1939, *SFN*

(caption): "Hong Kong started out in 1840 to be an English town. But the Chinese, seeking security, soon flocked in. Today 90 percent of the streets and houses are Chinese, and 99 percent of the population. In such streets as this there is not a sign of European influence and not a word of English is heard."

12. See *PML,* copy of letter to Harriette Janssen sent to Gerstle Mack, op. cit., November 24, 1938 (quoted). For refugees, compare Rowland, *Slipping the Lines,* op. cit., 243.

13. For legislature, see Powell, *My Twenty-Five Years in China,* op. cit., 69 (quoted). For law enforcement, see Cameron, *Illustrated History of Hong Kong,* op. cit., 49, 245. For Halliburton remark, see "Log of the *SD* #13," March 2, 1939, *SFN* (quoted). For cargo, see Haven, *Many Ports of Call,* op. cit., 216 (quoted). Also see Wilson, *Heyday,* 246-47. For beginnings of the protection racket, see 246.

14. For schools, see Cameron, *Illustrated History of Hong Kong,* supra, 242 (quoted) passim. For hospitals, see Cameron, 243 passim. Halliburton arrived at a time of changing attitudes toward the medical community. After 1920, writes medical historian Paul Unschuld, "the anachronistic attempt to combine a primitive variety of modern healing with Christian dogma and thereby render it useful for missions was replaced by the progressive alliance of the newest medical achievements with modern science, which for an untold number of Chinese signaled the philosophy of the future." Contributions from secular agencies such as the Rockefeller Foundation helped to build hospitals that "met the highest contemporary standards." But watchers still hoped that a synthesis of Western and Chinese medical practices would one day emerge. See Paul U. Unschuld, *Medicine in China—A History of Ideas* (Berkeley: University of California Press, 1985), 241-42, quoted at 242; for Western and Chinese medical philosophies, see 243 et seq. For Dartmouth, see Sykes, op. cit., 92 et seq.

15. For banks and banking, see Cameron, *Illustrated History of Hong Kong,* supra, 238 (quoted) passim. Also see Rowland, *Slipping the Lines,* op. cit., 243.

16. For transportation, see Cameron, *Illustrated History of Hong Kong,* supra, 235 (quoted) passim. For airplane service, see *Wings Over Hong Kong: An Aviation History; 1891-1998: A Tribute to Kai Tak* (Hong Kong: Pacific Century Publishers, 1998). Also consulted was *Information for Travellers,* supra.

17. For the "great typhoon" see Anthony Tully, with Bob Hackett and Sander Kingsepp, "The Great Hong Kong Typhoon, September, 1937," online. Also see S. Campbell (Hong Kong University of Science and Technology), "Typhoons Affecting Hong Kong—Case Studies," April 2005, *1937 Typhoon* at 51-55, online. For typhoon of 1906, see Cameron, *Illustrated History of Hong Kong,* op. cit., 209-10.

19. Rats, Lice, Morphine, and Misery

1. See *RHL,* November 2, 1938 (quoted). Full quote: "Hong Kong is a beautiful and interesting town—in fact I have no troubles of any sort." Sent to his "family" from the "American Consulate, Hong Kong." For Welch calling Hong Kong a "goddamn place," see *LHK* #1, October 24, 1938; for "lime hole" and "movies," see *LHK* #4, November 2, 1938. For appraisals of Shanghai and Hong Kong, see *LHK* #5, November 16, 1938 (quoted). Compare Leys, *After You, Magellan!,* (New York, The Century Company, 1927), 206. For "Coronas," see *LHK* #2, October 29, 1938. Also see *LHK* #7, November 25, 1938. For dislike of Orient, see *LHK* #1, October 24, 1938 (quoted). For "ricey," see *LHK* #4, November 2, 1938. For "good lays," see *LHK* #4, November 2, 1938, quoted. For city of

"refugees," see *RHL*, November 2, 1938, *His Story*, 410. For Mooney's opinion, see *PML*, December 31, 1938, letter to Alice Padgett (quoted).

2. For perceptions of cleanliness in the United States and China, compare Harry Franck, *Marco Polo, Junior—The True Story of an Imaginary American Boy's Travel-Adventures All Over China* (New York: Century, 1929), 44 et seq. For "brackish" water, see Halliburton, "Log of the *SD #9*," January 31, 1939, *SFN*. For food served to the European taste, see Leys, supra, 242.

3. For the history of disease, recommended is Kenneth F. Kiple, ed., *The Cambridge World History of Human Diseases* (Cambridge: Cambridge University Press, 1993). The following sources were also consulted: John Aberth, *The First Horsemen—Disease in Human History* (New York: Prentice Hall, 2007); R. S. Bray, *Armies of Pestilence—The Impact of Disease on History* (Cambridge, UK: Lutterworth, 1996); William H. McNeill, *Plagues and People* (New York: Anchor Doubleday, 1977). For "dysentery," see Welch, *LHK* #1, October 24, 1938. See *LHK*, November 2, 1938.

4. The Mexican and South American trips are covered in *RHL*, "Adventuring in New Worlds," 285–302. See Halliburton, "Log of the *SD #9*," January 31, 1939, *SFN* (quoted). For "insouciance," see Root, *HTMM*, 16. For these adventures, see, besides Halliburton's own books, Root, *HTMM*, beginning with the Panama Canal swim and its challenges to his health, 146 *et seq.*

5. For roughing it, see, for instance, the Devil's Island experience as recounted in Root, *HTMM*, 161–64.

6. Dr. Petersen was an osteopath. For the history and practice of osteopathy, see Wikipedia. For Halliburton's remarks on aging, see *Seven League Boots*, 207. *Zheng* is a Chinese catchall term for any malfunction of the body, and Halliburton might have heard it said. For Chinese medicine, see Unschuld, *Medicine in China*. For "scabies," see Welch, *LHK* #14, February 13, 1939. For Halliburton's use of medicines, see *RHL*, January 15, 1936, 375. For Macao, see *LHK*, November 2, 1938. For inoculations, see Yardley, *Chinese Black Chamber*, op. cit., 80.

7. For "dysentery" and "all" (crew members) having it, see Welch, *LHK* #1, October 24, 1938 (quoted). For Welch being vaccinated and inoculated, see *LHK*, November 2, 1938.

8. For "Dardanella," see Welch, *LHK*, November 2 1938, (quoted). For "a drink and a yarn," see *LHK* #2, October 29, 1938. For "universal gesture," see *LHK* #1, October 24, 1938 (quoted); for "tram," see *LHK* #5, November 16, 1938 (quoted). For Anderson, see *LHK* #1, October 24, 1938, and *LHK* #2, October 29, 1938. A significant contact, this Captain Anderson was almost certainly one-half of Anderson and Ashe of the American Bureau of Shipping, Marine Surveyors and Consulting Engineers. For squadron leaders, see *LHK* #14, February 13, 1939. For "Y" and hobnobbing with "limeys," see *LHK* #2, October 29, 1938.

9. For "scandalous," see *PML*, letter to Alice Padgett, December 31, 1938 (quoted). For the pervasiveness of opium and its use by all classes, see, for instance, Bly, *Around the World*, supra, 249, 254. According to his friend William Alexander (1909–1997), Paul referred to his father as a field anthropologist who got in trouble with his employers at the Smithsonian because he "got high with the Indians." Author's interview. For Paul's remarks, see *PML*, to Alice Padgett, December 31, 1938 (quoted). For James Mooney, see Lester George Moses, *The Indian Man: A Biography of James Mooney* (Urbana: University of Illinois Press, 1984). In general, see Peter Andreas, *Smuggler Nation—How Illicit*

Trade Made America (Oxford: Oxford University Press, 1913). For *Harrison Narcotic Act,* see 260; for rising concern about drug use in America, see 254–62, at 256; for early drug use legislation in San Francisco, see 257–58. For opium (one dollar a pipeful in Chinese currency) and its uses and abuses, see Powell, *My Twenty-Five Years in China,* op. cit., 285–86. Powell recalled smokers "filling (their) lungs with the sickeningly sweet fumes of opium"; see *My Twenty-Five Years in China,* 287. For drug use, see 285–86. For a discussion of the issue contemporary to Halliburton, consulted was H. G. W. Woodhead, "Morphia and Opium," in *Adventures in Far Eastern Journalism,* op. cit., 75–85. See also Hendrik DeLeeuw's *Flower of Joy,* op. cit. The once-popular book, which devotes a chapter to Hong Kong, is a sensationalist reporting of "the hell-holes of five continents" (sleeve). Also see James Holland, "World War Speed," *Secrets of the Dead* (Thirteen—Media With Impact, 2019), DVD.

10. Cf. Gene Gleason, *Tales of Hong Kong* (New York: Roy, 1967), 49. Compare Unschuld, *Medicine in China,* 99 passim. See Berthold Laufer, *Tobacco and Its Use in Asia,* Anthropology, Leaflet 18 (Chicago: Field Museum of Natural History, 1924), 39, 23–24 (quoted here). "Before tobacco became known in Asia," anthropologist Berthold Laufer wrote in the 1920s, "opium (not earlier than the eighteenth century) was taken internally, either in the form of pills, or was drunk as a liquid." The medical and recreational use of drugs has a long history. Users well known to both Mooney and Halliburton were Thomas De Quincy and Samuel Taylor Coleridge, both of whom *drank* opium. Poppy seeds or poppy straw could also be ingested. In *Confessions of an English Opium Eater,* written in the early 1800s, De Quincey reported how a tincture of what he called the *portable ecstasy* could turn one's darkest mood into the happiness of a summer day. Morphine itself could be extracted from the opiate alkaloids found in poppy pods. In some, opium induces Xanadu-type hallucinations and sleep, but in small doses the drug reduces pain, calms the nerves, and prevents diarrhea. Opium boiled in copper kettles was blended with tobacco; only later was it smoked in "its pure state." For Halliburton smoking Camels, see *Royal Road to Romance,* op. cit., 341. For Halliburton's mention of "dock peddlers" and so on, see "Cairo—the Capital of Sin," op. cit. (quoted). For Halliburton smoking, see newsreel of Alpine crossing (1930) and cantina scene from *India Speaks* (1932), retitled *The Bride of Buddha* (1940). "Editor" of Kurt Ludecke's 1937 exposé *I Knew Hitler,* Mooney had some idea that Adolf Hitler and members of his general staff regularly used mood enhancers. Drug use in Hollywood was well known. Movie producer David O. Selznick, for instance, regularly used Benzedrine ("bennies" or "speed") during the filming of *Gone With the Wind,* which, like the *Sea Dragon Expedition,* had involved long delays between shots. By the early 1930s, a pep pill version of the stimulant was easily obtained over the counter.

11. Despite the steady influx of refugees and the pressing war, everyday life in Hong Kong could seem "normal." As another observer offered, "The outbreak of war in September, 1939, did little or nothing to affect life in Hong Kong: indeed, if anything, the tempo of the cocktail parties increased. Although the Island was put on a war footing, there was little sense of urgency about the process." See Tim Carew, *The Fall of Hong Kong* (1960; repr., London: Pan Books, 1963), 17–18. For "showdown" with Japanese, see Halliburton, "Log of the *SD* #9," January 31, 1939, *SFN* (quoted). For lowered "British prestige," see " Log of the *SD* #4," December 27, 1938. For travel restrictions, bombing raids, and mood of the Colony, see *PML,* letter to Harriette Janssen, November 24, 1938 (quoted). For "reservoirs," see "Log of the *SD* #4," December 27, 1938 (quoted). For

military authorities "roused from their lethargy," see "Log of the *SD* #10," February 1, 1939, *SFN* (quoted).

12. For "Japs" taking Hong Kong, see Welch, *LHK,* November 2, 1938 (quoted).

13. For "morning papers" story and "limey" reporter, see Welch, *LHK* #2, October 29, 1938 (Saturday) (quoted).

14. For Welch being repelled by "Chinese food," see *LHK* #1, October 24, 1938 (quoted). For "filthy" Macao making Welch "ill," see *LHK,* November 2, 1938. Now the world's casino capital and a "glitzy Asian destination," Macao was a "seedy, vice-ridden backwater" in 1938. See Kelvin Chan, "Macao's Casinos Up the Ante," AP, Travel, *Wisconsin State Journal,* Sunday, August 23, 2015 (quoted). Also see en.macaotourism.gov.mo. Only Welch noted a visit to these "ruins."

15. For "Pearl River" venture, Canton, and "Scott," see Welch, *LHK,* November 2, 1938.

20. The Battle of the Books

1. See Welch, *LHK* #1, October 24, 1938 (quoted). Welch refused "to join the (*Canton River Company*) outfit," saying, "I can't see it yet." Halliburton might have paid him a better salary—besides, Welch was a captain, and he didn't want to linger in China. He could have gotten other jobs. While he was stationed in Shanghai, he was contacted by a *Standard Oil* representative who asked him to deliver a "yacht" from Manila to Hong Kong (or from Shanghai to Manila) the following April. Welch turned the offer down, saying, "I didn't commit myself to the job, just said that I would see what happened when I got back to the States." American enterprise flourished in Manila, and many American companies owned or leased large docks. Also, the *"Suiang"* that had brought Welch to Hong Kong needed *qualified help* aboard member ships in its line. *LHK* #1, October 24, 1938.

2. For being introduced by Halliburton as "my captain," see Welch, *LHK,* November 2, 1938 (quoted). For Welch "staying to himself," see *LHK* #2, October 29, 1938 (Saturday) (quoted).

3. For "Scott" incident, see Welch, *LHK* #5, November 16, 1938. For "China coast," see *LHK* #2, October 29, 1938 (Saturday) (quoted).

4. See Welch, *LHK* #1, October 24, 1938. For Welch peeking into Halliburton's room, see *LHK,* November 2, 1938 (quoted). For Mooney as a worshipper "at Venus' and Priapus' shrine," see his poem "Initiate" in Max, *HC,* 237. For "Ocean Tow," see "O, Captain, My Captain," chapter 9, note 8 above. For other works, see the bibliography.

5. See Welch, *LHK* #1, October 24, 1938 (quoted).

6. For Wetjen's career, see Wikipedia. Wetjen was married at the time to Edith Cecelia Eisenbrandt of Duluth, Minnesota. Welch's own story "Farewell to Sail" appeared in two parts in the September 1938 and October 1938 issues of *Man Magazine* (volume 4, numbers 4 and 5). "Farewell to Sail" was later published in *The "Man" Storyteller,* "a selection of the best Australian fiction by modern writers" (Sydney, Australia: K. G. Murray, 1945), 174–87. "Paradise Regained" appeared posthumously in *World's News* (Sydney, Australia), August 26, 1939, 8–9. See the bibliography for works by Welch who, at some point, had an agent named "Horn." See *LHK* #1, October 24, 1938. Welch's brother Edward, with whom he communicated, read Wetjen's stories. See *LHK* #1, October 24, 1938. Besides Albert Richard Wetjen, Welch knew best-selling author Jacland

Marmur (1901–70). He had also written letters to writers "Norby" and "Edgar." See *LHK* #1, October 24, 1938.

7. For quoted remarks on Wetjen, see E. Hoffman Price, *Book of the Dead: Friends of Yesteryear: Fictioneers and Others*, ed. Peter Ruber, with an introduction by Jack Williamson (Sauk City, WI: Arkham House, 2001). One might also name noted foreign correspondent, California storyteller, and football coach James "Jimmy" Hopper (1876–1956) as an influence on Wetjen. Hopper was also the author of early baseball classic *Coming Back With the Spitball* (1906) and of *Medals of Honor* (1929), about World War I combat heroes. For the career of Jack London, see James L. Haley, *Wolf—The Lives of Jack London* (New York: Basic Books, 2010). For the *Snark*, see 239–40 et passim.

8. For Welch being "put in first newspaper article," see *LHK* #6, November 24, 1938 (quoted). For moneymaking scheme and "Bell people," see *LHK* #1, October 24, 1938 (quoted). For writing to Herb Caen (1916–1997), see *LHK* #7, November 25, 1938; for writing "a good article" and "keeping notes," *LHK* #6, November 24, 1938; for keeping "a very accurate log," *LHK* #10, December 6, 1938; for Halliburton sounding "very anxious," *LHK* #1, October 24, 1938; for getting going on "the junk trip" once at sea, *LHK* #3, November 2, 1938.

9. For the fair opportunity, see Welch, *LHK* #2, October 29, 1938 (quoted). For the Sacramento State Fair, see *LHK* #3, November 2, 1938 (quoted).

10. For "British authorities," see Welch, *LHK* #13, January 29, 1939 (quoted); for "vain search," *LHK* #7, November 25, 1938 (quoted); for learning of the $1,500 kickback and "store list," *LHK* #5, November 16, 1938 (quoted); for "*American Club*," *LHK* #1, October 24, 1938. A locution introduced in the nineteenth century, "cumshaw" (from the Amoy word meaning "grateful thanks") is Pidgin English for a bribe, kickback, or tip. The word appears no fewer than six times in Welch's correspondence; see *LHK* #1, October 24, 1938; *LHK* #5, November 16, 1938; *LHK* #6, November 24, 1938; and *LHK* #13, January 29, 1939. The transcriber omitted only "personal stuff" and (Welch's probable need of) certain "clothes." See *LHK* #5, November 16, 1938, and *LHK* #13, January 29, 1939. Also, the letter dated November 16, 1938, contains a noted omission: "long paragraph re: Japs sinking junks, rape of women, British notes, etc." In addition, it seems Wetjen had mailed Welch an article he had written, and Welch had read it and returned it, presumably with some comments. See Welch, *LHK* #1, October 24, 1938, and *LHK* #3, November 2, 1938. For "good clean money," see *LHK* #13, January 29, 1939 (quoted). For "cumshaw" defined simply as "money," see Rowland, *Slipping the Lines*, op. cit., 243. For low cost of goods in this "shopper's paradise," see 245.

11. For *American Club*, see *LHK* #1, October 24, 1938.

21. The Master Shipbuilder and the Shipyard by the Peachy Garage

1. For Halliburton's report on the junks he inspected, see his letter to "Monica" at the Alber-Wickes agency, December 5, 1938, op. cit., Lowell Thomas Collection, Marist College Archives and Special Collections. Also see *RHL-P*, November 2, 1938 (quoted). Compare *RHL, His Story*, 410. For building a junk from scratch, see Welch, *LHK* #3, November 2, 1938: "We have decided to build the junk right here in a yard on the Hongkong side." For "sampan," "visiting dozens of junks," and "superb examples," see Halliburton, "Letters from the *SD* #2," January 18, 1939. In "Log of the *SD* #8," January 30, 1939, Halliburton reported, "We" decided to "build a new junk suitable to our

needs, regardless of the time it took." Whom he meant by "we" is disputed, but "Log of the *SD* #11," February 28, 1939, indicates that a committee decision was made. Said Halliburton, "My companions and I decided to build our own."

2. For "best junk-builder," see Halliburton, "Letters from the *SD* #2," January 18, 1939 (quoted). See "Log of the *SD* #11," February 28, 1939 (quoted). Confusion remains about the location of the shipyard. Root references "Fat Kau's Kowloon shipyard," while Prince mentions "Fat Kau's Shipyard in the Kowloon section of the peninsula." See Root, *HTMM*, 20; and Prince, *American Daredevil*, 215. Was it in Kowloon or Hong Kong? Or the New Territories? Welch clearly noted that the shipyard was in "Hong Kong proper." See *LHK* #3, November 2, 1938. Halliburton's tram ride and description of the terrain conform to views of the coastal island districts of Hong Kong Island. For shoreline activity, see Halliburton, "Log of the *SD* #12," March 1, 1939 (quoted).

3. For shipyard, see Halliburton, "Log of the *SD* #12," March 1, 1939 (quoted). For "joss sticks," see Welch, *LHK* #5, November 16, 1938.

4. What may be a photograph of Fat Kau in Halliburton's *His Story* belies that picture. Called "a Chinese shipwright," the short, graying gentleman in this image is hardly bald and perhaps fifty years old. He listens as Halliburton, who is easily half a foot taller, stands alongside him and points to the rigging. See *His Story*, image opposite 386. Another photograph shows Halliburton and this gentleman, who I am persuaded is Fat Kau, looking directly into the camera from the rail. Photograph in RHC, Princeton University Library. For Halliburton's description of Fat Kau, which does not match the shipwright in either photograph, see Halliburton, "Log of the *SD* #11," February 28, 1939 (quoted). Compare Root, *HTMM*, 20–21; also see Alt, who calls Fat Kau a "contractor" in *DDIB*, 333–34. For wives, see Halliburton, "Letters from the *SD* #3," January 27, 1939. For Welch on Fat Kau, see *LHK* #3, November 2, 1938 (quoted).

5. For "a group of foreign idiots," see Halliburton, "Log of the *SD* #11," February 28, 1939 (quoted). For price estimate, see Root, *HTMM*, 21 (quoted). For "deal," see "Log of the *SD* #8," January 30, 1939.

6. For "interpreter," see Welch, *LHK* #5, November 16, 1938 (quoted). For Halliburton on "interpreter," see "Log of the *SD* #11," February 28, 1939. I am assuming that this "smart as hell" messboy was the same person whom Halliburton identified as an "interpreter" and became one of the 'Chinese' messboys joining the crew of the *Sea Dragon*. If so, Captain Welch and Fat Kau might have struck a separate deal before Halliburton had even met Fat Kau. For language differences, see *RHL, His Story*, opposite 386. Halliburton and the Chinese shipwright in the photograph, perhaps Fat Kau, *appear ro be conversing*. The caption reads, "To a Chinese shipwright Richard explains essential points, while a crowd of curious bystanders looks on." For "deal," see "Log of the *SD* #8," January 30, 1939. For "Chinese shipbuilder" and Welch's "deal" with him, see *LHK* #3, November 2, 1938.

7. For "extras" and "bawdy prints," see Welch, *LHK* #7, November 25, 1938.

8. For "two small houses" and "space set aside," see Welch, *LHK* #4, November 2, 1938 (quoted). For "poop," see *LHK* #9, December 5, 1938, quoted. For Welch as "dictatorial," see *RHL*, November 2, 1938, *His Story*, 410. "Evil tempered" is cut from the original. For the Captain Ahlin incident and admonition about swearing, see *LHK* #5, November 16, 1938.

9. For "blueprints and drawings," see Welch, *LHK* #5, November 16, 1938 (quoted). For "blueprints" and contract, see Halliburton, "Letters from the *SD* #2," January 18,

1939. Certain mechanics of ship design and their practical application seemed commonsensical to veteran shipbuilders like Fat Kau. Vernacular words for nautical directives and gear were plentiful and various and through an interpreter he likely knew their meaning. For a glossary of terms, note *SeaTalk Nautical Dictionary*, online. For a photograph of the Chinese workers putting together the hull and frame, see *RHL, His Story*, opposite 386. Almost certainly the *Pang Jin* (or *Pan Jing*), a junk headed for the *New York World's Fair*, was built in the Fat Kau shipyard, and thus it might have been the *Sea Dragon*'s "sister" ship. In the only source noting the event, Captain Welch wrote, "There's another junk being built by an Australian in the same yard." Welch continued, "He is going to New York by way of the Cape of Good Hope and expects to be there in 145 days. He said he preferred that run to *mine* in winter months with the crew I had." See *LHK* #4, November 2, 1938. For the *Pang Jin*, Rex Purcell as told to George W. Polk, "The Cruise of the *Pan Jing*," *QST* (devoted to amateur radio), October 1939, 18, was consulted.

10. For "trial and error," see Halliburton, "Log of the *SD* #1," December 12, 1938 (quoted).

11. For rudders and their development in China, see Robert Temple, *The Genius of China—3000 Years of Science, Discovery & Invention*, with a foreword by Joseph Needham (Rochester, VT: Inner Traditions, 2007), 204–05. For Halliburton on junks, see "Log of the *SD* #1," December 12, 1938, *Cincinnati Enquirer*, January 29, 1939 (quoted). For "contract," see Halliburton, "Letters from the *SD* #2," January 18, 1939 (quoted).

22. Mr. Halliburton Builds His Dream Ship

1. See *PML*, November 24, 1938, letter to Harriette Janssen (quoted). For some Pacific cultures, notably the Tahitians, "the building of a canoe was a religious event, marked by prayers, ceremonies, and feasts." Trees were the children of Tana, god of the land, and they could only be toppled with Tana's permission. See Herb Kawainui Kane and Michael E. Long, "Discoverers of the Pacific" (folded insert), *National Geographic Society Magazine*, December 1974, at *Ships With Souls* (quoted). On November 2, Welch said of the keel, "[It] will be laid in a few days." *LHK* #3, November 2, 1938. For "jacal-wood," see Halliburton, "Log of the *SD* #11," February 28, 1939. "Jacal" may be chengal wood, otherwise known as "Malaysian teak," a durable (but scarce) hardwood resistant to rot, fungi, mildew, and worms. Along with the rudder and sails, a keel contributes to a ship's balance, assisting in horizontal rotation and turning lateral movement into forward motion. While no photographs that I have seen show fully the *Sea Dragon*'s underbelly, the junk's initial keel, not known to be either fixed or full, was a heavy, straight end-to-end girder-like beam. Author's collection. For "construction details," see *RHL-P*, November 10, 1938. For additional commentary on keels and rudders, see Rowland, *Slipping the Lines*, op. cit., 248, 250.

2. For "24 hours," see Halliburton, "Log of the *SD* #8," January 30, 1939, *SFN*. For "keel laid," see "Log of the *SD* #11," February 28, 1939 (quoted); for "demented idea," see "Log of the *SD* #11," February 28, 1939. For "primitive methods," see Welch, *LHK* #4, November 2, 1938 (quoted).

3. See Halliburton, "Log of the *SD* #2," December 13, 1938, *SFN*, January 18, 1939 (quoted).

4. See Halliburton, "Letters from the *SD* #2," January 18, 1939 (quoted).

5. For Halliburton's daily presence, see *RHL* November 10, 1938 (quoted). For put-

ting on "overalls" and being accused of doing "everything backwards," see Halliburton, "Log of the *SD* #8," January 30, 1939 (quoted). For progress, see Welch, *LHK* #5, November 16, 1938.

6. See Halliburton, "Log of the *SD* #11," February 28, 1939 (quoted).

7. Ibid. As conflicting cultures provided creative afflatus for Halliburton, one must wonder if his ignorance of primitive technology, likely pretended, was his way to tell a good story. By this point in his career, it is hard to believe that he was unaware of the extent to which other cultures' construction projects and methods differed from Western industrial models. In any event, Halliburton witnessed shipbuilding techniques performed by craftsmen whose inherited knowledge of such skills was becoming extinct. For "electric drill," see Welch, *LHK* #5, November 16, 1938. For "perfectly calm" shipwright, see Halliburton, "Log of the *SD* #11," February 28, 1939 (quoted).

8. For party, see Halliburton, "Log of the *SD* #11," February 28, 1939 (quoted).

9. For slow work and initial dislike of the drill, see Welch, *LHK* #5, November 16, 1938.

10. For "Chinese sails," see Welch, *LHK* #5, November 16, 1938 (quoted): "The mainsail is #2 duck, also the foresail. The mizzen will be twelve ounces. The mat sail we stow away until we get near the Farallone Islands." Canvas had to do: lightweight nylon or Dacron was at this time unavailable. For the moment, Welch decided to use the "Chinese sail (or batten lugsail) after all." He explained, "I like it better and the fore and after might not work with this rigging. The mainsail hoists forty five feet, thirty nine foot boom, the foresail is twenty seven by twenty nine hoist and looks over the bow at an angle of about fifteen degrees. The mizzen is a pocket handkerchief that sticks up like a sore thumb on the poop." Welch continued, "[I have] been outside in three of them and sailed in the bay in three more and I feel sure that I can handle them alright." For differences between "Chinese sails" and European ones, see Temple, *Genius of China*, 206. A photograph shows Halliburton in trench coat and tie next to the ultimately finished *Sea Dragon*; the batten mizzen-sail can be seen high on the poop. Photographs of the junk during an early trial run show its three masts: the mid-sail, stern-sail, and stem-sail. On this issue, William Taylor quotes Dale Collins: "Her three masts carried colored canvas sails. Typical mat sails had been shipped back to San Francisco via steam with the intention of changing the canvas for mat sails on arrival at Treasure Island . . . Crowell and Halliburton were to have a rendezvous at sea before arriving at San Francisco, trading the sails." As *Los Angeles Times* columnist Jack Smith reported (without confirmation), information received from a *lad living in Fresno*, the sails had been stored at the Schwabacher-Frey warehouse." See Taylor, *SSMTWOD*, 202–4, quoted at 204. For progress made, see Welch, *LHK* #6, November 24, 1938 (quoted). Scheduled to be installed about December 1, mid-December is a likelier date.

11. For Captain Welch's daily routine, beginning "at eight" and "feeling good," see *LHK* #5, November 16, 1938 (quoted). For "nuts," see *LHK* #2, October 29, 1938. For "firecrackers," "carte blanche" authority at shipyard, and "getting along" with workers, see *LHK* #5, November 16, 1938 (quoted). For Welch wishing he had never left San Francisco, see *LHK* #4, November 2, 1938 (quoted).

12. For "brand-new craft," see *PML*, letter to Alice Padgett, January 31, 1938 (quoted). For "junk we are going to build now," see Welch, *LHK* #3, November 2, 1938 (quoted). For the $1,000 bill, see *LHK* #6, November 24, 1938 (quoted). A federal note of this denomination, perhaps the 1934 issue with Grover Cleveland's profile, was a rare sight.

In today's money, $1,000 would be $8,000 to $10,000; in the Hong Kong of the 1930s, it would be equivalent to many times that amount. For the low cost of goods in Hong Kong, see Rowland, *Slipping the Lines*, op. cit., 245.

23. Flight of the Bumblebee

1. For living "in Hong Kong proper," see Welch, *LHK* #3, November 2, 1938: "*We are all* moving to Hongkong proper in a few days. The yard is over there and the run over in the ferry is too long to take every day. I am to go down at 10:00 a.m. and stay until about 10 at night to keep the thing going." For "French bed," see *LHK* #7, November 25, 1938 (quoted). For Welch as a "tyrant" and Von Fehren as "stubborn," see *RHL-P*, January 1, 1939. For cabin with Haliburton, Welch, and Petrich, see Rowland, *Slipping the Lines*, op. cit., 249. For "I am Richard Halliburton," see *LHK* #13, January 29, 1939. Halliburton contractor and friend William Alexander informed me that Halliburton often made the pronouncement when exasperated or challenged. For another side to the man—of Halliburton omitting "the egotistical 'I,'" see Potter, Cortese, *RHRR*, 181.

2. For "astonishing" progress, see *RHL*, November 21, 1938, *His Story*, 411–12 (quoted). For "construction decisions," see *RHL-P*, November 21, 1938. For *Pang Jin* departure, see Welch, *LHK* #7, November 25, 1938. See also "The Master Ship Builder," supra, at n9. For Halliburton's reaction, see "Letters from the *SD* #1," November 20, 1938 (quoted). For "thunder stolen," see *PML*, letter to Harriette Janssen, November 24, 1938, op. cit. (quoted). As for its arrival date, the *Hummel Hummel* docked in San Pedro on October 3, not on September 28, the day Halliburton's ship left San Francisco.

3. For "fuss," see Welch, *LHK* #5, November 16, 1938 (quoted). A tempting identification of "Emmerman" is decorated World War II German U-boat Naval Lieutenant Commander Carl Emmermann (1915–90). For the "primus stove," see 20; for inefficiency of stove, see 31. For mention of kitchen and stove by Halliburton, see *RHL-P*, January 1, 1939. For pistol and rifle, see Petersen, *Hummel Hummel*, 42. For "shakedown cruise," see 39. For "dream ship," see 20 et seq. For the first attempt, see 35. For river leg of the journey, see 27. For good Samaritans, see 34. For repairs, see 56–57.

4. For leaks, see 39. Also see 36–37, where one Captain Scurr (in Shanghai) noted that "a good pump" was necessary because "all junks leak like sieves." For the junk floating "like a cork," see Petersen, *Hummel Hummel*, supra, 38. For being detained by the Japanese, see 43–44, 53–56. For food stores aboard the junk, see 56–57 and 59–60; also see 45. For other supplies, see 51–52 and 53–54. For chickens, see 22 and 40. Safety harnesses are noted in passing; see 39. Petersen gave only the first names of the two Russians, Nick and Vic. For their full names, see the *Los Angeles Times*, October 5, 1938. Petersen said their English was poor See *Hummel Hummel*, 28–29. Yet compare Nick's English on page 31 with his English on page 29. Halliburton probably knew the two were political refugees. According to Petersen, both were "stateless," and immigration authorities directed them to return to China once celebration of the *Hummel Hummel* crossing ended. Nick and Vic had done "a little sightseeing" and were happy "to travel back in luxury with money in their pockets"; see 60. For junk sailing as a new experience, see Petersen, *Hummel Hummel*, 29, 34–35. Halliburton's concern over "immigration problems, passports and so forth" may seem odd since several mess-boys aboard the *Sea Dragon* were Chinese. See John Potter's recollections in Cortese, *RHRR*, 180. For ballast, see Petersen, *Hummel Hummel*, 29, 48. For food stores, see 56–57 and 59–60;

also see 45. Cf. Charles A. Borden, *Sea Quest—Global Blue-Water Adventuring in Small Craft* (Philadelphia: Macrae Smith, 1967), 235.

5. For departure date, see Petersen, *Hummel Hummel*, supra, 57. For progress, see 59. Dr. E. A. Petersen and his Chinese wife, Tani, set sail April 15, 1938; eighty-five days later they landed in San Pedro, California. Cf. Cortese, *RHRR*, 158. For the junk "laboring," see 41–42. For encounters with marine life, see 58–59, quoted at 58. For arrival in San Pedro, California, see *Life*, October 17, 1938, 56 (quoted). For Nick's comment, see Petersen, *Hummel Hummel*, 37. For Dr. Petersen's impressions, see 38 (quoted). For ocean fury, see 41–42. For Dr. Petersen comment, see 57 (quoted). For Dr. Petersen's assessment of their condition, during one leg of their voyage, see 39 (quoted). For voyage beginnings, see 36–37. For a summary of voyage, see Root, *HTMM*, op. cit., 158; for more detail, see Van Tilburg, *Chinese Junks on the Pacific*, supra, 29–31, who notes the departure date as April 28.

6. For Dr. Petersen's navigational skills and their improvement, see *Hummel Hummel*, 51 (quoted) and 68 (quoted). For equipment purchased and learning navigation on the way, see 21 (quoted). For the compass, see 27; for the compass proving deficient, see 37. For type of junk, see Van Tilburg, *Chinese Junks on the Pacific*, op. cit., 92 (quoted). For weather forecasts and forecasting, see Petersen, *Hummel Hummel*, 56. For sextant, see chapter 9, note 7 above. For the bulkheads, see Van Tilburg, *Chinese Junks on the Pacific*, supra, 92. For their importance to Halliburton on junk construction, see "Log of the *SD #2*," December 13, 1938, *SFN* (quoted). The three weeks between the *Sea Dragon*'s return to port and the commencement of its second attempt on March 4 are little known. What overhauls the junk received during this time, many presumed, are imperfectly known. If not completely removed, the bulkheads could have been repurposed. Bathroom features aboard the *Hummel Hummel* and the *Sea Dragon* require some comment. Dr. Petersen mentions that "the overhanging stern was open" and had "a convenient cross bar on each side of the rudder post where one could sit" and evacuate one's bowels. Not conceding to "the Chinese idea of outside plumbing," the *Hummel Hummel* had "a modern toilet below decks." Uncertainty persists regarding a similar upgrade to the *Sea Dragon*, though the junk did have a working "toilet" of some kind, perhaps a lidded chamber pot. As for sleeping quarters aboard the *Hummel Hummel*, "forward of the mainmast, a hatch led down into a sleeping compartment (which had) two bunks," evidently for use by Dr. Petersen and his wife. Tani. The two Russians presumably slept on deck. For plumbing and junk features, see Petersen, *Hummel Hummel*, 20 (quoted) et seq. The location of the diesel engine in the *Sea Dragon* is uncertain. Witness to the March 4 departure John Rowland mentioned "the diesel engine making the prow show a bone in her teeth." *Slipping the Lines*, op. cit., 251. At least one photograph shows, just to the bottom of the dragon-painted transom and at the waterline, what is likely the shaft for the propeller of the diesel engine.

7. For Japanese proverb and "regret," see Halliburton, "Log of the *SD #1*," December 12, 1938 (quoted.)

8. For other junk "sailing from Wenchow," see Welch, *LHK #5*, November 16, 1938 (quoted). Following his own inquiry into the matter, Halliburton enthusiast Ed Howell received a March 6, 1946, letter from the *San Francisco Public Library* that reads, "A check through reference does not disclose the name of the junk that raced against Richard Halliburton." For other junk with two months' head start, see *RHL-P*, September 28, 1938 (quoted).

9. For "other junk a total wreck," see Welch, *LHK* #7, November 25, 1938. For Halliburton's comment, see *RHL-P*, November 21, 1938 (quoted). The feature "Down East" appeared in *Motor Boating* (February 1951) and gave Potter's name as "Porter." For an alternative from Potter, see letter from John Potter to Edward Howell, January 15, 1946. After noting Halliburton's eagerness "to be awarded the concession," Potter said, "I recovered in short order and spent the rest of the year with a French journalist acting in the capacity of an interpreter while he covered the Sino-Japanese 'incident' in the China hinterland." In later years, Potter omitted any mention of a "race" and being awarded the concession. For Potter's later views, see Taylor, *SSMTWOD*, 210. Cf. Potter's recollections in Cortese, *RHRR*, 179–82. Welch noted the junk that had attempted to sail to the New York World's Fair, but he did not mention a race or prize money: "The other junk, bound to New York is away from Singapore. Made it to that place in twelve days and shipped another man before she sailed. She only has seven all told. I have twelve or will have if romance doesn't get any more of them and still I can't get a decent day's work done." See Welch, *LHK* #13, January 29, 1939.

10. For Halliburton cheering and Welch meeting a crew member from the injured ship, see Welch, *LHK* #7, November 25, 1938 (quoted).

24. Pandas and Other Distractions

1. For temperature, see *RHL-P*, November 21, 1938 (quoted). See Halliburton, "Log of the *SD* #11," February 28, 1939, *SFN* (quoted). For "driven mad," see "Log of the *SD* #11," supra (quoted). Halliburton often appeared dispirited. Crew member John Potter, calling the mission Halliburton's "dream," commented that the adventurer "hardly ever smiled and almost never laughed." See Cortese, *RHRR*, 181. For discussion of finances with Paul Mooney, see Root, *HTMM*, 32–33. Captain Welch overheard Mooney, who was also a "journalist," arguing with Halliburton about the articles, not about finances. See *LHK* #9, December 5, 1938. Also see Max, *HC*, 192. For Halliburton's activities, see Max, *HC*, 206. At the Foreign Correspondents Club, writes Jonathan Root, Halliburton "lounged languorously in a deep wicker chair on the veranda" while taking in the mutterings of fellow press members, or, as he had when he lived "in a house that overlooked New York Harbor," he idly "gaz[ed] out the window at the procession of great ocean liners that kept sailing in and out under (his) nose." See Root, *HTMM*, 15. For New York Harbor, see Halliburton, *Book of Marvels—The Occident*, op. cit., 58, 63 (quoted).

2. See *PML*, letter to Harriette Janssen, November 24, 1938, Kowloon Hotel, Hong Kong (quoted).

3. Ibid. (quoted). For "thousands of terrified refugees," see Halliburton, "Log of the *SD* #10," February 1, 1939 (quoted). For the authorship of the "Log of the *Sea Dragon*," see Max, *HC*, 2 passim. For the career of Saint Francisco Xavier (1506-1552), see Wikipedia. Also see Henry James Coleridge, *The Life and Letters of St. Francis Xavier*, 2 vols., 4th ed. (London: Burns and Oates, 1927). Mooney never mentioned the story or Saint Francis Xavier; it is only my suspicion that his was a dual mission. He remained secretive about his spiritual needs, as did his ethnologist father James, who tried to conciliate Catholic doctrine and ritual with the Indianist creeds he encountered. See Max, *HC*, 19 et seq.

4. For light show, see Halliburton, "Log of the *SD* #10," February 1, 1939 (quoted).

5. For magazines from America, see Yardley, *Chinese Black Chamber*, op. cit., 57. For Yardley at the Hong Kong Hotel, see 3 (quoted) and 4 (quoted). For Yardley as a writer

published by Bobbs-Merrill and a friend of its editor David Laurence Chambers, see Kahn, *Reader of Gentlemen's Mail*, 105–6 passim. For Halliburton's chumming with Patrick Kelly, see Max, *HC*, 206–7; also see Root, *HTMM*, 25–26 (quoted). For his depiction of Kelly, see Halliburton, "Log of the *SD* #16," March 19, 1939, *Boston Sunday Globe* (quoted).

6. For Mooney's experience with the pandas, see *PML*, copy of letter to Harriette Janssen sent to Gerstle Mack, November 24, 1938 (quoted). Photographs in the MEB Archive are of Paul and Richard with the pandas. See *His Story*, opposite 371, caption quoted. Of incidental interest, renowned trapper Frank "Fang and Claw" Buck fared equally well as a hunter and trapper, and his film *Bring 'Em Back Alive*—with its famous face-off between a python and a panther, dazzled moviegoers. According to Gordon Sinclair, Buck's animal adventures were of interest to Halliburton. See *Bright Paths to Adventure*, op. cit., 88–89.

7. For Mooney as pet lover and collector, see Max, *HC*, 27. The "one in America" was the famed Su Lin, or rather Mei Mei, as Su Lin, barely two years old, died on April 1, 1938, from an infection that developed after he ingested an oak twig. But for a moment, *Su Lin* had drawn daily crowds of sixty thousand while a featured guest of the Brookfield Zoo in Chicago. Like newspapers across the country, *Life* magazine carried the story of Su Lin's rise to stardom and sudden death. See "Su Lin, America's Favorite Animal, Dies of Quinsy in Chicago Zoo," *Life*, April 11, 1938, 14.

8. For early references to pandas see "Giant Panda," Wikipedia. For purchase of cub, see *RHL,* November 21, 412 (quoted) (oddly, such notice does not appear in the original letter). In 1929, Theodore Roosevelt's sons Kermit and Theodore Jr. had trekked deep into remote habitats and killed an aging female panda, which they brought back home to offer an eager public proof of their daring. Some people were curious to learn what panda meat tasted like and how a coat made from its fur might feel. In general, see Michael Kiefer, *Chasing the Panda* (New York: Four Walls Eight Windows, 2002), 166.

9. See Ruth Harkness, *The Lady and the Panda—an Adventure* (New York: Carrick Evans). Also see Ruth Harkness in Wikipedia for recent treatments. A fine treatment of her life is Vicki Constantine Croke, *The Lady and the Panda—The True Adventures of the First American Explorer to Bring Back China's Most Exotic Animal* (New York: Random House, 2006). In her last years "frequent bouts of malaria and heavy drinking took their toll" on Harkness; see 286.

10. See *PML*, letter to Harriette Jansen, November 24, 1938 (quoted).

11. For sailing times, see Welch, *LHK* #6, November 24, 1938 (quoted); and *LHK* #8, November 30, 1938. For Collins's observations, see "Royal Road Across the Pacific," op. cit., at 502. For Su Lin, see *New York Times*, April 2, 1938, 17. For Harkness's Peruvian adventure, see her *Pangoan Diary* (New York: Creative Age, 1942).

25. Big Men on Campus Join the Fraternity

1. Captain Welch quoted a news source when he wrote "blaze of publicity." See *LHK* #13, January 29, 1939. The article consulted was "Four Youths in a Junk," *SFN*, October 20, 1938 (quoted). For the Mark Hopkins and its stunning views, see Crowe and Bowen, *San Francisco Art Deco*, op. cit., 27. For Crowell's remarks, see *RHL-P*, November 10, 1938 (quoted). Also see *LA Times*, October 17, 1938.

2. For Welch's remarks, see *LHK* #13, January 29, 1939. He also notes that the collegians drank "rums." For Potter's remark, see "Four Youths in a Junk," supra (quoted). For

"greenest of green crews," see Halliburton, "Letters from the *SD* #3," January 27, 1939. For separate accounts from Potter and Torrey, see Cortese, *RHRR*, 177–82. Dartmouth College houses the "lads'" records. The MEB Archive also contains pertinent written and tape-recorded materials by Potter and Torrey. *Indian Mountain Boarding School* in Connecticut and *St. Andrews Prep School* in Middleton, Delaware, both of which George Barstow attended in the 1930s, are among the New England schools where Halliburton might have lectured.

3. See "Four Youths in a Junk," supra, quoted. Among other terms, *gaffer* and *grip* found their way from the navy to the circus and the Hollywood sound stage to denote personnel working with ropes and pulleys to operate curtains and set backings. For exciting train experience, see *RHL*, November 5, 1927, *His Story*, 277–78 (quoted). Halliburton conveyed Crowell's opinion in a letter to his parents. See *RHL-P*, November 10, 1938.

4. See *RHL-P*, November 10, 1938 (quoted). For "picnic," see Welch, *LHK* #2, October 29, 1938. Compare the comfort level with that of those aboard famed luxury yacht *Sunbeam*. From 1876 to 1877, the *Sunbeam* carried Lady Anna Brassey and her family, friends, pets, and servants. Brassey recounted her experiences in *A Voyage in the Sunbeam, Our Home on the Ocean for Eleven Months* (London: Longmans, Green, 1878); the best-selling book was often reprinted. See Alan Villiers, *The Making of a Sailor; the Photographic Story of Schoolships under Sail* (New York: William Morrow, 1938). For Seamen's Institute (associated with YMCA), which offered some practical training in ship-handling, see Leys, *After You, Magellan!* (op. cit.,), 150 and 219; Leys also mentions the *Royal Marine Commissary Department and Insular Marine Union*, 150 and 229. For the lads on the junk, see photograph showing from left to right, (should be) Paul Mooney, John Potter, Gordon Torrey, George Barstow, and Robert Chase in Williams, *Forgotten Adventures of Richard Halliburton*, op. cit., 151. For the topics a seamanship course might cover, see Robin Knox-Johnston's manual *Seamanship* (New York: W. W. Norton, 1987). For crew and "mooring lines," see Welch, *LHK* #6, November 24, 1938 (quoted). For "collective cool-headedness," compare J. R. L. Anderson, *The Ulysses Factor: The Exploring Instinct in Man* (New York: Harcourt Brace Jovanovich, 1970), 109 et passim. Also see "A Public School Teaches Art of Seamanship," *SFN*, May 4, 1939. For Collins's remarks, see Taylor, *SSMTWOD*, 209 (quoted). Also see Collins, "Royal Road Across the Pacific," op. cit., 502.

5. For "mothering" Barstow, see Welch, *LHK* #4, November 2, 1938; and *LHK* #5, November 16, 1938 (quoted). For "mothering" all the lads, and for the "mothering" Welch received, see *LHK* #12, December 21, 1938 (quoted). For "the bastards," see *LHK* #5, November 16, 1938 (quoted.) For "green hands," see Halliburton, "Log of the *SD* #11," February 28, 1939 (quoted).

6. As early as July 14, Halliburton told his parents, "Young Potter writes he would not come to Hong Kong until just before I start back." *RHL-P*, July 14, 1938. For the December arrival, see *RHL*, November 2, 1938; for training (same), quoted. For Torrey and Barstow arriving together, and for arriving too soon, see *RHL-P*, November 10, 1938 (quoted). For "profits," see *RHL-P*, November 21, 1938 (quoted). For Welch's first meeting with the lads, see *LHK* #9, December 5, 1938. For Potter's recollection of the arrival time, see Cortese, *RHRR*, 179. Also see "Four Youths in a Junk," supra. Captain Welch noted that Potter and Chase arrived on December 4; they had left Honolulu on December 3. See *LHK* #5, November 16, 1938. For Torrey's remarks, see Cortese, *RHRR*, 178. For Welch's remarks, see *LHK* #5, November 16, 1938. According to a "November"

news clipping (precise date and source unspecified, but likely from the *Honolulu Star*), Halliburton asked Potter and Chase to book passage on a clipper to Hong Kong. The article's byline reads, "Vacationists Here (in Honolulu) to Join Halliburton Pacific Junk Crew," and its features head photographs of "Mr. Potter" and "Mr. Chase." For Torrey's earliest known recollections (late 1938, when he mentions arriving "on Treasure Island next February"), including the name of the ship that took him "to the Orient," consulted was "Gordon E. Torrey," Rauner Special Collections, Dartmouth College Library, op. cit.

7. For Welch's remarks, including his description of Barstow as "honey," see *LHK* #6, November 24, 1938. For Welch's characterization of Barstow, see *LHK* #9, December 5, 1938 (quoted). For threat (of the lawsuit kind) from Barstow, see *LHK* #14, February 13, 1939. For information on the Barstow family, I am indebted to Patricia Suprenant, as well as to the *Providence Historical Society*. Letters to author, January 4 and 5, 2017. Also see Wikipedia, under Amos C. Barstow.

8. For "good able looking lads," see Welch, *LHK* #9, December 5, 1938 (quoted). For "lisping voices," see *LHK* #13, January 29, 1939 (quoted). Potter later said he arrived in November. See Cortese, *RHRR*, 180.

9. For Welch's defense, see *LHK* #9, December 5, 1938 (quoted). For state of the *Sea Dragon* as of December 12, 1938, see *RHL, His Story*, 411.

10. For the lads' attire and Torrey's remark, see *Hong Kong Daily News*, January 19, 1939 (quoted). Potter noted later that he conscientiously watched the junk being built. See Cortese, *RHRR*, 180. For Torrey's opinion, see Cortese, *RHRR*, 177–79. For Potter's remark, see *RHRR*, 181 (quoted).

26. *The Dark Side of Laughter*

1. For the Potter-Torrey or Chase film, which captures some of the lads' activities in the Colony, see Max, *HC*, 244. Chase carried a sixteen-millimeter color motion-picture camera, Potter a German twin-lens Rolleiflex. For Welch being annoyed by the cameras, see *LHK* #13, January 29, 1939. Knowledge of the lads' daily activities comes mainly from Potter's and Torrey's later recollections and from Captain Welch's letters. The lads might have joined the others at the Kowloon Hotel or have chosen the upscale Gloucester Hotel before staying aboard the *Sea Dragon*. For the Gloucester Hotel, a regular stop-off for crew members, see Rowland, *Slipping the Lines*, op. cit., 250–51. Hotel brochures from the period were also consulted.

2. Photographs of the crew at work reside at the MEB Archive, Rhodes. Paul Mooney participated in these sessions, providing an example of hard work to the others. Captain Welch ultimately singled him out as a surprise asset to the team. For training and the tiller, see Welch, *LHK* #13, January 29, 1939. The Hearst outtake film also includes frames of Mooney and other crew members straining at the ropes. See Hearst News Service, *Sea Dragon* newsreel outtakes, 1939, UCLA Film Archive.

3. See *RHL-P*, November 11, 1938, American Consulate, Hong Kong, (quoted). For gifts, see the same letter, whose contents are omitted from the published version except for a four-sentence notice of the gifts and Christmas. See *His Story*, 411. Mrs. Hutchison is often affectionately addressed as "Ammudder" (Grandmother) in the letters. For the chest, see Max, *HC*, 229–30. Given to Hangover House designer William Alexander by Paul's mother, Ione, and owned by Alexander until his death in 1997, the chest now belongs to a descendant of Alexander's. See Max, *HC*, 229.

4. For "caulkers," see Welch, *LHK* #9, December 5, 1938, quoted. For a "face like leather," see *LHK* #9, December 5, 1938. For "trick," see *LHK* #9, December 5, 1938, quoted. For departing on "Xmas day," see *LHK* #5, November 16, 1938 (quoted).

5. For "butterfly's wings," see Welch, *LHK* #7, November 25, 1938 (quoted); for "ensign" and "seven cannon," see *LHK* #10, December 6, 1938; for "picnic," "latest idea," "mad ship-building program," and "sane," see *LHK* #9, December 5, 1938 (quoted). For "cramped little shipyard," see Halliburton, "Log of the *SD* #11," February 28, 1939 (quoted). For "rusty cannons," probably from Fat Kau's shipyard, see "Log of the *SD* #1," December 12, 1938, *Cincinnati Enquirer*, January 29, 1939 (quoted). This article comprises materials also found in "Log of the *SD* #3," December 26, 1938. For Halliburton's guns, also see "Log of the *SD* #16," March 19, 1939. For another inspected junk, one with cannons, see Prince, *American Daredevil*, op. cit., 215. The following news was reported in the *South China Paper*: "Luxuriously equipped, the *Sea Dragon* will present an irresistible temptation to pirates should they be made aware of her movements. However, if they attack her they will find her a hard prize to take. Three shotguns, three rifles, and two revolvers provide her with a small armoury." For "armoury," see *South China Paper*, January 13, 1939, MEB Archive. Although Halliburton *armed* the junk, pirates remained a threat. On the evening of January 10, pirates boarded a junk near Sam Mun Island, stole livestock at gunpoint, and conveyed it to Hong Kong. See "Shipping News," *South China Morning Post*, Monday, January 16, 1939.

6. For Halliburton's "own quarters," see *RHL-P*, November 21, 1938 (quoted). Welch noted that he took one cabin and Halliburton took the other: "The gentlemen sleep in a forecastle forward, next to the engine. See *LHK* #5, November 16, 1938 (quoted). For "debutant," see *LHK* #4, November 2, 1938, (quoted). For "Marvel of China" and "special cabin," see *LHK* #9, December 5, 1938 (quoted). I know of no photographs that show the Halliburton cabin or the lads' quarters. The "God of Anger," as pictured in Fitch, "Life Afloat In China," op. cit., 676, could be *Tai Toa Fat*. For color schemes and "wooden eyes," see Halliburton, "Letters from the *SD* #2," January 18, 1939 (quoted). For Welch's rooms, see *LHK* #4, November 2, 1938. For mention of cabin, see *LHK* #10, December 6, 1938; for "honky-tonk crud," see *LHK* #12, December 21, 1938 (quoted). For limited space, see *PML*, letter to Alice Padgett, December 31, 1938. For cabin description, see Rowland, *Slipping the Lines*, op. cit., 249 (quoted). For alternative of a cabin built on deck, compare Romola Anderson and R. C. Anderson, *The Sailing-Ship—Six Thousand Years of History* (New York: Bonanza Books, 1963), 23, fig. 3. For "hooker" characterization, see *LHK* #5, November 16, 1938.

7. For "bunk in the galley," see *LHK* #10, December 6, 1938. For Kelly, "a Portuguese . . . 17 years old," see Welch, *LHK* #11, December 16, 1938.

8. For falling in love with Patrick Kelly, see Welch, *LHK* #11, December 16, 1938 (quoted). For "sailor's lament," "bloody Murder," and mail service, see *LHK* #12, December 21, 1938 (quoted).

9. For Welch's pride in the junk and for "Izzy's" at 848 Pacific in San Francisco, see *LHK* #10, December 6, 1938. For Welch's opinion of Mooney, see *LHK* #12, December 21, 1938. For "driv(ing) the jesus out of it," see *LHK* #3, November 2, 1938 (quoted).

10. For "spurts and whims," see Halliburton, "Log of the *SD* #12," March 1, 1939. For "unused bunk," see *RHL-P*, January 1, 1939 (quoted). Also see "Log of the *SD* #12," March 1, 1939. For "high poop," see Welch, *LHK* #13, January 29, 1939 (quoted). Occupancy of the fourth bunk is not indicated.

11. For high poops as "up-soaring castles," see "Log of the *SD* #1," *His Story*, 401. For the survey report, see Collins, "Royal Road Across the Pacific," 504 (quoted). For "survey certificate," see Welch, *LHK* #13, January 29, 1939 (quoted). For "western perspective" of poop, see Van Tilburg, *Chinese Junks on the Pacific*, op. cit., 92 (quoted). For "Chinese perspective," also see 92 (quoted). For Mooney comment, see *PML*, letter to Alice Padgett, December 31, 1938 (quoted).

12. For "carefully drawn paper plans," see "Log of the *SD* #12," March 1, 1939 (quoted). For Welch overworking, see *LHK* #13, January 29, 1939 (quoted). For Halliburton quip, see letter to "Monica" at the Alber-Wickes agency, December 5, 1938 (quoted), Lowell Thomas Collection, Marist College Archives and Special Collections.

13. For the "survey man," see December 6, 1938 (quoted). For Collins's description, see "Royal Road Across the Pacific," op. cit., 502 (quoted). For his having "Survey Report" in hand, 504. For Potter blaming Fat Kau, his September 10, 1987, letter to Zola Halliburton was consulted: "There were many things wrong with the junk almost none of which were [*sic*] attributable to Dick. The design and construction were completely in the hands of number one—i. e., the Chinese shipbuilder." For what the diesel installation involved, see Welch, *LHK* #5, November 16, 1938. For contracting with installers at "ten percent," see *LHK* #9, December 5, 1938. For his effort, Welch would receive a 10 percent kickback from the installers once the junk was moved to another shipyard. Unless rollers were employed, a dock crane or hoist with an extended arm or rotating boom had to lift the half-ton solid cast-iron engine, carry it horizontally to its new location, then lay it snugly into its bedding below deck. Clearly it sat in the rear of the ship. A photograph released by Wide World news shows an unadorned junk (presumably the *Sea Dragon*) launched into the water with the propeller opening in the lower stem (the upper part of which is without the high poop). The undated view is among the John Potter papers in the MEB Archive. Apparently, the junk was then returned to Fat Kau's shipyard. Chinese elements were retained, of course. The coastal regions whose styles of junk influenced the design, notes Van Tilburg, were "Guangzhou, Fujian, and Zhejiang." A hodgepodge, the design, he adds, "did not translate into a decipherable language, any known configuration," and that it "was an uncharacteristic and dangerously hybridized vessel, mixing technologies that were compatible only with the greatest care and attention." See Van Tilburg, *Chinese Junks on the Pacific*, op. cit., 97 (quoted). For the Soochow river-type junk, see Cortese, *RHRR*, 178 (where Torrey is quoted). For Polynesian models, see Kane and Long, "Discoverers of the Pacific," at *Ships With Souls* (quoted). For Ningpo model, see Halliburton, "Letters from the *SD* #2," January 18, 1939; also see Halliburton, "Log of the *SD* #12," March 1, 1939. For "up-climbing poops," see "Log of the *SD* #1," December 12, 1938, *Cincinnati Enquirer* (quoted). For Lieutenant Reed's high praise for the junk's design (as quoted in an unsigned letter to Wesley Halliburton dated April 13, 1939), see Taylor, *SSMTWOD*, 209–210. For Dale Collins' concerns, see Taylor, *SSMTWOD*, 210.

14. John Rowland estimated that the junk weighed fifty tons. See *Slipping the Lines*, 249. For "construction almost over," see Halliburton, "Letters from the *SD* #2," January 18, 1939 (quoted). For *Sea Dragon* "ready to sail," see Halliburton, "Log of the *SD* #11," February 28, 1939 (quoted). For merits of the engine, see "Letters from the *SD* #3," January 27, 1939. For "cow," crew, and tiller, see Welch, *LHK* #9, December 5, 1938, quoted. For "January 10" letter home, see *RHL-P*, December 12, 1938.

27. Preparing for the Real World

1. See Halliburton, "Log of the *SD* #12," March 1, 1939. Also see Welch, *LHK* #11, December 16, 1938 (quoted). After a trial run, Paul Mooney said, "[The junk] is back on the ways now while the final work of outfitting and *painting is done*." Italics mine. See *PML*, letter to Alice Padgett, January 31, 1938. Also noted in the letter is the long period without rain. Author's collection. For "automobiles," see Halliburton, "Letters from the *SD* #3," January 27, 1939 (quoted. For priest's ceremony, see "Letters from the *SD* #2," January 18, 1939 (quoted). Cf. Prince, *American Daredevil*, op. cit., 227. See Hearst News Service, *Sea Dragon* newsreel outtakes, 1939, UCLA Film Archive.

2. For this 'major' trial run, and for the comparison of the junk to "automobiles," see Halliburton, "Letters from the *SD* #3," January 27, 1939 (quoted.) For "high poop" see Welch, *LHK* #13, January 29, 1939 (quoted). For leaking "like a basket," see *LHK* #11, December 16, 1938. The trial runs and their number, frequency, and result are unclear. John Rowland, who was in Hong Kong at the time, suggested there was only one. But he ignored the first aborted sailing, which to him may have served as a trial run only. See *Slipping the Lines*, op. cit., 250. Root distinguishes the "first sailing" (February 4) from the "second sailing" (March 4); however, he evidently does not know the exact number of trial runs or "false starts" that occurred. A photograph in the January 15, 1939, *Hong Kong Sunday Herald* shows the junk "afloat for the first time," but there is no indication of when the photograph was taken, or the number of practice runs that ensued. *His Story* pictures the junk during separate trial runs, one with "motor," another under sail. However, no date for either run is given. Confusing the issue further is Halliburton's reference to the failed "first sailing" as a "shakedown cruise." The difference between a shakedown cruise and an actual sailing is comparable to that between a combat simulation and a live fire fight; thus a shakedown cruise is frightful but not life threatening. Late in January 1939, Captain Charles Jokstad recommended another "trial run." On January 13, 1939, the *South China Paper* published a story stating, "[A] two-day 'shakedown trip [will occur] in the near future" in which it is "likely that two police officers will be carried." No other source mentions this voyage. Nor do other sources mention the little side trips: "Yesterday (January 12) the *Sea Dragon* made one small trip. In tow, she travelled across the harbor to take on 2,000 gallons of oil at the Shell depot across the harbor." For the junk pictured, see *His Story*, opposite 410. For "shakedown cruise," see Halliburton, "Log of the *SD* #16," March 19, 1939, "Halliburton's Junk Battles Storm at Sea," *Boston Sunday Globe*, March 19, 1939—byline—"Adventurer and His Crew Put Back to Hong Kong to Rush Injured First Mate to Hospital." For late "trial run," see Jokstad, *Captain and the Sea*, op. cit., 189. See *South China Paper*, January 13, 1939 (quoted), MEB Archive. For crew assembled, see *RHL-P*, January 1, 1939 (quoted). In this letter Halliburton also noted, "We will take a good shakedown cruise first, before really going to sea." In the letter to subscribers dated January 27, he said that a shakedown cruise "outside the harbor" *had* occurred. Scenes of the "first" trial run (possibly on December 16, 1938), this one featuring the main crew, Fat Kau, and several of his wives, conform to sequences in the Hearst newsreel showing the sails reefed and the *Sea Dragon* making its way on diesel power. See Hearst News Service, *Sea Dragon* newsreel outtakes, 1939, UCLA Film Archive. Existing films of the launch give the mistaken impression that the junk shot down the ramp, the crew climbed aboard, and

crowds cheered while everyone sped off to Treasure Island. Several trial runs, each one of an increasing distance from the dock, preceded the *Sea Dragon*'s departure. Skipping useless details, Halliburton mentioned *at length* only a single "shakedown cruise" or "first" trial run, one designed as much "to test the engine as to test the crew and sails." See "Letters from the *SD* #3," January 27, 1939 (here quoted).

3. For "rough water" and "greenest of crews," see Halliburton, "Letters from the *SD* #3," January 27, 1939 (quoted.)

4. For "angry battles," see Halliburton, "Letters from the *SD* #2," January 18, 1939 (quoted). For assessment of crew, see "Letters from the *SD* #3," January 27, 1939 (quoted). For Mooney letter, see Cortese, *RHRR*, 160 (quoted); also see Taylor, *SSMTWOD*, 197 and (text) 198. Taylor notes that the recipient was Mooney's friend Lee Hutchings of Laguna Beach; again, it was not unusual for Paul to send a mimeographed copy or copies of his letters to multiple recipients, as he did with Harriette Janssen and William Alexander. Several crew members carried cameras. The "odious cameraman" could have been George Barstow. One wonders what an anti-Roosevelt *joke* entailed. In the recent presidential election, Franklin Roosevelt had only failed to carry Vermont and Maine. For Torrey's political opinions, see was "1937 Mint Bag," April 14, 1941 (quoted), Rauner Special Collections Library, Dartmouth College, op. cit. Written two years after the *Sea Dragon*'s disappearance, Torrey's comments were also dated nine months before the attack on Pearl Harbor, and they appear to reflect opinions developed from the time he arrived in the Orient. For Mooney's assessment of the crew, see Cortese, *RHRR*, 160 (quoted). For unbearable "element" (unspecified) of each crew member, compare *RHL-P*, January 1, 1939.

5. For "jewels," see *RHL*, December 12, 1938, *His Story*, 412 (quoted). For Potter's remarks, see Cortese, *RHRR*, 181 (Potter quoted). Also see *Motor Boating*, February 1951, op. cit.

6. For Halliburton as "mediator," see *RHL*, January 1, 1939, *His Story*, 413. For the leadership qualities of Thor Heyerdahl, see Anderson, *Ulysses Factor*, op. cit., 99 et seq. Mooney used the term "crazy cockle-shell" in a February 24, 1939, letter to friend and former roommate Eugene Hoenigsberg, a portion of which letter appeared in the *Brooklyn Eagle*, October 5, 1939. For "Jezebel," see *PML*, letter to Alice Padgett, December 31, 1938. Mooney qualified his description of the *Sea Dragon* thus: "However, that's how the Chinese paint theirs—with birds and dragons and proverbs on the stern and big eyes on the bows so the junk can find its way!" For Halliburton becoming an "autocrat in spirit," see *RHL-P*, May 28, 1938.

7. After their first meeting, Torrey seldom ran into Halliburton. Despite the adventurer's claim that he was regularly on hand during the building of the *Sea Dragon*, Torrey said that "promotional activities" kept him away, and that he spent little time with his crew. Torrey remarked, "[Halliburton] was obviously harassed by some personal matters and not particularly fun to be with or too sociable with us." See Cortese, *RHRR*, 177. For Potter's assessment, see *RHRR*, 181 (quoted); for Torrey's assessment, see 179 (quoted). Also see Sinclair, *Bright Paths to Adventure*, op. cit., 85–99, quoted at 93. After the disaster, Potter agreed with Torrey, noting, years later, that he had failed "to dissuade (Halliburton) from making this perilous sea voyage." See *JC*, p. 182.

8. For "clashes and quarrels," see *RHL-P*, January 23, 1939. See Cortese, *RHRR*, 178 (quoted). Earlier, Halliburton had seen fit to have the dismantled *Flying Carpet* "heavy-lifted" aboard the *Majestic* when he and Moye Stephens crossed the Atlantic. The plane

was again lifted aboard the SS *President McKinley* when they crossed the Pacific—but at considerable expense. Thus the idea was not farfetched. Root writes, "The *Flying Carpet's* wings were removed and crated with its fuselage and lashed down on the forward deck of the White Star liner *Majestic* which charged Richard $450 in freight plus $270 each (first class) for him and Stephens." See Root, *HTMM*, 178; for the SS *President McKinley*, see 203. For ditching Torrey, see Welch, *LHK* #13, January 29, 1939 (quoted).

9. For reference to "Captain Queeg" (from Herman Wouk's *The Caine Mutiny*), see Cortese, *RHRR*, 180. For Welch taking to the lads, and for "picnic," see Welch, *LHK* #9, December 5, 1938.

10. For Torrey's views, see Cortese, *RHRR*, 178 (quoted). Dale Collins, who thought six weeks' training was necessary "to train landsmen into competent deep-water sailormen," said that Welch "was the only man, who had previous practical experience in sailing ships." Collins added, "The handling of a large 3-masted junk in heavy seas should certainly require a thorough knowledge of seamanship from all hands as well as the master." See "Royal Road Across the Pacific," op. cit., 502. For Welch's opinion of Torrey, see *LHK* #13, January 29, 1939 (quoted).

11. Welch and Halliburton each took credit for the hiring of Sligh. For Sligh's service to other nations, see Halliburton, "Letters from the *SD* #2," January 18, 1939 (quoted). For Welch hiring a white cook, see *LHK* #11, December 16, 1938 (quoted). For additional background, see Root, *HTMM*, 32. For Welch hiring Sligh, see *LHK* #11, December 16, 1938. Halliburton said, "I've hired a fine cook and cook's boy—(crossed out "C" for, Chinese. American, and one more sailor. *RHL-P*, January 1, 1939, quoted. For "Chinese," the word "American" is substituted. Although in his dispatches Halliburton mentions Fat Kau by name, he never mentions the shipwright by name in his letters to his parents. See also the *Hong Kong Daily News,* January 19, 1939 (quoted). For Rowland, who was, as he explained, a "bull cook" or "pot walloping mechanic" and not a galley cook, see Rowland, *Slipping the Lines*, op. cit., 248. Halliburton wrote little about dietary logistics. He knew, however, that men working on a ship burned many calories. He also knew that food, whether in cans, barrels, or drums, can spoil. For "long passages," Joshua Slocum loaded goodly amounts of "potatoes, and salt cod and biscuits," as well as "plenty of coffee, tea, sugar, and flour." See Slocum, *Sailing Alone Around the World*, ed. and with an introduction and notes by Thomas Philbrick (New York: Penguin Books, 1999), 133–34, quoted at 133. For caveats on food and water consumption, see Knox-Johnston, *Seamanship*, 360–66, quoted at 363. For "not fit for human consumption," see John Wenlock Welch, who quotes Dick Wetjen, in his essay "Our Oceanic Ills," *U. S. Naval Institute Proceedings*, October, 1939, Vol. 65, 10, 440.

12. For the Lapp photograph, see Williams, *Forgotten Adventures of Richard Halliburton*, op. cit., 148. For Dr. J. M. Lapp, see Halliburton, *Royal Road to Romance*, op. cit., 153–54 (quoted); for *Gold Shell*, see 130. For Halliburton's age, see Max, *HC*, 246n9.

28. The Whole Town Is Talking

1. For "defects" and "final touches," see *RHL-P*, January 23, 1939 (quoted). The location of the "main dock" is uncertain. For "enthusiasm," see *RHL*, January 1, 1939, *His Story*, 412 (quoted).

2. For delays, see, in general, *RHL*, January 1, 1939. For the route, see Halliburton,

"Letters from the *SD #1*," November 20, 1938; for the route through Manila as "original intention," see "Letters from the *SD #3*," January 27, 1939. For Welch refusing to sail south, and Halliburton telling him about visiting Manila, see *LHK #12*, December 21, 1938. For trip to Manila, see *LHK #12*, December 21, 1938. For Andres de Urdaneta, see "From Asia to America: Conquering the Pacific," *National Geographic—History*, August–September 2015, 16–19. For another routing, compare that of the *Dove*; see Robin Lee Graham, *Dove*, with Derek L. T. Gill (New York: Harper and Row, 1972). For "all Hong Kong," see *RHL*, January 1, 1939, *His Story*, 412. For "splendid radio," see *RHL-P*, January 23, 1939 (quoted).

3. See *PML*, letter to (realtor) Alice Padgett, December 31, 1938 (quoted). The letter, marked *urgent*, was sent airmail, and much of it concerned the Laguna property. Wrote Paul Mooney, "By the way—perhaps you saw the (Hangover) house pictures in the *Architectural Record* or *Forum* or some such magazine, last October. How'd you like that gem in the text—'Being in earthquake country, sparsely settled and without adequate fire protection . . . ' Not just the way the C(hamber) of C(ommerce) would have put it! Or have all the city fathers of Laguna blown up in little bits by this time, on their own gas? I hope so. You see, you can't rub the hay off a real Lagunan just by sending him halfway round the world!"

4. For Halliburton's remarks, see "Letters from the *SD #2*," January 18, 1939 (quoted).

5. For Canton trip, see Halliburton, "Letters from the *SD #3*," January 27, 1939 (quoted). Also see Halliburton, "Log of the *SD #9*," January 31, 1939. For dating it to January 2, see *RHL-P*, January 1, 1939. Compare Powell, *My Twenty-Five Years in China*, op. cit., 314 passim. Of Halliburton's run-ins with brigands, especially in the stretch between Macao and Hong Kong, and for the mishap of the *Sui An*, see *Royal Road to Romance*, supra, 335–40.

6. Like Halliburton, Nellie Bly thought Canton was the soul of China. See Bly, *Around the World*, op. cit., 248. For Canton (Guangzhou) today, see "Chinese City (Shanghai) Known for Cars Moves to Limit Them," *New York Times*, International, September 5, 2012, A1 and A3. For the descriptions of war-ravaged Canton, see Halliburton, "Letters from the *SD #3*," January 27, 1939 (quoted).

7. Ibid., Halliburton, "Letters from the *SD #3*," January 27, 1939 (quoted). For Xu Xiake, see Tony Perrottet, "Traveler in the Sunset Clouds: The Indiana Jones of Imperial China Has Become a Modern Pop-Culture Celebrity," *Smithsonian*, April 2015, op. cit., 40–56. In 1641 Xu Xiake died of malaria shortly after his famed trip.

8. Ibid., Halliburton, "Letters from the *SD #3*," January 27, 1939 (quoted). For "10-hour ride," see "Letters from the *SD #3*," January 27, 1939 (quoted). For Canton trip, see Gordon Sinclair, "The Last Days of Richard Halliburton," in *Bright Paths to Adventure*, 85–99, 93–94 (quoted). For the historic beauty of Canton and its environs, see *Views of the Pearl River Delta: Macao, Canton and Hong Kong* (Urban Council of Hong Kong, 1996).

9. Ibid., Halliburton, "Letters from the *SD #3*," January 27, 1939 (quoted). For "queen city of South China," see Sinclair, *Bright Paths to Adventure*, supra, 93–94. For old Canton, see Halliburton, "Log of the *SD #13*," March 2, 1939 (quoted). Also see Halliburton, *Royal Road to Romance*, op. cit., 333–34 passim.

10. For fallen trade, see Halliburton, "Log of the *SD #4*," December 27, 1938 (quoted).

11. For assurances to his parents, see *RHL-P*, January 23, 1939 (quoted). For "vitamin pills" and "troubles," see Sinclair, *Bright Paths to Adventure*, op. cit., 93 (quoted).

29. History Is Made at Night

1. Halliburton well knew the *HMS Bounty* incident of April 1789, during which nine men under the leadership of Fletcher Christian assumed control of the ship and set Captain Bligh adrift in a launch with eighteen able-bodied seamen. What triggered the mutiny, say some, was not a few disgruntled sailors' outrage over a captain's alleged cruelty, but a few sexually repressed wantons' outburst of lust. See, in general, John Toohey, *Captain Bligh's Portable Nightmare* (New York: HarperCollins, 1998).

2. For Torrey, a letter from John Potter to Edward Howell, June 15, 1946, is quoted. For John Potter comment, see Cortese, *RHRR*, 181. For other comments, see Max, *HC*, 186–91. In a September 10, 1987 letter to Zola Halliburton, seventy-four-year-old Potter admitted to losses in his own hearing and vision; he also admitted to "the restraint inherent in (his) comments" during a taped interview with Michael E. Blankenship about the *Sea Dragon* and the people involved in the enterprise. Relevant items are to be found in the MEB Archive. For Welch "lecturing" the lads, see *LHK* #14, February 13, 1939 (quoted).

3. See Rowland, *Slipping the Lines*, op. cit., 242 (quoted), 250–51 (quoted).

4. Welch introduces the descriptive "the old complaint" in *LHK* #13, January 29, 1939. Rowland uses "Cupid's Revenge" in *Slipping the Lines*, op. cit., 251. He does not name the lad who had it. For Russians admitted to hospital, see *LHK* #13, January 29, 1938 (quoted). Syphilis itself can take many forms. DNA analysis has determined that it was imported to Europe from the Americas by the crew of Christopher Columbus. One theory states that Ponce de Leon's later search for the "Fountain of Youth" was in fact a search for a cure to all deadly venereal diseases. Except for the dedicated doctor, the main characters in an earlier British film *Damaged Goods*, which was already dated when it first appeared in 1919, are represented as ignorant or misinformed about VD. As with other VD propaganda films, the film attaches moral disgrace to the disease and recommends a proper marriage as the key to redemption. In 1940, making the scourge of syphilis better known to the public, Warner Brothers produced *Dr. Erhlich's Magic Bullet* starring Edward G. Robinson as the pioneering physician. By that time, syphilis (as chlamydia and gonorrhea) was renowned as a fairly common yet treatable disease. Penicillin revolutionized the treatment of infectious disease and eradicated most forms of VD, but in the pre-penicillin days, after 1910, one relied upon arsphenamine for a reduction of symptoms. In wartime the drug was hard to acquire. Also known as Salvarsan or compound 606, this first organic antisyphilitic chemosynthetic drug contained traces of arsenic and requires sustained weekly injections to effect a cure. Side risks for its use were skin rashes and liver damage. For leprosy in China, see Angela Ki Cei Leung, Angela, *Leprosy in China—A History*, Studies of the Weatherhead East Asian Institute (New York: Columbia University Press, 2009). Consulted was Irving Simons, *Unto the Fourth Generation: Gonorrhea and Syphilis; What the Layman Should Know* (New York: E. P. Dutton, 1940). Also see John Firth, "Syphilis—Its Early History and Treatment Until Penicillin and the Debate on its Origins," *Journal of Military and Veterans' Health*, History Issue, 20, no. 4 (November 2012). The key sources say little about medical provision or emergency care aboard the *Sea Dragon*, but the ship had a "big medicine chest." Besides quinine to alleviate muscle cramps, aching joints, and malaria, the chest presumably, carried "clap medicine." See *LHK* #14, February 13, 1939.

5. For idle hands, see Welch, *LHK* #13, January 29, 1939 (Sunday). For "clapped up," see *LHK* #14, February 13, 1939.

6. See Welch, *LHK* #14, February 13, 1939 (quoted). Torrey was "paid off." According to John Potter, Torrey (1913–1992) "remained in Hong Kong until the Japanese invaded the island (December, 1941) and (then he) escaped to the Philippines." Next, "the Japs started to catch up with him there (and he went next) to French Indo-China, only to be driven out once more. He finally ended up in India (Madras, I think) and when I was in India on Naval duty this past year (1945) I heard he was still working for Caltex Oil—his original employer in Hong Kong." Letter to Edward Howell, January 15, 1946.

7. For "married," "best bars," and "articles," see Welch, *LHK* #13, January 29, 1939 (quoted).

8. For crew work habits, see ibid. (quoted).

9. For crew defects, "brothel" reference, and Welch's assessment of the lads' work habits, see ibid. (quoted).

10. For "New Zealand aviator," see ibid. (quoted).

11. Captain Welch reported the incident days before the first crossing. See ibid. Without the benefit of Welch's letters to guide him, Root places the time of the Sligh incident after the first attempted crossing. See Root, *HTMM*, 32.

12. For "showdown, conference summary, whose topics were authority, "new contract," and "cumshaw," see Welch, *LHK* #13, January 29, 1939 (quoted).

13. Ibid., all quotes. For Torrey, see note 6 above.

14. Ibid., Welch, *LHK* #13, January 29, 1939 (quoted). For Mooney's help, see *RHL-P*, November 2, 1938 (quoted).

15. For photograph of junk, see *His Story*, opposite 386.

16. For *Pang Jin*, see Welch, *LHK* #7, November 25, 1938; and *LHK* #13, January 29, 1939 (quoted). For wanting to get home, see *LHK* #13, January 29, 1939 (quoted). For the "two logs," see *LHK* #6, November 24, 1938 (quoted).

30. *The Awful Truth—Captain Jokstad Inspects the* Sea Dragon

1. For "ministrations," "minor details," and "first 'sailing' date," see Halliburton, "Letters from the *SD* #3," January 27, 1939 (quoted). For "tightening," also see Welch, *LHK* #11, December 16, 1938. For sailing date and for adding "dragons," see *RHL-P*, January 1, 1939. The phrase "fixed sailing date" (late in January) appears in "Letters from the *SD* #4," February 16, 1939. His specialties masonry and iron manufacturing, Wesley Halliburton was a graduate of Vanderbilt who had done advanced study at the Massachusetts Institute of Technology.

2. For "autographs" and amusing the Chinese, see Welch, *LHK* #13, January 29, 1939 (quoted).

3. For junk qualities, see Halliburton, "Log of the *SD* #12," March 1, 1939, *SFN* (quoted). Compare his remarks to those he made about his airplane the *Flying Carpet*: "It is quite the most graceful and beautiful ship I've ever seen," he told friend Noel Sullivan. "Gold wings, with a scarlet body and a black forepart. Whenever we land, crowds of people gather around to admire it. My ever present concern has been to have the ship beautiful, otherwise, I could not love it nor enjoy my expedition." Halliburton letter to Noel Sullivan, December 15, 1930, Noel Sullivan Papers, Bancroft Library, University of

California—Berkeley. For "Royal Air Force officers," see Halliburton, "Letters from the *SD #2*," January 18, 1939 (quoted.)

4. For "special pride and joy," see Halliburton, "Log of the *SD #12*," March 1, 1939 (quoted); compare Halliburton, "Letters from the *SD #2*," January 18, 1939. For an extended treatment of these topics, see William Elliot Griffis (1843-1928), *China's Story in Myth, Legend, Art, and Animals* (Boston: Houghton Mifflin, 1911); reprinted, with it added "Brought Down to Date," 1935. If Halliburton consulted books on Chinese myth and graphic design, Griffis's book would have ranked high on his reading list. For "pictures obtained," see "Letters from the *SD #2*," January 18, 1939. Halliburton might have taken for his inspiration the stern of the Foochow pole junk brightly painted on the dust jacket (and again on the frontispiece) of Ivon A. Donnelly's classic *Chinese Junks and Other Nautical Craft*. The first and second editions, especially with the dust jacket, are rare, but the work has been reissued with an introduction by Gareth Powell. For "earnest conferences," see Halliburton, "Log of the *SD #12*," March 1, 1939 (quoted). For "ship's eyes," "traditional colors," and "circus-wagon," see "Letters from the *SD #2*," January 18, 1939 (quoted).

5. For publicity and "shameful excuses," see Welch, *LHK #13*, January 29, 1939. For Halliburton's preoccupation with finances, see Root, *HTMM*, 32. For appeals to America for money, see Halliburton, "Log of the *SD #16*," March 19, 1939. On March 13, 1939, the *Boston Sunday Globe* published an article conflating many of the events, making their precise day-by-day passage dependent on Welch's and Halliburton's own letters. Reads the article's byline: "Halliburton's Junk Battles Storm at Sea—Adventurer and His Crew Put Back to Hong Kong to Rush Injured First Mate to Hospital." For "weather" and "interviews," see *LHK #13*, January 29, 1939 (quoted). Mooney sustained his injury before the *Sea Dragon* departed on February 4—perhaps on January 28. For "replenishing," see "Log of the *SD #16*," March 19, 1939, *Boston Sunday Globe* (quoted). For food for twelve men, see Halliburton, "Letters from the *SD #4*," February 16, 1939 (quoted). Concerning the crew's dietary needs, details about the junk's inventory are lacking, yet food stores might have consisted of apples, bread, canned fish, and meat. As a protection against scurvy, large amounts of cabbage and vinegar—this revolutionary idea the proud dietary legacy of Captain Cook—might have been stowed. Besides "some rum," the ship held bottles, if not cases or barrels, of whiskey, and, to serve Halliburton's needs, a ready supply (I assume) of tomato juice. Numerous cartons of cigarettes were also necessary. For rum, "work," and "worry," see *RHL-P*, January 1, 1939 (quoted). For being a "mental wreck," see Halliburton, "Log of the *SD #10*," February 1, 1939, *SFN* (quoted). For the "slight skin itch," see *RHL-P*, February 23, 1939. For being "in bed," see Root, *HTMM*, 22 (quoted). For insect sting, see *RHL*, September 22 and October 1, 1922, *His Story*, 181-83.

6. See Welch, *LHK #13*, January 29, 1939 (quoted). For Lloyds, see Root, *HTMM*, 28, who referred to the document quoted in Collins, "Royal Road Across the Pacific," 504. Full text of the policy is not available. The insured figure of $10,000 was also the amount of a federal war-risk insurance policy payable to the Halliburton estate following proof of his death. An unidentified newspaper source (text located in the *Richard Halliburton Collection at Rhodes College*) notes the policy and includes a clause saying, "insurance companies holding policies for approximately $50,000 waived defense and signified willingness to pay after court certification of death." October 5, 1939. I note Lloyds' caveats. Evidently the company was not as knowledgeable about trans-Pacific routes

as it was about Atlantic ones. Consulted was *"Lloyd Routes Over the Seven Seas," Seven Seas* magazine, published by the North German *Lloyd* (from 1928). For "old Swede," one Anderson, see Welch, *LHK* #13, January 29, 1939. Welch mistakenly called Jokstad, who was Norwegian, a Swede whose ship was "docked nearby." For arrival time of SS *Pierce*, see Root, *HTMM*, 26 (quoted). For Jokstad's earlier run-in with Halliburton, see *Royal Road to Romance*, supra, 369–30. For Hong Kong, see Jokstad, *Captain and the Sea*, op. cit., 187 (quoted). Collins's remarks (quoted) are to be found in S. L. Kahn. *Commercial Appeal* (Memphis, Tennessee), after June 10, 1939 (undated clipping sent to Ralph B. Vawter by Wesley Halliburton) with letter from Wesley Halliburton to Vawter, December 6, 1940. Author's collection.

7. For investment by the lads, see Jokstad, *Captain and the Sea*, 187–88 (quoted). For Jokstad on Welch, see 189; for suggested route, see 188 (quoted). Halliburton said that Flagg, identified as "a recent graduate of Bowdoin College" and "a crew member of the President Pierce," had "missed his ship in Hong Kong." When "found (by Halliburton and company) on the beach," he was "practically shanghaied aboard the junk." See Halliburton, "Log of the *SD* #16," March 19, 1939, *Boston Sunday Globe* (quoted). My thanks to Caroline Moseley, George J. Mitchell Special Collections and Archives, Bowdoin College for sharing the file of John Benjamin Flagg, class of 1935, who prepared for Bowdoin at the Middlesex School in Concord, Massachusetts. "At Bowdoin he majored in history. He was a member of the freshman track and cross country squads, and ran on the varsity team as a sophomore. He was a member of the Delta Kappa Epsilon fraternity." See letter to author, February 13, 2013. Information about Flagg's work record was given to the Bowdoin archive by his mother, Ethel Marie Flinder Flagg. She said of her son, "When not at sea, he stayed either with me or with his sister in New York—no permanent address." Ben's father was Charles Flagg. Besides working for the President Lines, Ben had worked on the Isthmian, Calmar Steel Company, and American Export lines. Flagg is pictured alongside Halliburton and traveler and businessman John Rowland in Rowland, *Slipping the Lines*, op. cit., 480.

8. See Root, *HTMM*, 27 (quoted). The inspection, as reconstructed here, differs from Root's in its sequencing of events. For instance, Root places the routing information last and not first. As Halliburton's and Welch's letters do not duplicate, or even suggest, the substance of Captain Jokstad's remarks, Jokstad himself must be considered an only source. Almost certainly a correspondent of Jokstad's, Root might also have conducted a personal interview with him, who lived in San Mateo, near San Francisco where Root was employed at the *San Francisco Chronicle*. Root did consult with Thomas Wheeler of the American President Lines, but it is not clear whether Wheeler provided him with company records. Root's *Richard Halliburton—The Magnificent Myth* appeared in 1965; Jokstad's *The Captain and the Sea* appeared in 1967, the year, incidentally, that Root died. Significant portions of Root's analysis of the *Sea Dragon*'s design issues quote or reflect opinions given to him by Jokstad. Thus it seems reasonable to suppose that Jokstad read Root's book, and, in his own later account, offered revised recollections of the inspection. For the Jokstad inspection, see Root, *HTMM*, 26–28, quoted at 28.

9. For "cockroaches," see Jokstad, *Captain and the Sea*, supra, 187–88 (quoted). For basic background information about Jokstad, and for eating lunch together and strolling over to the ship, see Root, *HTMM*, 26–27. For the terse manner attributed to Jokstad, see Root, *HTMM*, 27 (quoted). For the "rat" omen, see Jokstad, *Captain and the Sea* ,187 (quoted).

10. For Jokstad and Welch, see Root, *HTMM*, 27 (quoted). For Jokstad and the crew, see Jokstad, *Captain and the Sea*, supra, 188. For "leg screws," see Root, *HTMM*, 27. For Jokstad on the same issue, see *Captain and the Sea*, supra, 188. Sails generally are guyed: they are held to the deck by ropes or cables called guy lines and secured to a chain-plate beneath the planking which is itself held to the hull by bolts. Bolts require nuts; screws do not. For "galvanized iron bolts," see Collins, who quoted the *Survey Report*, "The Royal Road Across the Pacific," op. cit., 504 (quoted). It is unclear what changes Welch made to the rigging, but it might have been reconfigured to allow for more speed. For variations on sail format, see Neville Wade, "Sailing a Model Square-Rigger," *Marine Modelling International*, April 2008, online.

11. For "tackle" and "endless fall," see Welch, *LHK* #9, December 5, 1938 (quoted). For "tiller," see Halliburton, "Log of the *SD* #12," March 1, 1939 (quoted). For Jokstad "waggling" the tiller, see Root, *HTMM*, 27 (quoted). For Jokstad's assessment of the rudder, see *Captain and the Sea*, 188 (quoted). For rudder issues, compare Paul Chow, *The Junk That Challenged the Yachts* (self-published, 2011), 90–91. For Welch's description of the tiller, see *LHK* #9, December 5, 1938 (quoted). See Wolfgang Asbach, *"The Hangzhou Trader – A 1/30 Scale Model of a Trading Junk from the Shanghai Area,"* Model Shipbuilder, No. 81–84, 1992–1993 *(Mariner's Museum Library)*, 3–11. For the rudder and the rigging, again see *LHK* #9, December 5, 1938. For a picture of the "balanced rudder," see Fitch, "Life Afloat In China," op. cit., 672.

12. For below deck inspection, see Root, *HTMM*, 27. For "backers," see Jokstad, *Captain and the Sea*, supra, 188. Also see Root, *HTMM*, 27, 28 (quoted). Jokstad recommended "backers," but he did not say whether these should be steel sheets; he later learned that Halliburton used wooden boards as backing. For the diesel engine, see Root, *HTMM*, 27 (quoted). For troubles with the engine aboard the *Snark*, see London, *Cruise of the Snark*, op. cit., 34–35 (quoted). For engine, see Rowland, *Slipping the Lines*, 251 (quoted).

13. For ballast, see Root, *HTMM*, 27 (quoted). Compare Jokstad, *Captain and the Sea*, supra, 188 (quoted). Root had not read Welch's letters, but he had interviewed Captain Jokstad. For interior, see Rowland, *Slipping the Lines*, op. cit., 249 (quoted). For "heavily built," see *RHL-P*, January 1, 1939; and Welch, *LHK* #5, November 16, 1938.

14. For recommended trial run, see Jokstad, *Captain and the Sea*, 189 (quoted). For Halliburton at the tiller, see Welch, *LHK*, January 28, 1939, quoted. For Halliburton's comments on the tiller, see "Log of the *SD* #13," March 2, 1939, *Boston Sunday Globe*, March 1, 1939 (quoted). For Halliburton taking the wheel on an earlier adventure, see *Royal Road to Romance*, op. cit., 136–37. For crew and tiller, see *Letters of the SD* #4, February 16, 1939. For difficulty of using tiller, see "Log of the *SD* #13 (quoted).

15. For Welch fighting hard to remove poop, see *LHK* #13, January 29, 1939. For being "very heavy with the engine," and for the "motor," see *LHK* #13, January 29, 1939 (quoted). For fighting hard to remove poop, and for "honky-tonk wagon," see *LHK* #13, January 29, 1939.

16. For Halliburton's commitment (provided in dialogue form), see Root, *HTMM*, 28 (quoted). For Halliburton taking his chances, see Jokstad, *Captain and the Sea*, op. cit., 188 (quoted).

17. For "cold feet," see Welch, *LHK* #13, January 29, 1939 (quoted). For "colony of woodpeckers" repairing the *Empress of Japan* (photographed), see "Shipping News," *South China Morning Post*, Monday, January 16, 1939 (quoted). For photograph of the

Sea Dragon alongside the *Empress of Japan*, see Getty Images 27 Richard Halliburton Pictures (and) Photos, online. For finances, see note 6 above. For "insurance policy" and "debts" cleared, see *RHL-P,* January 23, 1939 (quoted). For "first week's income," see *RHL-P,* February 23, 1939 (quoted).

31. A Gentleman Knows When It's Time to Leave

1. For festivities, see Gleason, *Tales of Hong Kong,* 49. Also see Alt, *DDIB,* 343 (quoted). For "Race Week" which followed the Chinese New Year, see Leys, *After You, Magellan!,* op. cit., 241. For "spirit festivals," see Ormond McGill and Ron Ormond, *Religious Mysteries of the Orient* (Cranbury, NJ: A. S. Barnes, 1976), quoted at 97. For the celebrations as a nuisance to Halliburton, see *RHL-P,* February 22, 1939. For Honolulu arrival time, and a stopover arranged by Halliburton representative in Hawaii Francis J. Brickner, see *Honolulu Star-Bulletin,* February 9, 1939 (Thursday).

2. For delays and Halliburton's dream day, see "Log of the *SD* #16," March 19, 1939, *Boston Sunday Globe* (quoted). For financial matters, see *RHL-P,* January 23, 1939. For money arriving, source not noted and exact time of arrival—after January 20 or after February 10 unclear, see Halliburton, "Log of the *SD* #16," March 19, 1939, *Boston Sunday Globe.* For money from Bobbs-Merrill, "friends" (possibly) and Barstow, see Alt, *DDIB,* 342–43. For Chambers, see Root, *HTMM,* 104 et passim. For ships "in the great port" and crowds that gathered, see "Log of the *SD* #16," March 19, 1939 (quoted). For Halliburton's thoughts on the Panama Canal swim, see *New Worlds To Conquer* (Indianapolis: Bobbs-Merrill, 1929), 94–95 (quoted).

3. The "docks" obscures the point of departure, presumably Bailey's. Evidently the *Sea Dragon* was alternately docked in Victoria and in Kowloon—these relocations after the junk left Fat Kau's shipyard. Collins notes that Bailey's Harbor was in Kowloon. Quoted in S. L. Kahn, *Commercial Appeal* (Memphis, TN), after June 10, 1939. Cf. Root, *HTMM,* 30. Earlier, Root notes that the SS Pierce was docked in Hong Kong "at the pier next to the one at which the *Sea Dragon* was moored" (26), probably in Kowloon. For the SS *Coolidge* occupying the other half of the wharf, see Halliburton, "Log of the *SD* #16," March 19, 1939 (quoted). For Bailey's Wharf and W. S. Bailey and Company, see "Hong Kong's Shipbuilding Industry," *Weekly Commercial News,* General Newspaper Specializing in Shipping, Foreign and Domestic Trade and Industries, August 9, 1913, 4, vol. 47, n. 6, online. Also see "The Industrial History of Hong Kong Group," May 14, 2019, online. For comment on bravery, see Collins, "Royal Road Across the Pacific," op. cit., 504 (quoted). Halliburton was not so naïve that he didn't know the worst that could befall him. See *RHL, His Story,* 405–6.

4. For references to early lecture appearances, see Root, *HTMM,* 100–101, quoted at 100.

5. Description of the apparel crew members wore that day is based on the photographic and film records. What gear they brought along might already have been stored aboard the junk. Compare Max, *HC,* 214. For crew, and animals, see Halliburton, "Letters from the *SD* #4," February 16, 1939. For possessions "shipped home on the large freighters," see Sarah Pratt, *SFN,* April 18, 1939.

6. Compare Root, *HTMM,* 28. For ceremony, see Halliburton, "Letters from the *SD* #4," February 16, 1939 (quoted). For the departure, see Collins, "Royal Road Across the Pacific," supra, 502 (quoted). For Halliburton's report of the departure, see "Letters

from the *SD #4*," February 16, 1939 (quoted). Also quoted was *RHL-P*, November 16, 1938. Also see Rowland, *Slipping the Lines*, op. cit., 251 (quoted).

7. At little or no cost, radio communication offered added security on a long voyage. Months earlier, on November 16, 1938, Welch had let his friend Dick Wetjen "bloody know" that, as he put it, "Being the bloody master I am entitled to buckshee radio messages." See *LHK #5*, November 16, 1938. Radiogram dated February 4, 1939, appears in *His Story*, 424. For the first leg of the voyage, see Halliburton, "Log of the *SD #16*," March 19, 1939, (quoted); Halliburton, "Letters from the *SD #4*," February 16, 1939, is nearly identical. See Root, *HTMM*, 28 (quoted). For "China coast" route, see *LHK #14*, February 13, 1939 (quoted). Consulted for the Collins testimonial was Kahn, *Commercial Appeal* (Memphis, TN), after June 10, 1939 (undated clipping sent to Ralph B. Vawter by Wesley Halliburton), (quoted). Collins reported the matter thus: "It was the intention of Captain Welch and Halliburton to sail up the China coast until about on the same latitude as northern Formosa, then sail eastward to Kiirun (Kelung)." See "The Royal Road Across the Pacific," op. cit., 502 (quoted). For Halliburton remarks, see "Log of the *SD #16*," March 19, 1939, *Boston Sunday Globe* (quoted). See also Halliburton, *Flying Carpet*, op. cit., 13 (quoted). Potter's recollections appeared in "Shipping News," *China Morning Post*, April 8 or April 15, 1939 (quoted); their source was *San Francisco* wire service for March 30.

8. For the first few hours, and absence of Japanese patrols, see Halliburton, "Log of the *SD #16*," March 19, 1939 (quoted); also see Halliburton, "Letters from the *SD #4*," February 16, 1939 (quoted). For "midnight," "bright spectacle," progress made, and "8000 miles" calculation, see "Letters from the *SD #4*," February 16, 1939 (quoted). For "fishing junks" and "fleet of fishers," see "Log of the *SD #16*," March 19, 1939 (quoted).

9. For "blissful dream voyage," see Halliburton, "Letters from the *SD #4*," February 16, 1939 (quoted). Also see Halliburton, "Log of the *SD #16*," March 19, 1939 (quoted). For route, see Collins, "Royal Road Across the Pacific," supra, 502–4, map insert at 503. For junk rolling, see Potter, "Shipping News," *China Morning Post*, April 8 or April 15, 1939, op. cit. (quoted). For being "slow under canvas," see Welch, *LHK #13*, January 29, 1939 (quoted). For junks in "dangerous seaways," nose-diving, and "bad steering," see *LHK #14*, February 13, 1939 (quoted).

10. For "auxiliary engine" and "fumes," see Halliburton, "Log of the *SD #16*," March 19, 1939 (quoted). For "decks awash," ship pitching and rolling, and crew "half-dead from seasickness," see Halliburton, "Letters from the *SD #4*," February 16, 1939 (quoted). For everything "not fastened" flying about, and the way in which one "wave pitched" the junk, see "Log of the *SD #16*," March 19, 1939 (quoted). For aerial, see Potter, "Shipping News," *China Morning Post*, April 8 or April 15, 1939, op. cit. (quoted). For "tiller and compass" and for "downpours" and being undiscouraged, see "Letters from the *SD #4*," February 16, 1939 (quoted). For Ben Flagg fixing the aerial, and wicked weather, see "Log of the *SD #16*, March 19, 1939 (quoted).

11. For the "slamming" and the behavior of the junk, see Welch, *LHK #14*, February 13, 1939 (quoted).

12. For "six o'clock" and sailing around the southern tip," see Halliburton, "Letters from the *SD #4*," February 16, 1939 (quoted). For "light-house," "Formosa" and Potter'" ruptured," see "Letters from the *SD #4*," February 16, 1939, quoted; also see *His Story*, 428 (for generalized chain of events). For Welch's decision, see Halliburton, "Log of the *SD #16*," March 19, 1939 (quoted). For Collins remarks, see "Royal Road Across the

Pacific," *supra,* 502–4, map insert at 503 (quoted). No bow thrusters existed at this time to assist the junk. Contributing to "blurred vision" may have been "mirages," optical illusions during winter from the play of light and light refractions on the water in the region. See G. S. P. Heywood, "Clouds and the Weather," *Hong Kong Naturalist,* April 1939, 119–28, at 128. For Potter's comments, see Taylor, *SSMTWOD,* 210 (quoted); also Potter interview (quoted) with William R. Taylor (1994), "A Beautiful Casket," *Richard Halliburton Biography—BOLD adventurer, globe trotter,"* online. Also consulted was letter of John Potter, March 26, 1994, MEB Archive. For Welch making good headway on the southern course, see *LHK #14,* February 13, 1939. For Flagg, see *LHK #14,* February 13, 1939. For "Potter prostrate," see Halliburton, "Log of the *SD #13,*" March 2, 1939, or "Log of the *SD #16,*" March 19, 1939 (quoted).

13. For Halliburton's account of Potter's condition—"high fever" and "flushed face"—see "Letters from the *SD #4,*" February 16, 1939. For rupture, see Halliburton, "Log of the *SD #13,*" March 2, 1939, *Boston Sunday Globe,* March 19, 1939. For Potter's assessment, consulted was his letter to Edward Howell, January 15, 1946 (quoted).

14. For "Lyman Island" and "Gap Rock," see Welch, *LHK #14,* February 13, 1939. For "treacherous" waters, see Halliburton, "Letters from the *SD #4,*" February 16, 1939 (quoted). For return, see Halliburton, "Log of the *SD #16,*" March 19, 1939. For Tai Toa Fat, see "Log of the *SD #16,*" quoted. For Potter "sneaked ashore," see *LHK #14,* February 13, 1939 (quoted). For "ambulance," see "Letters from the *SD #4,*" February 16, 1939 (quoted).

32. A Priest, a Whore, and a Siamese Cat Walk Into a Bar . . .

1. For "series of misadventures," see Halliburton, "Letters from the *SD #4,*" February 16, 1939 (quoted). Importantly, American followers of the *Sea Dragon Expedition* and its progress, experienced news delays: what was reported to them and what was occurring in so-called *real time* could be as many as five or six weeks apart. For instance, "Log of the *SD #16,*" appeared in the *Boston Globe* on March 19, 1939. The log recounted the first aborted crossing of February 4, giving a false impression of recency, as it coincided with actual news reports of the second crossing of March 4. Coincidentally, with "Log of the *SD #15,*" March 4, 1939, *SFN* suspended its publication of the *"Log of the Sea Dragon."* Although the story was now back-page news, newspapers continued to report the latest information about the *Sea Dragon.* "Log of the *SD #16*" was the syndicated *News-Press* version; nearly identical to the fourth number of the "Letters from the *Sea Dragon,*" it appeared March 19, 1939. Biographies of Halliburton, with the advantage of hindsight, have *not* been confused by the two sailings, and describe each departure as a distinct sailing event—what was once not so clear to readers of the newspapers at the time. See, for instance, Alt, *DDIB,* 346 et seq. For "all the news," see Welch, *LHK #14,* February 13, 1939 (quoted).

2. Ibid., Welch, *LHK #14,* February 13, 1939. Also see Halliburton, "Letters from the *SD #4,*" February 16, 1939.

3. Ibid., Welch, *LHK #14,* February 13, 1939 (quoted). For "torments" and "high fever," see Halliburton, "Letters from the *SD #4,*" February 16, 1939 (quoted). Also see Halliburton, "Log of the *SD #16,*" March 19, 1939. Cf. "Letters from the *SD #4,*" February 16, 1939. For good beginnings and worsening conditions, see *LHK #14,* February 13, 1939 (quoted). Also quoted is John Potter's letter to Edward Howell, January 15,

1946, op. cit. Cf. Taylor, *SSMTWOD*, 197–98. Potter held to the story. See "Comment" of WRT ("Bill Taylor") to Stuart Heaver, "Richard Halliburton—The Hero Time Forgot," *China Morning Post*, March 23, 2014 (online), which quotes John Potter's recounting of events in 1994. Also consulted was the *China Morning Post*, April 8 or 15, 1939, op. cit.

4. For "council" and "clap" see Welch, *LHK* #14, February 13, 1939 (quoted).

5. The radiogram appears in *His Story*, 424. For letter to his parents, see *RHL*, February 10, 1939, *His Story*, 424–25.

6. For "$500," see Root, *HTMM*, op. cit., 31 (quoted). Compare *RHL-P*, February 23, 1939 (quoted). For financial matters, see above, "A Gentleman Knows When It's Time To Leave," at note 2. Needed for a fuller picture of his implied financial distress are letters he sent to Wilfred Crowell and John Masterson. See *RHL-P*, February 23, 1939 (quoted). For the "race," see "Flight of the Bumblebee," chapter 23, note 9 above.

7. Consulted was Wesley Halliburton's letter to his son Richard, March 24, 1939, RHC, Rhodes.

8. See Welch, *LHK* #14, February 13, 1939, quoted. The names of Welch's companions are unknown. The "somebody" referenced is not identified. Chase was probably the "pure one" in this account, the only known written view of his nightlife in the Colony. From the context it seems reasonable to presume the "Kowloon belle" was "Chinese." Welch likened the "scuffle" to one he apparently witnessed "outside of the St. Francis (in San Francisco), if you get what I mean." He could have been referring to the Maritime Strike of May 1934 that involved longshoremen and by July had turned violent.

9. For "scabies," see Welch, *LHK* #14, February 13, 1939 (quoted). For "slight skin itch," see *RHL-P* (omitted in *His Story*). For Macao trip, see Rowland, *Slipping the Lines*, 249 (quoted). For "Nantucket doctor," see Root, *HTMM*, 103 (quoted). For "hay fever," see *RHL-P*, June 18, 1938.

10. For the keel, see *RHL*, February 23, 1939, *His Story*, 431 (quoted). Compare Root, *HTMM*, 30 (quoted). Quoted also is Kahn, *Commercial Appeal* (Memphis, TN), after June 10, 1939 (undated clipping sent to Ralph B. Vawter by Wesley Halliburton), op. cit., June 10, 1939. For the keel *not* being installed, see Collins, "Royal Road Across the Pacific," op. cit., 506. Also see Potter, who, writes Taylor, "scoffed at the practicality of such an undertaking on an already constructed boat of its size." See Taylor, *SSMTWOD*, 205–6, quoted at 206. Taylor, who quotes Collins, discusses the issue. In the letter dated February 22, Halliburton, adding a crude sketch of a "fin keel" he had drawn, assured his parents, "[Such a keel] has been put on." Ten photos (apparently lost) of the junk were sent to Wetjen; even if one of these shows the fin keel attached, or being attached, to the hull, it still remains possible that the keel was later removed later. A delicate issue was insurance, and the surveyor's insistence that, if the ship was to be underwritten, a strong keel had to be in place. If it is true Halliburton even procured a new keel, time and cost restraints may have induced him to forego its installation.

33. Proudly We Hail

1. For activities of Chase and Barstow, see Welch, *LHK* #14, February 13, 1939 (quoted). Of Barstow's guest, Welch says only that "they" went into the forecastle.

2. For "Chase" and "Flagg," and for "Johannsen" and "dinner over at the R.A.F.," see ibid. (quoted).

3. Quoted is letter from Paul Mooney to artist Eugene Hoenigsberg, *Brooklyn Eagle*,

October 5, 1939, op. cit. Mooney shared an apartment in Brooklyn with Hoenigsberg from 1934 to 1935. For Mooney's last known letter, to his mother Ione Gaut Mooney, see Max, *HC*, 223–24. For wartime conditions, also see *PML*, letter to Harriette Janssen, November 24, 1938; and letter to Alice Padgett, December 31, 1938. For "no turning back," see *RHL-P*, February 23, 1939 (quoted).

4. For the undated survey report, see Collins, "Royal Road Across the Pacific," op. cit., 504 (quoted). A surveyor from Lloyds, making his inspection before the junk's first attempt at a trans-Pacific crossing on February 4, noted that the ship was "solidly constructed." See Root, *HTMM*, 28. For ship insured for $10,000, also see Root, 28. For earlier assessment by Lloyds, see "The Awful Truth," Chapter 30, note 6 above. For talk with *Dollar Lines* "manager," see Welch, *LHK* #14, February 13, 1939. Collins probably had been told that the *Sea Dragon* would depart between February 20 or 25; he does not seem to have been informed that the junk left Hong Kong on March 4. Given the difference of ten days, between February 25 and March 4, that would have placed the rendezvous about 200 miles west of Midway rather than 1,200 miles west of Midway. For round-trip time of the SS *Coolidge*, see Collins, "Royal Road Across the Pacific," 501–2. Collins wrote that the liner departed from San Francisco on March 13 bound for Honolulu, then departed from Honolulu "bound for Yokohama via the Southern Track (Honolulu to 30 degrees North, 150 degrees East, rhumb line, thence direct to Yokohama." See "Royal Road Across the Pacific," 505.

5. For spring "arrival" and "profit," see *RHL*, March 3, 1939, *His Story*, quoted at 432. For weather, see *RHL-P*, February 23, 1939, *His Story*, quoted at 431. Also noted are "ghastly delays." For crew, see Welch, *LHK* #14, February 13, 1939. Also see *RHL-P*, February 23, 1939 (quoted). For Richard Davis, see Max, *HC*, 207, 208 (pictured); also see Max, *HC*, 211; and *His Story*, 425. Davis might have worked for foreign companies as a petroleum consultant, and he might have been on assignment for the very shipyard where the engine was fitted. But he too was "green" to the sea. On January 19, 1939, the *Hong Kong Daily News* published an article stating, "Adventure is his meat." Halliburton likely gave the press the information, and Davis likely offered his services to Halliburton in exchange for a ride to Honolulu. Slated to go on the first sailing, Davis bowed out on the second. Torrey perhaps swayed him to do so, but Davis's plans to marry provided the perfect excuse. Root notes that the two white (replacement) hires were "semi-professional seamen," one of whom jumped ship from the SS *Coolidge*, the other from the SS *Pierce*. Ben Flagg had been aboard the first aborted voyage. See Root, *HTMM*, 33. It seems that a second recruit, Ralph Granrud from the SS *Coolidge*, made the "rash decision" to join the *Sea Dragon* crew for reasons unknown, according to Dale Collins. See "Royal Road Across the Pacific," op. cit., 504. Granrud remains a mysterious stranger among the junk's crew. The *Errol Flynn Blog* notes a Ralph Granrud (from Tacoma, Washington, and first husband of blogger Randall's mother); Granrud captained boats for the actor in the mid-1930s and, later (as far as "Randall" could determine), he was lost at sea (going north from Australia). Granrud could have replaced Jack Rutherford of Huntington Beach, also of the *Coolidge*. Reporter Jack Smith noted that, even though "Halliburton offered him [Rutherford] fifty dollars a month" plus "a share in the adventure's profits," Rutherford's shipmates talked him out of joining the crew with "discouraging words and lots of *Johnny Walker Black Label*." With Huntington "the junk then would have had a crew of 13, and just to play it safe they signed on a cat." *Los Angeles Times* (March and April 1984). For an easy-access listing of the crew

members, see Alt, *DDIB*, 345. For midwife reference, see Root, *HTMM*, 33 (quoted). For the two pandas, see Cortese, *RHRR*, 179.

6. For "high spirits," see *RHL*, March 3, 1939 (quoted). Landsburg notes: "Richard was writing to calm his parents' nerves," and he saw in the phrase "one more—one last—good-bye letter" a possible "foreboding" from Richard. See Landsburg, "Richard Halliburton," op. cit., quoted at 24. For "diploma," see *RHL*, May 23, 1921, *His Story*, 79 (quoted).

7. Mailed to Midway Island—so Richard never received it, Wesley Halliburton's typed letter dated March 24, 1939 (quoted) is located in RHC, Rhodes.

8. Ibid., (quoted).

9. Consulted for events at the fair (and the world) were scrapbooks, the *SFN* and *NYT*. The PTA feature, for instance, appeared on January 25, 1939. Eleanor Roosevelt's daily activities were reported in "My Day," available online. Also consulted was the *New York Times*, March 3, 1939. The *Folies Bergere* also performed to packed houses at the California Auditorium. Consulted was the "Programme" for the spring 1939 production. Author's collection.

34. Don't Give Up the Trip

1. See Rowland, *Slipping the Lines*, op. cit., 248 (quoted); for harbor master, see 250 (quoted).

2. For party, see Rowland, supra, 251 (quoted). For Torrey's comments, see Landsburg, "Richard Halliburton," op. cit., 24 (quoted).

3. For departure date, see timeanddate.com. For weather conditions and weather predictions along the coast, the *E. China Daily Post* and the *Hong Kong Daily Press* were consulted. Compare Prince, *American Daredevil*, op. cit., 242 (quoted). The "Publisher's Note" in *His Story* indicates that the *Sea Dragon* "sailed from Hong Kong on March 5, 1939," ix. However, the letter dated March 3, 1939, indicates that the ship *will sail* "in a few hours," while the postal telegraph above notes March 4. For time differences—thirteen hours between Memphis and Hong Kong—the World Time Clock and Map was consulted. For temperatures, see the US Weather Bureau, the Monthly Weather Review, and the National Climatic Data Center; see also the *International Herald Tribune*. For weather conditions in earlier years and to gather insight into weather patterns, see w.underground.com. For "hurricane" or "typhoon," see "1939 Pacific Typhoon Season," Wikipedia. Cf. Root, *HTMM*, 34, and Max, *HC*, 214. For typhoon origins and trajectories on the Pacific, see Leys, *After You, Magellan!*, op. cit., 244-247. For Velman Ernest Fitch, see the *Morning Tribune* (Minneapolis), the *St. Paul (MN) Dispatch*, March 31, 1939, and *Daily Nonpareil* (Council Bluffs, Iowa), "Appeal To FDR To Institute Sea Search for Halliburton," April 13, 1939. Son of Minneapolis veterinarian Ernest Lyman Fitch (1891-1979) and his first wife, Louise (died 1935), "mysterious stranger" Velman (born 1917 in Council Bluffs, Iowa) was formerly enrolled as a student at the University of Minnesota. Velman possibly had some medical experience himself, or claimed that he did—a medic as well as a veterinarian (for Chow puppies) was needed aboard the junk. For Ernest Fitch, see *Minnesota* or *Minneapolis Dispatch*, July 11, 1961, which also notes that he was a DVM graduate of Iowa State College (1916) and started practice in Iowa in 1923. It still provokes wonder that others besides Fitch boarded the junk. Gordon Torrey's daughter, a resident of Bar Harbor, Maine, communicated to me by

telephone that it was common knowledge in the family and community that Gordon's younger brother Norman had joined the *Sea Dragon* crew. Whether he participated in the first attempted crossing, or the second (which is unlikely—he died in 1940) In any event, one must wonder what additional inductees to crew membership appeared at the last minute—and what crew member(s) might have defected at the last minute. Research continues. In the Dartmouth collection is an October 15, 1948, letter to the Alumni Records Office from Halliburton attorney J. Richard Townsend, asking the office "to locate Mr. Gordon E. Torrey and his brother Mr. Norman Torrey," certify whether the two brothers attended or graduated from Dartmouth, and "if so, [certify] whether they are still living." On August 20, 1940, the *Zanesville (OH) Signal* reported that the Japanese had detained yachtsmen Norman (Stuart) Torrey, an American, and James Petersen, an Australian, at Haines Island off of French Indo-China. Both men had come from or were going to Hong Kong. The *Mansfield (OH) News Journal* reported the same news on August 20, 1940. Other information found through www.findagrave.com notes that Norman S. Torrey (1917–40) is interred at the Ledgelawn Cemetery in Bar Harbor, Hancock County, Maine. A gravestone for one Norman S. Torrey with the same birth and death dates may be found at Mount Adams Cemetery, Deer Island, Hancock County, Maine. For denial of passage to Chinese sailors, see Cortese, *RHRR*, 180.

4. For Japanese activities, see *New York Times*, March 1–March 7, 1939, notably, March 3, 1939, 10 (quoted). Of incidental interest, the Chinese government in Peking imposed regulations for the export of certain goods, especially war-related materials. Exporters now required "certificates of sale exchange confirmed by the *Federal Reserve Bank*." These regulations were hard to enforce, and some considered them further incentive for smuggling. More Japanese troops, meanwhile, were sent to Northern China near Shanghai. There, following a two-day offensive, they captured the city of Hwaian. Soon Japanese armies would blockade commercial traffic throughout the mainland coast from Shanghai to Swatou.

5. For Halliburton being "in perfect health," see *RHL,* March 3, 1939, *His Story*, 431–32, quoted at 431. For "apprehension," see Root, *HTMM*, 34 (quoted). For residence in a new country, quoted is a letter from Richard Halliburton to Noel Sullivan, December 15, 1930, Noel Sullivan Papers, Bancroft Library, University of California—Berkeley. For spirits returning, see *RHL*, March 3, 1939, *His Story*, 431–32 (quoted). Halliburton demonstrated impatience on other occasions. Pilot Moye Stephens said that Halliburton grew "frantic" during the *Flying Carpet* Expedition when the military delayed their flight from Fez to Colomb Bechar. "You know," remarked Stephens to an interviewer, "he was very high strung and nervous (and) "being held up was just more than he could stand." See Taylor, *SSMTWOD*, 129 (quoted). Halliburton noted that, following one of the last trial runs, "*all*" the members of the outing, not just him, got "very sea sick." See *RHL-P*, January 23, 1939.

6. For onlookers, see Welch, *LHK* #13, January 29, 1939 (quoted). For forebodings, compare Root, *HTMM*, 33–35, at 34. Compare Alt, *DDIB,* 346 et seq. For "Southern Track," see Collins, "Royal Road Across the Pacific," op. cit., 505.

7. See Sinclair, *Bright Paths to Adventure*, op. cit., 85 (quoted). Also see Rowland, *Slipping the Lines*, op. cit., 251. Rowland disliked Sinclair, especially deploring his believability as a reporter, so he kept his distance; see 247–48. For Halliburton's lack of interest in other people, see *RHL*, July 20, 1926, *His Story*, 262 (quoted).

8. For message, see *RHL-P*, March 5, 1939 (quoted); compare *His Story*, 432. For

correspondence to "Monica," see Lowell Thomas Collection, Marist College Archives and Special Collections. A photo of the junk dated "March, 1939" and captioned "Departure for Doom" notes, "The vessel is using her engine power to head for the open sea." The prow faces the camera. Two figures are seen atop the mizzenmast whose sails are bundled near the deck.

9. Compare Alt, *DDIB*, 350. For the course, compare Chow, *Junk That Challenged the Yachts*, op. cit., 86. For Halliburton viewing the sky, see *RHL*, February 28, 1922, 136 (quoted). For quoted passage, see Gifford Pinchot, *To The South Seas* (Philadelphia: John C. Winston, 1930), 298.

10. For the March 9 message, see Rowland, *Slipping the Lines*, op. cit., 251 (quoted). For message, see *RHL-P*, March 13, 1939 (quoted); compare *His Story*, 432. For "exhilarating" trip, quoted is letter to Richard from Wesley Halliburton, supra, March 24, 1939, RHC, Rhodes. The note to "Monica" is in the Lowell Thomas Collection at Marist College's Archives and Special Collections. For Halliburton's attitude toward regular employment, compare Alt, *DDIB*, 103. For "respite," see *RHL*, February 1936, *His Story*, 376. For "the routine of a fixed existence," the theme he championed throughout his writing career, see his *Flying Carpet*, 13. The distance between points is as follows: Hong Kong 1,779 miles from Yokohama, Yokohama 2,551 miles from Midway Atoll, Midway Atoll 1,311 miles from Honolulu, and Honolulu is 2,393 miles from San Francisco. In total, this distance is certainly far less than "nine thousand" miles.

11. For radiogram, see *RHL-P* (quoted); compare *His Story*, 433. For Potter's remarks, consulted was the *China Morning Post*, op. cit., April 8 or April 15, 1939. For Amos Wood's remarks, see *Beachcombing the Pacific* (West Chester, PA: Schiffer, 1989), 209 (quoted).

12. For the radiograms, compare Taylor, *SSMTWOD*, 201–2. These appear in Collins's "Royal Road Across the Pacific," op. cit., 505. For "fruits and vegetables," see Root, *HTMM*, 266 (quoted).

13. Ibid., Taylor, *SSMTWOD*, 202. See Collins, "Royal Road Across the Pacific," supra, 505 (quoted). Also see Root, *HTMM*, 266 (quoted).

14. Besides readings of changing latitude and longitude, times of day and weather conditions (all numeric) were entered on the log's left-side pages and short notations (occasional) were entered on the log's right-side pages. Facsimiles of the *Coolidge*'s logs used here, now in the MEB Archive, were originally owned by "John Rust Potter"; they are signed "D. E. Collins," – Chief Officer, and "K. A. Ahlin"—Master."

15. Ibid., log of the SS *Coolidge* (quoted), MEB Archive. For the International Date Line and its *time travel* mystique, see Haven, *Many Ports of Call*, op. cit., 157: "Monday . . . Now, in mid-afternoon of the same day, we crossed the imaginary International Date Line and flew back into Sunday." For the *Beaufort Scale*, see Wikipedia. Also see Alt, *DDIB*, 351–52.

16. Ibid., log of the SS *Coolidge* (quoted), MEB Archive. For message from "Welch and Halliburton," see Collins, "Royal Road Across the Pacific," supra, 505 (quoted).

17. For the "General Characteristics" of the SS *Coolidge*, see Wikipedia. For text of message see Collins, "Royal Road Across the Pacific," op. cit., 505 (quoted).

18. See the log of the SS *Coolidge* (quoted), MEB Archive. For last messages, see Collins's report, "Royal Road Across the Pacific," 505. Compare commentary offered in Taylor, *SSMTWOD*, 201–2. For Charles Dunn, see *SSMTWOD*, Taylor's words quoted at 202. Dunn "thought the situation called for an *SOS* but he said he got the impression that Halliburton wouldn't allow it because he was too proud to do so." Dunn had contacted

columnist Jack Smith of the *Los Angeles Times* (March and April 1984). Dunn informed Smith that he had gotten "two ominous messages" from the *Sea Dragon* that essentially said, "Their food was gone and the junk was filling with water and in danger of breaking up." Dunn also noted that the "vainglorious Halliburton," to preserve his heroic image, would not allow an SOS to be sent.

35. Vanishing Point

1. For weather notes, see the log of the SS *Coolidge* (quoted), MEB Archive. For passenger reaction and the navy's opinion, consulted was S. L. Kahn, scrapbook of Ralph B. Vawter, op. cit. Author's collection. Compare Root, *HTMM*, 266.

2. For "normal weather conditions," see Collins, "Royal Road Across the Pacific," supra, 507 (quoted). For "timely precaution" of heaving to and securing hatches, see 506 (added from log "400 feet apart"). For "touch of rough weather," see Halliburton, "Letters from the *SD* #3," January 27, 1939 (quoted). Also see Root, *HTMM*, 267. For earlier secondary-source responses to the last radio message, see *His Story*, 433 and Cortese, *RHRR*, 164 (quoted). Charles A. Borden offers the following as Welch's last words: "Lee rail awash. We are virtually swamped." *Sea Quest*, op. cit., 234. For warm-front storms (which can occur in cold conditions) and cold-front storms (which can occur in warm conditions), see Eric Sloane, *Book of Storms—Hurricanes, Twisters and Squalls* (Mineola, NY: Dover, 2006), 48 et seq. For the effects of the storm on *the Coolidge*, see Root, *HTMM*, 266. Cf. Collins, "Royal Road Across the Pacific," 506 (quoted).

3. Welch casually mentioned "pump(ing) fuel over the side," but he did not directly mention a bilge pump or pumps. A regular discharge of water from the hull appears, however, in the Hearst newsreel of the junk during its preliminary maneuvers. See Welch, *LHK* #9, December 5, 1938. See Hearst News Service, *Sea Dragon* newsreel outtakes, 1939, UCLA Film Archive.

4. "Hard-tack," also called sea biscuit or pilot biscuit, is a brittle food "made from flour and water"; it can also be "hard tack bully beef." Either kind of biscuit is "a last resort when other perishable foods have been consumed." Of a different composition, "hard-tack bully beef" is likely pickled corn beef shipped in barrels or canned. "Bully beef" appears twice in Welch's story "Ocean Tow," which recounts his service in 1920 on the *Coringa*. For "salt beef and hard tack," see *RHL*, September 28, 1938, 401 (quoted). During one treacherous stretch through a violent storm in the North Atlantic, Welch noted that the crew had "pilot biscuits, bully beef and cold tea three times a day." As the tug was without an icebox, the crew fared "on salt and bully beef." John Potter noted that the ship was without a "fridge," so food stores were likely shelved. See Cortese, *RHRR*, 180. If meat was on board, it could rot in less than a day if it wasn't cured (as jerky) or canned in oil or brine. For the lads' consumption habits, see Welch, *LHK* #13, January 29, 1939 (quoted).

5. Consulted was the log of the SS *Coolidge*, MEB Archive.

6. Weather reports indicated to Dale Collins that the storm "covered an area of some 600 or 800 miles in diameter," an estimate that strikes me as hyperbolic. Also, a situation in which the storm was 800 miles across and the *Coolidge* was 400 miles from the junk would suggest that Collins inferred, from insufficient data, that the junk entered the very eye or center of the storm. See Collins, "Royal Road Across the Pacific," supra, 506. The storm's origin and precise range and intensity remain uncertain to this day.

Seismicity in the region might have been high, but it has not been proven that seismic-engendered waves caused the storm. The *Pacific Tsunami Warning Center* did not as yet exist, nor was the Honolulu Observatory operating. Still, until its encounter with the storm, the ship was making good time. The *Sea Dragon* was scheduled to reach Midway Island on April 8, and it was covering an average of 150 miles a day.

7. For the messages, see Collins, "Royal Road Across the Pacific," supra, 505; and Taylor, *SSMTWOD*, 201. For the *Coolidge's* radio capability, see Alt, *DDIB*, 353.

8. See Collins, "Royal Road Across the Pacific," 505 (quoted). Collins thought the text "facetious," according to the letter he sent to Edward Howell on April 22, 1946 (quoted). The April 30, 1939, issue of the *New York Herald Tribune* reported that Welch "told of danger with humorous bravado." Landsburg suggests that "Albert Wetjen" was aboard the *Coolidge* on March 23 based on a report he filed in the United States Naval Institute Proceedings "of June 1939." See Landsburg, "Richard Halliburton," op. cit., 26. Landsburg may have meant Collins, "Royal Road Across the Pacific"; no mention of the Halliburton junk appears in the June 1939 issue, volume 65/6/436. Available online: *Proceedings Archive U.S. Naval Institute*. For the junk being "over-engined," see *Commercial Appeal* (Memphis, TN), April 19, 1939 (quoted). For officers' report to press, see *El Dorado (AR) Times*, April 24, 1939 (quoted).

9. For *star* navigation, see Harold Gatty, *The Raft Book—Lore of the Sea and Sky* (New York: George Grady Press, 1943), 93 et seq., quoted at 94. Published shortly after Halliburton's time, the book discusses navigational instruments and "how to find your way to land—without instruments and without previous experience in navigation." Charts are enclosed. In 1931, Gatty served as flyer Wiley Post's navigator in the "Round the World Flight." For "star observations," see Welch, "Ocean Tow," 214. Certainly Captain Welch would have benefited from today's Global Positioning System. In his last message, he did ask for a (radio) direction finder. For weather conditions, see Douglas Myles, *The Great Waves* (London: Robert Hale, 1985), 189–90; for the rarity of sightings of seismic waves, see 143–44. The Associated Press report appeared in *SFN*, April 1, 1939 (quoted). According to the *Coolidge's* log, the storm ranked high on the Beaufort Scale, which measures wind velocity or wind-forces. As to the limitations of the international scale and the adjustments made to it since its inception and use, see Alfred Friendly, "Wind Scale," in *Beaufort of the Admiralty—The Life of Sir Francis Beaufort 1774–1857* (New York: Random House, 1977), 142–47, at 147. If the *Sea Dragon* headed into a storm of Force 6, as has been suggested, the craft would quickly have been "drawn and quartered, splintered and crushed." See Max, *HC*, 217–18 (quoted). Comparable to the Beaufort Scale, the Saffir-Simpson Hurricane Scale (devised in 1971) measures wind speeds and damage results with Categories 1–5. Category 5, with winds over 150 miles per hour (as recorded in Hurricane Patricia, October 23, 2015), is catalogued as the most severe. Were such a scale applied to the storm surge that struck the *Sea Dragon*, a Category 2 storm best fits the known evidence. Sustained winds exceeding seventy-five miles per hour and registered as dangerous may do minimal to extensive damage to property. See National Hurricane Center, online. Patricia's top wind speed was noted as 201 miles per hour. See Doyle Rice, "Patricia No. 1 on List of World's Strongest Storms," *USA Today*, October 24, 2015, 2B. Halliburton biographer John Alt compares "Halliburton's typhoon" with "Halsey's Typhoon," a violent storm in the Philippine Sea that struck the US Pacific Fleet of some 170 ships on December 17, 1944.

See Alt, *DDIB*, 348–49, at 348. Compare Robert Drury, *Halsey's Typhoon* (New York: Grove, 2007).

10. See Collins, "Royal Road Across the Pacific," 506. Also quoted is the log of the SS *Coolidge*, MEB Archive.

11. For *GEE* and *LORAN* see *Wikipedia*. For Alt's remarks, see *DDIB*, 353 (quoted). For Potter comment, see "Shipping News," *China Morning Post*, Saturday, April (possibly queried on March 30), byline, "Feared Lost—Anxiety for Safety Of *Sea Dragon*." Reports such as "Junk's Radio Silent" begin to appear in *SFN* from March 28, 1939. By comparison, note radio signaling in the search for Amelia Earhart, July 2, 1937, in Amelia Earhart article in *Wikipedia*, at "Radio signals." Also see "Speculation on disappearance."

12. See Collins, "Royal Road Across the Pacific," supra, 506 (quoted). Later, after July 1939, the *Coolidge* collided with the *Nissan Maru* on the Whangpoo River in China. Converted from a liner to a troopship, the *Coolidge* began its last voyage on October 6, 1942. Three weeks later, near Espiritu Santo in the New Hebrides just northeast of Australia, the ship was struck by mines planted there by the Japanese. The *Coolidge* then ran aground, its career in a variety of tough roles now ended. See SS *Coolidge*, with photographs, "*Coolidge* History Page, *Michael McFayden's Scuba Diving Web Site*. The writer mistakenly notes that the *Coolidge* saw the *Sea Dragon* enter the storm. For "mountainous waves," quoted is Ralph Lundquist, crew member of the *Ning-Po* (and now a mimeograph operator in the US Customs Building in San Francisco), "Sailor Thinks *Sea Dragon* Safe—Recalls Historical Voyage of *Ning-Po* to *Panama-Pacific Exposition*—Craft West Unreported 50 Days on Trip from Japan to California (1915)," *SFN*, April 5, 1939. After repairs in Yokohama, the *Ning-Po* sailed off into the Pacific, estimating landfall in California in thirty days. But for fifty days, and amid raging, ceaseless storms, the ship sent no received message as to its whereabouts. Rescued fortuitously by the liner the *Honolulan* and given food and water, the ship was redirected, and in days it headed in evident safety down the California coast toward Los Angeles. At midnight on February 19, 1913, the junk anchored inside the San Pedro breakwater. Two years later, the *Ning-Po* was brought to San Francisco and exhibited at the Panama Pacific International Exposition. For aerial, see chapter 31, note 10 above.

13. For the storm's effects on the *Coolidge*, see Root, *HTMM*, op. cit., 266. Cf. Collins, "Royal Road Across the Pacific," 506 (quoted).

14. For the *Coolidge* rescue mission and the *Jefferson Davis*, compare Taylor, *SSMT-WOD*, op. cit., 220. For the *Taney*, see *SFN*, March 28, 1939. For the *Coolidge's* estimated distance from the *Sea Dragon*, see Collins, "Royal Road Across the Pacific," supra, 506 (quoted). For the *Carpathia* rescue mission, see "Five Things You May Not Know about *Titanic's* Rescue Ship," *History in the Headlines*, Steven Cohen, April 2012, at "*Carpathia's* Rescue Preparations Were a Masterpiece of Multi-Tasking," online. Also see Daniel Allen Butler, "RMS Titanic Remembered—The *Carpathia* Responds," online. For other references, see *RMS Carpathia*, Wikipedia.

36. Live a Little, Die a Lot

1. See *SFN*, March 27, 1939 (quoted). For days, even weeks, news services reported the latest word on the whereabouts of the junk. Hundreds of newspaper clippings related to the event, many consisting of little more than a few sentences, and most identical in

content, were collected and placed into scrapbooks that now reside in the MEB Archive. Other newspaper references can be found among the materials in that collection. For Crowell contacting the Halliburtons, see Root, *HTMM*, 267 (quoted).

2. Ibid., *SFN,* March 27, 1939 (quoted). For the Navy ordering a search for the Halliburton junk and crew, see, for instance, *the St. Louis (MO) Post-Dispatch*, March 31, 1939. Quoted is John Potter in Cortese, *RHRR*, 180. For Senators McNary and Johnson, see "Legislators Seek Aid in Halliburton Hunt," *Commercial Appeal* (Memphis, TN), April 8, 1939.

3. Ibid., *SFN*, March 29, 1939 (quoted). Also see *SFN*, March 28, 1939 (quoted). "Fears for safety" of writer noted.

4. Ibid., *SFN*, March 28, 1939 (quoted). For stunt or "disappearing act," see Root, *HTMM*, 269. For Potter's remarks, consulted was the "Shipping News" in the *China Morning Post*, op. cit. (quoted).

5. Ibid., *SFN,* March 30, 1939 (quoted). Also consulted was Richard Crowell, "Author Lost: Richard Halliburton Missing 5 Days," *Nevada State Journal*, March 30, 1939; and Richard Crowell, "Report Concerning *Halliburton Trans-Pacific Chinese Junk Expedition.*" Issued by (Publicity) Manager of Expedition," *Nevada State Journal*, April 17, 1939.

6. See Potter, *China Morning Post,* supra, (quoted). For involvement of Mackay Radio Company, Pan American Airways, the Navy, and the Coast Guard, see SFN, March 27, 1939 (Monday), and March 28, 1939 (Tuesday). For S. W. Fenton, marine superintendent of the Mackay Radio Company, see Associated Press release, SFN, March 29, (1939) (quoted).

7. See United Press, March 29, 1939 (quoted).

8. Consulted was United Press (SAN FRANCISCO) and *SFN* for March 31, 1939. In another report the coast guard survey ship *Discoverer* (not Japanese fishermen, as in the initial report) sighted the waterlogged, limping junk off Cape Cook, Vancouver Island. The *Discoverer* tried to tow the ailing craft, which only keeled over and took on more water during the effort. Oblivious to world events, the people on board did not know they were at war with Germany, and remarked, as though it mattered, that one of their number was a German. For widening search for the junk, see *SFN*, April 1 and April 2, 1939 (quoted). For the *Empress of Canada*, also consulted was a United Press clipping, March 29, 1939, among the Robert Hill Chase documents at Dartmouth. For "fears felt" for crew, see *Boston Sunday Globe* (inset of Halliburton about to embark on "world tour" and *Sea Dragon* about to set sail from Hong Kong), April 2, 1939 (quoted).

9. For Richard Wetjen's remarks, see *SFN*, April 3, 1939 (quoted). He must have been referring to his receipt of the February 13 letter sent from Hong Kong. For reaching Midway, see *SFN*, April 4, 1939 (quoted).

10. For Lundquist story, see *SFN*, April 5, 1939 (quoted). For Jack London story, see the *New York Times,* January 10, 1908.

11. For searches by other vessels, *SFN*, April 5, 1939, and the Associated Press (quoted) were consulted. For the last sighting of the junk, see *Washington Star*, May 16, 1939 (quoted). For the "Jeff Davis" and the "Torak, out of Los Angeles," see *SFN*, April 4, 1939. For diminished hopes, see *SFN*, April 10, 1939. See *SFN*, April 6, 1939 (quoted).

12. For "400 miles north," see *SFN*, April 7, 1939 (quoted). For "navigators," see *SFN,* April 10, 1939 (quoted). For Petersen, see *SFN*, April 11, 1939 (quoted). For being adrift, compare Captain Devere Baker, *The Raft Lehi IV—69 Days Adrift on the Pacific Ocean* (Long Beach CA: Whitehorn, 1959).

13. For "cutter being sent," see *SFN*, April 12, 1939 (quoted). Compare Taylor, *SSMT-*

WOD, 209–10; Reed quoted at 210. Also quoted is the *Commercial Appeal* (Memphis, Tennessee), April 13, 1939. For ship being "lost," see the *Hong Kong Telegraph*, April 14, 1939 (quoted). For descriptions of Hawaii and Waikiki, see Pierrot, *Vagabond Trail*, op. cit., 17 (quoted).

14. For Sarah Pratt editorial, see *SFN*, April 18, 1939 (quoted). For "FEARS," see *SFN*, April 20, 1939 (quoted). For no wreckage found, see *SFN*, April 24, 1939. United Press (San Francisco), April 24, 1939, headlined "Searching Liner (the *Coolidge*) Reports No Hope for Halliburton—His Junk Crippled Said Last Word on March 23—All Drowned? Ship Was Heading Into Typhoon Area"; also headlined "Parents Won't Give Up Hope." The junk's last message that the lee rail was awash troubled Wesley Halliburton. On April 26, 1939, the *Commercial Appeal* (Memphis, Tennessee) noted, "[Wesley Halliburton] believes that the laconic last message from the junk, the *Sea Dragon*, evidenced his son's lack of fear that disaster impended. 'Why (he wonders), if the lee rail being awash signified danger, wouldn't the radio operator have broadcast a call for help?'" For "not a clue," see *SFN*, April 24, 1939 (quoted).

15. For search, and for Petrich, see *SFN*, April 24, 1939. For Collins's report of Welch's statements, see *New York Herald*, April 29, 1939 (quoted).

16. For the cartoon, and caption (quoted), see *SFN*, May 3, 1939. For the Andersons, see *SFN*, May 13, 1939 (quoted). John and Nellie Anderson are pictured in the *SFN* item announcing their rescue at sea. Their determination to finish the mission once reequipped and overhauled is noted. See *SFN*, May 15, 1939. Ernest Lombardi would also be undiscouraged. He planned a two-year cruise around the world on a small boat, and his plan received such publicity as had earlier gone to Halliburton. See *SFN*, May 11, 1939; also May 18, 1939. Twenty-seven-year-old George Priestly made another attempt to sail from San Francisco to Hawaii. See *SFN*, May 6, 1939. For the *Tai Ping*, also see Collins, "Royal Road Across the Pacific," op. cit., 507 et seq; and Haven, *Many Ports of Call*, op. cit., after 213.

37. Mistaken Identity

1. See *Washington Star*, May 16, 1939 (quoted). Newspaper clippings from scrapbook of Ralph B. Vawter, author's collection, were also consulted—notably, a retrospective of Halliburton's career by S. L. Kahn. For *Astoria* and Richmond K. Turner compare Alt, *DDIB*, 255–356. In its two-page typed report of the tragedy distributed "to editors" and interested parties (e.g., Ralph B. Vawter), the Bobbs-Merrill Company indicated that the search had been conducted by June 2, 1939. Ironically, the *USS Astoria* was itself lost when engaging the Japanese in the Battle of Savo Island on August 9, 1942. Its story is told through artifacts and exhibits housed at the Columbia River Maritime Museum in Astoria, Oregon.

2. For Washington "D.C. Aide," see *Washington Star*, May 17, 1939 (quoted). For search results, see a September 20, 1946, letter to Edward T. Howell Jr., Wilmington, Delaware, from Helene Philibert, Navy Department, Executive Office of the Secretary, Public Information, Washington 25, D.C. In her response, Philibert directed Howell to Collins's "Royal Road Across the Pacific." Accompanying the letter was *Hydrographic Bulletin* No. 2587, April 5, 1939, and under "Missing Vessels" the following was written: "The Chinese junk Seadragon [sic], length 75 feet, beam 20 feet, equipped with 100-horsepower Diesel engine, plus customary sails, and 50-watt Mackay radio, which

sailed from Hong Kong on March 4, 1939, with 15 persons aboard, was last reported in lat. 31010"N., lon. 155000"E, at 0500, March 24, 1939. Due to heavy weather prevailing since that time in that locality, fears for the safety of this craft are entertained. It is requested that vessels sighting this junk notify the Navy Department, Washington, D.C., by radio immediately." The *Hydrographic Bulletin* next reported that "the sloop *Show Me*, previously reported missing, arrived at Balboa, C.Z., on March 29, 1939." For search, see Alt, *DDIB*, op. cit., 355 (quoted).

3. Consulted was Philibert letter to Howell Jr., September 20, 1946, supra. Quoted is letter from Admiral Richmond K. Turner to Edward T. Howell Jr., May 2, 1946.

4. For "negative" results, see Turner, supra. For Leahy's concluding remarks, see S. L. Kahn retrospective in the scrapbook of Ralph B. Vawter (quoted), author's collection. It is unlikely that so large an undertaking was a combined effort to hunt for the junk and gather intelligence on Japan's expanding Pacific empire. The extensive search effort conducted two years earlier for Amelia Earhart had already spawned the belief that the Navy was combining the mission with surveillance. Earhart's mourners insisted that search teams actually knew where her plane had landed but searched elsewhere for the purpose of scouting Japanese activity in the region. Such suspicion naturally carried over into the Navy's late commitment to the Halliburton search. For the Earhart disappearance, see Jerry Adler, "The Lady Vanishes (Again)," New Clues/New Controversy, *Smithsonian*, January 2015, 32–41, especially at 39–41. Also see Randall Brink, *Lost Star—The Search for Amelia Earhart* (New York: W. W. Norton, 1994), op. cit., 181–82.

5. For *Navy* inquest, compare Root, op. cit., 269.

6. Consulted was the *Washington Star*, May 16 and May 17, 1939. Memphis' *Commercial Appeal* regularly ran news related to a favorite son. For Torrey contacted by Nelle Nance Halliburton, see Landsburg, "Richard Halliburton," op. cit., quoted at 24.

7. See *Boston Globe*, January 6, 1940. Cf. Taylor, *SSMTWOD*, 206–7. Records on Chase are housed at Dartmouth University, as are those of John Potter and Gordon Torrey. By family request the money went to Dartmouth College, to the college's chapter of Phi Gamma Delta, to Milton Academy, and to a boy's camp in Newfound Lake, New Hampshire. See *Dartmouth Alumni Magazine*, January 1940.

8. For the *Tai Ping* and the Andersons, see Dale Collins, "Royal Road Across the Pacific," op. cit ., 507 et. seq. Also see Haven, *Many Ports of Call*, op. cit., 213; quoted at 212 and 216. Photographs of the *Tai Ping* and its crew are opposite 214. See chapter 36, note 16 above. The Andersons' second departure occurred no earlier than June 14, when the couple gave one L. V. McAdams in Kobe, Japan, a signed and dated gouache watercolor of the *Tai Ping*. Author's collection.

9. Ibid., Haven, 213–15, quoted at 215. For other provisions, see Collins, "Royal Road Across the Pacific," op. cit., 508 (quoted).

10. For "sight of the girl," see Haven, supra, 218 (quoted); for Haven meeting Nellie, see 215 (quoted); for "lots of exercise," see 220 (quoted). For Anderson signaling a steamer, the caption on the reverse of ACME News photograph is quoted. Author's collection.

11. See *Ellesburg Daily Record* (Seattle, WA), October 4, 1939. Also see the *New York Times*, October 4, 1939. Time at sea was also given at 108 days. The 105 days could also have been counted from the time the ship met the SS *Coolidge* on July 13. See Haven, *Many Ports of Call*, op. cit., 221–22, quoted at 221; for the *axiom*, see 229 (quoted).

12. See Haven, supra, 222 (quoted).

38. The Summing Up

1. Cf. Root, *HTMM*, 270 et seq.; cf. Alt, *DDIB,* 357–58. For "changes of policy" and financial reckonings, see *RHL-P,* January 23, 1939 (quoted). If made to the estate or to any persons or institutions named as beneficiaries, Lloyds' payments, if any, are unknown to me. Wesley's remarks with pertinent newspaper clippings are in the Roma Borst Hoff Archive, Richard Halliburton Collection, Barret Library, Rhodes College, Memphis, Tennessee, op. cit. For the succession of tenants at Hangover House, see Max, *HC,* 225–226. Photographs of bookshelves reveal some of the book titles. For the property, consulted were letters of Paul Mooney and Wallace Scott. Author's collection.

2. For absence of furniture other than beds and boxes, see *PML,* letter to Harriette Janssen, November 24, 1938; also see *PML,* letter to Gerstle Mack, September 9, 1937. Halliburton also had a writing desk, and he added four (upholstered) chairs, two metal garden chairs, and dining-room furniture including eight (dinette) chairs. See *RHL-P,* July 14, 1938. William Alexander identified an oil on the living-room wall as one painted by friend Donald Forbes (1905–1951). Bertha Barstow had the best intentions for her only son, but Barstow family historian Patricia Suprenant believes that, had George Barstow II intervened (he died suddenly in 1934), he would have forbidden young George from joining the *Sea Dragon Expedition.* Bertha conveyed her wishes about the library to Berry College's founder, Martha Berry. In a letter dated November 25, 1939, she noted George's writing and musical interests. The letter is in the Bertha Barstow archive, Berry College, Rome, Georgia. For the Halliburton bequest, see the "Last Will and Testament of Richard Halliburton, Deceased," made out on August 10, 1937, signed before witnesses, and filed on October 12, 1939, RHC, Rhodes. *The New York Times* Sunday edition for October 15, 1939 was headlined "Halliburton Gift to Aid Princeton—Noted Author, Lost at Sea, Left $104,000 to Library." The paper noted that the funds were "to establish a Richard Halliburton geographical collection in the university library." Compare Taylor, *SSMTWOD,* 214–15.

3. See *Commercial Appeal* (Memphis, Tennessee), October 5, 1939 (quoted). Also see the *New York Herald Tribune,* October 5, 1939, which, like the *New York Times,* provided a sketch of Halliburton's career. Two months later, December 20, 1939, *The San Francisco Chronicle,* incidentally, reported that engineer Henry Von Fehren was declared legally dead two months later.

4. Ibid., *Commercial Appeal* (Memphis, Tennessee), quoted. Compare Root, *HTMM,* 270, 271. The jury ruled that the ship had sunk either March 23 or March 24. The uncertainty arose based on the junk's known proximity to the International Date Line, where ships lose or gain a day depending on whether they are bound east or west. *Commercial Appeal* (Memphis, Tennessee), October 5, 1939 (quoted). Also see Root, *HTMM,* op. cit., 271 (quoted). Chancery Court evidently said nothing about residuals from intellectual property. Royalties from Halliburton's books would amount to $4,000 to $5,000 in the ensuing years, yet those figures would have risen had he continued his lecture tours.

5. Numerous newspapers and magazines were consulted. Main sources include the *San Francisco Examiner*, the *New York Times*, and the *San Francisco News (SFN).* For *USS Squalus,* see *SFN,* May 23–24, 1939. For Eleanor Roosevelt's feelings toward San Francisco, see "My Day," *SFN,* March 21, 1939; and for Eleanor Roosevelt at the fair, see "My Day," *SFN,* March 23, 1939.

6. For Japan Day, see *SFN,* May 2, 1939.

7. For events, compare Julian Bryan, "What Hitler's Lightning War Will Do to England," *Look*, December 5, 1939 (Zorina cover), 10–17. Also featured is Gen. Hugh S. Johnson, "The United States Army Needs Guns," 20–21. America's readiness or unpreparedness for war is a regular theme in newspapers of the period. For news of the day, newspapers, magazines, and fair brochures were consulted. A postcard, dated October 10, 1939, marks the Crosby event. Author's collection.

8. For court ruling and its meaning, see Root, *HTMM*, 271 (quoted). Cf. Alt, *DDIB*, 357 et seq.

9. Bobbs-Merrill Company to editors, 1940 (quoted), from the scrapbook of Ralph Vawter, author's collection. Wesley Halliburton's remarks with pertinent newspaper clippings are in the Roma Borst Hoff Archive, Richard Halliburton Collection, Rhodes, op. cit.

39. Neptune's Realm

1. For Japanese cooperation, issues of *SFN* dating from March through June 1939 were consulted. In a noteworthy development, Japanese bombs had leveled Foochow, where Halliburton had thought to find a junk. See *SFN,* May 3, 1939. For Japanese intelligence operations at this time, see Ken Kotani, *Japanese Intelligence in World War II* (Oxford, UK: Osprey, 2009). A document in the possession of the *Ministry of Foreign Affairs* (MOFA) of its asking the *Japanese Army and Navy* for "flight permission" granted Halliburton to fly over Formosa in March 1931. Professor Kotani found no similar document from MOFA for the voyage of 1939. Letter from Ken Kotani to author, January 3, 2016. On the junk captured by Japanese warships or Chinese pirates, see Landsburg, "Richard Halliburton," op. cit., at 26. For Carveth Wells's view, see *My Candle at Both Ends*, op. cit., 108 (quoted).

2. For state-of-the-art coastal and monitoring systems, see Global Ocean Security Technologies (GOST), online. For the *USS Squalus*, the Navy enlisted the aid of a "diving bell" to rescue the crew, but the *Sea Dragon* was far too distant for the device to have practical merit. See *SFN*, May 24, 1939. Also tested at the time was an "electric robot" for solving ocean rescue problems. See *SFN*, May 6, 1939. The official search for Malaysia Flight MH 370 was suspended on January 17, 2017 (*Associated Press*).

3. For "temperament" of the *Sea Dragon*, see Halliburton, "Letters from the *SD* #3," January 27, 1939 (quoted).

4. For advancing storm conditions and their effect on the crew, see Welch, "Ocean Tow," op. cit., quoted at 209, 210, and 211.

5. See Collins, "Royal Road Across the Pacific," op. cit., 506 (quoted). Quoted also is the *Commercial Appeal* (Memphis, Tennessee), October 5, 1939.

6. For Rowland's comment, see *Slipping the Lines*, op. cit., 248 (quoted). For Torrey's remarks, see Cortese, *RHRR*, 177–78 (quoted).

7. See Charles A. Borden, *Sea Quest—Global Blue-Water Adventuring in Small Craft* (New York: David McKay, 1967), 234 (quoted).

8. A letter from Dale E. Collins to Edward T. Howell, April 22, 1939, is quoted. According to Captain Jokstad, "the first thing to give way" in so strong a gale would have been the rudder. See Jokstad, *Captain and the Sea*, op. cit., 190. For Garrett photograph, which appeared in *Click*, October 1939, see Max, *HC*, opposite 223.

9. See Nick Ward, with Sinead O'Brian, *Left For Dead—Surviving the Deadliest Storm*

NOTES TO PAGES 337-345

in Modern Sailing History (New York: Bloomsbury, 2007), 55–59, quoted at 63. Also see "Fastnet 1979—The Disaster That Changed Sailing," *Yachting World*, August 2009. Also see David Lynn, "Heavy Weather Sailing—Making a Series Drogue," online.

10. For killer waves, see Susan Casey, *The Wave–In Pursuit of the Rogues, Freaks, and Giants of the Ocean* (New York: Doubleday, 2010). For the "tall wave," see 16. For the use of mathematics to explain "monster waves," see 78–82. Also see YouTube.com: (1) "Ships in storm terrifying monster waves!"; (2) "10 Top Ship(s) in Storm Compilation Monster Waves." For Shackleton and others sighting such waves, see "List of Rogue Waves," Wikipedia, at note 7. For monster waves and tidal waves, compare L. Don Leet, *Causes of Catastrophe* (New York: McGraw-Hill, 1948), 169 et seq. For witnessed waves reaching eighty to one hundred feet, see Paul Theroux, "The Mount Everest of Surfing" and "Biggest Wave," in *Smithsonian*, July/August 2018, 26–39, at 33.

11. For Collins on Welch's handling of the junk, consulted was S. L. Kahn retrospective, scrapbook of Ralph B. Vawter, author's collection, op. cit.

12. For stove, see *RHL-P*, January 1, 1939. For "fumes," see "Log of the *SD* #13."

13. For Jokstad's blaming the rudder, see Jokstad, *Captain and the Sea*, 190.

14. For Walker's report, see "Believe San Diego Shipwreck Halliburton Boat," *Berkeley Daily Gazette*, February 10, 1945 (available online via Richard Halliburton Wikipedia article). For findings, see Max, *HC*, 216, 226. Dale Collins thought it conjecture that debris was identified as remnants of the *Sea Dragon*. Amos Wood, however, thought it likely that debris found off the shore of California was from the *Sea Dragon*. See Max, *HC*, 226.

15. See Jokstad, *Captain and the Sea*, 190 (quoted). Also noted is Edward T. Howell, letter to author, February 17, 2009. Unseen obstacles below the surface of the water, comments nautical theorist C. A. Marchaj, "may cause a *capsize* that is especially unexpected because the sudden increase in the heeling moment occurs without warning." See *Sailing Theory and Practice* (New York: Dodd, Mead, 1964), 343–44. For Halliburton on "some remote island," compare Landsburg, "Richard Halliburton," op. cit., 26–27. Compare Amelia Earhart and speculated landfalls on Milo Atoll or Jaluit Atoll. See *Amelia Earhart—The Lost Evidence* (History Channel, 2017), documentary.

16. For wreck simulation, see "Running Shipwrecks Simulation Backwards Helps Identify Dangerous Waves," *Phys. Org* Technical Computer Sciences, University of Michigan, online, October 1, 2007.

40. The Long Dusk

1. For Potter's remarks, see Cortese, *RHRR*, 180–81. For "madmen" image, see Welch, "Ocean Tow," op. cit., quoted at 211.

2. It is not known what lethal drugs, if any drugs at all, they had at their disposal. Had the crew requested it, MacTavish Chemists in Hong Kong could have confected all sorts of deadly toxins. For Jokstad's view, see *Captain and the Sea*, op. cit., 190.

3. For Halliburton spread-eagled (perhaps a wise rather than panic-stricken move), see Halliburton, "Letters from the *SD* #3," January 27, 1939.

4. For "dark water," see Chow, *Junk That Challenged the Yachts*, op. cit., 124–25 (quoted). For hoisting the rudder, see 91.

5. For "zero-moment point," see Sebastian Junger, *The Perfect Storm—A True Story*

of Men Against the Sea (New York: W. W. Norton, 1997), 40 (quoted). For drowning, compare 141–46.

6. For Earhart and Halliburton on island, compare Root, *HTMM*, 269. Also see chapter 39, note 17 above. Amelia Earhart would be officially declared dead on January 5, 1939. Conspiracy theorists who have studied her disappearance (her plane vanished in the Pacific on July 2, 1937) claim that at this time the Japanese were "disappearing" anyone whom they considered an enemy agent.

7. For "100-yard stare," see Frank Quirarte at ESPN.com, quoted in Michael Mc-Carthy, "'300' star (actor Gerard) Butler Survives Surfing Scare," *USA Today*, December 22, 2011, 3C.

8. See Richard Alfred Wetjen, *In Fiddler's Green, or The Strange Adventure of Tommy Lawn—A Tale of the Great Divide of Sailormen*, with illustrations by Ferdinand Huztl Horvath (Little Brown, 1931), xi (quoted). In his posthumous "Our Oceanic Ills," Welch notes Wetjen and *In Fiddler's Green,* op. cit. For the funeral of "Dick Wetjen," the Albert Richard Wetjen Papers at the University of California–Berkeley's Bancroft Library were consulted.

9. Quoted are Henry Van Dyke from essay "Camp-Fires and Guide-Posts," Rupert Brooke from poem "The Great Lover," and William Shakespeare, *The Tempest*, act 1, scene 2.

10. For Djuna Barnes, see *PML,* letter to Gerstle Mack, September 9, 1937. Paul's story appeared in *Washington Post*, "The Pen Pushers," March 7, 1920, and won first prize—one dollar.

11. For Lowell Thomas eulogy, see Root, *HTMM*, 272 (quoted); for Halliburton "in-sinuating himself into the legend of T. E. Lawrence (of Arabia)," see 190. For Nelle Nance running to the telephone, see Taylor, *SSMTWOD*, 212. Consulted was "Haunted House?" clipping on Associated Press photo of Hangover House. *San Francisco Examiner*, January 31, 1957, quoted. For "ghost ships," see Jack Phillips, "11 Mysterious 'Ghost Ships' Wash Up In Japan Containing 20 Dead Bodies," *Epoch Times*, November 30, 2015, online. The ships were wooden, probably Korean in origin, and similar to those Halliburton glimpsed along the China coast. Also see Hal McClure, "The Ghost of Laguna's Halliburton House," Orange County News, *Los Angeles Times*, January 20, 1957.

12. For Tower, see Root, *HTMM*, 276–277 (quoted). Also see Taylor, *SSMTWOD*, 231–32. For public image, and his "glorification of self-expression," see dust jacket blurb, *RHL, His Story,* op. cit.

13. Early in his career, Halliburton called himself "Don Quixote, Jr." *RHL,* May 14, 1923, *His Story*, 212. Also see *RHL*, 1921, 130. Compare Root, *HTMM*, 115. For Ulysses (or Odysseus), see Nikos Kazantzakis, *The Odyssey: A Modern Sequel*, first translated into English in 1958 by Kimon Friar.

14. For "third" book, see *RHL*, March 8, 1928, *His Story*, 282 (quoted).

Bibliography

Abend, Hallett, "Death Stalks Shanghai: A City Racked by War." *New York Times*, October 10, 1937.

——. *My Life in China—1926–1941*. New York: Harcourt, Brace and Company, 1943.

——. *Tortured China*. New York: Ives Washburn, 1930.

Aberth, John. *The First Horsemen—Disease in Human History*. New York: Prentice Hall, 2007.

Abora III Building and History. www.atlantisbolivia.org.

Accardi, Catharine A. *San Francisco's North Beach and Telegraph Hill*. Images of America. Charleston, SC: Arcadia, 2010.

Ackley, Laura P. *San Francisco's Jewel City—The Panama-Pacific International Exposition of 1915*. Berkeley, CA: Heyday, 2014.

Acqua Survey Abora III Final Update. November 26, 2007. www.youtube.com

Adams, Frederick Upham. *Conquest of the Tropics—The Story of the Development of a Great Concern: The United Fruit Company*. Garden City, NY: Doubleday, Page, 1914.

Adams, J. Donald. "One Hundred Years of Feminism." *New York Times Book Review*, July 9, 1933.

Adler, Jerry. "The Lady Vanishes (Again)." New Clues/New Controversy. *Smithsonian*, January 2015, 32–41.

Ainsworth, W. F., ed. *All Around the World: An Illustrated Record of Voyages, Travels, and Adventures in All Parts of the Globe; With Hundreds of Illustrations after Drawings of Gustave Dore, Berard, Lancelot, Jules Noel and Other Eminent Artists*. 3 vols. New York: Selman Hess, 1870.

Alber-Wickes Platform Service—Boston, Elbert A. Wickes, Manager, Lyceum Series Affiliated Service, 1926, 8 pp.

Aldington, Richard. *Lawrence of Arabia—A Biographical Enquiry*. Chicago: Henry Regnery, 1955.

Allen, Edward Frank. "'Richard Halliburton's Life of Adventure'—His Letters to His Father and Mother Are Both Engaging and Revealing." *New York Times*, June 30, 1940.

Allen, Frederick Lewis. *Only Yesterday—An Informal History of the 1920s*. 1931. Reprint. New York: Perennial, 1964.

Allen, Robert H. *The Classical Origins of Modern Homophobia*. Jefferson, NC: McFarland, 2006.

Alt, John H. *Don't Die In Bed—The Brief, Intense Life of Richard Halliburton*. Atlanta: Quincunx, 2013.

Amelia Earhart—The Lost Evidence. History Channel, 2017. Documentary.

American Experience. "The Boys of '36" (9-man University of Washington [State] Olympic medal-winning rowing team). 2016. PBS. Video.

American Experience. "The Chinese Exclusion Act." 2018. PBS. Video.

American Experience. "The Crash of 1929." 2012. PBS. Video.

Anderson, J. R. L. *The Ulysses Factor: The Exploring Instinct in Man*. New York: Harcourt Brace Jovanovich, 1970.

Anderson, Romola, and R. C. Anderson. *The Sailing-Ship—Six Thousand Years of History*. New York: Bonanza Books, 1963.

Anderson, Sherwood. *Sherwood Anderson's Memoirs*. New York: Harcourt, Brace, 1942.

Andrews, Allen. *The Mad Motorists—the Great Peking-Paris Race of '07*. Philadelphia: J. B. Lippincott, 1965.

Anema, Durlynn, *Harriet Chalmers Adams—Adventurer and Explorer*. Greensboro, North Carolina, 1997.

Angulo, Diana Hutchins. *Peking Sun, Shanghai Moon: Images from a Past Era* (1920–1940). Hong Kong: Old China House Press.

Another Hong Kong. Hong Kong: Emphasis Limited, 1999.

Appleby, Joyce. *Shores of Knowledge—New World Discoveries and the Scientific Imagination*. New York: W. W. Norton, 2013.

Art Program in the Continental Manner, Clifford C. Fischer Presents FOLIES BERGERE at the California Auditorium (at) Treasure Island, 1939. Souvenir. 24 pp., photographs and text.

Asbach, Wolfgang. *The Hangzhou Trader—A 1/30 Scale Model of a Trading Junk from the Shanghai Area, Model Shipbuilder*, No. 81–84, 1992–1993 (Mariner's Museum Library), 3–11.

Asbury, Herbert. *The Barbary Coast—An Informal History of the San Francisco Underworld*. New York: Alfred A. Knopf, 1933.

"Asia to America, From: Conquering the Pacific." *National Geographic—History*, August–September 2015,. 16–19.

Atherton, Gertrude. *Adventures of a Novelist*. New York: Liveright, 1932.

——. *An Intimate History of California*. New York: Blue Ribbon Books, 1936.

——. *My San Francisco—A Wayward Biography*. Indianapolis: Bobbs-Merrill, 1946.

——. *The Valiant Runaways*. New York: Dodd, Mead, 1898.

Austin, F. Britten. *A Saga of the Sea*. New York: MacMillan, 1929.

Baker, Capt. Devere. *The Raft Lehi IV—69 Days Adrift on the Pacific Ocean*. Long Beach, CA: Whitehorn, 1959.

Baker, W. A., and Tre Tryckare. *The Engine Powered Vessel—From Paddle-Wheeler to Nuclear Ship*. New York: Crescent Books, 1965.

Bancroft, Griffing. *Lower California: A Cruise; The Flight of the Least Petrel*. New York: G. P. Putnam's Sons, 1932.

Bankers Life Company (letter), Des Moines Iowa, Secretary G. W. Fowler, October 2, 1924.

Barczewski, Stephanie. *Antarctic Destinies—Scott, Shackelton and the Changing Face of Heroism*. London: Hambledon Continuum, 2007.

Barnes, Djuna. *Collected Poems with Notes towards the Memoirs*. Selected and edited by Phillip Herring and Osias Stutman. Madison: University of Wisconsin Press, 2005.

———. *Nightwood—The Original Version and Related Drafts*. Edited and with an introduction by Cheryl J. Plumb, Introduction vii–xxvi, Champaign, IL: Dalkey Archive Press, 1995.

Barrie, David. *Sextant—A Young Man's Daring Sea Voyage and the Men Who Mapped the World's Oceans*. New York: Harper/Collins, 2014.

Barron, Elwyn A., ed. *Deeds of Heroism and Bravery—The Book of Heroes and Personal Daring*. New York: Harper and Brothers, 1920.

Barrows, Edward Morley. *The Great Commodore—The Exploits of Matthew Calbraith Perry* Indianapolis: Bobbs-Merrill, 1935.

Barry, Iris. *Let's Go To The Movies*. New York: Payson and Clarke, 1926.

Barstow, Bertha Kellogg. Correspondence, 1939–1955. Berry College Archives, Rome, GA.

Barstow, George Eames II. *The Effect of Psychology on Americanism*. New York: Society of Applied Psychology, 1920.

Bartholet, Jeffrey. "What Ails the Taj Mahal." *Smithsonian*, September 2011, 44–57.

Battle Creek Sanitarium (subject), Menus, November 1927. menus.nypl.org.

Bauer, Paul J., and Mark Dawidziak. *Jim Tully—American Writer, Irish Rover, and Hollywood Brawler*. With foreword by Ken Burns. Kent, OH: Kent State University Press, 2011.

Bauerlein, Mark. *The Dumbest Generation—How the Digital Age Stupifies Young Americans and Jeopardizes Our Future*. New York: Jeremy P. Tarcher / Penguin, 2008.

Beaman, A. Gaylord. *A Doctor's Odyssey—A Sentimental Record of LeRoy Crummer: Physician, Author, Bibliophile, Artist in Living*. Baltimore, MD: John Hopkins Press, 1935. Copy inscribed by Myrtle [Crummer].

Becker, Elizabeth. *Overbooked—The Exploding Business of Travel and Tourism*. New York: Simon and Schuster, 2013.

Beech, Hannah (and others). "How China Sees The World." *Time*, June 17, 2013, 26–33.

Belfrage, Cedric. *Away from It All; An Escapologist's Notebooks*. New York: Literary Guild, 1937.

Bennett, Arnold. *How to Live 24 Hours a Day*. New York: George H. Doran, 1910.

"Berkeley Square—Resurrecting a West Adams Street Lost to the Freeway," *Historic Los Angeles* (2011), 19.

Bisland, Elizabeth. *In Seven Stages: A Flying Trip around the World*. 1891. Reprint. United Kingdom: Dodo, n.d.

———. *The Life and Letters of Lafcadio Hearn*. 2 vols. Boston: Houghton Mifflin, 1906.

Black and Gold (high school yearbook). R. J. Reynolds, High School, Winston-Salem, NC, 1936. (Richard Halliburton pictured on p. 178.)

Blaine, Capt. John. *The Boy Scout* series. 12 volumes. New York: Saalfield, 1916 et seq.

Blank, Hanne. *Straight—The Surprisingly Short History of Heterosexuality*. Boston: Beacon, 2012.

Blofeld, John. *The Wheel of Life—The Autobiography of a Western Buddhist*. Reprinted with a foreword by Huston Smith. Boston: Shambhala, 1988.

Bly, Nellie. *Around the World in Seventy-Two Days and Other Writings*. Reprinted with a foreword by Maureen Corrigan. New York: Penguin Books, 2014.

Blythe, Stuart O. "The Fair Is Ready to Open," *Golden Gate International Exposition* Premieres February 10 and 19." *California—Magazine of the Pacific*, February 1939, 5–9.

Bonner, Willard H. *Harp on the Shore—Thoreau and the Sea*. Completed and edited by George R. Levine. Albany: State University of New York Press, 1985.

Borden, Charles A. *Sea Quest—Global Blue-Water Adventuring in Small Craft*. Philadelphia: Macrae Smith, 1967.

Bowman, Heath, and Jefferson Bowman. *Crusoe's Island in the Caribbean*. Indianapolis: Bobbs-Merrill, 1939.

Boyle, David. *Toward the Setting Sun—Columbus, Cabot, Vespucci and The Race for America*. New York: Walker, 2008.

Boyle, T. Coraghessan. *The Road to Wellville*: *Story of John Harvey Kellogg Inventor of Cornflakes; A Novel*. New York: Viking, 1993.

Bradley, Capt. C., OBE. "A Master's Memories." *Sea Breezes*, July 1950, 32–34.

Brassey, (Mrs.) Lady Anna. *A Voyage in the Sunbeam, Our Home on the Ocean for Eleven Months*. London: Longmans, Green, 1878.

Bray, R. S. *Armies of Pestilence—The Impact of Disease on History*. Cambridge, UK: Lutterworth, 1996.

Breck, Flora E. *Jobs for the Perplexed*. New York: Thomas Y. Crowell, 1936.

Brink, Randall. *Lost Star—The Search for Amelia Earhart*. New York: W. W. Norton, 1994.

Broe, Mary Lynn, ed. *Silence and Power—A Reevaluation of Djuna Barnes*. Carbondale: Southern Illinois University Press, 1991.

Bronski, Michael. *A Queer History of the United States*. Boston: Beacon, 2011.

Brooke, Rupert. *The Collected Poems, with a Memoir by Sir Edward Marsh*. 3rd rev. ed. London: Sidgwick and Jackson, 1942.

Brown, Anthony. *Lloyd's of London*. New York: Stein and Day, 1973.

Bryan, Julian. "What Hitler's Lightning War Will Do to England." *Look*, December 5, 1939 (Zorina cover), 10–17.

Buhler-Wilkerson, Karen. *No Place Like Home—A History of Nursing and Home Care in the United States*. Baltimore, MD: Johns Hopkins University Press, 2003.

Buot, Francois. *Gay Paris—Un Histoire du Paris Interlope Entre 1900–1940*. Paris: Fayard, 2013.

Burr, Ty. *Gods Like Us—On Movie Stardom and Modern Fame*. New York: Pantheon Books, 2012.

Burrows, Anna. "The San Francisco Golden Gate Exposition." University of Maryland University Libraries Digital Collections. Digital.lib.umd.edu.

Burt, (Maxwell) Struthers. *Entertaining the Islanders*. New York: Charles Scribner's Sons, 1933.

Buruma, Ian, and Avishai Margalit. *Occidentalism—The West in the Eyes of Its Enemies*. New York: Penguin Books, 2004.

Butcher, Harold. "The Pittsburgh of Japan (Osaka)." *Travel*, August 1928, 18–24.

Butler, Daniel Allen. *Unsinkable: The Full Story of the RMS Titanic*. Boston: De Capo Press, 2012.

Byron, Robert. *The Byzantine Achievement*. London: George Routledge, 1929.

——. *Europe in the Looking Glass: Reflections of a Motor Drive from Grimsby to Athens*. 1926. Reprint. United Kingdom: Hesperus, 2012.

——. *The Road to Oxiana*. London: MacMillan, 1937; reprint New York, Oxford University Press, 2007.

Caen, Herb. *Herb Caen's San Francisco, 1976–1991*. Selected (features) by Irene Mecchi. San Francisco: Chronicle Books, 1992.

Caldwell, Wilber W. *Cynicism and the Evolution of the American Dream*. Washington, DC: Potomac Books, 2006.

Calhoun, Capt. C. Raymond, US Navy (Retired). *Typhoon: The Other Enemy; The Third Fleet and the Pacific Storm of December 1944*. Annapolis, MD: Naval Institute Press, 1981.

Cameron, Nigel. *An Illustrated History of Hong Kong*. Oxford: Oxford University Press, 1991.

Campbell, S. "Typhoons Affecting Hong Kong: Case Studies." (Report, Hong Kong University of Science and Technology, 2005).

Caputo, Philip. *The Longest Road: Overland in Search of America, from Key West to the Arctic Ocean*. New York: Henry Holt, 2013.

Caravan, June 1939 issue.

Card, James. *Seductive Cinema—The Art of Silent Film*. New York: Alfred A. Knopf, 1994.

Carey, David. "Gay Marriage Issue Traveled Long Road." Associated Press. *Wisconsin State Journal*, Nation & World, June 28, 2015.

Carlyle, Thomas. *On Heroes, Hero-Worship and the Heroic in History*. New York: Frederick A. Stokes, 1888.

Carpenter, C. Whitney, Jr. "The Peculiar Heathen Chinee." *Travel*, June 1919, 23–28, 48.

Carpenter, Frank G. *Around the World with the Children—An Introduction to Geography*. New York: American Book, 1935.

——. *Carpenter's Geographic Reader: Europe*. New York: American Book, 1902.

——. *Carpenter's World Travels: Familiar Talks about Countries and Peoples*. New York: Doubleday, 1927.

Carroll, Gladys Hasty. *To Remember Forever: The Journal of a College Girl, 1922–1923*. Boston: Little, Brown, 1963.

Carse, Robert. "The Devil's Crew." *Adventure Magazine*, December 15, 1934. 128 pp.

Carswell, John. *The Romantic Rogue—The Singular Life and Adventures of Rudolph Erich Raspe, Creator of Baron Munchausen*. New York: E. P. Dutton, 1950.

Casey, Robert. *Four Faces of Siva*. New York: Bobbs-Merrill, 1929.

Century of Hong Kong Roads and Streets. Hong Kong: Joint Publishing, 2000.

Chace, William M. "Where Have All the Students Gone?: The Demise of the English Department." *American Scholar*, Autumn 2009, 32–42.

Chapelle, Howard I. *American Small Sailing Craft—Their Design, Development, and Construction*. New York: W. W. Norton, 1951.

——. *The Search for Speed under Sail, 1700–1855*. New York: W. W. Norton, 1967.

—— *Yacht Designing and Planning for Yachtsmen, Students, and Amateurs*. New York: W. W. Norton, 1936.

"China," Foreign News *Time*, December 11, 1939, XXXIV 24, 17–18.

"China," World War II war front, *Time*, June 16, 1941, XXXVII 24, 24–25.

"Chinese Births on the Rise in USA—'Birth Tourism' Is a Booming Business Despite
   More Efforts at a Crackdown." *USA Today*, April 6, 2012. News 7A.

"Chinese City Known for Cars Moves to Limit Them." *New York Times*, International,
   September 5, 2012, A1, A3.

Chow, Paul. *The Junk That Challenged the Yachts*. Self-published, 2011.

Chuddacoff, Howard P. *How Old Are You?: Age Consciousness in American Culture*.
   Princeton, NJ: Princeton University Press, 1989.

Clark, Leonard. *A Wanderer Till I Die*. New York: Funk and Wagnalls, 1937.

Cleaton, Irene, and Allen Cleaton. *Books and Battles of the Twenties—American Literature, 1920–1930*. Boston: Houghton Mifflin, 1937.

Close, Upton. *In the Land of the Laughing Buddha—The Adventures of an American
   Barbarian in China*. New York: G. P. Putnam's Sons, 1924.

Clune, Frank. *Flight to Formosa*. Sydney, AU: Angus and Robertson, 1958.

———. *Scandals of Sydney Town*. Sydney, AU: Angus and Robertson, 1957.

Cocker, Mark. *Loneliness and Time—The Story of British Travel Writing*. New York:
   Pantheon Books, 1992.

Cohen, Steven. "Five Things You May Not Know about *Titanic's* Rescue Ship." *History
   in the Headlines,* April 2012. www.history.com.

Colcord, Lincoln. *An Instrument of the Gods and Other Stories of the Sea*. New York:
   MacMillan, 1922.

———. *The Game of Life and Death—Stories of the Sea*. New York: MacMillan, 1914.

Cole, Celia Caroline. "The Art of Staying Young—Escape and Renewal Are the Secrets
   of Youth." *Delineator*, January 1927, 29, 95.

Coleridge, Henry James. *The Life and Letters of St. Francis Xavier*. 2 vols. 4th ed. London: Burns and Oates, 1927.

Colley, Linda. *The Ordeal of Elizabeth Marsh—A Woman in World History*. New York:
   Pantheon Books, 2007.

Collins, Dale E., US Naval Reserve. "The Royal Road Across the Pacific." *US Naval
   Institute Proceedings*, April 1940, 501–12.

Collis, Maurice. *The Grand Peregrination*. London: Faber and Faber, 1949.

Connell, Evan S. *A Long Desire*. San Francisco: North Point, 1988.

Connolly, Kieron. *Dark History of Hollywood—A Century of Greed, Corruption, and
   Scandal behind the Movies*. London: Amber Books, 2014.

Connolly, Myles. *Mr. Blue*. New York: MacMillan, 1928.

———. *The Right to Romance*. Film based on story by Connolly. Directed by Alfred San-
   tell. Starring Ann Harding and Robert Young. RKO Radio Production, 1933.

Conover, Ted. *The Routes of Man—How Roads Are Changing the World and the Way We
   Live Today*. New York: Alfred A. Knopf, 2010.

Conrad, Barnaby. *San Francisco—A Profile with Pictures*. New York: Viking, 1959.

Conrad, Joseph. "An Outpost of Progress." In *Heart of Darkness and other Tales*, edited
   and with an introduction and notes by Cedric Watts. New York: Oxford University
   Press, 2002.

Conroy, F. Hillary, and Francis Conroy. *West across the Pacific: American Involvement
   in East Asia from 1898 to 1941*. With Sophie Quinn-Judge. Amherst, NY: Cambria,
   2008.

Cook, Andrew. *The Murder of the Romanovs*. Stroud, Glos., UK: Amberley, 2010.

Coolidge History Page. michaelmcfadyenscuba.info.

Corbin, Alain. *The Lure of the Sea—The Discovery of the Seaside in the Western World 1750-1840*. Translated by Jocelyn Phelps. London: Penguin Books, 1995.

Corbin, Alain et al. *A History of Virility*. Translated by Keith Cohen. New York: Columbia University Press, 2016.

Corradini, Robert E. *Narcotics and Youth Today*. New York: Foundation for Narcotics Research and Information, 1934.

Cortese, James. "Richard Halliburton—He Called the Road Royal." *Delta Review* 4, no. 5 (September 1967): 30-33, 48-51.

———. *Richard Halliburton's Royal Road*. Memphis: White Rose, 1989.

Cottle, Michelle. Notebook, "Race Baiting: All the Rage." *Newsweek*, February 20, 2012.

Cowan, Robert Ernest. *Forgotten Characters of Old San Francisco—1850-1870*. Los Angeles: Ward Ritchie Press, 1938. Includes, among its topics, the Emperor Norton.

Cowles, Virginia. *1913—An End and a Beginning*. Harper and Row Publishers, 1968.

Cram, Mildred. *Old Seaport Towns of the South*. Drawings by Allan G. Cram. New York: Dodd, Mead, 1917.

Crane, Louise. *China in Sign and Symbol*. With decorations by Kent Crane. Shanghai: Kelly and Walsh, 1926.

Crocker, Templeton. *The Cruise of the Zaca*. New York: Harper and Brothers, 1933.

Croke, Vicki Constantine. *The Lady and the Panda—The True Adventures of the First American Explorer to Bring Back China's Most Exotic Animal*. New York: Random House, 2006.

Cronkite, Walter. *A Reporter's Life*. New York: Alfred A. Knopf, 1996.

Crow, Carl. *400 Million Customers*. New York: Harper and Brothers, 1937.

———. "Japan's Hand in China." *World's Work*, September 1915, 529-44.

———. *Master Kung—The Story of Confucius*. New York: Harper and Brothers, 1938.

———. *My Friends, The Chinese*. Shanghai: Anderson Brothers Books, Printers and Stationers, 1938.

Crowe, Michael F., and Robert W. Bowen. *San Francisco Art Deco*. Chicago: Arcadia, 2007.

Crowell, Wilfred. "Author Lost: Richard Halliburton Missing 5 Days." *Nevada State Journal*, March 30, 1939.

———. "Report Concerning Halliburton Trans-Pacific Chinese Junk Expedition. Issued by (Publicity) Manager of Expedition." *Nevada State Journal*, April 17, 1939.

Cummings, Edna Mae. *Pots, Pans and Millions: A Study of Woman's Right to Be In Business, Her Proclivities and Capacity for Success*. Washington, DC: National School of Business Science for Women, 1929.

Cunarder, various issues.

Cutler, Leland, Letter. In *Golden Gate International Exposition* (spiral-bound volume). San Francisco: Division of Publications, 1938.

Dana, Richard Henry, Jr. *The Seaman's Friend—A Treatise on Practical Seamanship*. 1863. Reprint. Mineola, NY: Dover, 1997.

———. *The Seaman's Friend—Treatise on Practical Seamanship, with Plates; A Dictionary of Sea Terms; Customs and Usages of the Merchant Service; Laws Related to the Practical Duties of Master and Mariners*. Boston: Groom, 1863

———. *Two Years before the Mast*. Reprinted with a new introduction by John Seelye. New York: Signet, 2006.

——. *Two Years before the Mast—A Personal Narrative*. New Edition, With Subsequent Matter By the Author. Mineola, NY: Dover, 2007.

Darling, David. *Mayday—A History of Flight through Its Martyrs, Oddballs and Daredevils*. London: Oneworld, 2015.

Daugherty, Greg. "The Last Adventure of Richard Halliburton—the Forgotten Hero of 1930s America," *Smithsonian*, March 25, 2014.

David-Nell, Alexandra. *Magic and Mystery in Tibet*. New York: Claude Kendal, 1935.

Davis, Bette (1908–1989). www.imdb.com

Davis, Richard. "The Mysterious and Strange End of Richard Halliburton: What Happened to the Most Famous Man in America?"

Davis, Richard Harding. *The West from a Car-Window*. Illustrated (in part) by Frederic Remington. New York: Harper and Brothers, 1892.

——. *With the French in France and Salonika*. New York: Charles Scribner's Sons, 1916.

De Leeuw, Hendrik. *Flower of Joy*. New York: Lee Furman, 1939.

Deering, Mabel Craft. "Ho for the Soochow Ho." *National Geographic Magazine*, June 1927, 623–50.

Deffaa, Chip. "On the Trail of Richard Halliburton." *Princeton Alumni Weekly*, May 13, 1975, 8–12.

Delgado, James P. "Mapping Titanic—The New Frontier of Underwater Archaeology." *Archaeology*, May/June 2012.

Dew, Gwen. "I Photographed the Fall of Hong Kong." *Popular Photography*, February 1943, 28–29, 85–87.

Diamond, Jared. *Collapse—How Societies Choose to Fail or Succeed*. New York: Penguin, 2005.

Dickerson, A. F., "Light and Structure at *Golden Gate Exposition*" *Color: New Synthesis in the West*. Reprinted by General Electric from *Architectural Record*, 1939.

Dickstein, Morris. *Dancing In the Dark—A Cultural History of the Great Depression*. New York: W. W. Norton, 2009.

Dilling, Elizabeth "Mrs. Albert W." *The Red Network—A "Who's Who" and Handbook of Radicalism for Patriots*. Kenilworth, IL: self-published, 1934.

Diner, Stephen J. *A Very Different Age—Americans in the Progressive Era*. New York: Hill and Wang, 1998.

Dingle, Capt. Aylward E. *Seaworthy*. Boston: Houghton-Mifflin, 1930.

Dobie, Charles Caldwell. *San Francisco: A Pageant*. With illustrations by E. H. Suyham. New York: D. Appleton–Century, 1933.

——. *San Francisco's Chinatown*. With illustrations by E. H. Suyham. New York: D. Appleton–Century, 1936.

"Dollar's Career Is History of American President Lines." *San Francisco News*, February 15, 1939.

Dong, Stella. *Shanghai—The Rise and Fall of a Decadent City*. New York: William Morrow and Sons, 2000.

Donnelly, Ivon A. *Chinese Junks and Other Native Craft*. 1924. Reprinted with a foreword by Gareth Powell. China Economic Review Publishers, 2008.

Dreiser, Theodore. *Tragic America*. New York: Horace Liveright, 1931.

——. "What Are the Defects of American Democracy?" In *America Is Worth Saving*, 154–71. New York: Modern Age Books, 1941.

——. "What Is Democracy?" In *America Is Worth Saving*, 146–53. New York: Modern Age Books, 1941.

Drury, Robert. *Halsey's Typhoon*. New York: Grove, 2007.

Duffus, R. L. *The Tower of Jewels—Memories of San Francisco*. New York: W. W. Norton, 1960.

Duncan, Archibald, Esq. *The Mariner's Chronicle or Authentic and Complete History of Popular Shipwrecks*. Vol. 1. Black Apollo Press, 2004.

Durant, Will. *Fallen Leaves*. New York: Simon and Schuster, 2014.

Earhart, Amelia. *The Fun of It—Random Records of My Own Flying and of Women in Aviation*. New York: Brewer Warren and Putnam, 1932.

——. *The Last Flight*. New York: Harcourt, Brace, 1937.

Earle, David M. *All Man! Hemingway, 1950s Men's Magazines, and the Masculine Persona*. Kent, OH: Kent State University Press, 2009.

Easton, Emily. *Youth Immortal: A Life of Robert Herrick*. Boston: Houghton-Mifflin, 1934.

Eberhardt, Isabelle. *The Oblivion Seekers*. San Francisco: City Lights, 1972.

"Erle Palmer Halliburton," *The Encyclopedia of Oklahoma History*. Online.

"*Exposition* Edition." *San Francisco News*, Wednesday, July 6, 1938.

Fanthorpe, Lionel, and Patricia Fanthorpe. *Unsolved Mysteries of the Sea—An Eye-Opening Exploration of Lost Lands, Phantom Ships, and Dangerous Denizens of the Deep*. New York: MJF Books, 2004.

Fardon, George Robinson. *San Francisco in the 1850s: 33 Photographic Views*. With an introduction by Robert A. Sobieszek. Rochester, NY: Dover, 1977.

"Fastnet 1979—The Disaster That Changed Sailing." *Yachting World*, August 2009.

Faught, C. Brad. *Gordon—Victorian Hero*. Washington, DC: Potomac Books, 2008.

Feis, Herbert. *The China Tangle—American Effort in China from Pearl Harbour to the Marshall Plan*. Princeton, NJ: Princeton University Press, 1953.

Ferguson, Niall. "The End of Prosperity? A Noted Historian Looks At Parallels between This Financial Crisis and 1929 and Shows What Must Be Done to Avoid Depression 2.0," *Time*, October 13, 2008, 36–41.

Fermor, Patrick Leigh. *A Time of Gifts*. London: John Murray, 1977.

Finney, Ben R. *Hokule'a—The Way to Tahiti*. New York: Dodd, Mead, 1979.

Firth, John. "Syphilis—Its Early History and Treatment Until Penicillin and the Debate on Its Origins." *Journal of Military and Veterans' Health*, History Issue 20, no. 4 (November 2012).

Fisher, Irving. "How Much Is a Dollar?" *New York Herald Tribune*, January 27, 1929. 13 pages.

Fisher, J. T. *Dr. America—The Lives of Thomas A. Dooley*. Boston: University of Massachusetts Press, 1997.

Fisher, Margery, and James Fisher. *Shackleton*. Cambridge, MA: Riverside, 1958.

Fisher, Susanna. *The Makers of the Blueback Charts: A History of Imray Laurie Norie and Wilson Ltd*. Ithaca, NY: Regatta, 2001.

Fitch, George Ashmore. *My Eighty Years in China*. Taipei, Taiwan: Mei Ya, 1967. Privately printed by author Fitch.

Fitch, Robert F. "Life Afloat In China—Tens of Thousands of Chinese in Congested Ports Spend Their Entire Existence on Boats." *National Geographic Magazine*, June 1927, 665–86.

Fitzgerald, F. Scott. *A Short Autobiography*. Edited by James L. W. West III. New York: Scribner, 2011.

Fleming, Peter. *News from Tartary—A Journey from Peking to Kashmir*. New York: Charles Scribner's Sons, 1936.

——. *One's Company—A Journey to China*. New York: Charles Scribner's Sons, 1934.

——. *The Siege of Peking*. London: Rupert Hart-Davis, 1959.

Flynn, Errol. *Beam Ends*. New York: Longmans, Green, 1937.

Foot, Michael. *The Politics of Paradise—A Vindication of Byron*. New York: Harper and Row, Publishers, 1988.

Forbes, Alexander. *The Radio Gunner—A Fable of the Navy*. Boston: Houghton-Mifflin Cmpany, 1924.

Forbes, Edgar Allen. *Twice Around the World*. New York: Fleming H. Revell, 1912.

Ford, Ford Maddox. *Joseph Conrad—A Personal Remembrance*. New York, Ecco, 1924, 1989.

Forrai, Judith, "History of Different Therapeutics of Venereal Disease Before the Discovery of Penicillin," in *Syphilis—Recognition, Description, and Diagnosis* edited by Neuza Satomi Sato. London: InTech, 2001

Forsythe, Michael. *Hong Kong Journal*, "Where Humans Herd Themselves and Cattle Roam Free." *New York Times*. International, A4, A8.

Fox-Smith, Cicely, *Ocean Racers*. New York: Robert M. McBride, 1932.

Franck, Harry Alverson. *All About Going Abroad: With Maps and A Handy Travel Diary*. New York: Brentano's, 1927.

——. *East of Siam*. New York: D. Appleton-Century, 1939.

——. *Glimpses of Japan and Formosa*. 1924. Reprint. New York: Century, 1939.

——. *Marco Polo, Junior—The True Story of an Imaginary American Boy's Travel-Adventures All Over China*. New York: Century, 1929.

——. *Roving through Southern China*. New York: Century, 1925.

——. *A Vagabond Journey around the World—A Narrative of Personal Experience*. New York: Century, 1910.

——. *Wandering in Northern China*. New York: Century, 1923.

——. *Working My Way Around the World*. Rewritten (or edited) by Lena M. Franck from Harry A. Franck's *Vagabond Journey around the World*. New York: Century, 1918.

Freeman, Robert. "Echoes of History Reverberate From WWI: World Powers Are Again Forming Alliances, As They Did a Century Ago." Opinion, *Wisconsin State Journal*, November 16, 2014.

French, Paul. *A Tough Old China Hand—The Life, Times, and Adventures of an American in Shanghai*. Hong Kong: Hong Kong University Press, 2006.

Frere-Cook, Gervis, ed. *The Decorative Arts of the Mariner*. Boston: Little, Brown, 1966.

Freuchen, Peter. *Peter Freuchen's Book of the Seven Seas*.With David Loth. New York: Simon and Schuster, 1957.

Friendly, Alfred. *Beaufort of the Admiralty—The Life of Sir Francis Beaufort 1774–1857*. New York: Random House, 1977.

Fussell, Paul. *Abroad—British Literary Traveling between the Wars*. New York: Oxford University Press, 1980.

——. *The Great War and Modern Memory*. Oxford: Oxford University Press, 1975.

Garfield, Brian. *A Manifest Destiny—A True Romantic Saga of Theodore Roosevelt*. New York: Mysterious Press, 1989.

Garthwaite, Rosie. *How to Avoid Being Killed In a War Zone—The Essential Guide for Dangerous Places*. New York: Bloomsbury, 2011.

Gatty, Harold. *The Raft Book—Lore of the Sea and Sky*. New York: George Grady Press, 1943.

Gelernter, David. *1939—The Lost World of the Fair*. New York: Free Press, 2008.

Gentry, Curt. *San Francisco and the Bay Area—Present and Past*. With maps and photographs. Garden City, NY: Doubleday, 1962.

George, Ivy. Period photographs of *Hangover House*. Author's collection.

Gerend, Joseph. *Mistress of the Delta Land*. Memphis, TN: Brunner Printing, 1939.

"Gertrude Atherton and Ambrose Bierce." *California History—The Magazine of the California Historical Society*, Winter 1981.

Gibson, Charles R. *The Great Ball On Which We Live—An Interestingly Written Description of Our World, The Mighty Forces of Nature, & the Wonderful Animals Which Existed Before Man, All Described In Simple Language*. London: Seeley, Service, 1925.

Gilbert, Douglas. *Floyd Gibbons—Knight of the Air*. With an introduction by Floyd Gibbons. New York: Robert McBride, 1930.

Giles, Ray. *Sleep! The Secret of Greater Power and Achievement with 101 Tips from Famous People*. Indianapolis: Bobbs-Merrill, 1938.

Gillis, John R. *Youth and History*. Studies in Social Discontinuity. New York: Academic Press, 1974.

Gleason, Gene. *Tales of Hong Kong*. New York: Roy, 1967.

"Globe-Trotter, The." *The News Outline—An Elementary Story of the News* nos. 1–11 (September–December 1931): 1–48.

"Golden Gate Bridge." *Life,* May 31, 1937, 43–47.

"Golden Gate Exposition Opens With a Wild West Wallop." *Life,* March 6, 1939, 11–15.

Goldstein, Milton. *Old Mother Earth and Her Family—A Book of Geography for Young People*. New York: Platt and Munk, 1930.

Goodman, Mathew. *Eighty Days—Nellie Bly and Elizabeth Bisland's History-Making Race around the World*. New York: Ballantine Books, 2013.

Goodnow, Frank Johnson. "The Geography of China—The Influence of Physical Environment on the History and Character of the Chinese People." *National Geographic Magazine*, June 1927, 651–64.

Gould, Gerald. *The Collected Poems of Gerald Gould*. New York: Payson and Clark, 1929.

Grand Cherry Show—Takarazuka Girls on the Invitation of the *Golden Gate International Exposition* (music and theatre school with females playing roles of both genders). 1939. Unpaginated, 32 pp.

Graham, Robin Lee. *Dove*. With Derek L. T. Gill. New York: Harper and Row, 1972.

Graham, Stephen. *The Gentle Art of Tramping*. New York: D. Appleton, 1926.

Greatest Sailing Stories Ever Told, The—Twenty-Seven Unforgettable Stories. Lyons, 2002.

Green, Martin. *The Robinson Crusoe Story*. University Park: Pennsylvania State University Press, 1990.

Griffis, William Elliot. *China's Story in Myth, Legend, Art, and Animals*. Boston:

Houghton Mifflin, 1911. Reprinted with the addition of "Brought Down to Date."
    1935.

Gross, Alexander, F.R.G.S., ed. *Famous Guide to San Francisco and the World's Fair—
    Pictorial and Descriptive*. With 9 maps and 78 illustrations. San Francisco: San
    Francisco News Company, 1939.

Grundon, Imogen. *The Rash Adventurer: A Life of John Pendlebury*. With an introduc-
    tion by Patrick Leigh Fermor. UK: Libri, 2007.

Gugliotta, Guy. "The First Americans." *Smithsonian*, February 2013, 39–47.

Gurman, Joseph, and Myron Slager. *Radio Round-Ups—Intimate Glimpses of the Radio
    Stars*. Boston: Lothrop, Lee and Shepard, 1932.

Gwin, Peter. "The Mystery of Risk." *National Geographic*, June 2013, 30–45.

Gwulo: Old Hong Kong. gwulo.com.

Hacker, Arthur. *China Illustrated—Western Views of the Middle Kingdom*. With a fore-
    word by Frederic Wakeman. London: Tuttle, 2004.

Hagen, John Milton. *Holly-Would!* With illustrations by Feg Murray. New Rochelle,
    NY: Arlington House, 1974.

Haley, James L. *Wolf—The Lives of Jack London*. New York: Basic Books, 2010.

Hall, Capt. C. W. *Drifting Around the World—A Boy's Adventures by Land and Sea*.
    Boston: Lee and Shepard, 1882.

Hall, Edith. *The Return of Ulysses—A Cultural History of Homer's Odyssey*. Baltimore,
    MD: Johns Hopkins University Press, 2008.

Hall, James Norman, and Charles Nordhoff. *The Bounty Trilogy*. New York: Little,
    Brown, 1936.

——. *Mutiny on the Bounty*. Boston: Little, Brown, 1932.

Halliburton, Richard. *The Book of Marvels—The Occident*. Indianapolis: Bobbs-Merrill,
    1937.

——. "De Profundis." *Ladies Home Journal*, November 1929.

——. *The Flying Carpet*. Indianapolis: Bobbs-Merrill, 1925.

——. *The Glorious Adventure*. Indianapolis: Bobbs-Merrill, 1927.

——. "Humiliating the Matterhorn." In *New Horizons—The Discovery Series—Book
    Two*, edited by H. Augustus Miller and Bernice E. Leary, with maps by George
    Bell, 3–16. New York: Harcourt, Brace, 1937.

——. "I Killed the Czar—Assassin Confesses to Richard Halliburton." *Liberty Maga-
    zine*, July 27, 1935.

——. *India Speaks*. New York: Grosset and Dunlap, 1933.

——. "Monkey Business." *Ladies Home Journal*, October, 1929.

——. *New Worlds to Conquer*. Indianapolis: Bobbs-Merrill, 1929.

——. "Poor Richard Crusoe and Toosday." *Ladies Home Journal*, December 1929.

——. *Richard Halliburton—His Story of His Life's Adventure as Told in Letters to His
    Mother and Father*. Indianapolis: Bobbs-Merrill, 1940.

——. "Richard Halliburton 'Prince of Lovers' Talks About Women and Love." Inter-
    view with Dorothy Dayton. *Love Magazine*, March 1930, 36–41.

——. *The Royal Road to Romance*. Indianapolis: Bobbs-Merrill, 1925.

——. *Second Book of Marvels—The Orient*. Indianapolis: Bobbs-Merrill, 1938.

——. *Seven League Boots*. Indianapolis: Bobbs-Merrill, 1935.

——. "Seven Volunteers for Death" (Admiral Hobson in Santiago, Cuba, 1898).
    Reader's Digest, January, 1935, 106–7.

———. "The SS *Halliburton.*" *Ladies Home Journal,* June 1929.

———. Unpublished letter to Captain Simmons (author of *Sinbads of Science*), post-marked May 5, 1928. Author's collection.

———. Unpublished letter to Hannah Watterson, November 2, 1927. Author's collection.

———. Unpublished letter to J. Watson Webb Jr., August 30, 1938 (Chancellor Hotel), 12 pp. with enclosures. Author's collection.

———. Unpublished letter to Mary Clarke, July 26, 1927 (Victoria Hotel, London), with envelope and photograph (with words on reverse, presumably by Clarke). Author's collection.

———. "Upon a Peak in Darien." *Ladies Home Journal,* July 1929.

———. "The Wickedest City in the World," also entitled "Cairo—the Capital of Sin," #38 (in Bell Syndicate series of articles, 1935). Unpublished. Richard Halliburton Collection, Princeton University Library, Princeton, NJ.

Halliburton, Richard (ghost), "Does Adventurer's Ghost Prowl His Strange Home?" *Richmond Times,* January 20, 1957.

Hamill, John. *The Strange Career of Mr. Hoover—Under Two Flags.* New York: William Faro, 1931.

Hammond, Tom. *Showdown in Memphis: An Epic Tale of the Forties.* Memphis, TN: Retrospective Productions, 1997.

Hark, Ina Rae, ed. *American Cinema of the 1930s—Themes and Variations.* New Brunswick, NJ: Rutgers University Press, 2007.

Harkness, Ruth. *The Lady and the Panda—an Adventure.* New York: Carrick and Evans.

———. *Pangoan Diary.* New York: Creative Age, 1942.

Harmetz, Aljean. *Round Up All the Usual Suspects: The Making of Casablanca; Bogart, Bergman, and World War II.* New York: Hyperion, 1992.

Harrington, Lyn. *The Grand Canal of China.* Folkestone, Great Britain: Bailey Brothers and Swinfen, 1974.

Harrington, Peter. *Travel & Exploration.* London, 2000 et seq. Various catalogues noting rare books on subject.

Harrisburg Telegraph, July 10, 1937.

Harrison, Marguerite. *Marooned in Moscow—The Story of an American Woman Imprisoned In Soviet Russia.* New York: George H. Doran, 1921.

———. "Pekin—Grand Hotel de Pekin," Letter to Laura Owens," October 18, 1922. Author's collection.

Hart, Dorothy. *Thou Swell, Thou Witty—The Life and Lyrics of Lorenz Hart.* New York: Harper and Row, 1976.

Haven, Violet Sweet. *Gentlemen of Japan—A Study of Rapist Diplomacy.* Chicago: Ziff-Davis, 1944.

———. *Many Ports of Call.* New York: Longmans, Green, 1940.

Hawks, Ellison. *The Book of Natural Wonders.* New York: Tudor, 1935.

Hayes, Claire W. *The Boy Allies of the Army* series. New York: A. L. Burt, 1915 et seq.

Hearn, Lafcadio. *Inventing New Orleans.* Edited by S. Frederick Starr. Oxford: University Press of Mississippi, 2001.

Hearst News Service. *Sea Dragon* newsreel outtakes. 1939. UCLA Film Archive.

Heat-Moon, William Least. *Here, There, Elsewhere: Stories from the Road.* New York: Little, Brown, 2013.

Heaver, Stuart. "Richard Halliburton—The Hero Time Forgot." *China Morning Post*, March 23, 2014.

Henry, William A., III. *In Defense of Elitism*. New York: Doubleday, 1994.

Heyerdahl, Thor. *Kon Tiki*. New York: Rand McNally, 1950.

——. *The Ra Expeditions*. New York: HarperCollins, 1993.

Heywood, Graham S. P. "Clouds and the Weather." *Hong Kong Naturalist*, April 1939, 119–28.

——. *Rambles in Hong Kong*. With a new introduction and commentary by Richard Gee. Hong Kong: Oxford University Press, 1992.

Hillyer, V. M. *A Child's Geography of the World*. New York: Appleton-Century-Crofts, 1929.

"Hitlerism in America Grows Stronger and Bolder." *Look*, March 28, 1939.

Hoefler, Paul L. *Africa Speaks: A Story of Adventure; The First Trans-African Journey by a Motor Truck from Mombosa on the Indian Ocean to Lagos (Nigeria) on the Atlantic*. Chicago: John C. Winston, 1931.

——. *Africa Speaks*. 1930. Directed (and produced) by Walter Futter. Narrated by Lowell Thomas. Cinematography by Paul Hoefler. 75 minutes. Reissued on DVD.

Hoff, Roma Borst. "My Royal Road to Romance" (unpublished) with original narrative and documents, including correspondence between Professor Roma B. Hoff (1926–2017) and Wesley Halliburton Sr. (1870–1965), 1939–1963. Roma Borst Hoff Archive. Richard Halliburton Collection. Barret Library, Rhodes College, Memphis, TN.

Holland, James. "World War Speed." *Secrets of the Dead*. Episode Thirteen. Media with Impact, 2019. DVD.

Hollister, Mary Brewster. *Beggars of Dreams*. New York: Dodd, Mead, 1937.

Holman, C. Hugh. "The Epic Impulse." In *The Loneliness at the Core—Studies in Thomas Wolfe*, 155–67. Baton Rouge: Louisiana State University Press, 1975.

Holme, Cedric Geoffrey (photography), and William Gaunt (text). *Touring the Ancient World With a Camera*. London: Studio Limited, 1932.

"Hong Kong—Britain's Far-Flung Outpost in China." With 16 illustrations. *National Geographic Magazine*, March 1938, 349–60.

Hong Kong Naturalist. 1930–41. Hong Kong Journals Online. https://hkjo.lib.hku.hk

"Hong Kong's Shipbuilding Industry," *Weekly Commercial News*. Online.

Honolulu Star-Bulletin, 1938–39 issues.

Hookham, Hilda. *A Short History of China*. London: Longman, 1969.

Horne, Charles F., ed. *Great Men and Famous Women: A Series of Pen and Pencil Sketches of the Lives of More Than 200 of the Most Prominent Personages in History*. 8 vols. New York: Selmar Hess, 1894.

Horney, Karen. *Neurosis and Human Growth—The Struggle toward Self-Realization*. New York: W. W. Norton, 1950.

Horowitz, Helen Lefkowitz. *Rereading Sex—Battles over Sexual Knowledge and Suppression in Nineteenth-Century America*. New York: Alfred A. Knopf, 2002.

Hounchell, Whitley "Cementing a Reputation—The Story of Erle P. Halliburton," November 2013. whitleyhounchell.wordpress.com.

"House Richard Halliburton Built, The." *Look*, December 19, 1939.

"House Richard Halliburton Built and Never Lived In, The." *Look*, December 19, 1939.

Howard, Harvey J. *Ten Weeks with Chinese Bandits*. Dodd, Mead, 1932.

Howe, Herbert. Unpublished letter to author, November 15, 1996.

Hudson, Kenneth, and Ann Nicholls. *Tragedy on the High Seas—A History of Ship-wrecks*. New York: A and W, 1979.

Hunter, Stanley Armstrong. *Temple of Religion and Tower of Peace at the Golden Gate International Exposition*. San Francisco, 1940. 96 pp. with photographs.

Hutchinson, Paul. "New China and the Printed Page (*and* Among the People of Cathay)." *National Geographic Magazine*, June 1927, 687–722.

Huxley, Aldous. *Jesting Pilate—The Diary of a Journey*. With 22 illustrations of scenes in India, Burma, the Pacific and America. London: Chatto and Windus, 1926.

Huysmans, Joris-Karl. *Against the Grain (A rebours)*. Translated by Margaret Mauldon. Edited and with an introduction and notes by Nicholas White. Oxford: Oxford University Press, 1998.

——. *Against the Grain (A Rebours)*. With an introduction by Havelock Ellis. New York: Dover, 1969.

In the Wake of the Zaca, produced and directed by Luther Greene, edited by Carrie Ledereer, 2005, DVD, 53 min., EP 103. www.zaaca.com.

India Speaks, or *The Bride of Buddha*. Norman Houston, 1940. UCLA Film and Television Archive.

Information for Travellers Landing in Hong Kong. London: Thomas Cook and Son, 1919. 41 pp.

Ingles, Fred. *A Short History of Celebrity*. Princeton, NJ: Princeton University Press, 2010.

Intercollegiate Student Tours, Europe. Hamburg-America, 1932. 32 pp.

Irwin, Inez Haynes. "Women's Clubs." In *Angels and Amazons—A Hundred Years of American Women*. Garden City, NY: Doubleday, 1934.

Isaac, Benjamin. *The Invention of Racism in Classical Antiquity*. Princeton, NJ: Princeton University Press, 2004.

Jackson, Joseph Henry. *A Trip to the San Francisco Exposition with Bobby and Betty*. New York: Robert M. McBride, 1939.

Jamali, Arash. "Investigation of Propeller Characteristics with Different Locations of the Rudder." Report No. X-10/253. Master's thesis, Chalmers University of Technology (Goteburg, Sweden), 2010.

James, George Wharton. *Our American Wonderlands*. Chicago: A. C. McClurg, 1916.

James, William. *The Varieties of Religious Experience—A Study in Human Nature*. Edited and with an introduction by Martin E. Marty. New York: Penguin Books, 1982, 1985.

Jameson, W. C. *Amelia Earhart—Beyond the Grave*. Lanham, MD: Taylor Trade, 2016.

"Japanese Empire, The." *Fortune*, September 1936.

Jenkins, Mark Collins. *The Book of Marvels—An Explorer's Miscellany*. Washington, DC: National Geographic Society, 2009.

Jenks, Tudor. *The Century World's Fair Book for Boys and Girls, Being the Adventures of Harry and Philip with Their Tutor, Mr. Douglass, at the World's Columbian Exposition with Off-Hand Sketches by Harry and Snap-shots by Philip and Illustrations by Better-Known Artists and Reproductions of Many Photographs*. New York: Century, 1893.

"Jeweled Radiance," *Magic of Night*. *San Francisco Examiner*, Wednesday, February 15, 1939. Section 3, 1–12.

Jokstad, Capt. Charles. *The Captain and the Sea*. New York: Vantage, 1967.

Jones, Christopher. *New Heroes in Antiquity—From Achilles to Antinoos*. Cambridge, MA: Harvard University Press, 2010.

Jones, Nigel. *Rupert Brooke—Life, Death and Myth*. London: Head of Zeus, 1999; 2014.

Joseph Conrad: Including "An Approach to His Writings," "A Biographical Sketch," "A Brief Survey of His Works," and a "Bibliography." Garden City, NY: Doubleday, 1926.

Juanita McCown Hight's Conversation (1934) with Richard Halliburton. In *Golden Days—Reminiscences of Alumnae, Mississippi State College for Women*. Jackson: University Press of Mississippi, 2008.

June, Laura. "The Life of Djuna Barnes, Stunt Reporter and Shocking Modernist." *Literally*. June 3, 1915. Online.

Junger, Sebastian. *The Perfect Storm—A True Story of Men Against the Sea*. New York: W. W. Norton, 1997.

Kahn, David. *The Reader of Gentlemen's Mail: Herbert O. Yardley and the Birth of American Intelligence*. New Haven: Yale University Press, 2004).

Kahn, S. L. Clipping (undated) sent to Ralph B. Vawter by Wesley Halliburton. *Commercial Appeal* (Memphis, TN), after June 10, 1939.

Kane, Herb Kawainui, and Michael E. Long. "Discoverers of the Pacific" (folded insert). *National Geographic Society Magazine*, December 1974.

Kane, John Francis, ed. *Picturesque America*. New York: Garden City, 1937.

Kazantzakis, Nikos. *Japan China*. With an epilogue by Helen Kazantzakis. New York: Simon and Schuster, 1963.

Keats, John. *You Might As Well Live—The Life and Times of Dorothy Parker*. New York: Paragon House, 1970.

Keay, John. *Empire's End—A History of the Far East from High Colonialism to Hong Kong*. New York: Scribner, 1997.

———. *A History of China*. New York: Basic Books, 2009.

Keels, Thomas H., and Elizabeth Farmer Jarvis for Chestnut Hill Historical Society. *Images of America—Chestnut Hill* [Philadelphia]. Charleston, SC: Arcadia, 2002.

Kemble, John Haskell. *San Francisco Bay—A Pictorial Maritime History*. Cambridge, MD: Cornell Maritime, 1957.

Ki Cei Leung, Angela. *Leprosy in China—A History*. Studies of the Weatherhead East Asian Institute. New York: Columbia University Press, 2009.

Kiefer, Michael. *Chasing the Panda*. New York: Four Walls Eight Windows, 2002.

King, Thomas F. "The Islands of the Japanese Mandate in 1937." The Earhart Project, TIGHAR #0391ECB, 2015.

Kiple, Kenneth F., ed. *The Cambridge World History of Human Diseases*. Cambridge: Cambridge University Press, 1993.

Kipling, Rudyard. *Captains Courageous*. With a new introduction by Marilyn Sides. New York: Signet Classics, 2004.

——— (Kipling as commissioner, Boy Scouts in England). *Land and Sea Tales for Scouts and Scoutmasters*. Garden City, NY: Doubleday, Doran, 1929.

Kirkland, Frances, and Winifred Frances. *Girls Who Made Good*. New York: Richard R. Smith, 1930.

Kirkland, Lucian Swift. "What Japan Thinks of Us." *Travel*, August 1923, 10-15, 37-38.

Kirsh, Adam. "The 'Five-Foot Shelf' Reconsidered." *Harvard Magazine* 103, no. 2 (November–December 2001).

Kjellstrom, Rolf. *Eskimo Marriage: An Account of Traditional Eskimo Courtship and Marriage.* Stockholm: Nordiska Museets, 1973.

Klapholz, Jesse. "The History and Development of Microphones." *Sound and Communication*, September 1986.

Knepper, Max. *Sodom and Gomorrah.* Los Angeles: End Poverty League, 1935.

Knox, James. *Robert Byron.* London: John Murray, 2003.

Knox-Johnston, Robin. *Seamanship.* New York: W. W. Norton, 1987.

Kotani, Ken. *Japanese Intelligence in World War II.* Oxford, UK: Osprey, 2009.

Kramer, Alan. *Dynamic of Destruction—Culture and Mass Killing in the First World War.* Oxford: Oxford University Press, 2007.

Kroger, Grove. "The Horizon Chaser and *Hangover House.*" *Laguna Beach Art Magazine*, October 31, 2014.

Kuhn, Annette. *Cinema, Censorship and Sexuality, 1909–1925.* London: Routledge, 1988.

Kuo, Huei-Ying. *Networks beyond Empires—Chinese Business and Nationalism in the Hong Kong-Singapore Corridor, 1914–1941.* Leiden, Netherlands: Brill Academic, 2014.

Kushner, Malcolm. *Public Speaking for Dummies.* 2nd ed. Hoboken, NJ: Wiley, 2004.

"Kweilin Incident," *New York Times*, August 25, 1938.

Labor, Earle. *Jack London—An American Life.* New York: Farrar, Straus and Giroux, 2013.

La Farge, Henry, ed. *Lost Treasures of Europe—A Pictorial Record.* With 427 photographs. New York: Pantheon Books, 1946.

Lamb, W. Kaye. *Empress to the Orient.* Vancouver Maritime Museum, 1991.

Landsburg, Alan. "Richard Halliburton." In *In Search of Missing Persons*, 9–29. New York: Bantam Books, 1978.

Lane, Allen Stanley. *Emperor Norton—Mad Monarch of America.* Caldwell, ID: Caxton Printers, 1939.

Lanham, Edwin Moultrie. *Sailors Don't Care.* New York: Jonathan Cape and Harrison Smith, 1930. P. x (quoted). Earlier (banned) version: Paris: Contact Edition, 1929, printed in Dijon by Darantiere.

Lardner, John. "The Lindbergh Legends." In *The Aspirin Age, 1919–1941*, edited by Isabel Leighton. New York: Simon and Schuster, 1949.

Laskowski, Lenny, and the Princeton Language Institute. *Ten (10) Days to More Confident Public Speaking.* New York: Grand Central, 2001.

Last Voyages: Cavendish, Hudson, Raleigh; The Original Narratives. Introduced and edited by Philip Edwards. Oxford: Clarendon, 1988.

Laufer, Berthold. *Tobacco and Its Use in Asia.* Anthropology, Leaflet 18. Chicago: Field Museum of Natural History, 1924.

Laurence, Ray. *Roman Passions—A History of Pleasure in Imperial Rome.*(London: Continuum, 2009.

La Varre, William. *Gold, Diamonds, and Orchids—An Amazing Story of a Year's Expedition into a Lost World.* New York: Fleming H. Revell, 1935.

Le Blond, Mrs. Aubrey. "Tales of a Mountain Climber." *Strand Magazine—An Illustrated Monthly* 34, no. 204 (January 1908): 619–27.

Le passage du Grand-Saint Bernard (par Richard Halliburton). July 19 (beginning),

1935. Centre Valaisan du Film, Patria. Film Archive du Kantons Wallis. B/w, 10 minutes.

Leach, William. *Country of Exiles—The Destruction of Place in America.* New York: Pantheon Books, 1999.

Leet, L. Don. *Causes of Catastrophe.* New York: McGraw-Hill, 1948.

Leider, Emily Wortis. *California's Daughter: Gertrude Atherton and Her Times.* Stanford, CA: Stanford University Press, 1991.

Lengel, Edward G. *Inventing George Washington—America's Father, in Myth & Memory.* New York: Harper/Collins, 2011.

Lengyel, Emil. *Turkey.* New York: Random House, 1941.

Leonard, Arthur Lee. *Lost Road.* New York: Macaulay, 1934.

Lesieutre, Susie Seefelt. "Joy Camps: The Camp Craft Camps for Girls." *A Northern Wisconsin Adventure, Wisconsin Magazine of History,* Summer 2015, 36–49.

Lethem, Jonathan. "Dickens: Greatest Animal Novelist of All Time?" *Believer,* March 2003. also Internet.

Levenstein, Harvey. *Seductive Journey—American Tourists in France from Jefferson to the Jazz Age.* Chicago: University of Chicago Press, 1998.

Levine, Lawrence W. *Highbrow Lowbrow—The Emergence of Cultural Hierarchy in America.* Cambridge, MA: Harvard University Press, 1988.

Levy, Alexander. "Halliburton House called Hangover," *California Arts and Architecture Magazine,* November 1937.

Lewis, Wyndham. "If I Were a British Agent." in *America, I Presume,* 203–28. New York: Howell, Soskin, 1940.

Leys, James F. *After You Magellan.* New York: The Century Company, 1927.

Life, October 17, 1938 (Carole Lombard cover).

Lindley, Capt. Augustus. *The Log of the Fortuna—A Cruise in Chinese Waters, Containing Tales of Adventure in Foreign Climes, by Sea and by Shore.* London: Cassell, Peter and Galpin, n.d., ca. 1870–80.

Linkletter, A[rthur] G. *America! Cavalcade of a Nation.* A. L. Vollman, 1940. 16 pp.

——. *I Didn't Do It Alone—An Autobiography of Art Linkletter As Told to George Bishop.* Jameson Books, 1980.

Lipsky, William. *San Francisco's Panama-Pacific International Exposition.* Arcadia, 2005.

Lollar, Michael. "'Little Friend' Part of Halliburton Lore—Adventurer's Dad Accepted Girl as Surrogate Grandchild." *Commercial Appeal* (Memphis, TN), May 30, 2012.

London, Charmian Kittredge (Mrs. Jack London). *Jack London in the South Seas.* London: Mills and Boon, 1917.

London, Jack. *From Coast to Coast with Jack London Written by Himself from Personal Experiences.* Erie, PA: A-No-1, 1917.

——. *The Cruise of the Snark.* 1911. Reprint. London: Mills and Boon, n.d.

Long, Elgen M., and Marie K. Long. *Amelia Earhart—The Mystery Solved.* New York: Simon and Schuster Paperbacks, 1999.

Lorch, Fred W. *The Trouble Begins At Eight—Mark Twain's Lecture Tours.* Ames: Iowa State University Press, 1968.

Los Angeles County, California. Los Angeles Chamber of Commerce brochure). 1931. 64 pp.

Lower East Side. Tenement Museum. www.tenement.org.

Lowry, Malcolm. *Ultramarine*. Philadelphia: J. B. Lippincott, 1962.

Lunbeck, Elizabeth. *The Americanization of Identity*. Cambridge, MA: Harvard University Press, 2014.

Lyceum World, Instruction, Entertainment Enlightenment for Lecturers, Entertainers, Concert Companies, Bureaus, Committees, Chautauquas and Audiences, Arthur E. Gringle, Editor, Indianapolis, Indiana, 10, no. 5 (August 1917).

Lynn, David. "Heavy Weather Sailing—Making a Series Drogue." Online.

Lytton, Earl of. *Antony—A Record of Youth*. With a foreword by J. M. Barrie. London: Peter Davies, 1935.

MacCannell, Dean. *The Tourist—A New Theory of the Leisure Class*. Berkeley: University of California Press, 1999, 1976.

MacDonald, William. "Modern Life and Its Hazards." Review of C. Delisle Burns' *Modern Civilization on Trial*, by C. Delisle Burns. *New York Times Book Review*, Sunday, August 16, 1931. P. 1.

MacFadden Exhibit. Souvenir Program. *New York's World Fair* 1939, Communications Building. 8 p.

Mack, Gerstle. *The Land Divided: A History of the Panama Canal and Other Isthmian Canal Projects*. Alfred A. Knopf, 1944.

——. *1906—Surviving San Francisco's Great Earthquake and Fire*. Chronicle Books, 1981.

MacKerras, Colin. *Western Images of China*. Oxford: Oxford University Press, 1989.

"Magic in the Night," Official Souvenir—*The Golden Gate International Exposition*. San Francisco: Crocker, 1939. Unpaginated.

Mahan, Capt. Alfred Thayer. *The Influence of Sea Power Upon History, 1660–1783*. 1890. Reprint. Mineola, NY: Dover, 1987.

Malkin, Richard. *Marriage, Morals, and War—A Candid Inquiry into Morals in Wartime*. New York: Arden Book Company, 1943.

Mandell, Richard D. *The Nazi Olympics*. New York: MacMillan, 1971.

Marchaj, C. A. *Sailing Theory and Practice*. New York: Dodd, Mead, 1964.

Marquis, Alice Goldbarb. *Hopes and Ashes—The Birth of Modern Times, 1929–1939*. New York: Free Press, 1986)

Marrero, Frank. *Lincoln Beachey—The Man Who Owned the Sky*. San Francisco: Scollwall, 1997.

Marshall, Jim. "The West Throws a Party," *Magnificence in San Francisco* (Golden Gate International Exposition). *Collier's—The National Weekly*, February 18, 1939, 21–23; 64–65.

Martyr, Weston. *The Perfect Ship and How We Built Her*. New York: Ives Washburn, 1928.

Mathews, Basil. *Wilfred Grenfell: The Master Mariner; A Life of Adventure on Sea and Ice*. New York: George H. Doran, 1924.

Matthews, Owen. *Glorious Misadventures—Nikolai Rezanov and the Dream of a Russian America*. New York: Bloomsbury, 2013.

Matzen, Robert. *Fireball—Carole Lombard and the Mystery of Flight 3*. Pittsburgh: Paladin Communications, 2013.

Maurois, Andre. *Byron*. Translated from the French by Hamish Miles. New York: D. Appleton, 1930. Halliburton's favorite biography of the poet.

Max, Gerry. "He Must Have Died In Summer." *Thomas Wolfe Review* (Fall 1991): 63–69.

——. *Horizon Chasers*. Jefferson, NC: McFarland, 2007.

——. "The Royal Road to Romance in the USA: Thomas Wolfe, Richard Halliburton, Eco-Tourism, and Eco-Poetry." *Thomas Wolfe Review* (2014): 80–94.

McCarthy, Michael. "'300' Star [actor Gerald] Butler Survives Surfing Scare." *USA Today*, December 22, 2011. 3C.

McClure, Hal. "The Ghost of Laguna's Halliburton House." Orange County News, *Los Angeles Times*, January 20, 1957.

McCormick, Elsie. *Audacious Angles on China*. Shanghai: Chinese American Publishing, 1922.

McDonald, Kevin. "Timbuktu: Mali's Past Under Threat—Bloodied but Unbowed, Malian Heritage Weathers a Storm of Conflict." *World Archaeology Magazine* #58, April/May 2013, 26–31.

McElvaine, Robert S. *The Great Depression: America, 1929–1941*. New York: Three Rivers, 1994.

McGill, Ormond, and Ron Ormond. *Religious Mysteries of the Orient*. Cranbury, NJ: A. S. Barnes, 1976.

McGovern, William Montgomery. *To Lhasa in Disguise—A Secret Expedition through Tibet*. New York: Century, 1924.

McHugh, Fionnuala. "'I Feel Lost in Hong Kong': Why Travel Writer Paul Theroux Finds the City Impenetrable." *Post Magazine,* November 30, 2014.

McNeill, William H. *Plagues and People*. New York: Anchor Doubleday, 1977.

Meacham, Jon. "The American Dream: A Biography." *Time*, July 2, 2012, 26–39.

Meade, Marion. *Bobbed Hair and Bathtub Gin—Writers Running Wild In the Twenties*. New York: Doubleday, 2004.

Mencken, H. L. "Types of Men." in *A Mencken Chrestomathy*. New York: First Vintage Books Edition, 1982.

Mertz, Henriette. *Pale Ink—Two Ancient Records of Chinese Explorations in America*. Chicago: Swallow, 1953. 2nd rev. ed., 1972.

Meyer, Richard J. *Ruan LIng-Yu—The Goddess of Shanghai*. With DVD. Hong Kong: Hong Kong University Press, 2005.

Meyers, Jeffrey. *Somerset Maugham—A Life*. New York: Alfred A. Knopf, 2004.

Milfred, Scott, editorial page director. "Resolve to Live With Spirit Like This," Opinion, *Wisconsin State Journal*, January 1, 2008. p. A10.

Millard, Candice. *The River of Doubt—Theodore Roosevelt's Darkest Journey*. New York: Broadway Books, 2005.

Miller, Lee. G. *An Ernie Pyle Album—Indiana to Ie Shima*. New York: William Sloane Associates, 1946.

Miller, Nathan. *New World Coming*. New York, 2003.

Miller, Scott. *The President and the Assassin: McKinley, Terror, and Empire at the Dawn of the American Century*. New York: Random House, 2011.

Miller, William D. *Memphis—During the Progressive Era, 1900–1917*. Memphis, TN: Memphis State University Press; Madison, WI: American History Research Center, 1957.

Milligan, Barry. *Pleasures and Pains: Opium and the Orient in Nineteenth-Century British Culture*. Charlottesville: University Press of Virginia, 1995.

Mizener, Arthur. *Scott Fitzgerald and His World*. New York: J. P. Putnam's Sons, 1972.

Montgomery, Roselle Mercier. *Ulysses Returns and Other Poems*. New York: Brentano's, 1925.

Moody, Alton B., *Navigation Afloat—A Manual for the Seaman*. New York: Van Nostrand Reinhold, 1980.

Moore, David William. *Scoot McKay*. Philadelphia: David McKay, 1939.

Morden, William. *Across Asia's Snows and Deserts—On the Marco Polo Trail from Bombay to Peking*. New York: G. P. Putnam, 1927.

Morella, Joseph, and George Mazzei. *Genius and Lust: The Creativity and Sexuality of Cole Porter and Noel Coward*. New York: Carroll and Graf, 1995.

Morgan, Ainsworth. *Man of Two Worlds—The Novel of a Stranger*. Indianapolis: Bobbs-Merrill, 1933.

Morgan, Alfred P. *Wireless Telegraphy and Telephony*. New York: W. Henley, 1922.

Morris, Charles E., III. "Richard Halliburton's Bearded Tales." *Quarterly Journal of Speech* 95, no. 2(May 2009): 123–47.

Morris, Joe Alex. *What a Year!* New York: Harper and Brothers, 1956.

Morrissey, Thomas L. *Odyssey of Fighting Two*. Self-published, 1945.

Mortimer, Gavin. *Chasing Icarus—The Seventeen Days in 1910 That Forever Changed Aviation*. Walker, 2010.

Moses, Lester George. *The Indian Man: A Biography of James Mooney*. Urbana: University of Illinois Press, 1984.

Moulin, Gabriel. *San Francisco Peninsula—Town & Country Homes, 1910–1930*. Sausalito, CA: Windgate, 1985.

Mrantz, Maxine. *R. L. Stevenson—Poet in Paradise*. Honolulu: Aloha Graphics, 1977.

Munsterberg, Hugo. *Psychology and Industrial Efficiency*. Boston: Houghton-Mifflin, 1913.

———. *Psychotherapy*. New York: Moffatt, Yard, 1909.

Myles, Douglas. *The Great Waves*. London: Robert Hale, 1985.

Nason, Leonard H. *The Incomplete Mariner*. Garden City, NJ: Doubleday, Doran, 1929.

National Security Agency. *The Many Lives of Herbert O. Yardley*. Washington, DC: National Security Agency, 2012.

Nazi Games, The: Berlin 1936. 2016. PBS. Video.

Needham, Noel Joseph. *Science and Civilization in China*. 6 vols. (7th vol. later). Cambridge: Cambridge University Press, 1956–78.

Negroni, Christine. *The Crash Detectives—Investigating the World's Most Mysterious Air Disasters*. New York: Penguin Books, 2016.

New Century, 1900–1914—A Changing World. History of the 20th Century. Andromeda Oxford Ltd.: Chancellor, 1993.

Nevins, Allan. "The Troubled Life of Lafcadio Hearn." *Mentor*, March 1925, 47–50.

Newby, I. A. *Jim Crow's Defense—Anti-Negro Thought in America, 1900–1930*. Baton Rouge: Louisiana State University Press, 1965.

Newman TravelTalks, (E. M. Newman), 1929 Twenty First Season 1930, South America and Europe, Unpaginated, 22 pp.

Nilsson, Jeff. "Edward Snowden's Forgotten Predecessor." *Saturday Evening Post*, January 17, 2014.

Nivelon, Francis. *The Rudiments of Genteel Behavior*. Facsimile reprint of the unique edition of 1737. London: Paul Hoberton, 2003.

Nock, Samuel. "I Knew Hitler" by Kurt G. W. Ludecke. *Saturday Review of Literature* 17, no. 6 (December 4, 1937).

Nott, G. William. "The Charm of Old New Orleans." *Mentor*, March 1925, 27–46, with photographs. Article on Lafcadio Hearn in New Orleans follows.

Nourse, Mary Augusta. *The Four Hundred Million—A Short History of the Chinese.* Indianapolis: Bobbs-Merrill, 1935.

———. *Kodo: The Way of the Emperor; A Short History of the Japanese.* Indianapolis: Bobbs-Merrill, 1940.

———. *Official Metropolitan Guide*, Association of New York City (brochure). Week beginning June 24, 1928. 48 pp.

O'Bar, Jack. "The Origins and History of the Bobbs-Merrill Company." *University of Illinois Graduate School of Library and Information Sciences Occasional Papers.* 172, December 1985. University of Illinois.

Official Daily Program—World Championship Rodeo. Livermore Day (noted), May 21, 1939.

Official Guide Book Golden Gate International Exposition on San Francisco Bay. Includes foldout map, 25 cents. Rev. ed. 1939. 116 pp.

Official Souvenir Route Book Commemorating the History Making Transcontinental Route of the Tom Mix Circus Ocean to Ocean Border to Border. 1936.

Olds, Jacqueline, and Richard S. Schwartz. *The Lonely American—Drifting Apart in the Twenty-First Century.* Boston: Beacon, 2008.

Optic, Oliver. *Little by Little,* or *The Cruise of the Flyaway—A Story for Young Folks.* Chicago: M. A. Donohue, n.d.

O'Shea, M. V., and J. H. Kellogg. *The Body in Health.* The Health Series of Physiology and Hygiene. New York: MacMillan, 1916.

Overstreet, R. Larry. "The Greek Concept of the 'Seven Stages of Life' and Its New Testament Significance." *Bulletin of Biblical Research* 19, no. 4 (2009): 537–63.

Parish, James Robert, and William T. Leonard. *Hollywood Players—The Thirties.* New Rochelle, NY: Arlington House, 1976.

Parkinson, R. B. *The Tale of Sinuhe and Other Egyptian Poems, 1940–1640 BC.* Oxford World's Classics, an imprint of Oxford University Press, 1997.

Patel, Samir S. "Something in the Water." *Archaeology,* November/December 2007. P. 13.

Peat, Neville. *Shackleton's Whiskey.* London: Preface—The Random House Group, 2012.

Pederasty (subject), Wikipedia.

Peh-T'i Wei, Betty. *Old Shanghai.* Hong Kong: Oxford University Press, 1993.

Peplow, S. H, and M. Barker. *Hong Kong About and Around.* 2nd ed. with map. Hong Kong: Commercial, 1931.

Perrottet, Tony. "Traveler in the Sunset Clouds: The Indiana Jones of Imperial China Has Become a Modern Pop-Culture Celebrity." *Smithsonian,* April 2015, 40–56.

Petersen, E[than] Allen. *Hummel Hummel.* New York: Vantage, 1952.

———. *In a Junk Across the Pacific.* London: Elek, 1954.

Pettegrew, John. *Brutes in Suits—Male Sensibility in America, 1890–1920.* Baltimore, MD: John Hopkins University Press, 2007.

Phillips, Cabell. *From the Crash to the Blitz, 1929–1939—The New York Times Chronicle of American Life.* New York: MacMillan, 1939.

Phillips, Jack. "11 Mysterious 'Ghost Ships' Wash Up In Japan Containing 20 Dead Bodies." *Epoch Times*, November 30, 2015. Online.

Phillips-Birt, Douglas. *Fore & Aft Sailing Craft and the Development of the Modern Yacht*. London: Seely, Service, 1962.

Pierrot, George F. *The Vagabond Trail: Around the World in 100 Days*. With a foreword by Lowell Thomas. New York: D. Appleton–Century, 1935.

Pierson, Melissa Holbrook. *The Place You Love Is Gone—Progress Hits Home*. New York: W. W. Norton, 2006.

Pigafetta, Antonio. *Magellan's Voyage—A Narrative Account of the First Circumnavigation*. Translated and edited by R. A. Skelton. New York: Dover, 1994.

Pinchot, Gifford. *To the South Seas*. Philadelphia: John C. Winston, 1930.

Pinsky, Robert. *Thousands of Broadways—Dreams and Nightmares of the American Small Town*. Chicago: University of Chicago Press, 2009.

Pollack, Howard. *The Ballad of John LaTouche*. Oxford: University of Oxford Press, 2017.

"Potter, Col. Wilson" (obituary). *New York Times*, Thursday, June 13, 1946.

Powell, Edward Alexander. *Asia at the Crossroads—Japan, Korea, China, Philippine Islands*. New York: Century, 1922.

Powell, John B. *My Twenty-Five Years in China*. New York: MacMillan, 1945.

Predmore, Richard L. *The World of Don Quixote*. Cambridge, MA: Harvard University Press, 1967.

Preston, Diana. *A First Rate Tragedy—Robert Falcon Scott and the Race to the South Pole*. Boston: Houghton Mifflin, 1998.

Price, E. Hoffman. *Book of the Dead: Friends of Yesteryear; Fictioneers and Others*. Edited by Peter Ruber. With an introduction by Jack Williamson. Sauk City, WI: Arkham House, 2001.

Prince, Cathryn J. *American Daredevil: The Extraordinary Life of Richard Halliburton, the World's First Celebrity Travel Writer*. Chicago: Chicago Review Press, 2016.

Purcell, Rex, as told to George W. Polk. "The Cruise of the *Pan Jing*." *QST* (devoted to amateur radio), October 1939. P. 18.

Pu-u, Hu. *A Brief History of Sino-Japanese War (1937–1945)*. Taipei, Taiwan: Chung Wu Publishing, 1974. Contains list of reference books, 343–47.

Quick, William J. "Reading F. Scott Fitzgerald's Tax Records." *American Scholar*, (Autumn 2009), 96–101.

Rand, Edward A. *All Aboard for Lakes and Mountains—A Trip to Picturesque Localities in the United States*. Illustrated. Chicago: M. A. Donahue, 1885.

———. *From the Golden Gate through Sunrise Lands—A Trip through California across the Pacific to Japan, China and Australia*. Oriental, 1894.

Red Revolution—In America and in Russia: (I) Logan, Malcolm, These Terrible Reds, How Strong Is Communism in America? How Should It Be dealt With?; (II) Home Office of the Revolution, White, William C. White, A Singularly Revealing Article on the Organization and Real Purposes of Communist Uprisings All Over the World, *Scribner's Magazine*, June 1930, 649–65.

"Rediscovered: Notes on Wetjen by E. Hoffman Price." Up and Down These Mean Streets: The Official Website of Don Herron. DonHerron.com.

Register of Commissioned Officers, Cadets, Midshipmen and Warrant Officers of the United States Naval Reserve, July, 1941. Online.

Reid, William Jameson. *Through Unexplored Asia*. With illustrations by L. J. Bridgman. Boston: Dana Estes, 1899.

Reinhardt, Richard. *Treasure Island—San Francisco's Exposition Years*. San Francisco:
Scrimshaw, 1973.

Reinsch, Paul Samuel. *An American Diplomat in China*. (London: George Allen and
Unwin, 1922).

Remember When . . . a Nostalgic Look Back in Time. Birmingham, AL: SeekPublishing, 20.

Returning the Free China. Directed by Robin Greenberg. New Zealand International
Film Festival (NZIFF), 2015.

Rice, Doyle. "Patricia No. 1 on List of World's Strongest Storms." *USA Today*, October
24, 2015. 2B.

Richard Halliburton Pictures (and) Photos, Getty Images 27. Online.

"Richard Halliburton Speaker for Soroptimist International in Savoy Theatre, April
20," *San Diego News*, Sunday, April 4, 1937.

Richards, Audrey I. *Hunger and Work in a Savage Tribe—A Functional Study of Nutrition among the Southern Bantu*. With an introduction by Prof. B. Malinowski.
London: George Routledge and Sons, 1932.

Richards, Robert. "Halliburton Letters Depict the Man Himself, Noted Adventurer
Reveals What He Thought About Life—First as a Youngster, and Later as a Veteran
of Wanderlust." *Press-Scimitar*, July 19, 1940.

Ripley, Robert L. *Believe It or Not*. New York: Simon and Schuster, 1929.

Robb, Brian, ed. *Twelve Adventures of the Celebrated Baron Munchausen*. London:
Peter Lunn, 1947.

Robinson, William Albert. *10,000 Leagues Over the Sea*. New York: Brewer, Warren
and Putnam, 1932.

Root, Jonathan. *Halliburton—The Magnificent Myth*. New York: Coward-McCann, 1965.

Ross, Alex. "Berlin Story—How the Germans Invented Gay Rights—More Than A Century Ago." *New Yorker*, January 26, 2015, 73–77.

Rowe, J. G. *Crusoe Island*. New York: Cupples and Leon, 1927.

Rowland, John. *Slipping the Lines—Adventures around the World in Peace and War*.
North Battleford, SK, Canada: Turner-Warwick, 1993.

"Running Shipwrecks Simulation Backwards Helps Identify Dangerous Waves," *Phys.
Org*. Technical Computer Sciences, University of Michigan, October 1, 2007.

Rupert, Greenberry G. *The Yellow Peril, or Orient vs. Occident as Viewed by Modern
Statesmen and Ancient Prophets*. 1911. On-demand reprint of original. Nabu, 2012.

Russell, Jan Jarboe, ed. *They Lived to Tell The Tale: True Stories of Modern Adventure
from the Legendary Explorers Club*. Guilford, CT: Globe Pequot, 2007.

San Francisco in the 1930s—The WPA Guide to The City By the Bay. With an introduction by David Kipen. Berkeley: University of California Press, 2011.

San Francisco News, 1937–39.

San Francisco 1939. Copy of booklet with signed letter written by John Cuddy, managing director of the exposition, dated March 6, 1939. San Francisco: Californians
Inc., 1939. Unpaginated.

Saslow, James M. *Pictures and Passions—A History of Homosexuality in the Visual
Arts*. New York: Penguin Putnam, 1999.

Savage-Landor, Arnold Henry. *Across Coveted Lands*. 2 vols. New York: Charles Scribner's Sons, 1903.

——. *China and the Allies* (Boxer rebellion focus). 2 vols. New York: Charles Scribner's Sons, 1901.

——. *Everywhere—The Memoirs of an Explorer.* 2 vols. New York: Frederick A. Stokes, 1924.

Sayer, Geoffrey Robley. *Hong Kong, 1841–1867: Birth, Adolescence and Coming of Age.* Hong Kong: Hong Kong University Press, 1937.

Schell, Orville. "Can the U.S. and China Get Along?" *New York Times,* Op-Ed, July 10, 2015. A25.

Scheurer, Timothy E. *Born in the U.S.A.—The Myth of America in Popular Music from Colonial Times to the Present.* Jackson: University of Mississippi Press, 1991.

Schilling, James von. *The Magic Window—American Television, 1939–1953.* New York: Haworth, 2003.

Schipske, Gerrie. *Early Aviation in Long Beach.* Mount Pleasant, SC: Arcadia, 2009.

Schubert, Marie. *Minute Myths and Legends.* New York: Grosset and Dunlap, 1934.

Scull, Andrew, *Madness in Civilization—From the Bible to Freud, From the Madhouse to Modern Medicine.* London: Thames and Hudson, 2015.

Schult, Joachim. *The Sailing Dictionary.* 2nd ed. Dobbs Ferry, NY: Sheridan House, 1992.

Schultz, Barbara H., *Flying Carpets / Flying Wings.* Lancaster, CA: Little Buttes, 2010.

Schwartz, David M. "On the Royal Road to Adventure with 'Daring Dick.'" *Smithsonian,* March 1989. 159–78.

Scott, John Murphy (Paul Mooney's nephew). Unpublished letters to the author.

SeaTalk Nautical Dictionary. Online.

Seelye, John. *War Games—Richard Harding Davis and the New Imperialism.* University of Massachusetts Press, 2003.

Sertima, Ivan Van. *They Came Before Columbus—The African Presence in Ancient America.* New York: Random House, 1976.

Seven Seas, various issues from 1920s and 1930s.

Severin, Tim. *The China Voyage—Across the Pacific by Bamboo Raft.* Reading, MA: Wesley-Addison, 1994.

Shea, Gail Hynes. "Treasure Island Fair: Golden Gate International Exposition." Shaping San Francisco's Digital Archive at Found. www.foundsf.org.

Sheff, David. "Halliburton-By-The-Sea." *New York Times,* May 2, 2004.

Shenk, Joshua Wolf. *Lincoln's Melancholia—How Depression Challenged a President and Fueled His Greatness.* (New York: Mariner Books, an imprint of Houghton Mifflin, 2005.

Sherwood, Martyn. *The Voyage of the Tai-Mo-Shan.* 1935. Reprint. London: Geoffrey Bles, 1946.

Shlaes, Amity. *Coolidge.* New York: Harper/Collins, 2013.

Simons, Irving. *Unto the Fourth Generation: Gonorrhea and Syphilis; What the Layman Should Know.* New York: E. P. Dutton, 1940.

Sinclair, Gordon. *Bright Paths to Adventure.* With illustrations by Stanley Turner. Toronto: McClelland and Stewart, 1945.

Sinful Cities of the Western World. New York: Julian Messner, 1934.

Skemer, Don. "A New View of Richard Halliburton's *Sea Dragon.*" PUL Manuscripts News. March 17, 2014. blogs.princeton.edu.

Skrenda, Alfred, and Isabel Juergens. *Minute Wonders of the World—Describing by*

Means of Picture and Text 144 of the Natural and Man-Made Wonders of the World. New York: Grosset and Dunlap, 1933.

Sloane, Eric. *The Book of Storms—Hurricanes, Twisters, and Squalls*. Mineola, NY: Dover, 2006.

——. *Weather Book*. Mineola, NY: Dover, 2005.

Slocum, Joshua. *Sailing Alone Around the World*. Edited an with an introduction and notes by Thomas Philbrick. New York: Penguin Books, 1999.

Smith, Holly Austin. *Walking Prey—How America's Youth Are Vulnerable to Sex Slavery*. New York: Palgrave MacMillan, 2014.

Smith, James R., *Guidelines—Golden Gate International Exposition—San Francisco World's Fair I & II; also San Francisco's Final World's Fair*.

——. *San Francisco's Lost Landmarks*. Sanger, CA: Word Dancer, 2005.

Smith, Joseph Russell. *California: Life, Resources, and Industries*. Sacramento: California State Department of Education, 1936.

——. *World Folks*. Chicago: John C. Winston, 1939.

Smith, Richard Gordon. *Travels in the Land of the Gods (1898–1907)—The Japan Diaries of Richard Gordon Smith*. Edited by Victoria Manthorpe. New York: Prentice Hall, 1986.

Smyth, Admiral W. H. *The Sailor's Word Book*. London: Conway, 1991.

Smyth, H. Warrington. *Mast and Sail in Europe and Asia*. Illustrated. New York: E. P. Dutton, 1906.

Snow, Edward Rowe. *The Vengeful Sea*. New York: Dodd, Mead, 1957.

Snow, Phillip. *The Fall of Hong Kong: Britain, China and the Japanese Occupation*. New Haven: Yale University Press, 2003.

Sobol, Dava, and William J. H. Andrewes. *The Illustrated Longitude—The True Story of the Lone Genius Who Solved the Greatest Scientific Problem of His Time*. New York: Walker, 1995.

Soiland, Albert (Honorary Commodore). *Transpacific Ocean Races and Transpacific Yacht Club: Facts, Fancies and Some Gossip about One of the Most Unique and Interesting Yacht Clubs in the World, and the Races It Sponsor*s. Los Angeles: privately printed, 1937. Dedicated to Clarence W. MacFarlane.

Souhami, Diana. *Selkirk's Island—The True and Strange Adventures of the Real Robinson Crusoe*. New York: Harcourt, 2001.

Sowers, Phyllis Ayer. *Let's Go 'Round the World with Bob and Betty*. With illustrations by Robert Von Neuman. New York: Grosset and Dunlap, 1934.

Spedding, Charles T. (for many years purser of the *Aquitania*). *Reminiscences of Transatlantic Travelers*. Illustrated. Philadelphia: J. P. Lippincott, 1925.

Speed Service "Map Showing Routes, Ports of Call, and Services of Java-China Japan Line." 1930.

Spence, Jonathan D. *The Chan's Great Continent—China in Western Minds*. New York: W. W. Norton, 1998.

Springer, Arthur. *Red Wine of Youth—A Life of Rupert Brooke*. New York: Bobbs-Merrill, 1952.

Spurling, Hilary. *Pearl Buck in China—Journey to the Good Earth*. New York: Simon and Schuster, 2010.

Stansell, Christine. *American Moderns—Bohemian New York and the Creation of a New Century*. New York: Metropolitan Books, an imprint of Henry Holt, 2000.

Starr, Kevin. *Golden Gate: The Life and Times of America's Greatest Bridge.* New York: Bloomsbury, 2010.

———. *Inventing the Dream—California through the Progressive Era.* New York: Oxford University Press, 1985.

Steer, Dugald A. *Dr. Ernest Drake's Dragonology—The Complete Book of Dragons.* Cambridge, MA: Candlewick, 2003.

Stein, R. Conrad. *The Story of the Child Labor Laws.* Chicago: Children's Press, 1984.

Stewart, James B. "An Aftershock with Present." Business Day, *New York Times*, Saturday, August 13, 2011, B1, B5.

Stevens, Thomas. *Around the World on a Bicycle.* 2 vols. New York: Charles Scribner's, 1888.

Stoddard, Lothrop. *The Rising Tide of Color against White World Supremacy.* New York: Charles Scribner's Sons, 1920.

Stone, Irving. *Sailor on Horseback—The Biography of Jack London.* New York: Houghton-Mifflin, 1938.

Stone, Peter. *The Lady and the President—the Life and Death of the S. S.* President Coolidge. Yarram, Victoria, Australia: Oceans Enterprises, 2004.

Strand, Ginger. *Killer on the Road—Violence and the American Interstate.* Austin: University of Texas Press, 2012.

Stuart, Tristram. *The Bloodless Revolution—A Cultural History of Vegetarianism from 1600 to Modern Times.* New York: W. W. Norton, 2006.

"Sunk." *Outside Magazine*, April 13, 2013, 78–83.

Sutton, Bettye. "1930-1939" *American Cultural History.* Lonestar College-Kingwood Library, 1999.

Sweeney, Gladys Acevedo, and John H. Horan (Pennsylvania State University). "Separate and Combined Effects of Cue-Controlled Relaxation and Cognitive Treatment of Musical Performing Anxiety." *Journal of Counseling Psychology* 29, no. 5 (1982): 488–97.

Sykes, Charles J. *The Hollow Men: Politics and Corruption in Higher Education.* Washington, DC: Regnery Publishing, 1990.

Takeo Club. The Wreck of the *S.S. Hoover* Part II. www.takaoclub.com.

Talley, Jeannine. *Women at the Helm.* Racine, WI: Mother Courage, 1990.

Taussig, Michael. *Shamanism, Colonialism and the Wild Man—A Study in Terror and Healing.* Chicago: University of Chicago Press, 1987.

Taylor, J. Hudson, ed. *China's Millions.* London: Morgan and Scott, 1876-77. Includes "A Hong Kong Junk From the Diary of G. W. Clarke."

Taylor, William "Bill." Interview by Jean Feraca. *The Ideas Network on Tape.* 0520e. May 20, 1994.

Taylor, William R. *A Shooting Star Meets the Well of Death.* Abbeville, SC: Moonshine Cove, 2013.

Temple, Robert. *The Genius of China—3000 Years of Science, Discovery & Invention.* With a foreword by Joseph Needham. Rochester, VT: Inner Traditions, 2007.

———. "The Mount Everest of Surfing" and "Biggest Wave." *Smithsonian*, July/August 2018, 26–39.

———. "Raw Material—Thomas Hart Benton and America's Modern Era." *Smithsonian*, December 2014, 58-67.

———. *The Tao of Travel—Enlightenments from Lives on the Road.* Boston: Houghton Mifflin Harcourt, 2011.

Thomas, Lowell. *Good Evening Everybody: An Autobiography; From Cripple Creek to Samarkand*. New York: Avon, 1976).

——. *Tall Stories—The Rise and Triumph of the Great American Whopper*. With illustrations by Herb Roth. New York: Funk and Wagnalls, 1931.

——. "Thomas Lawrence the Man." *Asia: The American Magazine of the Orient*, August 1920, 670–76.

——. *Thrilling Moments in Thrilling Lives*. Sun Oil Company, 1936. 37 pp.

——. *A Trip to New York with Bobby and Betty*. Dodge, 1936.

——. *With Lawrence in Arabia*. New York: Century, 1924.

Thomas, Margaret Loring. *George Washington Lincoln Goes Around the World*. With illustrations by Willy Pogany. New York: Thomas Nelson and Sons, 1927.

Thompson, Neal. *A Curious Man—The Strange and Brilliant Life of Robert "Believe It or Not" Ripley*. New York: Crown Archetype, 2013.

Thomson, H. C. *The Case for China*. New York: Charles Scribner's Sons, 1933.

Time, June 24, 1929.

Time, June 28, 1931.

Time, July 26, 1937.

Time, April 17, 1939.

Time, November 21, 1960.

Time, July 3, 2006.

Time, June 17, 2013.

Time-Life Books, Editors of. *This Fabulous Century*. Vol. 4, *1930–1940—The Thirties*. New York: Time, 1969.

"Timeline" (for Karel Frederik Mulder [1901–78]). *International Institute for Asian Studies* 58 (Autumn 2011).

Today at the Fair, Complete Magazine of Events, *Exposition* Premiere Souvenir. Sunday, February 19, 1939. 16 pp.

Toohey, John. *Captain Bligh's Portable Nightmare*. New York: HarperCollins, 1998.

"Transpacific Yacht Races" (subject), Wikipedia and Wikiwand.

Travel, August 1928, 6.

"Treasure Island of 1939—Coloroto." *Popular Mechanics Magazine*, May 1938,. 649–56, 128A–29A.

Tuchman, Barbara W. *Stilwell and the American Experience in China, 1911–1945*. New York: MacMillan, 1970.

Tully, Anthony, Bob Hackett, and Sander Kingsepp. "The Great Hong Kong Typhoon—September 1937," in *Rising Storm: The Imperial Japanese Navy and China, 1931–1941*. www.combinedfleet.com.

Twain, Mark. *Mark Twain on the Move—A Travel Reader*. Edited by Alan Gribben and Jeffrey Alan Melton. Tuscaloosa: University of Alabama Press, 2009.

"The Two Coasts of China: Asia and the Challenge of the West." Episode 1, *The Pacific Century*. 10 Episodes. Annenberg/CPB Collection, 1992.

Underwater Universe. Documentary TV series (2011–).

Unschuld, Paul U. *Medicine in China—A History of Ideas*. Berkeley: University of California Press, 1985.

"Vagabond Trail to Romance." *Milwaukee Journal*, Sunday Magazine, September 2, 1934.

Van Allen, Elizabeth J. *James Whitcomb Riley—A Life*. Bloomington: Indiana University Press, 1999.

Van Dyke, Henry. *Little Rivers—A Book of Essays in Profitable Idleness*. Charles Scribner's Sons, 1895.

——. *The Works of Henry Van Dyke*. 16 vols. Avalon ed. New York: Charles Scribner's Sons, 1920.

Van Loan, Derek. *Design and Build Your Own Junk*. Arcata, CA: Paradise Cay, 2006.

Van Loon, Henrik Willem. *Van Loon's Geography*. New York: Simon and Schuster, 1932.

Van Tilburg, Hans Konrad. *Chinese Junks on the Pacific—Views from a Different Deck*. New Perspectives on Maritime History and Nautical Archaeology. Gainesville: University Press of Florida, 2013.

Varende, Jean De La. *Cherish the Sea: A History of Sail*. Translated from the French by Marvin Saville. New York: Viking, 1958.

Vawter, Ralph B. (Buffalo, New York). "My Scrapbook, 1939–1940." Author's collection.

Views of the Pearl River Delta: Macao, Canton and Hong Kong. Urban Council of Hong Kong, 1996.

Villiers, Alan. *The Making of a Sailor; the Photographic Story of Schoolships under Sail*. New York: William Morrow, 1938.

——. *Posted Missing—The Story of Lost Ships without a Trace in Recent Years*. New York: Charles Scribner's Sons, 1958.

Wade, Mary Hazelton. *Our Japanese Cousin*. With illustrations by L. J. Bridgman. Boston: L. C. Paige and Cooper, 1901.

Wade, Neville. "Sailing a Model Square-Rigger." *Marine Modelling International*, April 2008.

Walker, Fred M. *Ships & Shipbuilders—Pioneers of Design and Construction*. With a foreword by Trevor Blakeley. Annapolis, MD: Naval Institute Press, 2010.

Wallace, David. *Lost Hollywood*. LA Weekly Books, 2001.

Wallace, Max. *The American Axis—Henry Ford, Charles Lindbergh and the Rise of the Third Reich*. New York: St. Martin's, 2003.

Walter, Ellery. *Russia's Decisive Years*. New York: G. P. Putnam's Sons, 1932.

——. *The World on One Leg*. New York: G. P. Putnam's Sons, 1928.

Ward, Julian. *The Art of Travel Writing*. London: Taylor Francis, 2015.

Ward, Nick. *Left For Dead—Surviving the Deadliest Storm in Modern Sailing History*. With Sinead O'Brian. New York: Bloomsbury, 2007.

Ware, Susan. *Holding Their Own—American Women in the 1930s*. American Women Series. Boston: Twayne, 1982.

Warinner, Emily V. *Voyager to Destiny*. Indianapolis: Bobbs-Merrill, 1956.

Watt, Louise. "Pollution Woes Inspire China." Associated Press. *Wisconsin State Journal*, December 14, 2015.

Watt, O. M., ed. *Reed's Nautical Almanac*. London: Thomas Reed, 1938.

Weatherford, Jack. *Savages and Civilization—Who Will Survive?* New York: Fawcett Columbine, 1994).

Weber, Bruce. "Henry Worsley, a British Adventurer Trying to Cross Antarctica, Dies at 55." *New York Times*, Obituaries, January 26, 2016.

Weigall, Arthur. *Personalities of Antiquity*. Garden City, NY: Doubleday, Doran, 1928.

Weintraub, Aileen. *The Pacific Ocean—The Largest Ocean*. New York: Rosen Publishing Group's PowerKids Press, 2001.

Welch, John (Architect). *A Six Days' Tour of the Isle of Man—A Passing View of Its Present, Natural, Social and Political Aspect by a Stranger*. Douglas, 1836.

Welch, John "Jack" Wenloch (or Wenlock). "Farewell to Sail." In *The "Man" Storyteller*, 174-87. Sydney, Australia: K. G. Murray, 1945. First published in *Man Magazine* 4, nos. 4 and 5 (September and October 1938).

———. "Ocean Tow." *Proceedings Magazine*, US Naval Institute, vol. 63/2/408 (February 1937): 206-14 (Coringa crew photographed). Online.

———. "Our Oceanic Ills," "By the late Ensign John Wenlock Welch, U. S. Naval Reserve," *Naval Institute Proceedings*, October, 1939, Vol. 65, 10, 440.

———. "Paradise Regained." *World's News* (Sydney, Australia), August 26, 1939, 8-9.

———. "Signaling and the U. S. Merchant Marine." *Proceedings Magazine*, US Naval Institute, vol. 62/1/395 (January 1936): 79-81. Online.

Weller, Earle, and Jack James. *Treasure Island: The Magic City; The Story of the Golden Gate International Exposition, 1939-1940*. San Francisco: Pisani, 1941.

Weller, George. "The Passing of the Last Playboy"—Richard Halliburton Was Not a Phony, He Had an Appetite for Action, and Made His Living Having a Good Time. *Esquire*, April 1940, 58, 111-12.

Wells, Carolyn. *The Lover's Baedeker and Guide to Arcady*. With twenty illustrations and a cover by A. D. Blashfield and maps by George W. Hood. New York: Frederick A. Stokes, 1912.

Wells, Carveth. *Adventure*. New York: Robert M. McBride, 1931.

———. *Around the World with Bobby and Betty*. New York: Robert M. McBride, 1939.

———. *Exploring America with Conoco and Carveth Wells* (the Man They Call "Radio's Truthful Liar"). Station WEAF, New York, 1933 (through April 15) (Conoco Travel Bureau—America's Foremost Free Travel Service), 16-panel brochure. Author's collection.

———. *Exploring the World with Carveth Wells*. New York: Robert M. McBride, 1934.

———. *My Candle at Both Ends—The Autobiography of John Carveth Wells*. London: Jarrolds, 1950.

———. *North of Singapore*. New York: Robert M. McBride, 1940.

———. Unpublished letter to Elizabeth Cleveland, September 28, 1933. Author's collection.

Wells, H. G. *The King Who Was a King*. Garden City, NY: Doubleday, Doran, 1929.

———. *The New America: The New World*. London: Cresset, 1935.

———. *The Way the World Is Going—Guesses and Forecasts of the Next Few Years*. London: Ernest Benn, 1928.

Wells, Linton. "Richard Halliburton—His Story of His Life's Adventures." *Saturday Review* 22, no. 10 (June 29, 1940).

Wetjen, Albert Richard. *The Chronicles of Shark Gotch*. The World's Work (1913) Ltd. *The Master Thriller* Library. London: Kingswood, 1913.

———. *In Fiddlers' Green, or The Strange Adventure of Tommy Lawn—A Tale of the Great Divide of Sailormen*. With illustrations by Ferdinand Huztl Horvath. Little Brown, 1931.

———. *The Way of the Sea*. New York: Century, 1928.

——. *Youth Walks On the Highway*. With illustrations by John Alan Maxwell. New York: Heron, 1930.

Whalen, Grover. *A Trip to the World's Fair*. Dodge, 1938.

When and Where, at the Fair, Official Bulletin for Special Events, Monday and Tuesday, July 29 and 30, 1940. 4 pp.

White Countess, The. A Merchant-Ivory Production, 2005, DVD, 136 minutes.

Wild China—Valley of the Giant Panda. National Geographic Travel. DVD. 2015.

Wilde, Oscar. *The Uncensored Picture of Dorian Gray. Edited by Nicholas Frankel*. Cambridge, Massachusetts: Harvard University Press, 2011.

Williams, R. Scott. *The Forgotten Adventures of Richard Halliburton—A High-Flying Life from Tennessee to Timbuktu*. History Press, 2014.

Williamson, C. N., and A. M. Williamson. *The Golden Silence*. With illustrations by George Brehm. Garden City, NY: Doubleday, Page, 1911.

Willis, Bruce, and Adrian Brody. *Unbreakable Spirit*. Directed by Xiao Feng. Released August 2018.

Wilson, Ben. *Heyday—The 1850s and the Dawn of the Global Age*. New York: Basic Books, 2016.

Wilson, Derek. *The World Encompassed: Drake's Great Voyage, 1577–80*. New York: Harper and Row, 1977.

Winchester, Simon. *The Man Who Loved China—The Fantastic Story of the Eccentric Scientist Who Unlocked the Mysteries of the Middle Kingdom*. New York: Harper/Collins, 2008.

Wings Over Hong Kong: An Aviation History 1891–1998; A Tribute to Kai Tak. Hong Kong: Pacific Century, 1998.

"Wireless Issue, The—10 Ways Your Phone Is Changing the World." *Time*, August 27, 2012.

Wolfe, Thomas. *The Four Lost Men*. Edited by Arlyn Bruccoli and Matthew J. Bruccoli. Columbia: University of South Carolina Press, 2008).

——. *The Starwick Episodes*. Edited and with an introduction by Richard S. Kennedy. Baton Rouge: Louisiana State University Press, 1989).

Wolff, Daniel. *How Lincoln Learned to Read—Twelve Great Americans and the Educations That Made Them*. New York: Bloomsbury USA, 2009.

Wolff, Geoffrey. *The Hard Way Around—The Passages of Joshua Slocum*. New York: Alfred A. Knopf, 2010.

Wood, Amos. *Beachcombing the Pacific*. West Chester, PA: Schiffer, 1989.

Wood, Frances. *Did Marco Polo Go to China*. Boulder, CO: Westview, 1995.

Wood, Max. *Sailing Tall*. Lanham, MD: Rowman and Littlefield, 2004.

Woodhead, H. G. W. *Adventures in Far Eastern Journalism—A Record of Thirty-Three Years' Experience*. Tokyo: Hokuseido, 1935.

Woods, Gregory. *Homintern—How Gay Culture Liberated the Modern World*. New Haven: Yale University Press, 2016.

Worcester, G. R. G. *The Junks and Sampans of the Yangtze*. Annapolis, MD: United States Naval Institute, 1971.

"World Explorer Discovers Hollywood." *Motion Picture*, April 1933 (Jean Harlow cover).

Wormser, Richard. *Hoboes—Wandering in America, 1870–1940*. New York: Walker, 1994.

Worsley, F. A., *Shackleton's Boat Journey*. With a narrative introduction by Sir Edmund Hillary. New York: W. W. Norton, 1977.

Wright, William. *Harvard's Secret Court—The Savage 1920 Purge of Campus Homosexuals*. New York: St. Martin's, 2005.

Wu, Judy Tzu-Chun. *Doctor Mom Chung of the Fair-Haired Bastards—The Life of a Wartime Celebrity*. Berkeley: University of California Press, 2005.

Yardley, Herbert Osborn. *The Chinese Black Chamber—An Adventure in Espionage*. London: New English Library, 1983.

Yarwood, A. T., Samuel Marsden (1765–1838), *Australian Dictionary of Biography*, 1967, Vol. 2.

Yates, Helen E. *Shopping and Sightseeing in Hong Kong*. Hong Kong: Cathay, 1952.

Yin, James Yin, Ron Dorman, and Young Shi. *The Rape of Nanking—An Undeniable History in Photographs*. Chicago: Innovative, 1996.

Young, Lt. Stephanie. "*Itasca* and the Search for Amelia Earhart." July 2, 2012.

Young, William H., and Nancy K. Young. *The 1930s*. American Popular Culture through History Series. Westport, CT: Greenwood, 2002.

Zeitz, Joshua. *Flapper—A Madcap Story of Sex, Style, Celebrity and the Women Who Made America Modern*. New York: Crown, 2006.

Zug, James. *American Traveler—The Life and Adventures of John Ledyard, the Man Who Dreamed of Walking the World*. New York: Basic Books, 2005.

———. "Sea of Dreams." *Dartmouth Alumni Magazine*, July–August 2014. Online.

Zweig, Paul. *The Adventurer—The Fate of Adventure in the Western World*. Princeton, NJ: Princeton University Press, 1974.

Zweig, Stefan. *The World of Yesterday: An Autobiography*. New York: Viking, 1943.

Index